Springer Tracts in Advanced Robotics

Volume 31

Editors: Bruno Siciliano · Oussama Khatib · Frans Groen

AF789294

Manuel Ferre, Martin Buss, Rafael Aracil,
Claudio Melchiorri and Carlos Balaguer (Eds.)

Advances in Telerobotics

Springer

Professor Bruno Siciliano, Dipartimento di Informatica e Sistemistica, Universitá di Napoli Federico II, Via Claudio 21, 80125 Napoli, Italy, E-mail: siciliano@unina.it

Professor Oussama Khatib, Robotics Laboratory, Department of Computer Science, Stanford University, Stanford, CA 94305-9010, USA, E-mail: khatib@cs.stanford.edu

Professor Frans Groen, Department of Computer Science, Universiteit vanAmsterdam, Kruislaan 403, 1098 SJ Amsterdam, The Netherlands, E-mail: groen@science.uva.nl

Editors

Manuel Ferre
Universidad Politécnica de Madrid
Dpto. de Automática, Ingeniería
Electrónica e Informática Industrial
28006 Madrid, Spain
E-mail: m.ferre@upm.es

Martin Buss
Technische Universität München
Institute of Automatic Control
Engineering (LSR)
D-80290 Munich, Germany
E-mail: mb@tum.de, M.Buss@ieee.org

Rafael Aracil
Universidad Politécnica de Madrid
Dpto. de Automática, Ingeniería
Electrónica e Informática Industrial
28006 Madrid, Spain
E-mail: aracil@etsii.upm.es

Claudio Melchiorri
University of Bologna
Laboratory of Automation and Robotics
Department of Electronics, Computer
Science and Systems
40136 Bologna, Italy
E-mail: cmelchiorri@deis.unibo.it

Carlos Balaguer
Universidad Carlos III de Madrid
Robotics Lab.
Depto. de Ingeniería de Sistemas y
Automática
28911 (Leganés) Madrid, Spain
E-mail: balaguer@ing.uc3m.es

Additional material to this book can be downloaded from http://extras.springer.com

ISSN print edition: 1610-7438
ISSN electronic edition: 1610-742X
ISBN-10 3-642-42592-5 Springer Berlin Heidelberg New York
ISBN-13 978-3-642-42592-9 Springer Berlin Heidelberg New York

This work is subject to copyright. All rights are reserved, whether the whole or part of the material is concerned, specifically the rights of translation, reprinting, reuse of illustrations, recitation, broadcasting, reproduction on microfilm or in any other way, and storage in data banks. Duplication of this publication or parts thereof is permitted only under the provisions of the German Copyright Law of September 9, 1965, in its current version, and permission for use must always be obtained from Springer. Violations are liable for prosecution under the German Copyright Law.

Springer is a part of Springer Science+Business Media
springer.com

© Springer-Verlag Berlin Heidelberg 2007
Softcover re-print of the Hardcover 1st edition 2007

The use of general descriptive names, registered names, trademarks, etc. in this publication does not imply, even in the absence of a specific statement, that such names are exempt from the relevant protective laws and regulations and therefore free for general use.

Typesetting: Digital data supplied by editor.
Data-conversion and production: SPS, Chennai, India
Cover-Design: WMXDesign, Heidelberg
Printed on acid-free paper SPIN: 12103775 89/3180 5 4 3 2 1

Editorial Advisory Board

Herman Bruyninckx, KU Leuven, Belgium
Raja Chatila, LAAS, France
Henrik Christensen, Georgia Institute of Technology, USA
Peter Corke, CSIRO, Australia
Paolo Dario, Scuola Superiore Sant'Anna Pisa, Italy
Rüdiger Dillmann, Universität Karlsruhe, Germany
Ken Goldberg, UC Berkeley, USA
John Hollerbach, University of Utah, USA
Makoto Kaneko, Hiroshima University, Japan
Lydia Kavraki, Rice University, USA
Sukhan Lee, Sungkyunkwan University, Korea
Tim Salcudean, University of British Columbia, Canada
Sebastian Thrun, Stanford University, USA
Yangsheng Xu, Chinese University of Hong Kong, PRC
Shin'ichi Yuta, Tsukuba University, Japan

STAR (Springer Tracts in Advanced Robotics) has been promoted under the auspices of EURON (European Robotics Research Network)

Foreword

At the dawn of the new millennium, robotics is undergoing a major transformation in scope and dimension. From a largely dominant industrial focus, robotics is rapidly expanding into the challenges of unstructured environments. Interacting with, assisting, serving, and exploring with humans, the emerging robots will increasingly touch people and their lives.

The goal of the new series of *Springer Tracts in Advanced Robotics (STAR)* is to bring, in a timely fashion, the latest advances and developments in robotics on the basis of their significance and quality. It is our hope that the wider dissemination of research developments stimulates exchanges and collaborations among the research community and contributes to further advancement of this rapidly growing field.

The edited volume by Manuel Ferre, Martin Buss, Rafael Aracil, Claudio Melchiorri and Carlos Balaguer is focused on the most recent advances in telerobotics, a technology that deals with the inclusion of a human operator in the control loop of a remote robot. Telerobotics encompasses an area at the crossroads of several scientific disciplines such as mechatronics, control, communication, computers, sensor-based recognition, multimodality and even teleoperation through Internet.

The material is organised in twenty-eight chapters by well-recognised authors in the field, which are grouped in three main parts on human system interfaces, control, and applications. The introduction by the editors provides a useful reading guide throughout the contents of the book, thanks to an effective table of topics and keywords for each part. A number of problems and solutions of today's research on telerobotics are addressed, with emphasis on methods, techniques, experimental results, and developments. Remarkably, the volume is accompanied by a collection of videos illustrating several practical applications of telerobotics.

As the first focused STAR volume in the broad area of telerobotics, this title constitutes a fine addition to the series!

Naples, Italy
January 2007

Bruno Siciliano
STAR Editor

Preface

The main purpose of this book is to provide readers with recent advances in the field of Telerobotics. It describes methods, experimental results, applications, and developments, highly relevant for scientists, researchers, and students in Teleoperation.

In its broadest sense Telerobotics may be defined as the technology that deals with the inclusion of a human operator in the control loop of a remote robot. Many scientific disciplines are to contribute to the area of telerobotics: mechatronics, control, communication, engineering, computer science, speech/gesture/image recognition, psychology (psychophysics), etc. Multimodality – vision, audio, haptics – and teleoperation through the Internet are important contemporary issues.

This book is structured in three parts: I. Human System Interfaces, II. Control, and III. Applications. Chapters in part I concentrate on human interface technology which allows a human operator to close the control loop of a remote robot. Topics related to control algorithms - in particular for the case with time delay in the communication network - are the focus in part II, concentrating on bilateral control methods. Part III presents a variety of advanced applications in surgery, space, and other fields relevant to everyday life. The book is complemented by a CD containing fifteen research videos, which make the contents of the book even more descriptive. This collection of videos can also be found at *www.lsr.ei.tum.de/telerobotics*.

The editors would like to thank all the authors for their valuable contributions to this book. The quality and freshness found in each chapter are due to the excellent work carried out by the authors. In addition we would like to offer our special thanks to the junior researchers who have made large parts of their PhD theses available in the chapters of this volume.

Spain, Germany, Italy
January 2007

Manuel Ferre
Martin Buss
Rafael Aracil
Claudio Melchiorri
Carlos Balaguer

Contents

Introduction to Advances in Telerobotics
Manuel Ferre, Martin Buss, Rafael Aracil, Claudio Melchiorri, Carlos Balaguer 1

Part I: Human System Interfaces

1 The Human Role in Telerobotics
Rafael Aracil, Martin Buss, Salvador Cobos, Manuel Ferre, Sandra Hirche, Martin Kuschel, Angelika Peer 11

2 Design, Control, and Evaluation of a Hyper-redundant Haptic Device
Marc Ueberle, Nico Mock, Martin Buss 25

3 A Novel Parallel Haptic Interface for Telerobotic Systems
José María Sabater, Rafael Aracil, Roque J. Saltaren, Luis Payá 45

4 Exoskeletons as Man-Machine Interface Systems for Teleoperation and Interaction in Virtual Environments
Massimo Bergamasco, Antonio Frisoli, Carlo Alberto Avizzano 61

5 Stereoscopic Image Visualization for Telerobotics. Experiments with Active Binocular Cameras
Alexandre Bernardino, José Santos-Victor, Manuel Ferre, Miguel A. Sanchez-Urán 77

6 Stereoscopic 3-D Acquisition, Processing, and Display for Telerobotic Applications
Fergal Shevlin, Barry McCullagh, David Eadie, Manuel Navas-Herreros, Christophe Rabaud 91

7 User Voice Assistance Tool for Teleoperation
Oscar Reinoso, César Fernández, Ramón Ñeco 107

8 Computer Vision Body Modeling for Gesture Based Teleoperation
Manel Frigola, Alberto Rodriguez, Josep Amat, Alícia Casals 121

9 Enhanced Teleoperation Through Virtual Reality Techniques
Abderrahmane Kheddar, Ee-Sian Neo, Riichiro Tadakuma, Kazuhito Yokoi 139

Part II: Control

10 Bilateral Control Architectures for Telerobotics
Sandra Hirche, Manuel Ferre, Jordi Barrio, Claudio Melchiorri, Martin Buss 163

11 Experimental Results on Bilateral Control Using an Industrial Telemanipulator
Manuel Ferre, Jordi Barrio, Claudio Melchiorri, Juan M. Bogado, Pedro L. Castedo, Juan M. Ibarra 177

12 Human Perceived Transparency with Time Delay
Sandra Hirche, Martin Buss 191

13 Environment Estimation in Teleoperation Systems
Luigi Biagiotti, Claudio Melchiorri 211

14 Power Scaling in Port-Hamiltonian Telemanipulation over Packet Switched Networks
Cristian Secchi, Stefano Stramigioli, Cesare Fantuzzi 233

15 Adaptive Synchronization of Bilateral Teleoperators with Time Delay
Nikhil Chopra, Mark W. Spong 257

16 Bilateral Control of Teleoperation Systems Through State Convergence
José M. Azorín, Rafael Aracil, Nicolás M. García, Carlos Pérez 271

17 Re-configurable Control Scheme for Guiding Telerobotics
Adrián Mora, Antonio Barrientos 289

18 Teleprograming: Capturing the Intention of the Human Operator
Miguel Hernando, Ernesto Gambao 303

Part III: Applications

19 DLR's Advanced Telerobotic Concepts and Experiments for On-Orbit Servicing
Detlef Reintsema, Klaus Landzettel, Gerd Hirzinger 323

20 Underwater Telerobotics for Collaborative Research
Pere Ridao, Marc Carreras, Emili Hernandez, Narcis Palomeras 347

21 Robot Assisted Force Feedback Surgery
Tobias Ortmaier, Barbara Deml, Bernhard Kübler, Georg Passig, Detlef Reintsema, Ulrich Seibold 361

22 Telerobotic Control by Virtual Fixtures for Surgical Applications
Ming Li, Ankur Kapoor, Russell H. Taylor 381

23 Technologies for a Telesurgery Laboratory Implementation
Jesús Manuel Gómez-de-Gabriel, Víctor Fernando Muñoz-Martinez, Francisco Jesús Domínguez-Fernández, Javier Serón-Barba 403

24 Proprio and Teleoperation of a Robotic System for Disabled Persons' Assistance in Domestic Environments
Carlos Balaguer, Antonio Giménez, Alberto Jardón, Raúl Correal, Santiago Martínez, Angelo M. Sabatini, Vincenzo Genovese 415

25 Internet-Based Tele-Laboratory: Remote Experiments Using the SNRP Distributed Network Architecture
Raúl Marín, Raúl Wirz, Pedro J. Sanz, Josep Fernández 429

26 Force Reflecting Teleoperation Via IPv6 Protocol with Geometric Constraints Haptic Guidance
Emmanuel Nuño, Adolfo Rodríguez, Luis Basañez 445

27 Telerobotics for Aerial Live Power Line Maintenance
Rafael Aracil, Manuel Ferre 459

28 Advanced Telerobotics: Dual-Handed and Mobile Remote Manipulation
Martin Buss, Kwang-Kyu Lee, Norbert Nitzsche, Angelika Peer, Bartlomiej Stanczyk, Ulrich Unterhinninghofen 471

Index 499

List of Contributors

Josep Amat
Technical University of Catalonia (UPC)
Automatic Control and Computer Engineering Department
08028 Barcelona, Spain
josep.amat@upc.edu

Rafael Aracil
Universidad Politécnica de Madrid
Dpto. de Automática, Ingeniería Electrónica e Informática Industrial
28006 Madrid, Spain
aracil@etsii.upm.es

Carlo A. Avizzano
Scuola Superiore Sant'Anna
PERCRO
56127, Pisa, Italy
avizzano@sssup.it

José M. Azorín
Universidad Miguel Hernández
Dpto. Ingeniería de Sistemas Industriales
03202 Elche-Alicante, Spain
jm.azorin@umh.es

Carlos Balaguer
Universidad Carlos III de Madrid
Robotics Lab.
Depto. de Ingeniería de Sistemas y Automática
28911 (Leganés) Madrid, Spain
balaguer@ing.uc3m.es

Antonio Barrientos
Universidad Politécnica de Madrid
Dpto. de Automática, Ingeniería Electrónica e Informática Industrial
28006 Madrid, Spain
barrientos@etsii.upm.es

Jordi Barrio
Universidad Politécnica de Madrid
Dpto. de Automática, Ingeniería Electrónica e Informática Industrial
28006 Madrid, Spain
jbarrio@etsii.upm.es

Luis Basañez
Technical University of Catalonia
Institute of Industrial and Control Engineering
08028 Barcelona, Spain
luis.basanez@upc.edu

Alexandre Bernardino
Instituto Superior Técnico
Instituto de Sistemas e Robótica
1049-001 Lisbon, Portugal
alex@isr.ist.utl.pt

Massimo Bergamasco
Scuola Superiore Sant'Anna

PERCRO
56127, Pisa, Italy
bergamasco@sssup.it

Luigi Biagiotti
University of Bologna
40136 Bologna, Italy
lbiagiotti@deis.unibo.it

Juan M. Bogado
Universidad Simón Bolivar
1080A - Valle Sartenejas (Baruta)
Miranda, Venezuela
jbogado@usb.ve

Martin Buss
Technische Universität München
Institute of Automatic Control
Engineering (LSR)
D-80290 Munich, Germany
mb@tum.de, M.Buss@ieee.org

Marc Carreras
University of Girona
Institute of Informatics and Applications
17071 Girona, Spain
marcc@eia.udg.es

Alícia Casals
Technical University of Catalonia
(UPC)
Automatic Control and Computer
Engineering Department
08028 Barcelona, Spain
alicia.casals@upc.es

Pedro L. Castedo
Universidad Politécnica de Madrid
Dept. Electrónica, Automática e
Informática Industrial
28012 Madrid, Spain
castedo@elai.upm.es

Nikhil Chopra
University of Illinois at Urbana-
Champaign
Coordinated Science Lab (CSL)
Urbana, Illinois 61801. USA
nchopra@uiuc.edu

Salvador Cobos
Universidad Politécnica de Madrid
Dpto. de Automática, Ingeniería
Electrónica e Informática Industrial
28006 Madrid, Spain
cobosalvador@etsii.upm.es

Raúl Correal
Universidad Carlos III de Madrid
Robotics Lab.
Depto. de Ingeniería de Sistemas y
Automática
28911 (Leganés) Madrid, Spain
rcorreal@ing.uc3m.es

Barbara Deml
University of the Armed Forces Munich
Human Factors Institute
D-85577 Neubiberg, Germany
barbara.deml@unibw.de

Francisco J. Domínguez
University of Málaga
Dept. Ingeniería de Sistemas y
Automática
29013Málaga, Spain
frandominguez@isa.uma.es

David Eadie
Trinity College
Dept. of Computer Science
Dublin 2, Ireland
david.eadie@cs.tcd.ie

Cesare Fantuzzi
Univ. of Modena and Reggio Emilia
42100 Reggio Emilia, Italy
fantuzzi.cesare@unimore.it

Cesar Fernandez
Universidad Miguel Hernández
Dpto. Ingeniería de Sistemas Industriales
03202 (Elche) Alicante, Spain
c.fernandez@umh.es

Josep Fernández
Technical University of Catalonia (UPC)
Automatic Control and Computer Engineering Department
08028 Barcelona, Spain
josep.fernandez@upc.edu

Manuel Ferre
Universidad Politécnica de Madrid
Dpto. de Automática, Ingeniería Electrónica e Informática Industrial
28006 Madrid, Spain
m.ferre@upm.es

Manel Frigola
Technical University of Catalonia (UPC)
Automatic Control and Computer Engineering Department
08028 Barcelona, Spain
manel.frigola@upc.edu

Antonio Frisoli
Scuola Superiore Sant'Anna
PERCRO
56127, Pisa, Italy
a.frisoli@sssup.it

Ernesto Gambao
Universidad Politécnica de Madrid
Dpto. de Automática, Ingeniería Electrónica e Informática Industrial
28006 Madrid, Spain
gambao@etsii.upm.es

Nicolás M. García
Universidad Miguel Hernández
Dpto. Ingeniería de Sistemas Industriales
03202 Elche-Alicante, Spain
nicolas.garcia@umh.es

Vincenzo Genovese
Scuola Superiore Sant'Anna
Arts Lab
Pisa 56127, Italy
genovese@mail-arts.sssup.it

Antonio Giménez
Universidad Carlos III de Madrid
Robotics Lab.
Depto. de Ingeniería de Sistemas y Automática
28911 (Leganés) Madrid, Spain
agimenez@ing.uc3m.es

Jes t'us M. Gómez
University of Málaga
Dept. Ingeniería de Sistemas y Automática
29013Málaga, Spain
degabriel@ctima.uma.es

Emili Hernandez
Universitat de Girona
Institute of Informatics and Applications
17071 Girona, Spain.
emilihb@eia.udg.es

Miguel Hernando
Universidad Politécnica de Madrid
Dpto. Electrónica, Automática e Informática Industrial
28012 Madrid, Spain
miguel.hernando@upm.es

Sandra Hirche
Tokyo Institute of Technology
Fujita Laboratory, Dept. of Mechanical and Control Engineering
152-8552 Tokyo, Japan
s.hirche@ieee.org

Gerd Hirzinger
German Aerospace Center (DLR)
Institute of Robotics and Mechatronics
D-82234Wessling, Germany
erd.Hirzinger@dlr.de

Juan M. Ibarra
Instituto Politécnico Nacional
Centro de Investigacin y de Estudios Avanzados (CINVESTAV)
07300,México DF, México
jibarra@ctrl.cinvestav.mx

Alberto Jardón
Universidad Carlos III de Madrid
Robotics Lab.
Depto. de Ingeniería de Sistemas y Automática
28911 (Leganés) Madrid, Spain
ajardon@ing.uc3m.es

Ankur Kapoor
The Johns Hopkins University
Department of Computer Science
Baltimore, MD 21218, USA
kapoor@cs.jhu.edu

Abderraman Khedar
AIST/CNRS Joint Japanese-French Robotics Laboratory
Intelligent Systems Research Institute
AIST Central 2, Umezono 1-1-1, 305-305-8568, Tsukuba, Japan.
kheddar@ieee.org

Bernhard Kubler
German Aerospace Center (DLR)
Institute of Robotics and Mechatronics
D-82234Wessling, Germany

Martin Kuschel
Technische Universität München
Institute of Automatic Control Engineering (LSR)
D-80290 Munich, Germany
Martin.Kuschel@tum.de

Kwang-Kyu Lee
Technische Universität München
Institute of Automatic Control Engineering (LSR)
D-80290 Munich, Germany
kk.lee@tum.de

Klaus Landzettel
German Aerospace Center (DLR)
Institute of Robotics and Mechatronics
D-82234Wessling, Germany
Klaus.Landzettel@dlr.de

Ming Li
The Johns Hopkins University
Department of Computer Science
Baltimore, MD 21218, USA
liming@cs.jhu.edu

Raúl Marín
University of Jaume I
Departament of Computer Science and Engineering
12006 Castelló, Spain
rmarin@icc.uji.es

Santiago Martínez
Universidad Carlos III de Madrid
Robotics Lab.
Depto. de Ingeniería de Sistemas y Automática
28911 (Leganés) Madrid, Spain
smartinez@ing.uc3m.es

Barry McCullagh
Trinity College
Dept. of Computer Science
Dublin 2, Ireland
Barry.McCullagh@cs.tcd.ie

Claudio Melchiorri
University of Bologna
Laboratory of Automation and Robotics
Department of Electronics, Computer Science and Systems
40136 Bologna, Italy
cmelchiorri@deis.unibo.it

Nico Mock
Technische Universität Berlin
Control Systems Group
D-10587 Berlin, Germany
nico.mock@iee.tu-berlin.de

Adrián Mora
Universidad Politécnica de Madrid
Dpto. de Automática, Ingeniería Electrónica e Informática Industrial
28006 Madrid, Spain
amora@etsii.upm.es

Victor F. Muñoz
University of Málaga
Dept. Ingeniería de Sistemas y Automática
29013Málaga, Spain
victor@ctima.uma.es

Manuel Navas-Herreros
Dyoptyka Ltd.
Sth. Cumberland St.,
Dublin 2, Ireland
mnavas78@yahoo.es

Norbert Nitzsche
Technische Universität München
Institute of Automatic Control Engineering (LSR)
D-80290 Munich, Germany
norbert.nitzsche@mytum.de

Emmanuel Nuño
Technical University of Catalonia
Institute of Industrial and Control Engineering
08028 Barcelona, Spain
emmanuel.nuno@upc.edu

Ramón Ñeco
Universidad Miguel Hernández
Dpto. Ingeniería de Sistemas Industriales
03202 (Elche) Alicante, Spain
ramon.neco@umh.es

Tobias Ortmaier
German Aerospace Center (DLR)
Institute of Robotics and Mechatronics
D-82234Wessling, Germany
tobias.ortmaier@dlr.de

Narcis Palomeras
Universitat de Girona
Institute of Informatics and Applications
17071 Girona, Spain
npalomer@eia.udg.es

Georg Passig
German Aerospace Center (DLR)
Institute of Robotics and Mechatronics
D-82234Wessling, Germany
georg.passig@dlr.de

Luis Payá
Universidad Miguel Hernández
Dpto. Ingeniería de Sistemas Industriales
03202 (Elche) Alicante, Spain
lpaya@umh.es

Angelika Peer
Technische Universität München
Institute of Automatic Control Engineering (LSR)
D-80290 Munich, Germany
Angelika.Peer@tum.de

Carlos Pérez
Universidad Miguel Hernández
Dpto. Ingeniería de Sistemas Industriales
03202 Elche-Alicante, Spain
carlos.perez@umh.es

Christophe Rabaud
Dyoptyka Ltd.
Sth. Cumberland St.,
Dublin 2, Ireland
crabaud@dyoptyka.com

Oscar Reinoso
Universidad Miguel Hernández
Dpto. Ingeniería de Sistemas Industriales
03202 (Elche) Alicante, Spain
o.reinoso@umh.es

Detlef Reintsema
German Aerospace Center (DLR)
Institute of Robotics and Mechatronics
D-82234Wessling, Germany
Detlef.Reintsema@dlr.de

Pere Ridao
Universitat de Girona
Institute of Informatics and Applications
17071 Girona, Spain
pere@eia.udg.es

Alberto Rodriguez
Technical University of Catalonia (UPC)
Automatic Control and Computer Engineering Department
08028 Barcelona, Spain
alberto.rodriguez@upc.edu

Adolfo Rodríguez
Technical University of Catalonia
Institute of Industrial and Control Engineering
08028 Barcelona, Spain
adolfo.rodriguez@upc.edu

José María Sabater
Universidad Miguel Hernández
Dpto. Ingeniería de Sistemas Industriales
03202 (Elche) Alicante, Spain
j.sabater@umh.es

Angelo Sabatini
Scuola Superiore Sant'Anna
Arts Lab.
Pisa 56127, Italy
a.sabatini@arts.sssup.it

Roque J. Saltarén
Universidad Politécnica de Madrid
Dpto. de Automática, Ingeniería Electrónica e Informática Industrial
28006 Madrid, Spain
rsaltaren@etsii.upm.es

Miguel A. Sanchez-Urán
Universidad Politécnica de Madrid
Departamento de Ingeniería Eléctrica
28012 Madrid, Spain
mas@ie.upm.es

José Santos-Victor
Instituto Superior Técnico
Instituto de Sistemas e Robótica
1049-001 Lisbon, Portugal
jasv@isr.ist.utl.pt

Pedro J. Sanz
University of Jaume I
Departament of Computer Science and Engineering
12006 Castelló, Spain
sanzp@icc.uji.es

Cristian Secchi
Univ. of Modena and Reggio Emilia
42100 Reggio Emilia, Italy
secchi.cristian@unimore.it

Ulrich Seibold
German Aerospace Center (DLR)
Institute of Robotics and Mechatronics
D-82234Wessling, Germany
Ulrich.Seibold@dlr.de

Javier Serón
University of Málaga
Dept. Ingeniería de Sistemas y Automática
29013Málaga, Spain
javierseron@ctima.uma.es

Fergal Shevlin
Dyoptyka Ltd.
Sth. Cumberland St.,
Dublin 2, Ireland
fshevlin@dyoptyka.com

Ee-Sian Neo
AIST/CNRS Joint Japanese-French Robotics Laboratory
Intelligent Systems Research Institute
AIST Central 2, Umezono 1-1-1,
305-305-8568, Tsukuba, Japan.
rio.neo@aist.go.jp

Mark W. Spong
University of Illinois at Urbana-Champaign
Coordinated Science Lab (CSL)
Urbana, Illinois 61801. USA
mspong@uiuc.edu

Bartlomiej Stanczyk
Technische Universität München
Institute of Automatic Control
Engineering (LSR)
D-80290 Munich, Germany
stanczyk@tum.de

Stefano Stramigioli
University of Twente
7500 AE Enschede, Nederland
S.Stramigioli@ieee.org

Riichiro Tadakuma
AIST/CNRS Joint Japanese-French
Robotics Laboratory
Intelligent Systems Research Institute
AIST Central 2, Umezono 1-1-1,
305-305-8568, Tsukuba, Japan.
r-tadakuma@aist.go.jp

Russell H. Taylor
The Johns Hopkins University
Department of Computer Science
Baltimore, MD 21218, USA
rht@cs.jhu.edu

Marc Ueberle
Technische Universität München
Institute of Automatic Control
Engineering (LSR)
D-80290 Munich, Germany
ueberle@tum.de

Ulrich Unterhinninghofen
Technische Universität München
Institute of Automatic Control
Engineering (LSR)
D-80290 Munich, Germany
ulrich.unterhinninghofen@tum.de

Raúl Wirz
University of Jaume I
Departament of Computer Science and
Engineering
12006 Castelló, Spain
wirz@icc.uji.es

Kazuhito Yokoi
AIST/CNRS Joint Japanese-French
Robotics Laboratory
Intelligent Systems Research Institute
AIST Central 2, Umezono 1-1-1,
305-305-8568, Tsukuba, Japan.
kazuhito.yokoi@aist.go.jp

Introduction to Advances in Telerobotics

Manuel Ferre[1], Martin Buss[2], Rafael Aracil[1], Claudio Melchiorri[3], and Carlos Balaguer[4]

[1] Universidad Politécnica de Madrid
Dpto. Automática, Ingenierá Electrónica e Informática Industrial
C/. José Gutierrez Abascal 2, 28006 Madrid, Spain
aracil@etsii.upm.es, m.ferre@upm.es

[2] Technische Universität München
Institute of Automatic Control Engineering (LSR)
D-80290 Munich, Germany
M.Buss@ieee.org

[3] University of Bologna
Department of Electronics, Computer Science and Systems
Via Risorgimento 2, 40136 Bologna, Italy
cmelchiorri@deis.unibo.it

[4] University Carlos III de Madrid
Depto. de Ingeniería de Sistemas y Automática
Avenida de la Universidad, 30. 28911 (Leganés) Madrid, Spain
balaguer@ing.uc3m.es

The objective of atelerobotic system is to connect humans and robots in order to reproduce operator actions at a distance. We must take numerous matters into account in order to efficiently produce such interaction, most of which are related to human interaction, distributed control, and information flow between operator and remote robots. All these topics are dealt with in depth in this book.

Early developments in Telerobotics were carried out in the areas of nuclear, underwater and space applications [1, 2] with the common aim of reducing risk to human life in each of these fields. During the 1950's and 1960's Ray Goertz planted the seed of Telerobotics at the Argon National Laboratory when he achieved the safe handling of radioactive materials, a process required both for power generation and for military usage [3]. This greatly spurred on all teleoperation-related technologies and as a result many telerobotic systems have been developed since then [1, 4, 5, 6, 7, 8, 9].

Specific features may be observed in the underwater field due to the hostility of the environment, high pressure, poor visibility and corrosion being the most relevant of these. Successful usage in rescue missions and seabed works have continuously fostered the development of new teleoperated manipulators for underwater vehicles [1, 10]. Their first well known application dates back to 1966 in the South of Spain in the decommissioning of a nuclear weapon by the U.S. Army. Since then, new developments have been introduced in order to improve the performance of underwater tasks.

M. Ferre et al. (Eds.): Advances in Telerobotics, STAR 31, pp. 1–7, 2007.
springerlink.com © Springer-Verlag Berlin Heidelberg 2007

Telerobotic technology has also been greatly boosted through the area of Space Investigation. The Flight Telerobotic Servicer (FTS) program [11] and the NASA/NBS Standard Reference Model (NASREM) [12]have been some relevant NASA contributions. The FTS was an ambitious program focussed on assisting and reducing crew for extravehicular activities. This program started in 1987 and was called off in 1991. NASREM architecture describes teleoperation systems as a hierarchical structure based on a 7-level task division, which at the same time is built in three blocks dedicated to information processing, task modelling and operator performance. A summary of NASA's main contributions to Telerobotics can be found in the 21st Century Guide to Robotics [13]. Other relevant studies in Telerobotics have been carried out by the German Aerospace Center [14] and by the Japanese Space Agency, NASDA [15].

Whilst the above mentioned form the traditional core of Teleoperation some advances have been extrapolated to other industrial fields. The major drawbacks, however, have been the particularity of designs for its respective telerobotic task. New fields of application have emerged during the 80s and 90s, such as surgery [16], live-line maintenance [17, 18, 19], rescue [20], education, people assistance, mining, etc. [7, 21], all characterized by reducing the risk for human operators to an acceptable or non-existent level. The main challenges lie in linking the local operator and the remote robot sites efficiently, since achieving top level performance is a crucial factor for any successful telerobotic system. It is vital to achieve such proficiency since new teleoperation systems are applied as an alternative to the traditional manual procedure of tasks in the above mentioned fields. Moreover, the complexity of the teleoperated system is a further essential factor to be considered in its design. A balance between complexity and performance must be taken into account in order to obtain an efficient and feasible telerobotic system.

Advances in human interfaces and in the area of control are therefore necessary to develop powerful and useful Telerobotics. The topics analyzed in this book are specifically aimed at researchers interested in the development of such new applications. The chapters are organized into three main parts focused on the design of human interfaces, the area of control and the description of new developments in Telerobotics. The following sections set out how the contents have been organised.

1 Human Interfaces

The first part of the book, entitled "Human Interfaces", analyses the interaction of a human operator in a teleoperation control loop. The main topics concern human perception of a remote environment and the transmission of operator actions. These topics and keywords are summarized in table 1.

The first chapter is an introduction about human interfaces within Teleoperation, providing a general overview of the human role in Telerobotics. The human perception of remote environments is analyzed first, and then the generation

Table 1. Chapter main topics and keywords for Part I: Human Interfaces

Chap.	Main topic	Keywords
1	Human interaction	Telepresence, multimodal interfaces remote control
2	Guidance	Haptic interfaces, serial kinematics, hyper-redundant kinematics
3	Guidance	Haptic interfaces, parallel kinematics, telemanipulation parameters
4	Exoesqueleton	Anthropomorphic interfaces, Human body modelling
5	3D perception	Stereoscopic vision, binocular disparity, automatic vergence control
6	3D perception	3D display, adaptive optics, real-time image processing
7	Voice interfaces	Voice commands, natural language, manipulation assistance tools
8	Advance interfaces	Gesture recognition, computer vision, body modelling
9	Advance interfaces	Virtual reality, augmented reality, graphic programming teleoperation

of operator command is dealt with. The next two chapters cover the guidance of telemanipulators by means of haptic devices. One of the most basic tasks of remote control is telerobot guidance as it represents the capability of remote manipulation in reproducing operator hand movements. Moreover, remote contact forces reflected on the operator significantly increase operator dexterity for remote telemanipulation. Serial haptic master devices are examined in chapter two, describing and analyzing in detail a hyper-redundant device with 10 degrees of freedom. As a result, good performance in different types of remote environments has been observed. Chapter three examines parallel haptic devices showing a comparison between a well known serial interface and a new parallel haptic interface in which the performance of each is measured according to telemanipulation parameters. Chapter four focuses on exoskeleton devices whose anthropomorphic design allows the identification and reproduction of accurate human behaviour. The following two chapters deal with stereoscopic vision of remote locations in order to reproduce human perception of depth. It is fundamental for the operator to be provided with 3D information in order to manipulate remote objects more precisely. Chapter five describes an advanced stereoscopic video interface that includes an automatic vergence control in order to minimize image disparity, which allows extensive movements in the remote site. Chapter six presents a new 3-D display developed by StereoFocus. This device is based on the application of adaptive optics in order to display image depth properly. The next two chapters describe examples of applications of artificial intelligence in Telerobotics. Chapter seven studies voice command generation for controlling a remote robot, allowing high level commands to be sent during remote task execution. Natural language orders can be processed in real time, but some constraints have to be taken into account such as those described by the authors. In chapter eight we find techniques for identifying and translating human gestures by remote control based on computer vision. The human body is modelled and its gestures are used to interact with the telerobot. The last chapter in this section analyses the contribution that virtual reality technologies have made to Telerobotics and presents an exposition of current applications are shown.

2 Control

This part is made up of nine chapters, the first seven focussing on bilateral control and the last two analyzing specific architectures for teleprogramming and reconfigurable systems. The goal of a bilateral telerobotics system is to design a master controller that reproduces the slave robot interaction to the operator as accurately as possible. Early telemanipulators demonstrated that bilateral controllers significantly improved remote task performance. However, bilateral systems give rise to numerous problems related to stability in the event of communication delays, the balance of energy flowing between master and slave, the human perception of the remote environment, and the properties of the mechanical device which prevent the development of ideal mechanisms with no damping or bandwidth limits. The topics analyzed in this part are summarized in table 2.

Table 2. Chapter keywords of part II

Chap.	Keywords
10	2-channel models, 4-channel models, delays, scattering, passivity
11	Experimental evaluation of bilateral models,
12	Transparency, environment impedance evaluation, haptic perception
13	Intrinsically passive controllers, compliant environments, human perception
14	Power scaling, port-Hamiltonian modelling, controller design
15	Robot network, passive systems, multi-agents systems
16	State-space modelling, state convergence, controller design
17	Position vs. rate control, reconfigurable systems
18	Teleprogramming, predictive simulation, virtual force generation

The first chapter in this part constitutes an introduction to bilateral control summarizing the main modelling techniques applied in Telerobotics. The architecture of the two- and four- channel models are described in the first sections and passivity is introduced and evaluated in a testbed. The following chapter presents many experiments that implement the above control schemes. The results of these experiments allow better comprehension of such bilateral architectures. Chapter twelve analyzes transparency and impedance concepts in the event of communication delays, and centres on the subject of time delay transparency analysis from the point of view of human perception. Some psychophysical insights such as just noticeable difference (JND) are also discussed. In chapter thirteen we discuss the modelling of remote environments to maximize the degree of transparency obtained, looking at linear and nonlinear contact models in describing compliant environments. Intrinsically passive controllers are used for bilateral control in this study. Chapter fourteen models a bilateral controller as a port-Hamiltonian and packet switched network with the goal of guaranteeing the stability of a scaled-down/up master-slave system in the event of delays and the loss of packets in the communication channel. In chapter fifteen we extend the passive theory to networks that link master and slave devices with different configurations, examining the synchronization and passivity of many devices. In chapter sixteen we are introduced to a new methodology for designing bilateral controllers, defining master and slave according to their state-space description. State error between both devices is calculated, so controllers are designed in order to cancel out such error. Chapter seventeen describes a reconfigurable system that permits the position and rate control of a telerobot. Both types of controllers are compared under different conditions, thereby obtaining criteria concerning their best performance configuration. Finally, the last chapter is focussed on teleprogramming techniques, which are used in event of long communication delays. A predictive simulator is described and a test bed is used to evaluate the teleoperation performance.

3 Applications

The third and final part of this book concentrates on showing the latest developments in Telerobotics. New systems are being designed due to advances in teleoperation technologies related to new human interfaces that allow more powerful human-machine interaction, better network protocols that enable real time control for distributed systems and more powerful computers that execute faster processing. New applications in Telerobotics mean this technology is continually growing and extending its use into new areas. The main obstacle to the usage of Telerobotics in new fields is cost, since it is a two-fold system which implies the development of both operator and robot sites. The fields of application and keywords for the chapters in part III are summarized in table 3.

The ten chapters which make up this final section provide an excellent overview into the current state of the art. Advanced robotics in orbital space applications is dealt with in chapter nineteen which summarizes pioneer developments in robot assisted on-orbit assembly and servicing for the near future, covering its application in the areas of inspection, and maintenance or repair of infrastructure and systems. Key robotics technologies are analysed regarding their future contribution to teleoperated on-orbit servicing applications. Chapter

Table 3. Fields and keywords of chapters in Part III

Chap.	Field	Keywords
19	Space	On-Orbit Servicing, outer space and maintenance tasks DLR's telerobotics technologies
20	Underwater	Underwater robots, distributed architecture, navigation experiments
21	Telesurgery	Minimally invasive robotic surgery, sensorized instruments force feedback evaluation
22	Telesurgery	Telesurgery laboratory, minimally invasive surgery, teleoperation interface
23	Telesurgery	Telesurgery tools, motion constrain, modelling tasks
24	Assistance	Disabled people, teleassistance robot, domestic environments
25	Education	Education platforms, distributed architecture, remote robot experiments
26	Telemanipulation	IPv6 protocol, geometric reasoning, virtual force feedback
27	Telemanipulation	Aerial live power line works, telemaintenance
28	Telemanipulation	Bimanual telemanipulation, extensive telepresence, collaborative tasks

twenty describes the new control architecture of an underwater vehicle showing its control in a network. The following three chapters are focussed on telesurgery applications. Chapter twenty-one describes a prototypical force reflecting robotic surgery system based on two surgical robots which are applied for minimally invasive surgery followed by a description, in. chapter twenty-two, of a complete telerobotic system for laparoscopy and minimally invasive surgery. The following chapter analyses movement constraints that improve surgical procedures, such as suturing or tissue cutting. Chapter twenty-four informs about teleoperated robot assistance for disabled people, demonstrating how a robot with the proper user interface can become a powerful tool for assisting people in this situation. Chapter twenty-five gives a further example of the application of Telerobotics in new fields in describing a telelaboratory together with the results of several courses in Telerobotics. Chapter twenty-six explains how an industrial robot is integrated into a teleoperation architecture with powerful tools being developed for geometric reasoning. Chapter twenty-seven described an telerbotic application for live-line maintenance. It is a clear example of tasks where operator has a medium risk and teleoperation is an alternative to classical procedures. The last chapter deals with advanced topics on teleoperation such as dual-handed, extensive telepresence and collaborative telemanipulation. These topics have been studied in some test beds.

References

1. H.L. Martin and D.P. Kuban. *Teleoperated Robotics in Hostile Environments.* Robotics International of SME, 1985.
2. J. Vertut and P. Coiffet. *Robot Technology, Volume 3A: Teleoperation and Robotics: Evolution and Development.* Prentice-Hall, 1986.
3. R. Goertz. Manipulator Systems Development at ANL. In *Proceedings of the 12th Conference on Remote Systems Technology, ANS*, pages 117–136, 1964.
4. M.C. Becquet. *Teleoperation: Numerical Simulation and Experimental Validation.* Kluwer Academic Publishers, 1992.
5. M. Jamshidi and P. Eicher. *Robotics and Remote Systems for Hazardous Environments.* Prentice-Hall, 1993.
6. Y. Shinohara, U. Hozumi, and F. Yoshio. Development of telerobotic manipulators for reactor dismantling work. *Advances Robotics*, 8(1):95–120, 1994.
7. T. Schilling. *Telerobotic Applications.* Professional Engineering Publishing, 2000.
8. T.B. Sheridan. *Telerobotics, Automation, and Human Supervisory Control.* MIT Press, Cambridge, Massachusetts, 1992.
9. T.B. Sheridan. Telerobotics. *Automatica*, 25(4):487–507, 1989.
10. J. Yuh. Design and control of autonomous underwater robots: A survey. *Autonomous Robots*, 8:7–24, 2000.
11. H.G. McCain, J.F. Andary, D.R. Hewitt, and D.C. Haley. The space station freedom flight telerobotic servicer: the design and evolution of a dexterous space robot. *Astronaut*, 24:45–54, 1991.
12. J.S. Albus, R. Lumia, and H.G. McCain. Nasa/nbs standard reference model for telerobot control system architecture (nasrem). *NASA document SS-GSFC-0027*, 1987.
13. NASA-JPL *et al.* *21st Century Complete Guide to Robotics.* CDROM, Progressive Management, 2004.
14. G. Hirzinger. Robots in space - a survey. *Advanced Robotics*, 9(6):625–651, 1995.
15. K. Machida, T. Mikama, S. Komada, and K. Akita. Precise ev robot: Flight model and telerobotic operation for ets-vii. In *Proc. of 1996 IEEE Int. Conf. on Intelligent Robots and Systems*, pages 1550–1557, 1996.
16. G.H. Ballantyne, J. Marescaux, and P.C. Giulianotti. *Primer of Robotic and Telerobotic Surgery.* Lippincott Williams and Wilkins, 2004.
17. R. Aracil, M. Ferre, M. Hernando, E. Pinto, and J.M. Sebastian. Telerobotic system for live-power line maintenance: Robtet. *Control Engineering Practice*, 10:1271–1281, 2002.
18. EPRI - Electric Power Research Institute. Live-line repair with tomcat. *EPRI Journal Robotics*, 1:1–29, 1987.
19. Y. Maruyama. Robot applications for hot-line maintenance. *Industrial Robots*, 27:357–365, 2000.
20. R.J. Anderson, D.L. Shirey, and W.D. Morse. A telerobot control system for accident response. In *Proc. of the 2000 IEEE Int. Conf. on Robotics and Automation*, pages 2000–2006, 2000.
21. L.E. Parker and J. Draper. Maintenance and repair. In Shimon Nof, editor, *Handbook of Industrial Robotics*, pages 1023–1036. Wiley, 2 edition, 1999.

References

Part I

Human System Interfaces

1

The Human Role in Telerobotics

Rafael Aracil[1], Martin Buss[2], Salvador Cobos[1], Manuel Ferre[1], Sandra Hirche[2], Martin Kuschel[2], and Angelika Peer[2]

[1] Universidad Politécnica de Madrid
Dpto. de Automática, Ingeniería Electrónica e Informática Industrial
C/. José Gutierrez Abascal 2, 28006 Madrid, Spain
aracil@etsii.upm.es, m.ferre@upm.es, cobosalvador@etsii.upm.es

[2] Technische Universität München
Institute of Automatic Control Engineering (LSR)
D-80290 Munich, Germany
{mb,hirche,martin.kuschel,angelika.peer}@tum.de

Summary. This chapter introduces the main topics of a telerobotic system. It describes the architecture of such a system from a general point of view and emphasizes the interaction between a human operator and a robot that performs the task in the remote environment. Furthermore it focuses on multi-modal human system interfaces and explains the main features of haptic, auditory, and visual interfaces. Finally important issues for the measurement and evaluation of the attribute telepresence are described.

1.1 Introduction

Telerobotic systems allow human operators to properly interact with a telerobot to telemanipulate objects located in a remote environment. This means that human actions are extended to remote locations allowing the execution of complex tasks and avoiding risky situations for the human operator [1].

In a telerobotic system the human operator plays an important role. He *perceives* information from the remote environment through the human system interface and *acts* accordingly by sending commands to the remote devices. Thus the human system interface has two important functions; first, it has to excite the operator senses so as to show the status of the executed task in the remote environment and second, it has to process the operator commands in order to properly control remote devices. Multi-modal[1] commands are generated by the operator at his or her working site by means of the human system interface using motion, force, voice or symbolic inputs. Such commands are transmitted to the telerobot in order to perform the remote task. Sensors are placed at the remote site to gather data from the task which is then transmitted back to the operator and displayed by the multi-modal human system interface.

Designing of multi-modal human system interfaces is one of the key challenges in telerobotics. Vision, audition, and haptics are senses excited by the multi-modal

[1] The term *multi-modal* refers to the different human senses.

M. Ferre et al. (Eds.): Advances in Telerobotics, STAR 31, pp. 11–24, 2007.
springerlink.com © Springer-Verlag Berlin Heidelberg 2007

interface. It is thereby important to take into account human perception capabilities so as to obtain a better interaction.

Telepresence is one of the key factors that enhances performance of a telerobotic system. Telepresence means that the information about the remote environment is displayed to the operator in a natural manner, which implies a feeling of presence at the remote site. A good degree of telepresence guarantees the feasibility of the required manipulation task.

The following sections describe the key points of a telerobotic system. Section two shows a general structure of such a system and explains its main components and functions. Section three is focused on the design of multi-modal human system interfaces. Main features of human senses and capabilities are briefly described. It allows defining criteria for a better design of visual, auditory and haptic devices. Section four deals with the concept of telepresence and other performance measurements. Lastly, conclusions of this chapter are summarized in section five.

This chapter is an introduction to the topics of this part of the book. Chapters 2 and 3 describe serial and parallel haptic interfaces respectively. Chapter 4 focuses on exoskeletons. Chapters 5 and 6 describe two different stereoscopic video systems that reproduce human binocular vision. Chapter 7 deals with voice command generation for telerobotics. Chapter 8 describes how to process operator gestures in order to remotely control a robot. Finally, chapter 9 contains a review of the technology in virtual reality applied to telerobotics.

1.2 General Structure of a Telerobotic System

A telerobotic system is comprised of two main parts; the operator environment and the remote environment, as visualized in Fig. 1.1. Both environments are linked by a communication channel that transmits commands from the operator to the remote devices and sends back information of the remote task to the operator. The operator environment is made up of a multi-modal human system interface, which the operator uses in order to control the remote devices. The remote environment consists of teleoperated devices, sensors and objects that take part in the teleoperation task. Each environment contains processing modules which have double functions: first, to transform data transmitted by the communications channel and second, to execute the corresponding local control loops.

A central issue related to the design and operation of telerobotic systems is the degree of coupling between the human operator and the remote robot. It is generally classified as *weak* or *strong*. If the operator gives symbolic commands to the robot by pushing buttons and watching the resulting action in the remote environment, its coupling is rather *weak*. Some degree of "intelligence" is required for a remote robot to execute such symbolic commands. The coupling is comparably *strong* for the kinesthetic modality in a bilateral teleoperation scenario. Commonly, the motion (and/or force) of the human operator is measured, communicated and used as a set-point for teleoperator motion (and/or force) controller. On the other hand, forces (motions) of the teleoperator in the remote environment are sensed, communicated and fed back to the human operator through the multi-modal human system interface. The degree of coupling is thereby related to the control distribution between operator and remote robot controller. Literature on telerobotics distinguishes among *shared*, *cooperative*, *supervisory*, and *bilateral control*. A comprehensive review of the control modes in telerobotics can be found in [2].

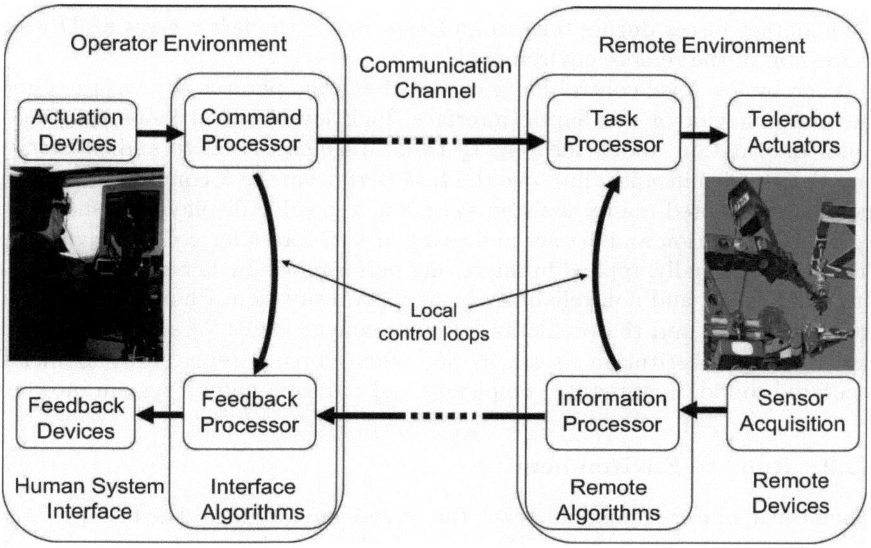

Fig. 1.1. Main modules of a telerobotic system

1.2.1 Operator Environment

The human system interface plays an important role in a telerobotic system. It provides input devices that are used to generate operator commands and display devices that are used to monitor the interaction between remote robot and environment. Telerobot commands are generated by input devices that identify the operator actions. According to the control mode, commands have to be processed to a greater or lesser degree before they are transmitted to the remote environment. For example, when an operator executes a guidance task using a master-slave system, i.e. with strong coupling, motion (force) commands are continuously processed. They could be scaled or transformed to different coordinates. This is an example for a rather simple processing. More complex processing would be required if commands were symbolic, like e.g. "picking an object". Symbolic commands have to be transformed - on operator or remote site - to the corresponding sequence of remote device actions.

Simultaneously, multi-modal sensor information is received from the remote environment. This multi-modal feedback consists of 2D or 3D visual, mono/stereo acoustic, haptic (force, motion, tactile, temperature) and symbolic information which is generated by feedback information processors and displayed by the corresponding interface devices. The purpose of a feedback device is to excite the operator's senses in order to show him the remote task status. Force feedback master-arms and stereoscopic screens are typical examples of devices used as human system interfaces. The former informs the operator about applied contact forces during telemanipulation, while the latter gives a 3D visual impression of the remote environment.

Whereby low level control loops executed at the operator site ensure a good tracking behavior of the haptic interface, high level control loops show additional information about the remote task. *Augmented reality* and *predictions* may thereby significantly improve the task performance. A common example of such an augmented reality assisted system is a graphic display that shows safe region for operation and arrows indicating virtual forces for collision avoidance. Prediction is usually applied in improving performance by lowering the effect of long time delays and non-reliability in signal transmission. Photo-realistic scene prediction [3, 4] and the prediction of environment forces [5] are typical examples. Sensory substitution as e.g. in [6], where force is replaced by artificially generated sound, may reduce complexity and cost of a human system interface.

1.2.2 Remote Environment

When the operator commands reach the remote environment, the task processor transforms them into actions. Once again, the complexity of data processing depends on the type of command and the degree of coupling. Complex data processing is required when the operator and the telerobot are weakly coupled, i.e. in cases where the robot has some degree of autonomy or when the robot only receives symbolic commands. Simple data processing is required when the operator and the telerobot are strongly coupled.

The information captured by sensors is used in obtaining data from the remote task and sending them to the operator environment via the communication channel. Computer vision recognition and object localization algorithms are good examples of sensor processing. They obtain information from the objects located at the remote site and thereby define spatial positions of telemanipulated objects.

Local control loops that are executed at the remote site, ensure the motion (force) tracking of the robot. Trajectories are provided by the operator or generated from symbolic commands. Several researchers have looked into human skill and expertise modelling so as to supplement control from the local teleoperator, e.g. [7, 8, 9, 10, 11, 12]. The main concept is to have an intelligent teleoperator that performs tasks by demonstration. Such operator can acquire expert control knowledge (skills) from measured data and apply skills in performing tasks in semi-autonomous teleoperation control.

1.3 Multi-modal Human System Interfaces for Telerobotic Applications

The interaction with a remote robot is done through the human system interface, which transmits operator's actions and excites human senses according to the information received from the remote environment. Multi-modal human system interfaces refer to the perceptual modalities of human beings, such as visual, auditory, and haptic [2] modalities. Thus designing new devices the human sensing ability must be taken into account.

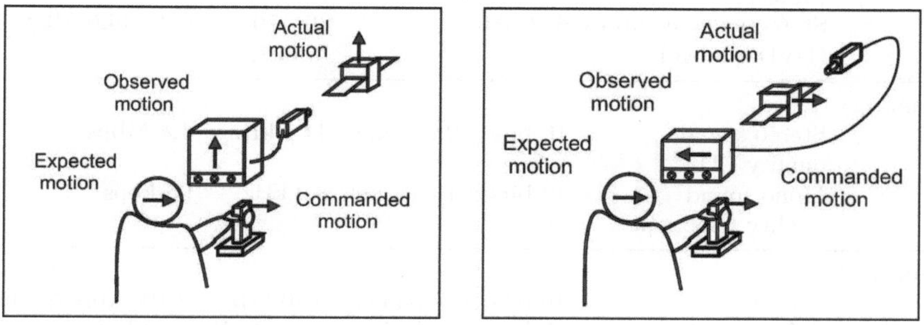

Fig. 1.2. Examples of visual disadjustment due to kinematic transformations (left) or observational reasons (right)

Furthermore operator *teleproprioception* have to be considered. Operator telepropioception implies coherence between operator's commands and their execution. Fig. 1.2 describes some examples that show disadjustments between commanded and observed motions. Such disadjustments are due to kinematic transformations, observational reasons or relative movements between object and camera, which make the guidance references incoherent to the given visual references. Moreover aspects such as information redundancy and stimulus fidelity of the information provided to the human operator are essential in obtaining an accurate perception of the remote environment. The following sections review the human sensing abilities and provide a classification of state-of-the-art human system interfaces.

1.3.1 Sense of Vision

The sense of vision informs us about shapes, colours and distances of the objects that can be seen by the human eye. The retina consists of a large number of photoreceptor cells. The two main types are rods and cones. They are excited by light and transmit signals to the brain through the optic nerve. The brain

[2] Haptics refers to the feeling of force, motion, and vibration. It is divided into kinesthetic, proprioreceptive, and tactile submodalities.

Table 1.1. Main features of signals transmitted in a telerobotic system

Channel	Type of signals	Sample size (pixels or n. of bits)	Samples per second	Bandwidth
Visual				
	TV video (PAL/NTSC)	720x480 - 720x576	25 - 30 frames/s	165,9 Mbps
	TV Video compres. (DVD quality)	720x480 - 720x576	25 - 30 frames/s	5,2 Mbps
	Stereo video (uncompressed)	640x480	30 - 70 frames/s	147 - 344 Mbps
	Stereo video compres. (DVD quality)	640x480	30 - 70 frames/s	6,3 - 14,6 Mbps
Auditory				
	Stereo sound quality CD	16 bits x 2 channels	44,1 kHz	1,4 Mbps
	Mono sound quality telephone	12 bits x 1 channel	8,0 kHz	96 kbps
Haptics				
	Tactile	10 bits (per point)	0-10 kHz	0-100 kbps/point
	Soft contact forces	10 bits (per DoF) (6 DoF)	0,1-1,0 kHz	6-60 kbps
	Hard contact forces	10 bits (per DoF) (6 DoF)	10-100 Hz	0,6-6 kbps

processes this information in several layers in order to properly interpret visual excitation.

Numerous studies and experiments have been performed to compare monoscopic effectiveness of images versus stereoscopic images [13, 14, 15]. These studies show that stereoscopic images are better than monoscopic images when performing a telemanipulation task. Other studies, such as [16, 17] highlight the complexity of the mechanisms applied in stereoscopic image perception. These works demonstrate that monoscopic data such as shadows and reflections could be as important as stereoscopic data [18].

Human visual perception has three mechanisms to perceive spatial information, which are binocularity, motion parallax and image realism. Binocularity is due to having two points of view which are 6 to 7 centimeters apart. Spatial information received by this mechanism has a predominant effect for closer objects, which is less than 1 meter. Many visual interfaces such as head-mounted displays, shutter glasses, parallax barrier, etc. imitate this effect providing users with different images for each eye. The motion parallax effect is predominant for objects farther than 1 meter and refers to relative movements between objects. As everybody knows nearer objects move faster than farther objects. An example can be clearly stated out when driving: trees located next to the car move faster than the mountains in the background that can be seen without any motion. As a consequence, a monoscopic camera in motion informs about spatial positions because different points of view of the scene are provided. Finally the third mechanism, the image realism is related to our manner of perceiving environments, which is based on texture gradients, object projections, light reflections, shadows and so on. Features of these data imply transmitting high quality images that require a large bandwidth. According to table 1.1 the minimum bandwidth is 5,2 Mbps. It corresponds to monoscopic compressed images. The maximum bandwidth is 344 Mbps, which corresponds to stereoscopic images with 70 frames per second and a resolution of 640*480 pixels per image.

Visual Interfaces

Teleoperation visual interfaces show images from the remote site. Cameras observe the scene of the remote environment and the captured images are displayed on the corresponding interfaces that provide visual information to the human operator.

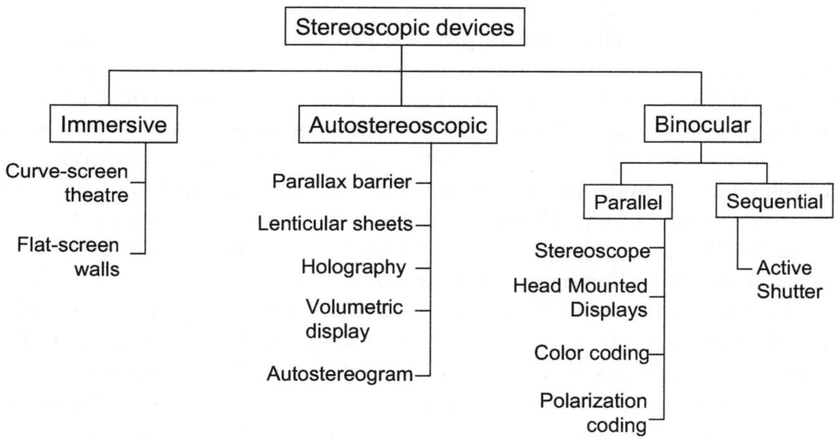

Fig. 1.3. Stereoscopic device classification

Simulating full human binocular vision requires better technology than what is currently available. No current display can meet all the specifications required in order to reproduce human depth perception properly. Therefore, a great variety of stereoscopic devices are available for specific applications, as is shown in Fig. 1.3. Stereoscopic devices can be divided into the following categories: binocular, autostereoscopic and immersive (according to its purpose). Binocular devices require an additional component such as glasses or a helmet in order to show a different image to each eye. Head Mounted Displays and systems based on shutter or polarized glasses are representatives of these devices. Autostereoscopic devices show a different image to each eye without needing any additional device, such as lenticular sheets or parallax barrier. Immersive devices make use of broad scenes where the sensation of depth is attained by covering the whole visual field, such as flat-screen walls and curve-screen theatres.

The number of images displayed per second is an important parameter for visual devices. Common visual interfaces usually show 25 or 30 frames per second, while stereoscopic devices based on shutter glasses display more than 60 frames per second. This high frequency is to avoid flicking problems upon visualization of stereoscopic images and it implies a large bandwidth as shown in table 1.1.

1.3.2 Sense of Hearing

Auditory cues are also important for teleoperation interfaces since they increase the situation awareness, attract visual attention and covey a variety of complex information without overwhelming the visual system. Auditory cues are particularly usefull when the visual channel is saturated or in the case of a limited field of view. Since the response to the auditory stimulus is fast (30-40 ms faster than for visual signals) sounds are also very suitable for alarms and sporadic messages from the computer interface.

Sound can be described by its physical properties frequency, intensity, and complexity. These properties correspond to the perceptual analogues pitch, loudness, and timbre. Designing auditory displays it is important to consider physical as well as perceptual properties. It is known that humans are able to hear sounds with frequencies from 20 Hz to 22.000 Hz, whereby the absolute sensitivity varies with frequency. Humans perceive sound intensity on a logarithmic scale, which spans over a range of 110-120 dB from just detectable sounds to sounds that cause pain. The physical measure "‘intensity"’ is not linear dependent on the perceptual measure "‘loudness"’ of a sound: The same increase in intensity can result in different increments in loudness, depending on the frequency of the signal. As the intensity of a signal is correlated to the loudness of a signal, the frequency correlates to the pitch. While for periodic signals the perceived pitch of a signal is directly dependent on the frequency of a signal, for nonperiodic signals the perceived pitch is affected by several stimulus attributes as e.g. harmonicity and loudness. Also the perception of timbre, which enables us to distinguish between different speakers or instruments depends on a number of physical parameters as spectral content and temporal envelope. Well known are also masking effects, which appear when multiple acoustic sources are presented to the listener simultaneously or in rapid succession.

Another important characteristic is the spatial acuity of the auditory system. While humans are able to distinguish sounds which are displaced by one degree from the median plane, this ability decreases drastically for sound sources located directly to the side of the human such that lateral displacements of about 10 degrees can be just detected. The human perception of spatial hearing is based on the evaluation of binaural, spectral, anechoic and dynamic cues as well as reverberation and the prior knowledge about the environment.

Auditory Interfaces

As already mentioned designing auditory interfaces it is important to consider physical as well as perceptual properties. E.g. it is known that the intensity of everyday sound has a range of about 80 to 90 dB. Thats why typical sound systems use 16 bits to represent the pressure of an acoustic signal (see table 1.1). Further it is known that speech information is characterized by frequencies of 200 to 5000 Hz. Hence acoustic interfaces must be designed to cover at least this range of frequencies.

In order to enable a high immersion signals should be replayed by providing an accurate spatial information. Spatial auditory cues can be generated either by using headphones or loudspeakers. While headphones allow a more precise control of the different auditory cues, a loudspeaker-system doesn't interfere with the human's head.

Depending on the different level of spatial immersion diotic, dichotic and spatialized headphone displays can be distinguished. Diotic displays present identical signals to both ears, dichotic displays simulate frequency independent interaural time and intensity differences and spatialized audio displays coupled with a head-tracking system can provide several spatial cues available in the real world. So called Head-Related-Transfer-Functions (HRTFs) are used to describe how an acoustic signal is transformed on the way from its source to the ear drum of the listener.

A similar impression to that generated with headphones can also be produced by using speakers arranged around the listener. In this case the signals of all speakers must be controlled in such a way, that the sum of all signals generates the appropriate spatial cues. Since all signals influence each other, the signals for each ear cannot be manipulated independently and complex calculations are necessary in order to get a real spatial impression. Even here nonspatial, stereo displays and spatial displays can be distinguished. While nonspatial displays correspond to the diotic, stereo displays (using only two speakers) correspond to the dichotic headphones. Most commercial available stereo headphones are based on this last mentioned two speaker system, which is able to control the lateral sound location. Using more than two speakers the spatial simulation increases. The most common Surround sound systems available at the market are 5.1, 6.1 and 10.1 systems.

1.3.3 Sense of Touch

The sense of touch is another complex sense which can be divided into two main components: the tactile and the kinesthetic component. Integration of both is known as haptics, which is a Greek word meaning "science of touch".

The tactile receptors are located directly under the skin and the stimulation of such receptors has a high frequency (up to 10 kHz). They participate in the first contact when the interaction with an object occurs. Furthermore they make it possible to perceive texture, geometry, and temperature of manipulated objects. Four different mechanoreceptors are distinguished according to the velocity of adaptation as well as the size of the receptive fields: FA II receptors (Pacinian Corpuscles) which are acceleration sensitive, FA I receptors (Meissner Corpuscles) which are velocity dependent, SA I receptors (Merkel's cells) which react on pressure, and SA II receptors (Ruffini endings) which are sensitive on stretching the skin. More information about tactile receptors and the tactile sensation of human beings can be found in [19, 20].

The kinesthetic excitation is related to the receptors located in muscles, sinews, and joints. They inform us about pose and motion, contact forces, weight and object deformability. This interaction has a lower frequency (less 1 kHz), and its reproduction is thus more realistic. A good description of this stimulus can be found in [19, 20].

To summarize it can be stated that bandwidth and location of receptors are the main differences between tactile and kinesthetic sensation. This has to be taken into account when developing new haptic interfaces.

Haptic Interfaces

Haptic interfaces cover a very extensive variety of devices. Their classification could be based on two main criteria which are predominant feedback component and device portability. The predominant feedback component criterion permits classifying haptic devices into two categories, which are kinesthetic predominant devices and tactile predominant devices. According to the other criterion which is portability, haptic interfaces can be classified as portable and non-portable devices. Non-portable haptic interfaces are devices that are bolted to a desk, a wall, the ceiling, or the floor. Portable haptic interfaces are devices that are worn by the operator. An overview of typical types of haptic interfaces can be found in [21, 22, 23, 24].

Table 1.2. Classification of haptic interfaces

	Portable	Non-portable
Kinesthetic predominant	exoskeletons	joysticks, pen/string-based systems, robot-like systems
Tactile predominant	gloves with vibrotactile and temperature feedback	pin actuators, general vibrotactile and temperature devices

Table 1.2 shows the most important representatives according to the classification mentioned above. Exoskeletons are usually worn devices mounted to the arm or leg. They reproduce the human body motions and the feedback is predominantly kinesthetic. Such a kind of interface is presented in chapter 4. Non-portable and kinesthetic predominant devices are devices that have a serial or parallel kinematic configuration and are manipulated by the operator's fingers

or hands. Chapter 2 and 3 describe such systems in more detail. Gloves are common interface devices that highlight tactile interaction. Examples of them are described in [25, 26]. Finally, tactile and non-portable displays provide information about object features such as surface structure, geometry and temperature. An overview of the principles of vibrotactile and electrotactile displays can be found in [27] and [28]. Examples for temperature feedback systems are presented in [29] and [26].

1.4 Measuring and Evaluating Telepresence

A telepresence system allows a human operator to operate in a remote environment using the superior motor and sensor skills of a robot and the unrivalled cognitive skills of a human being. Teleoperation provides the means to successfully achieve telepresence.

Technically as well as philosophically telepresence is a difficult concept. Hence, defining an *ideal* telepresence system results in many different, partly contradictory, conditions. For example, reducing feedback may improve the performance of a fatigue operator. However, as common ground a telepresence system should enable the operator to feel *immersed* and *involved* in the remote environment. Immersion and involvement are psychological states that depend on the display of the remote environment. High *immersion* means that the user is enveloped by all stimuli necessary to provide a congruent picture of the remote environment. High *involvement* means that the user is provided by all stimuli essential to interact with the remote environment [30]. Both conditions are accommodated by reflecting a high extent of sensory information to the human operator and by enabling her/him to naturally explore and manipulate the remote environment [31].

A basis that structures different evaluation methods is given by the distinction between *objective* and *subjective* telepresence. Objective telepresence is defined by the pure capability of the human operator to successfully complete a given task in the remote environment. Subjective telepresence is more strict emphasizing that the operator must feel as if physically being present in the remote environment [32]. Objective performance measures are *task completion time* or *reaction time* to a remote stimulus. Subjective performance can be measured by *presence questionnaires* asking the human operator about her/his individual feelings [30, 33, 34].

Another way to evaluate the performance is offered by the *transparency* paradigm. A telepresence system is transparent if it exactly reproduces the remote environment. Hence the operator can 'look through' the telepresence system sensing only the remote environment [35]. That results in a number of performance measures for the different channels of the telepresence system. The quality of haptic telepresence, for example, can be judged by comparing positions and forces at operator and teleoperator side [36]. Another possibility is to compare the displayed mechanical impedance with the impedance of the remote environment [37]. For the other modalities transparency criteria also result in comparisons between the operator and the teleoperator site.

1.5 Conclusions

Telerobotics implies linking a human operator and a robot in order to execute a remote task. The coupling between the operator and the robot is considered as strong, when most of the remote control loops are closed by the operator, or considered as weak, when symbolic commands are sent by the operator to be processed by the remote robot control loops.

Teleoperation multi-modal human system interfaces have a double function; first, to process the operator commands and second, to excite operator sense with the information coming from the remote environment. Visual, auditory and haptic interfaces as described are natural manners in controlling a remote task.

Many factors as e.g. acting and human sense capabilities have to be taken into account upon designing a multi-modal interface properly. The goal is to achieve the maximum possible degree of telepresence in order to increase the performance of the telerobotic system.

References

1. T.B. Sheridan. Telerobotics. *Automatica*, 25(4):487–507, 1989.
2. M. Buss and G. Schmidt. Control Problems in Multimodal Telepresence. In *Advances in Control: Highlights of the 5th European Control Conference ECC'99*, pages 65–101, 1999.
3. T. Burkert, J. Leupold, and G. Passig. A photorealistic predictive display. *Presence: Teleoperators and Virtual Environments*, 13(1):22–43, 2004.
4. N. Chong, S. Kawabata, K. Ohba, T. Kotoku, K. Komoriya, K. Takase, and K. Tanie. Multioperator teleoperation of multirobot systems with time-delay: part 1 - aids for collision-free control. *Presence: Teleoperators and Virtual Environments*, 11(3):277–291, 2002.
5. N.A. Tanner and G. Niemeyer. Improving perception in time delayed telerobotics. *International Journal of Robotics Research*, 24(8):631–644, 2005.
6. M. Massimino and T. Sheridan. Sensory Substitution for Force Feedback in Teleoperation. *Presence: Teleoperators and Virtual Environments*, 2(4):344–352, 1993.
7. M. Buss. *Study on Intelligent Cooperative Manipulation.* PhD thesis, University of Tokyo, Tokyo, 1994.
8. M. Buss and H. Hashimoto. Motion Scheme for Dextrous Manipulation in the Intelligent Cooperative Manipulation System—ICMS. In V. Graefe, editor, *Intelligent Robots and Systems (IROS'94). Post-Conference Book*, pages 279–294, Elsevier Science, Amsterdam, 1995.
9. N. Delson and H. West. Robot programming by human demonstration: Subtask compliance controller identification. In *Proceedings of the IEEE/RSJ International Conference on Intelligent Robots and Systems IROS*, pages 33–41, Yokohama, Japan, 1993.
10. H. Friedrich, J. Holle, and R. Dillmann. Interactive Generation of Flexible Robot Programs. In *Proceedings of the IEEE International Conference on Robotics and Automation*, pages 538–543, Leuven, Belgium, 1998.

11. Y. Kunii and H. Hashimoto. Tele-teaching by human demonstration in virtual environment for robotic network system. In *Proceedings of the IEEE International Conference on Robotics and Automation*, Albuquerque, New Mexico, pages 405–410, 1997.
12. J. Yang, Y. Xu, and C. Chen. Hidden markov model approach to skill learning and its application to telerobotics. In *Proceedings of the IEEE International Conference on Robotics and Automation*, pages 396–402, Atlanta, Georgia, 1993.
13. H. Baier, M. Buss, F. Freyberger, and G. Schmidt G. Interactive Stereo Vision Telepresence for Correct Communication of Spatial Geometry. In *Advanced Robotics*, 17(3):219–233, 2003.
14. R.E. Cole and D.L. Parker. Stereo TV improves manipulator performance. In *Proceedings of the SPIE*, 1083:18–27, Bellingham, WA, 1990.
15. M. Ferre, R. Aracil, and M. Navas. Stereoscopic Video Images for Telerobotic Applications. *Journal of Robotic Systems*, 22(3):131–146, 2005.
16. J. Hsu, Z. Pizlo, D.M. Chelberg, C.F. Babbs, and E.J. Delp. Issues in the design of studies to test the effectiveness of stereo imaging. *IEEE Trans. Syst. Man Cybernetics*, Part A 26(6):810-819, 1996.
17. L.B. Stelmach, W.J. Tam, and D.V. Meegan. Perceptual Basis of Stereoscopic Video. In *Proceedings of the SPIE*, 3639:260–265, 1999.
18. W.S. Kim, S. Won, E. Stephen, T. Mitchell, H. Blake, and S. Lawrence. Quantitative evaluation of perspective and stereoscopic displays in three-axis manual tracking tasks. *IEEE Trans Syst Man Cybernetics*, 17(1):418-425, 1987.
19. E.B. Goldstein. Sensation and Perception. Wadsworth Publishing Company, 2002.
20. E.B. Goldstein. Cognitive Psychology. Wadsworth Publishing Company, 2004.
21. G. Burdea. Force and Touch Feedback for Virtual Reality. John Wiley and Sons Inc., 1996.
22. S.D. Laycock and A.M. Day. Recent developments and applications of haptic devices. *Computer Graphics Forum*, 22(2):117–132, 2003.
23. J. Martins and J. Savall. Mechanisms for haptic torque feedback. In *Proceedings of the First Joint Eurohaptics Conference and Symposium on Haptic Interfaces for Virtual Environment and Teleoperator Systems*, pages 611–614, 2005.
24. C. Yongblut, R.E. Johnston, S.H. Nash, R.A. Wienclaw, and C.A. Will. Review of virtual environment interface technology. IDA Paper P-3186, Institute of Defense Analysis (IDA), 1996.
25. M. Benali-Khoudja, et al. Tactile interfaces: a state-of-the-art survey. In *Proceedings of 35th International Symposium on Robotics (ISR 2004)*, Paris, pages 1-9, 2004.
26. A. Kron and G. Schmidt. Multi-fingered tactile feedback from virtual and remote environments. In *Proceedings of the 11th Symposium on Haptic Interfaces for Virtual Environment and Teleoperator Systems (HAPTICS)*, pages 16–23, 2003.
27. K.A. Kaczmarek and P. Bach-y-Rita. Tactile displays. *Virtual environments and advanced interface design*, Oxford, England, Oxford University Press, 1993.
28. K. Shimoga. A Survey of Perceptual Feedback Issues in Dextrous Telemanipulation: Part II. Finger Touch Feedback. In *Proceedings of IEEE Virtual Reality Annual International Symposium*, New York, pages 271–279, 1993.
29. S. Ino, T. Izumi, M. Takahashi, and T. Ifukube. Psychophysical study on tactile sense produced by grasping forhand with sensory feedback. *Systems and Computers in Japan*, 24(13):89–97, 1993.
30. B.G. Witmer and M.J. Singer. Measuring Presence in Virtual Environments: A Presence Questionnaire. *Presence: Teleoperators and Virtual Environments*, Presence 7(3):225–240, 1998.

31. T.B. Sheridan. Musings on Telepresence and Virtual Presence. *Presence: Teleoperators and Virtual Environments*, 1(1):120–126, 1992.
32. D. Schloerb. A Quantitative Measure of Telepresence. *Presence: Teleoperators and Virtual Environments*, 4(1):64–80, 1995.
33. W.A. IJsselsteijn, H. de Ridder, J. Freemanand, and S.E. Avons. Presence: Concept, determinants and measurement. In *Proceedings of the SPIE*, 3959:520–529, 2000.
34. M. Usoh, E. Catena, S. Arman, and M. Slater. Using Presence Questionaires in Reality. *Presence: Teleoperators and Virtual Environments*, 9(5):497-503, 2000.
35. B. Hannaford. A Design Framework for Teleoperators with Kinesthetic Feedback. *IEEE Trans. on Robotics and Automation*, 5(4):426–434, 1989.
36. Y. Yokokohji and T. Yoshikawa. Bilateral Control of Master-Slave Manipulators for Ideal Kinesthetic Coupling - Formulation and Experiment. *IEEE Trans. on Robotics and Automation*, 10(5):605-620, 1994.
37. D.A. Lawrence. Stability and Transparency in Bilateral Teleoperation. *IEEE Trans. on Robotics and Automation*, 9:624–637, 1993.

2

Design, Control, and Evaluation of a Hyper-redundant Haptic Device

Marc Ueberle[1], Nico Mock[2], and Martin Buss[1]

[1] Technische Universität München
Institute of Automatic Control Engineering (LSR)
D-80290 Munich, Germany
ueberle@tum.de, M.Buss@ieee.org
[2] Technische Universität Berlin
Institut für Elektrische Energietechnik
D-10587 Berlin, Germany
nico.mock@iee.tu-berlin.de

Summary. The design and control concept of a serial hyper-redundant haptic interface with 10 actuated degrees-of-freedom (DOF) is discussed. The main motivation for the redundant DOF is the avoidance of interior singularities to increase the workspace while reducing the overall device size. Furthermore, the wrist singularities are eliminated resulting in an orientation workspace of 360° around each axis. Hardware experiments evaluating the closed loop performance and the capability of the inverse kinematics solution to avoid singular configurations confirm the applicability of the proposed design and control concept for human haptic interaction.

2.1 Introduction

Haptic interfaces are force feedback devices enabling bidirectional human system interactions via the sense of touch. Being able to exert and react to the operator's motions and interaction forces they are used to mimic the dynamic behavior of virtual or remote environments in virtual reality and telepresence systems. Although in recent years haptic devices have been successfully implemented in various task domains including e. g. medical and surgical VR systems [1], telesurgery, rehabilitation, tele- and micromanipulation [2], telemaintenance, virtual prototyping, scientific visualization [3], and education [4], their enormous application potential does not seem to be exhaustively investigated and exploited.

The exploration of novel applications is often impaired by the unavailability of qualified haptic hardware. Scenarios involving operations in large regions are e. g. little studied mostly due to the fact that commercially available haptic devices suffer from a comparatively small workspace. Among the reasons for this is the design rationale of most interfaces to provide low dynamic properties. A large workspace seems to be contrary to this requirement because it usually leads to large and heavy interfaces with reduced mechanical stiffness. A possible solution to this problem is the introduction of actuated kinematic redundancies. These can be exploited for the avoidance of interior singularities allowing for

M. Ferre et al. (Eds.): Advances in Telerobotics, STAR 31, pp. 25–44, 2007.
springerlink.com © Springer-Verlag Berlin Heidelberg 2007

significantly larger workspace without increase of the device size. Besides the increase of workspace the redundant kinematical DOF offer a potential for operator collision avoidance and improvement of the dynamic properties and output capability. Even though redundancies have been successfully introduced in industrial robot designs, see e.g. [5], this concept is only little explored and employed for haptic interface kinematics. The only kinematically redundant haptic devices known to the authors are exoskeleton constructions, off-the-shelf redundant industrial robots, and the 7 DOF DLR light-weight robots [6] which are, however, not specifically designed for haptic applications.

In order to offer a haptic device allowing high force tasks in large volumes we designed the hyper-redundant system ViSHaRD10 (**Vi**rtual **S**cenario **Ha**ptic **R**endering **D**evice with **10** actuated DOF). This design is based on the experiences obtained from a non-redundant haptic device ViSHaRD6 previously developed [7]. Both devices employ a force-torque sensor for active force feedback control to shape the device dynamics. Directed towards versatility and extensibility the objective is to provide a testbed and experimental environment for a rapid evaluation of haptic applications.

This paper is organized as follows. Sec. 2.2 briefly reviews the requirements and state of the art of haptic hardware also discussing the use of kinematic redundancies. Sec. 2.3 presents the system design of ViSHaRD10 and Sec. 2.4 basic control strategies along with the inverse kinematics solution technique. Finally, hardware experiments for closed-loop performance and the capability of the inverse kinematics solution to avoid singular configurations are presented in Sec. 2.5.

2.2 Kinesthetic Device Design

2.2.1 Requirements

An 'ideal' haptic device provides for a large variety of haptic applications a completely transparent interface to the remote or virtual environment, i. e. the user cannot detect any difference to the interaction with real objects. The transparency and versatility of haptic devices is affected by a number of design criteria characterizing its performance.

- *Dynamic properties:* The haptic interface should be able to display a large dynamic range of impedances. This requires a good backdrivability to render unconstrained motion and a high closed loop stiffness to mimic solid contact situations. In case the task involves a highly dynamical interaction of the operator with the device it is important that backdrivability and stiffness can be provided at a large bandwidth. It has to be noted, that the bandwidth requirements for these two tasks differ substantially. The render of low impedances needs the control system to suppress disturbance forces due to the operator's motion input and the device natural dynamics. As a result the bandwidth of the disturbances typically corresponds to the bandwidth of the user input which has been reported by Tan *et al.* [8] to be on the order of 20 to 30 Hz for the human finger. Contact situations raise the bar for bandwidth requirements drastically. Hitting a solid wall results in a very sharp force response. Accordingly, a realistic render of such interactions requires the interface force bandwidth to correspond to the bandwidth of the human's somatosensory system which is according to Tan *et al.* around 1 kHz. An increase of the backdrivability in the high frequency domain can be achieved by a reduction of the mechanical stiffness, inertia, and friction. The force bandwidth is, however, widened by increasing the viscous friction and stiffness as well as by lowering the inertia and eliminating backlash. In case of open loop controlled devices the bandwidth characteristics are purely determined by the mechanical design, whereas active closed loop control can shape the frequency response characteristics. As the bandwidth requirement at contact situations is more stringent than at free space operations the maximization of the open loop force bandwidth is the common design guideline for both, open loop and closed loop controlled haptic interfaces. Needless to say that this does not hold for interfaces with application domains primarily including interactions with soft environments (e. g. soft tissue interaction in telesurgery).

- *Output capability:* The device output capability by means of maximum force, velocity, and acceleration defines limits for the haptic interactions that can be rendered. Moreover, a low output capability is likely to reduce the robustness of closed loop control due to actuation saturation.
- *Workspace:* The decision for the number of the DOF and the size of the workspace mainly affects the range and variety of applications the interface can be applied for.
- *Extensibility:* Besides the workspace and the output capability the most influential factor for the versatility of the device is its extensibility. The addition of complex end-effectors requires sufficient mounting space and torque capability to compensate for the payload.

It is intuitively clear that these requirements are contrary and one has to balance amongst them. In particular workspace enlargement is a considerable technological challenge because the unavoidable increase in hardware size usually entails a drastic impairment of the dynamic device properties. It is therefore increasingly difficult to satisfy the requirements on backdrivability, stiffness, and bandwidth with open loop control when the demands for workspace size and output capability are rising.

2.2.2 State of the Art

In the last decade a great number of kinesthetic haptic interfaces has been developed at educational and research institutions. A comparatively wide but far from complete overview on recent kinesthetic haptic devices is given by Martin and Savall [9]. More detailed but less comprehensive comparisons of haptic hardware can be found in the reviews of Laycock and Day [10], Youngblut *et al.* [11], and Burdea [12]).

Significant research effort has been devised to the development of kinesthetic feedback devices characterized by a high degree of specialization for certain task domains, in particular medical applications. This includes for instance the Laparoscopic Impulse Engine (Immersion) and the VEST-system VSOne (Select-IT VEST Systems AG) [13] with kinematical designs and output capability matching exactly the requirements of minimally invasive surgical simulation. This approach aims at the maximization of the device performance for the intended application but, in turn, narrows the range of task domains it can be used for.

Viewing general purpose haptic interfaces suitable for a wide application range it can be observed that the vast majority is based on *hybrid* kinematical designs. Unlike pure *serial* designs, where the end-effector is linked to the base via one single opened kinematical chain, and *fully-parallel* designs, where the connection is established by multiple parallel kinematical chains (see chapter 3), hybrid mechanisms contain serial-chain and parallel-chain modules connected either in series or in parallel. Various hybrid designs can be considered as extensions of existing serial or parallel mechanisms. For instance the small orientation workspace of many parallel devices motivated the design of hybrid haptic interfaces with an orientation stage connected in series to a parallel position stage as for example the interfaces described by Tsumaki *et al.* [14] and Hayward *et al.* [15]. Other motivations for hybrid kinematics include the coupling of several existing devices in parallel to increase the number of DOF [16], or the incorporation of four or five bar link mechanisms in serial designs to move actuators closer to the base [17].

The most wide class of general purpose haptic interfaces that achieved a very convenient development status are passive designs providing kinesthetic feedback in comparatively small workspaces at moderate force levels. These devices are characterized by highly lightweight mechanical designs requiring no active force feedback control to provide a good backdrivability. The lack of force sensing capability greatly reduces the complexity of the control design and the hardware cost. The low mass, size, and force capacity is beneficial in terms of safety aspects and human-friendliness. The commercially most successful passive devices for general tool based applications are the PHANToM devices (SensAble Technologies) developed at MIT [17]. They are available in a variety of sizes and 3 or 6 actuated DOF and allow interactions through a finger sled or a stylus. Offering low dynamic properties they are able to render free-space in a high-quality. Another passive design with disturbance forces very close to the human perceptual threshold is the Freedom 6S (MPB Technologies), the commercial version of Freedom 7, see [15]. One of the disadvantages of these passive design approaches are a comparatively limited control stiffness due to the low physical damping present in the joints.

As these purely passive devices do not provide force measurement capability they cannot be used for applications that fundamentally require the render of admittances where motion is a response to force input and not vice versa. An example for such an application is a bone drilling training scenario where the surgeon should learn to apply a dedicated constant force to the drill [1]. Interfaces with mildly increased dynamic properties compared to common passive designs but additional force sensing are for instance the commercially available DELTA Haptic Device (FORCE dimension) described in [18] and the VIRTUOSE 6D (Haption). Both provide force feedback in 6 DOF with increased (but still moderate) force capability compared to the 6 DOF PHANTOM devices.

A common deficiency of the majority of currently available haptic devices is their comparatively small workspace and low force capability forbidding for example large ergonomic studies, the display of stiff immovable walls during assembly and disassembly simulations, or the accommodation of heavy end-effectors such as exoskeleton devices providing kinesthetic feedback to the operator's hand or tactile interfaces. One of the reasons is that the design rationale to have low dynamic properties is contrary to other requirements such as versatility or large workspace.

For haptic realization of tasks requiring a large workspace and high force capability often serial off-the-shelf industrial robots are used [19,20]. These robots are, however, not optimized for interactions with humans; the force capability exceeds by far the strength of a human and the mechanical stiffness is much larger than required for haptic applications. Consequently, these devices show major deficiencies regarding dynamic properties and safety aspects. Interfaces with human matched force capability and workspace, that are devices filling the gap between passive designs and industrial robots, are uncommon and rarely available. Two of the very few examples are the serial devices Excalibur [21] with very high peak stiffness and *Haptic*MASTER (FCS Control Systems) described in [22] showing good performance regarding deceleration capability. Both provide 100 N continuous force in 3 DOF but in a rather limited workspace.

Another approach to provide a large workspace combined with high force are exoskeleton constructions with jointed linkages fixed to the operator (see chapter 4). The workspace is exceptionally high in case of wearable devices as for example the L-EXOS device described by Frisoli *et al.* [23]. Portable interfaces have, however, the disadvantage that their weight has to be supported by the operator strongly tightening the requirement for a lightweight design. Furthermore, because wearable haptic interfaces are not grounded to the environment, the net force exerted by the device to the operator equals zero. As a consequence, external forces cannot be rendered adequately. Exoskeletons grounded to the environment, as for instance the SARCOS dexterous master [24], do not show this deficiency but, in turn, limit the mobility of the operator. For example arm exoskeletons with shoulder joint attached to the environment require the operator's shoulder to remain at a fixed position. A general drawback of both, grounded and portable exoskeleton devices, is that they tend to be quite complex and encumbering. The fact that they are firmly attached to the operator is highly disadvantageous in terms of ergonomics and safety aspects.

Another line of research targets at an unlimited workspace by mounting haptic devices on a movable platform [25,26]. Compared to portable exoskeletons this approach has the advantage that the device weight has not to be supported by the operator. At the downside, this solution drastically increases system complexity.

Moreover, the render of high stiffness is challenging due to the typically more compliant coupling of the system to the environment.

2.2.3 Kinematical Redundancies

A maximization of the translational working volume in respect to the length of the kinematical chain requires a serial mechanism with a revolute joint at the base. Indeed, such designs can yield a spherical workspace with radius equal to the length of the robot arm. If the kinematical design is *non-redundant* (i. e. has equal number of actuated joints and DOF at the end-effector) large areas of the workspace will, however, not be available for haptic simulations due to the existence of *interior singularities*. Singularities are positions in space where the robot looses a DOF. Whereas in common industrial robot applications it is frequently allowed to drive the robot through such singularities it is necessary to circumvent these locations in haptic systems for the following reason: around these positions the dynamic properties of the robot degrade because high joint velocities only produce small end-effector velocities in certain directions. This results in a significant impairment of the end-effector output capability regarding acceleration and velocity. Due to the fact that the device can be moved by the human operator at will motions along these directions cannot be avoided by trajectory planning methods. Singularities are also the reason for the limitation of the angular workspace. It is in general true that an angular workspace of 360° around each axis is not achievable for non-redundant 6 DOF robots [27].

To illustrate the effect of interior singularities the workspace of ViSHaRD6, a 6 DOF haptic interface [7], can be viewed. Fig. 2.1 shows the singularity free translational working volume allowing arbitrary orientations of the end-effector in the range of 360°, 90°, and 360° for the angle of roll, pitch, and yaw, respectively. Although the working area is comparatively large an area in the center of the workspace is not available for haptic interaction due to interior singularities. On account of this the unpropitious workspace shape of non-redundant serial designs with revolute joints at the base often diminishes the benefit of their improved workspace volume when compared to mechanisms free of interior singularities as for instance serial 3 DOF linear axes designs or the parallel DELTA mechanism introduced by Clavel [28].

A possible solution to this problem is the introduction of actuated kinematical redundancies. Such mechanisms allow for a change of their internal configuration without changing the position and orientation of the end-effector. This kind of motion is called *null space movement* or *selfmotion*. A well directed control of the selfmotion can contribute to increase the overall system performance. The following summarizes some of the well known attractive features available in redundant kinematical designs.

- *Workspace:* The redundant DOF can be used to avoid interior singularities. This can drastically increase the workspace while simultaneously reducing the device size.
- *Dynamic properties:* The selfmotion can be controlled to maximize inertial and bandwidth performance criteria as well as to reduce friction forces

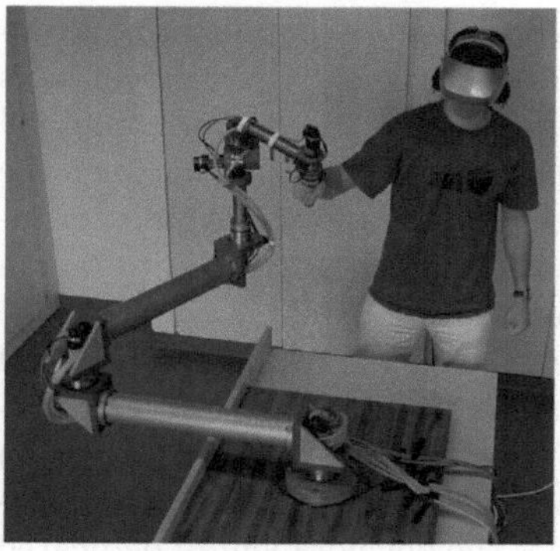

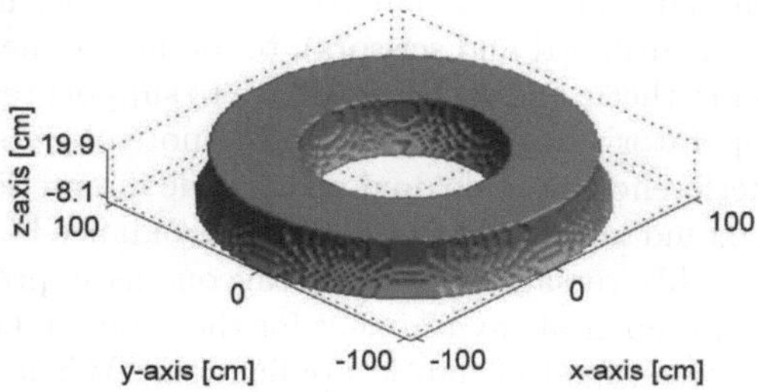

Fig. 2.1. VISHARD6 prototype (left) and its singularity free translational workspace for an orientation workspace of 360°, 90°, and 360° for the angle of roll, pitch, and yaw, respectively (right)

at the end-effector. The potential for improvement of the dynamic properties is exceptionally large when dealing with redundant designs forming a macro/micro system comprised of a large (macro) robot carrying a small (micro) robot [29].
- *Output capability:* Also feasible is the maximization of performance criteria affecting the output capability as e.g. force/velocity transmission or acceleration capability.
- *Collision avoidance:* Redundancies offer an increased potential for collision avoidance with the environment and human operator. This can for example be exploited for the prevention of user interference or link interference at dual-arm haptic devices.

In practice, however, kinematical redundancies are rarely used for haptic devices. A possible explanation is the increased cost and complexity of the mechanical design. To control the redundant DOF a computational augmentation is unavoidable. Also, the introduction of additional joints seems to be contrary to the objective of low inertia and high stiffness. The potential reduction of the device size relaxes these disadvantages. Moreover, the removal of the interior singularities allows to operate in workspace regions with increased stiffness and output capability.

2.3 Design of ViSHaRD10

2.3.1 Design Rationale

The driving motivation for the design of the new hyper-redundant haptic interface is the vision of a general-purpose haptic interface that can be used in a large variety of application domains. Distinct advantages compared to existing solutions include an unlimited orientation workspace free of singularities; large translational workspace; human matched force capability; high payload capability to accommodate various (actuated) application specific end-effectors such as surgical tools (e. g. drills [1] and scissors), to mount tactile stimulation actuators for combined kinesthetic and tactile feedback to support realistic direct interaction between the operator's finger or hand and remote objects; offer redundancy to avoid user interference; provide dual-arm haptic interaction with full 6 DOF capability (again redundancy facilitates collision avoidance between the two arms).

This versatility is advantageous as it provides a benchmarking testbed and experimental environment for the rapid and cost-effective development and evaluation of novel haptic applications. When the task domain requires a certain workspace or force capability the device can be constrained to these specifications by appropriate controller design; this includes the development of dedicated inverse kinematics algorithms incorporating the specific needs of these applications. Once the new haptic application has been rudimentarily developed using ViSHaRD10 and the feasibility is verified, a tailored, highly specialized haptic display with exactly matching mechanical properties can be developed.

2.3.2 Standard 7 DOF Anthropomorphic Arm Designs

The first considerations regarding the kinematic design of a redundant interface have been focused on the class of standard kinematical designs consisting of a 3-jointed spherical shoulder, a single elbow joint, and a 3-jointed spherical wrist. These arms can be described as anthropomorphic after [30]. Exemplarily, a configuration with a wrist in roll-pitch-roll configuration is illustrated in Fig. 2.2. The strength of these mechanisms is the size of the workspace which is optimum for 7 DOF robots in terms of the ratio of the arm length to the working volume. The translational workspace is a sphere with an interior singularity at the center. The angular workspace is 360° around each axis since singularities in the wrist can be avoided by rotating the elbow around the line from the shoulder to the wrist. A kinematic analysis of the design shown in Fig. 2.2 is presented in [31]. Among the drawbacks we identified for 7 DOF anthropomorphic arms are:

- *Gravitational load:* Only the first joint axis is designed to be vertical for arbitrary positions and orientations of the end-effector. As a consequence high motor torque is required to compensate for gravitational load.
- *Interior singularity:* The singularity in the center of the workspace impairs the dexterity and thus the performance of the device when moving the end effector close to the shoulder. An elimination of this singularity requires at least two additional redundant joints placed between the shoulder and the wrist.
- *Safety:* The most critical deficiency is the selfmotion of anthropomorphic 7 DOF arms, the rotation of the upper and forearm. Especially in case of operating with the end-effector close to the shoulder the elbow-orbit may deeply intrude into the operator's workspace bearing the risk for severe conflicts between the elbow joint and the user. The safety aspect can be solved by using a 4-jointed roll-yaw-pitch-roll wrist as described in [32]. Then, the position of the elbow can be controlled to prevent collisions with the operator because singular wrist configurations can be avoided with the selfmotion of

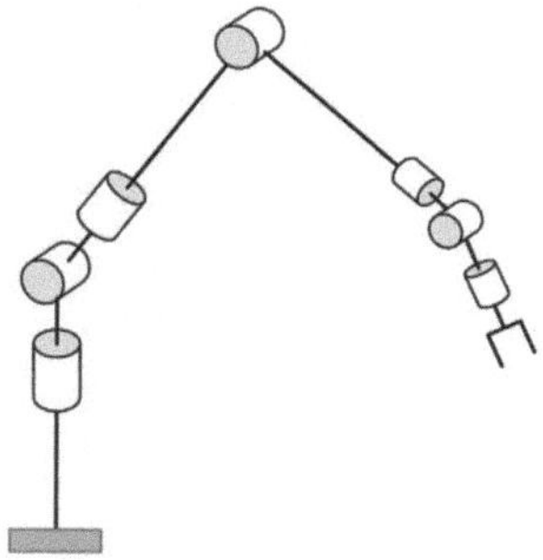

Fig. 2.2. Typical 7 DOF anthropomorphic robot arm

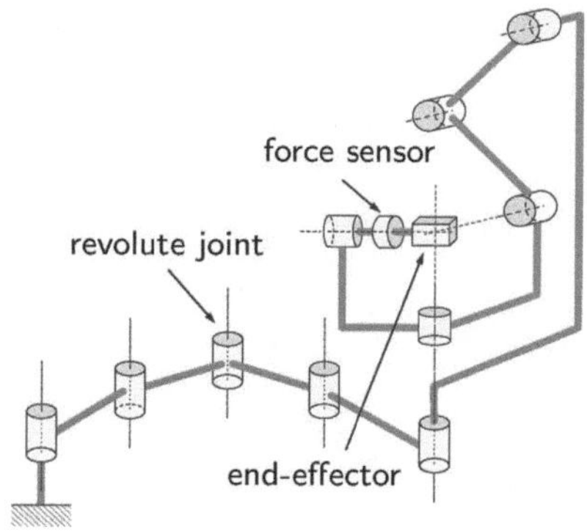

Fig. 2.3. Hyper-redundant haptic device VISHARD10

the redundant wrist. This 8 DOF solution, however, intensifies the deficiency regarding gravitational load significantly.

2.3.3 Haptic Device Description

In order to circumvent these deficiencies of anthropomorphic arms we finally decided for a mechanism without a 3-jointed spherical shoulder, see Fig. 2.3. The kinematic structure of the hyper-redundant design with 10 DOF, VISHARD10, is depicted in Fig. 2.4, the link length design is summarized in Tab. 2.1. It shows the reference configuration with all joint angles θ_i defined to be zero. In Fig. 2.5 the prototype is shown in a typical operational configuration.

The first five joints are arranged in a SCARA configuration with vertical axes avoiding the need for an active compensation of gravity. This segment is assigned for the positioning of the end effector in the x-y-plane. Although a SCARA segment with three links is sufficient for the elimination of the singularity in the center of the planar workspace we decided for an arm with 4 revolute joints (4R) for two reasons: First, it is well known that the 4R arm provides improved dexterity compared to the 3R arm [33] and second, the avoidance of user interference is much simpler to achieve.

Alternatively we weighed up the use of a 2 DOF linear axes design for the positioning in the x-y-plane as this also provides an interior singularity free workspace. This mechanism, however, suffers from a significantly reduced workspace; two prismatic joints with a length of one meter are required to achieve a workspace of $1 \times 1\,\text{m}^2$ whereas an 4R arm with a total length of two meters can provide a sphere with radius 2 m. Moreover, as the base of the 4R arm is less bulky it is better suited for two arm simulators with two haptic devices.

Joints 6 and 7 are assigned to adjust the height of the end-effector. One simple inverse kinematics solution for them is the imitation of a prismatic joint by means of not changing the end-effector position in the x-y-plane. The decision for two revolute joints over one prismatic joint is amongst other reasons due to

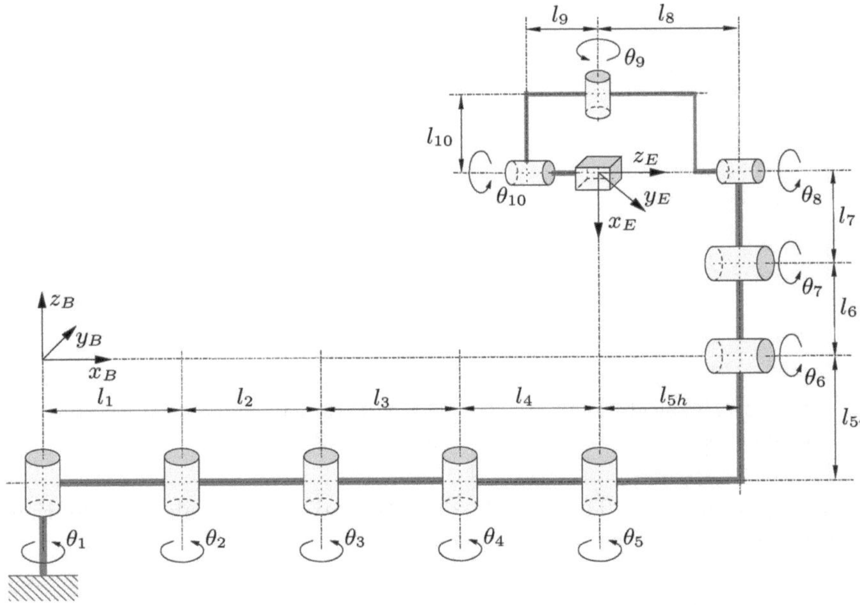

Fig. 2.4. Kinematical model of **ViSHaRD10**

the fact that off-the-shelf prismatic joints matching our requirements regarding low friction, mass, stiffness, velocity, and zero backlash could not be found on the market and modularity of the rotational joint components is kept.

Joint 5 is used to prevent singular configurations in the wrist formed by joints 8, 9, 10. Despite tending to an increased wrist size we decided for a yaw instead of roll orientation for joint 8 to obtain decoupling of the wrist configuration from the end-effector height. This mechanism has a singularity when the axes of joint 5 and 9 as well as of joint 8 and 10 have the same orientation. Whereas the first condition cannot be avoided when holding the end-effector in a horizontal orientation the second can be avoided by a rotation of joint 5.

The axes of joint 5, 8, 9, 10 intersect at one point which is located 5 cm in front of the force-torque sensor (assuming that the motion of joint 6 and 7 is controlled accordingly). This enables the operator to grip the end-effector at the point where the angular DOF are mechanically decoupled from the translational ones as for example desired for simulations involving direct haptic interactions with the finger or hand. Alternatively, the user can hold the device at an other location of the end-effector to simulate for instance the exploration of a remote environment with the tip of a tool. The benefit of such a decoupling of the positioning from the orientation mechanism is twofold. First, it results in reduced natural dynamics of the orientational DOF and second, it allows for taking different output capability requirements of these DOF into account; the maximum torque exerted to the end-effector is usually bounded by the low torque capability of the human wrist

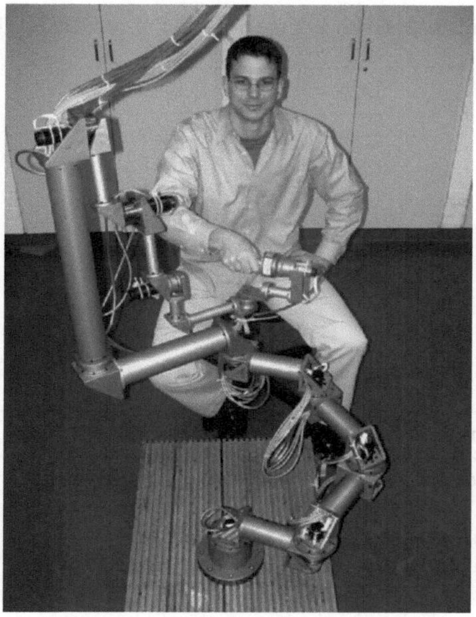

Fig. 2.5. Haptic display ViSHARD10

whereas the high torque capability of the human shoulder joints allows for comparatively high forces.

The mechanical realization of this kinematical design is without joint angle limits and possibility for collision between parts of the structure. One important goal pursued with the kinematical design of ViSHARD10 has been to provide the option to partition the inverse kinematics problem into two separate problems: the inverse kinematics for the positioning and the orientation stage. This can decrease the computational power required for the redundancy resolution significantly. The decoupling of the translational from the rotational movement is achieved when controlling joint 6 and 7 to mimic the operation of a prismatic joint. The distance between joint axis 10 and joint 9 is 0.1 m providing sufficient mounting room for end-effectors of moderate size. The link length design (see table 2.1) gives an overall system size similar to ViSHARD6 (see Fig. 2.1). The dexterous workspace, however, which is a cylinder with $\varnothing 1.7\,\mathrm{m} \times 0.6\,\mathrm{m}$, is significantly larger.

The actuation torque is provided by DC-motors coupled with harmonic drive gears offering zero backlash. The moment stiffness of all gears in the SCARA segment is increased by additional bearing support in order to avoid damage due to deflection in the harmonic drive component sets. The motors and the gears have been selected to meet the target specifications summarized in Tab. 2.2. The motors are actuated by PWM-amplifiers supplying a control of the motor current at a bandwidth of 2.5 kHz. In order to permit force feedback control the device is equipped with a six-axis JR3 force-torque sensor providing a bandwidth of 8 kHz at a comparatively low noise level. The joint angles are measured by

Table 2.1. Link length design of VISHARD10

Link i	Length
$l_1 = l_2 = l_3 = l_4$	0.25 m
$l_{5h} = l_8$	0.47 m
l_{5v}	0.71 m
$l_6 = l_7$	0.212 m
l_9	0.15 m
l_{10}	0.15 m

Table 2.2. Target specifications of VISHARD10

Property	Value
workspace	cylinder ⌀1.7 m × 0.6 m 360° for each rotation
peak force	170 N
peak torque	pitch, yaw: 13 N m roll: 4.8 N m
translational velocity	> 1 m/s
maximum payload	7 kg
mass of moving parts	≈ 24 kg

digital magneto resistive encoders with a resolution of 4 096 counts per revolution, resulting in a comparatively high position resolution when multiplied with the gear ratio varying from 100:1 to 160:1.

The VISHARD10 is currently employed for a dual arm telemanipulation system (see chapter 28) as well as for a virtual reality multi-user training system [34]. The successful application of the device for the render of haptic interactions with deformable objects such as soft tissue is reported in [35]. One VISHARD10 prototype has been extended by end-effectors providing tactile feedback to the operator [36]. It is used at the Max Planck Institute for Biological Cybernetics in Tübingen for psychophysical research. The focus is on the investigation of tactile suppression effects [37].

2.4 Control of ViSHaRD10

2.4.1 General Control Scheme

The haptic simulation of a human's bilateral interaction with a remote or virtual environment requires the control of the motion-force[1] relation between operator

[1] Throughout this chapter force stands for both, linear force and torque, while motion in terms of a generalization of position, velocity, and acceleration refers to both, translational and angular motion quantities.

and robot. This can be achieved by either controlling the interaction force of the device with the operator (impedance display mode) or the device motion (admittance display mode).

In the former approach, also called impedance display mode, the robot acts as impedance and the human as admittance. In other words, the robot accepts the human's motion input and is controlled to give force output according to the impedance of the remote or simulated environment. Impedance controlled devices do not necessarily require force measurement, frequently simple open loop force control schemes are used. Force feedback, however, reduces the disturbance forces due to the natural device dynamics significantly. Impedance control usually used for light and highly backdrivable devices provides good results in rendering low impedances. A drawback of these algorithms is that a dedicated shaping of the closed loop impedance error due to the natural device dynamics is hard to obtain as it strongly relies on the accuracy of the dynamic model of the interface. Moreover, forming the mass and inertial properties of the impedance error requires either acceleration measurement, which is rarely provided by haptic hardware, or force sensing[2]. For robots with considerable anisotropic[3] dynamics the operator will usually get a somewhat peculiar impression at the display of low impedances.

Admittance control is particularly well suited for robots with hard nonlinearities and large dynamic properties compared to the remote or virtual environment being emulated. In this display mode forces are measured and motion is commanded, i. e. the robot acts as admittance and the human as impedance. Accordingly, a force sensor is required for admittance control. Contrary to haptic displays driven in the impedance mode all admittance control implementations aim at forming the closed-loop mass and inertia. The high gain inner control loop closed on motion allows for an effective elimination of nonlinear device dynamics as for instance friction. It is thus possible to render an isotropic closed-loop dynamic behavior in order to provide the operator a more 'natural feeling'. The drawbacks are a reduced capability for the display of low impedances and a decreased closed-loop bandwidth of the force feedback. A more detailed analysis of haptic control schemes is given in [38].

In order to provide an effective compensation of disturbances due to friction and to be able to shape the inertial and mass properties of the impedance error an operation in the admittance display mode is preferred as illustrated in Fig.2.5. The interaction force $\boldsymbol{h}_{\text{ext}}^T = \left[\boldsymbol{f}_{\text{ext}}^T \; \boldsymbol{\mu}_{\text{ext}}^T\right]$ of the operator is measured by a force-torque sensor. The vectors $\boldsymbol{f}_{\text{ext}}$ and $\boldsymbol{\mu}_{\text{ext}}$ denote the interaction force and torque, respectively. According to the master dynamics the deviation between the reference force $\boldsymbol{h}_{\text{r}}$ and the measured force is related to the reference end-effector velocity $\dot{\boldsymbol{x}}_{\text{r}}$. An algorithm for inverse kinematics resolution calculates the reference joint velocities $\dot{\boldsymbol{q}}_{\text{r}}$. Alternatively, the mapping of the end-effector to the

---

[2] In fact, shaping the closed-loop inertia by model-based force feedback control aims 'in its heart' at the control of the device acceleration and can therefore also be classified as admittance control.

[3] Physical properties vary in different directions.

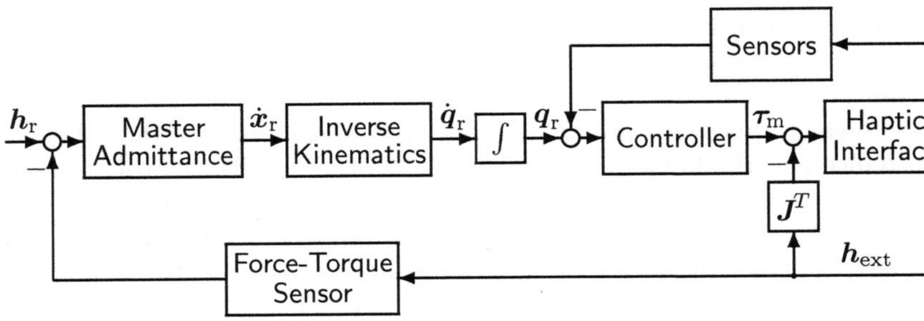

Fig. 2.6. Admittance control scheme

joint motion can be realized at the position or acceleration level. The joint angles $\boldsymbol{q}_r$ are the reference input to a conventional computed torque position control scheme with acceleration feedforward [39].

2.4.2 Master Admittance

An increase of the master admittance decreases the deviation between the haptic interface closed loop dynamics and the target impedance rendered by the relation between $\boldsymbol{q}$ and $\boldsymbol{h}_r$. One has, however, to consider, that the minimum closed loop mass and inertia of the haptic interface is bounded by stability (see Sec. 2.5). A common worst case scenario is free space motion of the teleoperator producing $\boldsymbol{h}_r = 0$ for arbitrary device motions. In that case the minimum mass and inertia has to be solely provided by the master admittance.

The implementation of the ViSHaRD10 master admittance is illustrated in Fig. 2.7. The rendering of the inertia and rotational damping is based upon the well known Euler's dynamical equations of rotation. The indices $_B$ and $_E$ indicate the corresponding quantity to be defined respective the coordinate system $\{B\}$ and $\{E\}$ defined in Fig. 2.4. The rotation matrices ${}^E_B\boldsymbol{R}$ and ${}^B_E\boldsymbol{R}$ map vectors from the base to the end-effector coordinate system and vice versa. The virtual mass, inertia, translational, and rotational damping is defined by the matrices $\boldsymbol{M}_{trans}$, $\boldsymbol{M}_{rot}$, $\boldsymbol{K}_{D,trans}$, and $\boldsymbol{K}_{D,rot}$, respectively. Possible extensions to this master dynamics are e. g. the implementation of virtual spring forces to constrain the workspace of the haptic interface and the teleoperator.

2.4.3 Inverse Kinematics

The mapping

$$\dot{\boldsymbol{x}} = \boldsymbol{J}\dot{\boldsymbol{q}} \tag{2.1}$$

relates the joint velocity $\dot{\boldsymbol{q}} \in \mathbf{R}^n$ to the end-effector velocity $\dot{\boldsymbol{x}} \in \mathbf{R}^m$, where $\boldsymbol{J}$ is the $m \times n$ Jacobian matrix of the manipulator. If $n > m$ the manipulator is said to be redundant with respect to the end-effector task. Then, a solution to the inverse kinematics problem, the calculation of $\dot{\boldsymbol{q}}$ from $\dot{\boldsymbol{x}}$, is not uniquely determinable as there are fewer equations than unknowns. A large number of

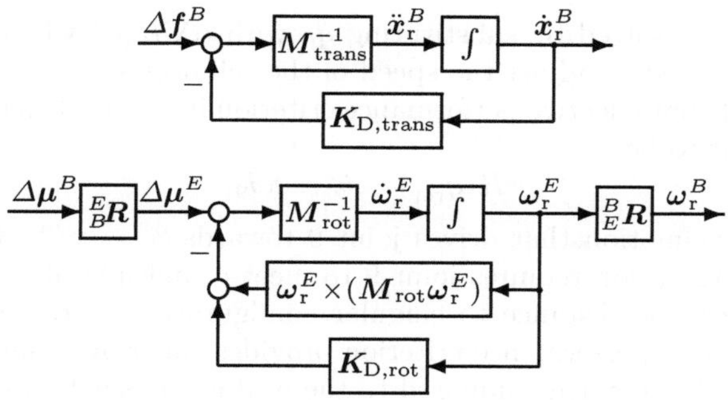

Fig. 2.7. Implementation of the master admittance

inverse kinematics solution approaches has been proposed in the literature. An excellent discussion of the fundamental properties of standard techniques has been given by Wampler [27].

The current implementation of the VISHARD10 inverse kinematics employs a combination of the inverse function approach for the translational degrees of freedom and Pseudoinverse control as proposed by Liegeois [40] for the rotational motion. Alternative techniques along with simulation experiments for performance comparison are presented in [41]. The inverse function is defined by controlling the SCARA segment and joint 6 and 7 to mimic the operation of three prismatic joints:

$$\theta_1 = \arccos\frac{y}{2l_1} - \frac{3}{2}\pi, \qquad \theta_2 = -2\arccos\frac{y}{2l_1} + 2\pi, \tag{2.2}$$

$$\theta_3 = \arccos\frac{x}{2l_3} - \theta_1 - \theta_2, \qquad \theta_4 = -2\arccos\frac{x}{2l_3} + 2\pi, \tag{2.3}$$

$$\theta_6 = -\arccos\frac{z}{2l_6}, \qquad \theta_7 = -2\theta_6, \tag{2.4}$$

where x, y, z is the end-effector position respective the coordinate system $\{B\}$ defined in Fig. 2.4.

For the solution of the wrist motion the Moore-Penrose generalized inverse $J^{\#}$ of the Jacobian matrix is applied. By setting $q_{\text{rot}}^T = \left[\theta_5^*\ \theta_8^*\ \theta_9\ \theta_{10}\right]$ with $\theta_5^* = \theta_5 + \sum_{i=1}^{4}\theta_i$ and $\theta_8^* = \theta_8 + \sum_{i=6}^{7}\theta_i$ the inverse kinematics for $\dot{q}_{\text{rot}}$ is decoupled from the translational movement:

$$\dot{q}_{\text{rot}} = J^{\#}_{\text{rot}}\omega + [I - J^{\#}_{\text{rot}} J_{\text{rot}}]\dot{q}_{\text{rot},0} \quad \text{with} \quad \dot{q}_{\text{rot},0} = \alpha\nabla H(q_{\text{rot}}). \tag{2.5}$$

Here ω is the rotational Cartesian velocity command and $J_{\text{rot}} \in \mathbf{R}^{3\times4}$ the Jacobian relating q_{rot} to ω. The first term is the minimum norm joint velocity solution and the second term the homogeneous solution where $[I - J^{\#}_{\text{rot}} J_{\text{rot}}]$ projects an arbitrary joint velocity vector $\dot{q}_{\text{rot},0}$ onto the nullspace of J_{rot}. It is also common to solve (2.5) applying a weighted Jacobian matrix offering e. g. the opportunity to ease motions of certain joints. The homogeneous solution is used to improve the device performance by choosing $\dot{q}_0$ to optimize a performance criterion $H(q)$, a scalar function of the joint angles. Using gradient projection the redundancy is solved by substituting $\dot{\boldsymbol{q}}_0$ with $\alpha \nabla H(\boldsymbol{q})$ where the negative scalar α can be used to adjust the speed of the selfmotion.

A very simple but effective performance criterion in terms of singularity avoidance turned out to be

$$H(\boldsymbol{q}_{\mathrm{rot}}) = \theta_9^2 - \pi\theta_9 \tag{2.6}$$

resulting in a selfmotion that drives joint 9 towards $\theta_9 = \pi/2$. Recalling, that a singular configuration requires joint 9 to meet a multiple of π, this joint position maximizes the distance to singular configurations. Practical experience confirmed that this performance criterion provides superior results by means of singularity avoidance when compared to the well established manipulability index introduced by Yoshikawa [42] or the condition number of the wrist Jaobian. The disadvantage of the manipulability index is the fact, that it is a function of θ_8^* and θ_9. Thus, the nullspace motion can move joint 9 away from $\theta_9 = \pi/2$. The motion of joint 8 has, however, no impact on the distance to a singular configuration: Singular configurations require the end-effector to be in a horizontal orientation. In these orientations $\dot{\theta}_8^*$ does not lie in the nullspace of $\boldsymbol{J}_{\mathrm{rot}}$ causing θ_8^* to remain fixed to the critical angle. The distance to the singularity is therefore solely determined by the position of joint 9.

2.5 Hardware Experiments

The experimental results presented in the following are to evaluate the closed loop performance of the device regarding the dynamic range of achievable admittances. A common benchmark experiment comprising a high range of admittances and a very sudden transition between them is the haptic interaction with a virtual wall. The dynamic range of admissible admittances is then described by the minimum mass/inertia and maximum stiffness parameters producing no instability during vigorous end-effector motions induced by a human operator.

The virtual walls have been oriented perpendicular to the axes of the base coordinate system to keep x, y, and z within $[-0.3\,\mathrm{m}\ 0.3\,\mathrm{m}]$ and to limit the yzx-Euler angles to $\alpha \in [0°\ 360°]$, $\beta \in [-70°\ 70°]$, and $\gamma \in [0°\ 360°]$. Outside the walls the device has been commanded to render the dynamic behavior of a pure mass/inertia without damping, whereas in the inside a linear stiffness and damping has been added to the target admittance. The damping parameters have been selected to obtain a damping ratio of $\zeta = 1/\sqrt{2}$ to minimize the transition time.

The determination of the minimum target mass and inertia has been constrained to the investigation of diagonal matrices. It has been found to be around $\boldsymbol{M}_{\mathrm{trans}} = \mathrm{diag}(5, 7, 2)\,\mathrm{kg}$ and $\boldsymbol{M}_{\mathrm{rot}} = \mathrm{diag}(0.01, 0.01, 0.005)\,\mathrm{kg\,m^2}$. The significant dependence of the minimum mass and inertia on the direction is in accordance with the inertial device characteristics. This points to a potential benefit of more advanced inverse kinematics solution approaches optimizing the VISHARD10 mass and inertial properties.

The render of stiff walls has shown to cause no difficulties. The experimental results indicate a realizable stiffness of 2 000 N/mm in x-, 3 000 N/mm in y-, 800 N/mm in z-, 10 kN m/rad in α- and β-, and 1.8 kN m/rad in γ-direction.

Please note, that these values constitute the maximum target stiffness of the virtual model. This is different from the stiffness perceived by the operator as it is reduced by the flexibility of the mechanical device structure and the compliance of the inner position control.

In order to study the capability of the wrist inverse kinematics solution to avoid singular configurations the virtual walls constraining the angular workspace have been removed. These experiments clearly confirmed the benefit of redundant joints in haptic hardware design; it was possible to drive the end-effector to arbitrary orientations while effectively preventing singular configurations.

Difficulties with the avoidance of wrist singularities have only been encountered when rotating the end-effector from a vertical orientation ($x_E \parallel z_B$) exactly around joint axis 9 into the horizontal orientation as illustrated in Fig. 2.8. In this case $\dot{\theta}_9$ does not lie in the nullspace of $\boldsymbol{J}_{\mathrm{rot}}$ resulting in zero selfmotion. As a consequence, the wrist is driven into a singular configuration. In practice, however, the end-effector rotation is rarely exactly parallel to the axis of joint 9. Hence, the singularity avoidance is typically successful, but the selfmotion can induce undesirable high accelerations of joint 5 for fast end-effector motions. This can be avoided by placing a virtual wall keeping θ_9 within the bounds $\left[0°+\zeta \;\; 180°-\zeta\right]$ with $0° < \zeta < 45°$, where ζ is a measure for the distance of the 3R wrist from a singular configuration. Because the side criterion given in (2.6) causes the selfmotion to drive this wall away from the current end-effector orientation, contacts with the virtual wall are likely to occur only in case of critical end-effector trajectories. The wall will then, however, avoid undesirable fast selfmotions.

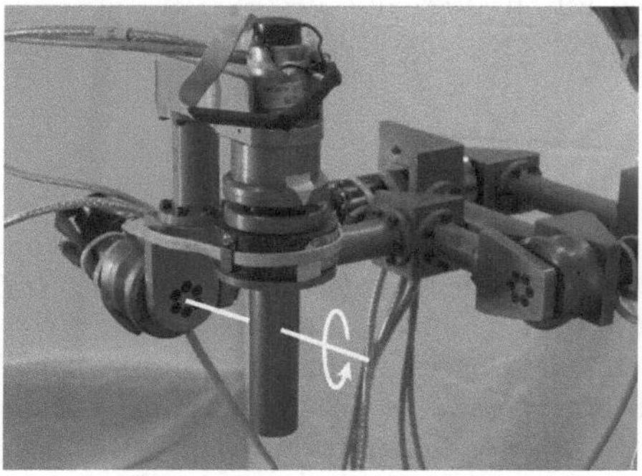

Fig. 2.8. Critical end-effector motion: rotation around the axis of joint 9 drives the wrist towards a singular configuration producing high accelerations of joint 5

The emphasize of the above discussion on the critical end-effector trajectories may cloud the fact that these motions rarely occur in practice. At all other interactions a robust singularity avoidance has been achieved. Therefore it can be concluded, that the pseudoinverse control approach can indeed offer an unlimited orientational workspace. Moreover, as the resultant selfmotion is simple to anticipate this solution appears to be well suited for human-system interaction.

2.6 Conclusion

The introduction of actuated kinematical redundancies in kinesthetic haptic hardware designs has been proposed to overcome performance limitations of non-redundant interfaces due to internal singularities. We presented the kinematical design of a highly versatile haptic interface VISHARD10 with 10 actuated DOF offering an unlimited rotational and large translational workspace free of singularities as well as a high force capability. The goal of this prototype is to provide a benchmarking testbed for the development and feasibility studies of novel haptic applications. We described the basic control strategy of the prototype along with the inverse kinematics solution technique. For the orientational degrees of freedom the potential for an effective singularity avoidance has been confirmed by hardware experiments. The experiments for the evaluation of the dynamic range of achievable admittances show good results regarding the display of stiff environments and low inertia. Future work will include further in-depth studies of redundancy exploitation techniques specific to the goal of haptic human-device interaction, the implementation of tactile actuators, and bimanual collaborative telemanipulation tasks in a large workspace.

Acknowledgements

This work is part of the TOUCH-HapSys project financially supported by the 5th Framework IST Programme of the European Union, action line IST-2002-6.1.1, contract number IST-2001-38040. For the content of this paper the authors are solely responsible for, it does not necessarily represent the opinion of the European Community.

References

1. H. Esen, K. Yano, and M. Buss. A virtual environment medical training system for bone drilling with 3 DOF force feedback. In *Proc. IEEE/RSJ Int. Conf. on Intellig. Rob. and Syst.*, pages 3631–3636, 2004.
2. M. Buss and G. Schmidt. Control Problems in Multi-Modal Telepresence Systems. In P.M. Frank, editor, *Advances in Control: Highlights of the 5th European Control Conf. ECC'99 in Karlsruhe, Germany*, pages 65–101. Springer, 1999.
3. J.D. Brederson, M. Ikits, C.R. Johnson, and C.D. Hansen. The visual haptic workbench. In *Proc. 5th PHANToM Users Group Workshop*, 2000.
4. C. Richard, A.M. Okamura, and M.R. Cutkosky. Getting a feel for dynamics: using haptic interface kits for teaching dynamics and controls. In *ASME IMECE 6th Annu. Symp. on Haptic Interfaces*, Dallas, Texas, 1997.
5. Y. Nakamura. *Advanced Robotics: Redundancy and Optimization.* Addison-Wesley, 1991.
6. G. Hirzinger, N. Sporer, A. Albu-Schäffer, M. Hähnle, R. Krenn, A. Pascucci, and M. Schedl. DLR's torque-controlled light weight robot iii - are we reaching the technological limits now? In *Proc. IEEE Int. Conf. Rob. Automat.*, pages 1710–1716, 2002.

7. M. Ueberle and M. Buss. Design, control, and evaluation of a new 6 DOF haptic device. In *Proc. IEEE/RSJ Int. Conf. on Intellig. Rob. and Syst.*, pages 2949–2954, 2002.
8. H.Z. Tan, M.A. Srinivasan, B. Eberman, and B. Cheng. Human factors for the design of force-reflecting haptic interfaces. In *Proc. ASME Winter Annu. Meeting: Dyn. Syst. and Contr., DSC-55(1)*, pages 353–359, 1994.
9. J. Martin, J. Savall. Mechanisms for haptic torque feedback. In *Proc. First Joint Eurohaptics Conf. and Symp. on Haptic Interfaces for Virtual Environment and Teleoperator Systems*, pages 611–614, 2005.
10. S.D. Laycock and A.M. Day. Recent developments and applications of haptic devices. *Computer Graphics Forum*, 22(2):117–132, 2003.
11. C. Youngblut, R.E. Johnston, S.H. Nash, R.A. Wienclaw, and C.A. Will. Review of virtual environment interface technology. IDA Paper P-3186, Institute for Defense Analysis (IDA), 1996.
12. G. Burdea. *Force and Touch Feedback for Virtual Reality.* John Wiley & Sons, 1996.
13. U. Kühnapfel, H.K. Çakmak, B. Chantier, H. Maaß, G. Strauss, C. Trantakis, E. Novatius, J. Meixensberger, K. Lehmann, H.. Buhr, M. Lawo, and G. Bretthauer. HapticIO: haptic interface-systems for virtual-reality training in minimally-invasive surgery. In *Proc. Int. Statustagung Virtuelle und Erweiterte Realität*, 2004.
14. Y. Tsumaki, H. Naruse, D.N. Nenchev, and M. Uchiyama. Design of a compact 6-dof haptic interface. In *Proc. IEEE Int. Conf. on Rob. Automat.*, pages 2580–2585, 1998.
15. V. Hayward, P. Gregorio, S. Greenish, M. Doyon, L. Lessard, J. McDougall, I. Sinclair, S. Boelen, X. Chen, J.-P. Demers, and J. Poulin. Freedom-7: A high fidelity seven axis haptic device with application to surgical training. In A. Casals and A.T. de Almeida, editors, *Experimental Robotics V*, Lecture Notes in Control and Information Sciences 232, Springer-Verlag, pages 445–456, 1998.
16. H. Iwata. Pen-based haptic virtual environment. In *Proc. IEEE Virtual Reality Annual Int. Symp.*, pages 287–292, 1993.
17. T. Massie and J. Salisbury. The PHANTOM haptic interface: A device for probing virtual objects. In *Proc. ASME Winter Annu. Meeting: Dyn. Syst. and Contr.*, volume 55, pages 295–301, 1994.
18. S. Grange, F. Conti, P. Rouiller, P. Helmer, and C. Baur. Overview of the Delta Haptic Device. In *Eurohaptics*, 2001.
19. C.L. Clover, G.R. Luecke, J.J. Troy, and W.A. McNeely. Dynamic simulation of virtual mechanisms with haptic feedback using industrial robotics equipment. In *Proc. IEEE Int. Conf. on Rob. Automat.*, pages 724–730, 1997.
20. J. Hoogen, R. Riener, and G. Schmidt. Control aspects of a robotic haptic display for kinesthetic knee joint simulation. In *IFAC J. on Control Eng. Practice*, volume 10, pages 1301–1308, 2002.
21. R. Adams, M. Moreyra, and B. Hannaford. Excalibur - a three axis force display. In *Proc. ASME Int. Mech. Eng. Congr. Exhib.*, 1999.
22. R.Q. Van der Linde, P. Lammertse, E. Frederiksen, and B. Ruiter. The HapticMaster, a new high-performance haptic interface. In *Eurohaptics'02*, pages 1–5, 2002.
23. A. Frisoli, F. Rocchi, S. Marcheschi, A. Dettori, F. Salsedo, and M. Bergamasco. A new force-feedback arm exoskeleton for haptic interaction in virtual environments. In *Proc. First Joint Eurohaptics Conf. and Symp. on Haptic Interfaces for Virtual Environment and Teleoperator Systems*, pages 195–201, 2005.

24. S. Jacobsen, F. Smith, D. Backman, and E. Iversen. High performance, high dexterity, force reflective teleoperator II. In *ANS Topical Meeting on Robotics and Remote Systems*, 1991.
25. N. Nitzsche and G. Schmidt. A mobile haptic interface mastering a mobile teleoperator. In *Proc. IEEE/RJS Int. Conf. on Intellig. Rob. and Syst.*, pages 391–3917, 2004.
26. N. Nitzsche, U.D. Hanebeck, and G. Schmidt. Motion compression for telepresent walking in large target environments. *Presence*, 13(1):44–60, 2004.
27. C. Wampler. Wrist singularities: theory and practice. In O. Khatib, J. Craig, and T. Lozano-Pérez, editors, *The Robotics Review 2*, pages 173–189. MIT Press, 1992.
28. R. Clavel. *Conception d'un robot parallèle rapid à 4 degrés de liberté.* PhD thesis, École Polytechnique Fédérale de Lausanne, 1991.
29. A. Sharon, N. Hogan, and D.E. Hardt. High bandwidth force regulation and inertia reduction using a macro/micro manipulator system. In *Proc. IEEE Int. Conf. on Rob. Automat.*, pages 126–132, 1988.
30. J.M. Hollerbach. Optimum kinematic design for a seven degree of freedom manipulator. In *2nd Int. Symp. of Robotics Research*, pages 215–222, 1984.
31. K. Kreutz-Delgado, M. Long, and H. Seraji. Kinematic analysis of 7 dof anthropomorphic arms. In *Proc. IEEE Int. Conf. on Rob. Automat.*, pages 824–830, 1990.
32. R.L. Williams II. Local performance optimization for a class of redundant eight-degree-of-freedom manipulators. In *Proc. IEEE Int. Conf. on Rob. Automat.*, pages 992–997, 1994.
33. F.B.M. Duarte, J.A.T Machado, and L. Horváth. A trajectory planning algorithm for redundant manipulators. In *Proc. IEEE Int. Symp. Ind. Electr.*, pages 1002–1007, 1999.
34. H. Esen, K. Yano, and M. Buss. Training strategies allowing haptic interaction in multi-user virtual environments. In *Proc. IEEE Int. Conf. on Robotics & Automation*, 2007. in review.
35. H. Esen, K. Yano, and M. Buss. Interaction with virtual deformable objects for surgery simulation using a hyper redundant haptic display. In *Eurohaptics'06*, pages 519–522, 2006.
36. M. Fritschi, M.O. Ernst, and M. Buss. Integration of kinesthetic and tactile display - modular design concept. In *Eurohaptics'06*, pages 607–612, 2006.
37. M.P. Vitello, M.O. Ernst, and M. Fritschi. An instance of tactile suppression: Active exploration impairs tactile sensitivity for the direction of lateral movement. In *Eurohaptics'06*, pages 351–356, 2006.
38. M. Ueberle, N. Mock, and M. Buss. VisHaRD10, a novel hyper-redundant haptic interface. In *Proc. Int. Symp. Haptic Interfaces for Virtual Environment and Teleoperator Systems*, pages 58–65, 2004.
39. P.K. Khosla and T. Kanade. Treal-time implementation and evaluation of the computed-torque scheme. *IEEE Trans. Rob. Automat.*, 5(2):245–253, 1989.
40. A. Liégeois. Automatic supervisory control of the configuration and behaviour of multibody mechanisms. *IEEE Tans. on Syst. Man, and Cybern.*, 7(12):868–871, 1977.
41. M. Ueberle and M. Buss. Control of kinesthetic haptic interfaces. In *Proc. IEEE/RSJ Int. Conf. on Intellig. Rob. and Syst. Workshop on Touch and Haptics*, 2004.
42. Tsuneo Yoshikawa. Manipulability of robotic mechanism. *The Int. J. of Robotics Research*, 4(5):3–9, Summer 1985.

3

A Novel Parallel Haptic Interface for Telerobotic Systems

José María Sabater[1], Rafael Aracil[2], Roque J. Saltaren[2], and Luis Payá[1]

[1] Universidad Miguel Hernandez, DISI, Elche (Alicante), 03202 Spain
{j.sabater,lpaya}@umh.es
[2] Universidad Politecnica de Madrid, DISAM, ETSII, Madrid, 28006 Spain
{aracil,rsaltaren}@etsii.upm.es

Summary. This chapter presents a novel parallel structure based on a modification of the 6 degrees of freedom (d.o.f.) Gough-Stewart platform where the linear actuators have been replaced by cable-driven pantographs. This device is used as a haptic interface. The goal of this design is to be used as a first prototype of a haptic device that tries to explote the particular characteristics of parallel structures to get a high bandwidth haptic interface able to be used as an impedance or as an admittance display. The remainder of the chapter is as follows; first some considerations about using parallel structures as haptic devices are made; then the geometrical model and the kinematic analysis are shown. Next, the hardware and software architectures used on the system, and the control schemes implemented on a multi-axis board are detailed. The designed control setup of the interface allows the implementation and experimentation of several bilateral control schemes. Experimental results with this device inside a classical force-position bilateral control are also shown at the end of this chapter.

3.1 Introduction

The main objective in a telerobotic system is to develop "a partnership between man and machine that seeks to exploit the capabilities of both to do a task better than either can do alone" [1]. To establish this relationship, a mechanical interface that allows the operator to interact with the remote robot will be needed. When this mechanical interface has the ability to render tactile or kinesthetic forces to the operator is named a haptic interface. "Touching" can be defined as the feeling when the skin and muscles are stimulated by mechanical, thermic, chemical or electrical means [2], so the goal of the haptic interface is to provoke this needed stimuli to the operator. Besides, other objective is to be transparent to the operator, i.e. the user must feel only the forces of the remote zone, and never the weight and forces of the own mechanical system, so the own inertia, friction and weight of the device must not be felt when using the interface [3]. Nevertheless, a deeper discussion of human psychophysics is complicated by a number of factors, such as the temporal and spatial characteristics of the receptor involved, their saturation or the knowledge of the operator [4]. As Kalawsky [5]states:

Clearly, anyone wishing to construct a device to communicate the sensation of remote touch to a user must be fully aware of the dynamic range of the touch

M. Ferre et al. (Eds.): Advances in Telerobotics, STAR 31, pp. 45–59, 2007.
springerlink.com © Springer-Verlag Berlin Heidelberg 2007

receptors, with particular emphasis on their adaptation to certain stimuli. It is only too easy to disregard the fundamental characteristics of the human body.

Leaving the important psychophysical aspects out of this chapter, we can establish that there are several classifications of haptic interfaces; depending on the technology of the actuators (electric, hydraulic, pneumatic, etc.), the control architecture (impedance system or admittance system), its portability (desk-grounded interfaces, exoskeletons, hand gloves, etc.) or depending on its kinematic chain. Attending to the last criteria, haptic devices can be classified as spherical interfaces (joysticks), serial kinematic interfaces (anthropomorphic, cartesian, etc.) or parallel kinematic devices.

By other side, the increase of the computational power of new processors has renewed researchers interest on parallel robots. Parallel Kinematic Mechanisms (PKM's) are constituted of an end effector connected to the base by a number of kinematic chains (legs). The special characteristics of this kind of robots make them suitable for many applications, as for example, machine tools, robot manipulators or climbing service robots. Especially interesting is the application of parallel structures as haptic devices. Several examples, like the Gough based platforms for surgery applications (CRIGOS [6]), 3 degrees of freedom spherical mechanisms (SHaDE [7]), cable driven mechanisms (Videt [8]) or mechanisms with legs of several links (Delta [9], HapticMaster [10]) have been presented during last years. All these devices try to explote the special characteristics of parallel structures, like low inertia, high rigidity, compactness, precise resolution and high load/power ratio, compared with serial mechanisms. However, to our knowledge, there are not many research works in the literature that analyze the performance of these devices inside a teleoperation scheme.

Physiologically, haptic feedback relates to two cognitive senses: the tactile sense that gives an awareness to stimuli on the surface of the body, and the kinesthetic sense that provides information on body position and movement. Bidirectionality is the most prominent characteristic of the haptic channel. Haptic perception always involves exchange of (mechanical) energy - and therefore of information - between the body and the world outside.

To characterize the performance of a haptic display is not either an easy task, due to its bidirectional nature; it is capable of both reading and writing input to and from a human user. Several indexes can be used to compare between different devices with some similar characteristics, but frequently those indexes point on a very particular feature. Those indexes can be found on specialized literature. Some of them are used to characterize the static performance of parallel devices, as the Ozaki (1996) global pay-load index (GPI) [11], the global conditioning index (GCI) used by Gosselin and Angeles [12] or the constant orientation workspace (COW) defined by Yoon and Ryu [13]. Other researchers have presented measures of the dynamic performance of various haptic devices, for example Moreyra and Hannaford [14] suggested a method to characterize and experimentally measure the dynamic performance of a haptic device. Carignan and Cleary [15] measured the quality of haptic devices in terms of impedance accuracy and resolution. Colgate and Brown [16] suggested the dynamic range of achievable impedance (Z-width) as a measure of performance. Other indexes have been thought for more general features of the haptic interfaces, like the resolution, precision, bandwidth or structural response. Nevertheless, the wide variety of the existent designs makes difficult to establish some comparatives.

3.2 Using Parallel Devices as Haptic Interfaces

In an ideal haptic device, the desirable features for force display are summarized as:

- High workspace,
- transparency and backdrivability,
- high control and sensing bandwidth to render tactile and kinesthetic forces,
- high ratio power-mass to minimize the operator fatigue,
- low inertia, and
- high dynamic range.

Some of the desired conditions for an ideal haptic display can be achieved using PKM's. In general, parallel mechanisms show the next "advantages":

- High ratio load/power. In some cases, the actuators connect the base and the end effector directly, being part of the structure of the mechanism. That way, parallel robots have a great capacity of manipulating loads that are higher than its own weight.
- High stiffness and rigidity at relatively low weight.
- High operation velocity.

Nevertheless, some "disadvantages" for its use as haptic devices must be considered (and overcome if possible):

- The forward kinematic problem of parallel mechanisms is an open problem for many of them, and commonly there are multiple solutions for this problem (known as assembly modes). Nevertheless, if the haptic display is thought to be an impedance display (rendering forces to the operator), we can avoid this obstacle easily, as explained later.
- The workspace is usually small and difficult to analyze, but some solutions can be found when indexing and mapping workspace of the haptic master and the slave [17].
- The singularities of a parallel manipulator must be calculated in order to avoid jams on the mechanism [18].

If we thought about serial and parallel mechanisms, we can observe that some "duality" exists between them, the stronger features in one structure appear to be the weaker in the other, and viceversa.

3.3 Performance Parameters of a Haptic Interface

For any robotic or mechanical system, such as a haptic device, there are several essential criteria for describing the system, e.g. inertia, friction, weight and backlash. However, the duality of the haptic channel to drive and to be driven causes discrepancies such as where these measures should be taken from. For example, in [19] Hayward expressed it with the following question: "is inertia measured as seen from the actuators, or from the output device itself?".

Next, some indexes and features to characterize the performance of haptic devices are briefly reviewed. This measures are discussed with respect to its importance inside a teleoperation scheme. For example, one factor that is commonly not taken into account on some papers is the bandwidth, nevertheless the interaction of human and machine implies that the bandwidth of the device is of great importance.

Bandwidth — On the design process of a haptic device, the sensing and motor-control bandwidth plays an important role. Sensing bandwidth refers to the frequency with which tactile or kinesthetic stimuli are sensed, and control bandwidth refers to the rapidity with humans can respond. The sensing and control loops are asymmetric, meaning that input (or sensing) bandwidth is much larger than output (or control) bandwidth.

Constant Orientation Workspace — The COW is defined as the 3D region that can be obtained by the end-effector when the mobile platform is kept at a constant orientation. Therefore, large values of COW will give wide range of motion.

Instantaneous Peak Force — The instantaneous forces rendered to the operator must be according with the sensing human capacity. The maximum continuous force an operator can support without feeling fatigue is near 20 N. For example, the peak output force of the well-known Phantom is 10 N, and continuous force (without actuator overheating) is only 1.5 N. The measurement of peak force is relatively easy since it involves the use of a sufficiently accurate load cell and interposing it between the end effector and a stiff reference, thus under isometric conditions.

Resolution — It has been observed many times that resolution is a critical detailed feature of haptic devices, comparing to precision that matters less. The resolution of the system is considered from the output point of view, and represents the smallest deviation from system equilibrium which can be detected by sensor under study.

Global Condition Index — Salisbury and Craig [20] defined the dexterity of a robotic manipulator as the kinematic accuracy associated with it. Mathematically, the quality of performance of a robot with respect to the force and velocity transmission can be addressed by using the Jacobian matrix of the robot. The condition number ranges from $1 \leq \kappa < \infty$, and hence, the reciprocal of the condition number -and referred as the dexterity of the manipulator- is used instead. The local dexterity (CI) is defined as the inverse of the condition number of the Jacobian matrix:

$$CI = \frac{1}{\kappa(\mathbf{J})}; \qquad 0 \leq CI \leq 1 \quad \kappa(\mathbf{J}) = \|\mathbf{J}\|\|\mathbf{J}^{-1}\| \tag{3.1}$$

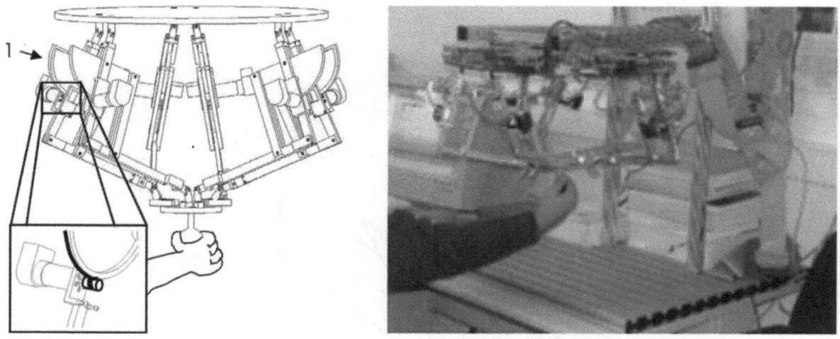

(a) Cad model (b) Designed prototype

Fig. 3.1. CAD model and prototype of designed device

where $\|\mathbf{J}\|$ is the Euclidean 2-norm of the Jacobian matrix. When the CI approaches 1 the matrix $\mathbf{J}$ is said to be well-conditioned, i.e., is far from singularities.

Global dexterity is then evaluated by integrating the local dexterity over the workspace (w), i.e.:

$$GCI = \frac{\int_w \left(\frac{1}{\kappa} dw\right)}{\int_w dw}. \tag{3.2}$$

3.4 Magister-p: Analysis of a 6-U$\underline{\text{R}}$S Parallel Haptic Device

Next, the analysis of a parallel haptic device is done. The mobile base of the designed joystick is the inferior one. That way the resulting design becomes an ergonomic device, showing to the operator a free collision workspace (as the mechanical structure stays overhead), and at the same time, the offset torques required to "mask to the operator" the gravity effects over the own links are decreased. Besides that, the device shows an equilibrium position when no forces are acting over the system. Because of the special mechanical configuration of the joints, such a haptic device has several advantages; it has 6 programmable degrees of freedom, large workspace, low inertia, high bandwidth, and a comfortable using. The pantograph on each kinematic chain with a U$\underline{\text{R}}$S configuration (Universal-Rotational-Spherical) where the underline indicates that the joint is actuated, and the using of a non slip-cable transmission between the actuator axis and the rotation of one of the links of the pantograph (marked with 1 in Fig. 3.1) increase the mentioned advantages of parallel devices.

3.4.1 Kinematics

The kinematic analysis of the proposed 6-U$\underline{\text{R}}$S platform is based on a modification of the algorithms of the 6-U$\underline{\text{P}}$S platform, and has been presented on [21], so here we make a brief resume.

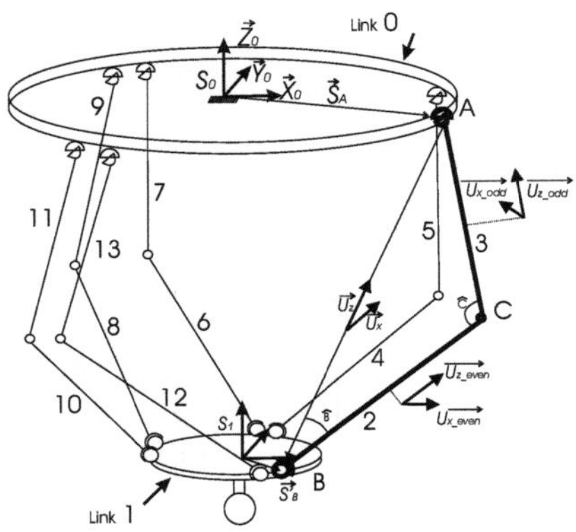

Fig. 3.2. Geometrical model

The kinematic analysis of the platform has been carried out considering only 13 bodies, because the adding of the transmission links of the pantographs does not give additional information to the kinematic model and extends the vectorial equations. This assumption must be reviewed for a dynamical modelling. Fig. 3.2 shows the used model.

Using the Euler parameters to represent the orientation, each body needs seven generalized coordinates, leading to 91 generalized coordinates for the 13 bodies to completely define the display. The sum of the constraints imposed by the spherical, the rotational and the universal joints, and the constraints imposed by the normalization of the Euler parameters, give a total of 85 constraints. The difference are the degrees of freedom of the mechanism.

The motion of the device is defined by the time variation of angle $\hat{C}$ between the even and odd links. This is equivalent to impose actuators constraints to the rotational joints, so the six degrees of freedom will be completely determined.

To solve the kinematics of the mechanism, a reference system must be assigned to each link.

The upper base, that is fixed, is named as "*link 0* "and its reference system $\mathbf{S_0}$ (considered inertial). The $\mathbf{S_0^{A_n}}$ vectors localize the application points of the universal joints A_n and are referenced to $\mathbf{S_0}$ by the expressions:

$$\mathbf{S_0^{A_n}} = \begin{bmatrix} r_{base}\cos\left((n-1)\frac{\pi}{3}+\frac{\delta_0}{2}\right) \\ r_{base}\sin\left((n-1)\frac{\pi}{3}+\frac{\delta_0}{2}\right) \\ 0 \end{bmatrix}^T \quad n=1,3,5 \qquad \mathbf{S_0^{A_n}} = \begin{bmatrix} r_{base}\cos\left((n-2)\frac{\pi}{3}-\frac{\delta_0}{2}\right) \\ r_{base}\sin\left((n-2)\frac{\pi}{3}-\frac{\delta_0}{2}\right) \\ 0 \end{bmatrix}^T \quad n=2,4,6 \tag{3.3}$$

where δ_0 defines the angle between two consecutive joints and r_{base} is the separation radius of the joints from the origin of the reference frame.

The lower base or joystick is named as "*link 1*" and its reference system S_1. The $\mathbf{S_1'^{B_n}}$ vectors localize the application points of the spherical joints B_n.

Inverse Kinematics

To find the unique solution of the inverse kinematic problem (IKP), the distance between the A_n and B_n anchor points is used.

Given the position and orientation of the *"link 1"* joystick, by a vector $\mathbf{q_1} = [\mathbf{r_1}, \mathbf{p_1}]^T$ where $\mathbf{r_1} = [x_1,\ y_1,\ z_1]^T$ is the cartesian position and $\mathbf{p_1} = [e_0,\ e_1,\ e_2,\ e_3]^T$ are the euler parameters, or $\mathbf{q_1} = [\mathbf{r_1},\ \alpha_1,\ \beta_1,\ \gamma_1]^T$ if the orientation is given with the 313 euler angles. The distance between the universal and spherical joints can easily be obtained by:

$$\mathbf{r_{A_nB_n}} = \mathbf{r_1} + \mathbf{A_1}\mathbf{s_1}'^{\mathbf{B_n}} - \mathbf{s_0^{B_n}} \tag{3.4}$$

where $\mathbf{A_1}$ is the rotation matrix given the orientation of *"link 1"*. Getting the norm of $\mathbf{r_{A_nB_n}}$, the solution is shown by (3.5) (see Fig. 3.2).

$$\hat{C}_n = \arccos\left(\frac{\left(\overline{BC}\right)^2 + \left(\overline{AC}\right)^2 - norm\left(\mathbf{r_{A_nB_n}}\right)^2}{2\left(\overline{BC}\right)\left(\overline{AC}\right)}\right) \tag{3.5}$$

Forward Kinematics

To solve the forward kinematics (FKP) of a U<u>R</u>S platform is to establish the relations between the command variables of the angles $\hat{C}$ and the final position of the end effector. A solution to the FKP problem is obtained using multibody formulation in [21], but it is not needed here.

3.4.2 Jacobian Analysis

In order to obtain the forces rendered to the operator, the relation between the joint forces exerted by the actuators and the cartesian force rendered at the end-effector (*"link1"*) must be known. This relation must be evaluated for each pose of the device. The are many ways to get this relationship, like differential geometry or screws theory. Here we use an easy way that starts by assigning the reference systems to the even and odd links, and an auxiliary reference system composed by the vectors $\mathbf{u_z}$ and $\mathbf{u_x}$.

$$\mathbf{u_z^n} = \frac{\mathbf{r_{A_nB_n}}}{\left\|\mathbf{r_{A_nB_n}}\right\|}, \quad \mathbf{u_x^n} = \frac{\mathbf{u_z^n} \times \mathbf{s_0^{An}}}{\left\|\mathbf{u_z^n} \times \mathbf{s_0^{An}}\right\|} \tag{3.6}$$

Vectors $\mathbf{u_{z\_even}}$ are part of the reference systems of the even links; 2, 4, 6, 8, 10 and 12, and are obtained by a rotation of the vector $\mathbf{u_z}$ around vector $\mathbf{u_x}$.

$$\mathbf{u^n_{z\_even}} = \left(\mathrm{Rot}(\mathbf{u_x}, (\hat{B}_n))\right) \times \mathbf{u_z} \tag{3.7}$$

To obtain the vectors $\mathbf{u_{x\_even}}$, two possible situations must be considered. Fig. 3.3 shows that the third component of $\mathbf{u_{z\_even}}$ can be positive or negative (considering null the z coordinate of the B_n points), meaning that the vector

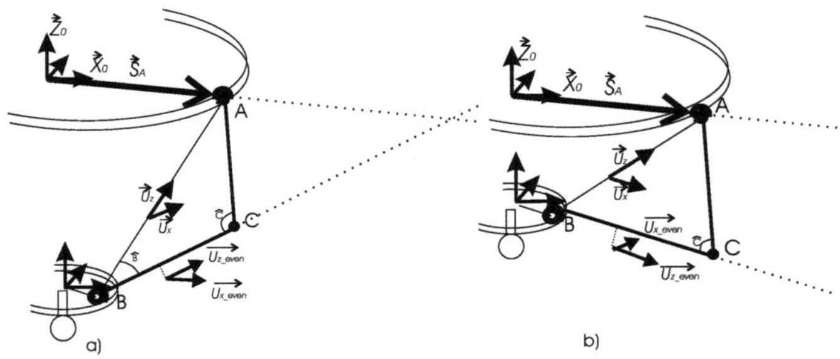

Fig. 3.3. Two possible configurations of the URS leg

$\mathbf{u_{x\_even}}$ lays above or beyond the vector $\mathbf{s_0^{An}}$ (whose third component is zero). So the expressions to get $\mathbf{u_{x\_even}}$ are

$$\mathbf{u^n_{x\_even}} = \frac{\mathbf{u^n_{z\_even}} \times \mathbf{s_0^{An}}}{\left\| \mathbf{u^n_{z\_even}} \times \mathbf{s_0^{An}} \right\|}, \text{ if } \mathbf{u(3)_{z\_even}} \geq 0 \tag{3.8}$$

or

$$\mathbf{u^n_{x\_even}} = \frac{\mathbf{s_0^{An}} \times \mathbf{u^n_{z\_even}}}{\left\| \mathbf{s_0^{An}} \times \mathbf{u^n_{z\_even}} \right\|}, \text{ if } \mathbf{u(3)_{z\_even}} < 0 \tag{3.9}$$

$$\mathbf{u^n_{y\_even}} = \mathbf{u^n_{z\_even}} \times \mathbf{u^n_{x\_even}} \tag{3.10}$$

Then, if $[\mathbf{f}, \mathbf{n}]^T$ are the forces and torques on the "*link 1*", and $\mathbf{N_n}$ are the torques generated by the actuators on the rotational joints,

$$\mathbf{f} = \sum_{n=1}^{6} \mathbf{u_{y\_even}} \times \overline{BC} \times \mathbf{N_n} \tag{3.11}$$

$$\mathbf{n} = \sum_{n=1}^{6} \left(s_1^{Bn} \mathbf{u_{y\_even}} \right) \times \overline{BC} \times \mathbf{N_n} \tag{3.12}$$

Writing in matrix form the previous equations, expression (3.13) is obtained.

$$\begin{bmatrix} \mathbf{f} \\ \mathbf{n} \end{bmatrix} = \left(\mathbf{J}^T\right)^{-1} \mathbf{F} = \begin{bmatrix} \mathbf{u^1_{y\_even}} & \mathbf{u^2_{y\_even}} & \cdots & \mathbf{u^n_{y\_even}} \\ \mathbf{s_1^{B1}} \times \mathbf{u^1_{y\_even}} & \mathbf{s_1^{B2}} \times \mathbf{u^2_{y\_even}} & \cdots & \mathbf{s_1^{Bn}} \times \mathbf{u^n_{y\_even}} \end{bmatrix} \mathbf{F} \tag{3.13}$$

3.4.3 Performance Indexes of This Interface

With the previous analysis and the built prototype, the performance of the haptic device is calculated and summarized on table 3.1. Some of these values are compared with the ones of the Phantom Premium 1.5 Prototype.

Table 3.1. Performance indexes of 6-U<u>R</u>S and Phantom Premium 1.5

	Magister-P	Phantom Premium 1.5
Kinematic chain	Serial	Parallel
Dof	6	6
Linearity		
at low range	no linear	almost linear
Bandwidth	320 Hz	1000 Hz
Workspace	approx. sphere of 15 cm	19.5×27×37.5 cm
Resolution	1 mm	0.03 mm
Peak Force	99.9 N	8.5 N
GCI at orientation (0,0,0)	0.2516	unknown

As it can be observed, some parameters are worse than on our parallel device, mainly because of manufacturing and mounting problems. The resolution is one of these parameters. The bandwidth of the present prototype is also worse, but it arrives until 320 Hz, that is the bandwidth beyond which the human fingers cannot discriminate two consecutive force input signals. Nevertheless, the parallel device is able to render forces higher than a serial device (the value of 99.9 N is obtained on a vertical axis). Unfortunately, the GCI value of the serial display is not obtained in the comparative. Nevertheless, the value of this parameter indicates the global dexterity of the device at a constant orientation. The maximum value is for an orientation of (0,0,0) (i.e. when the both platforms are parallel), and shows that at different orientations, the dexterity of the device decreases.

3.5 Hardware and Software Setup

Some of the more relevant characteristics of the designed device are the ability of working as impedance or as admittance display, and to present an open control programmable architecture that allows implementing any bilateral control scheme. To get those characteristics, an open hardware architecture and several software tools have been developed.

This section shows the operation of the Magister-P and the system architecture of the whole interface.

3.5.1 Hardware Scheme

The system hardware architecture has been built on a general purpose multiaxis board (dSpace model DS1103PPC), that is equipped with a real time Motorola PowerPC 604e processor and with a slave-DSP subsystem based on the Texas Instruments TMS320F240 DSP microcontroller. The board is allocated on the ISA bus of the workstation computer. All the simulation software and the visualization interface is programmed over the workstation PC. Communications between the PC-Workstation and the PowerPC 604e microprocessor are made

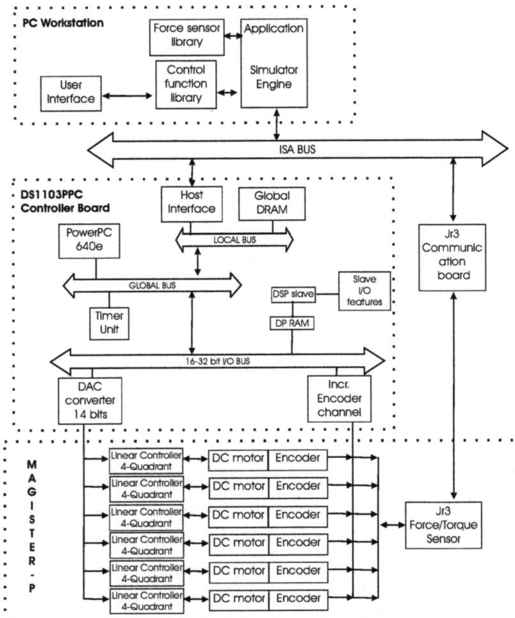

Fig. 3.4. Magister-P Open Control Architecture

using the dSpace Clib libraries, which control the access to the DS1103PPC memory. The sampling of the optical encoders, the calculation of the control signal, and the management of the linear servo amplifiers are carried out in the PowerPC microprocessor. Fig. 3.4 shows the scheme of the designed architecture for the control of the haptic interface.

The six 24V DC motors are managed by 4-Quadrants linear controller power amplifiers. Those amplifiers are set on current control configuration mode (torque control), so they supply a torque that is proportional to the command voltage.

The forces and torques applied by the operator are obtained by a six d.o.f. force/torque sensor (JR3 sensor) located on the end-effector (joystick link), and the signal is read by the own communication board.

3.5.2 Software Tools

An ANSI-C library (MBlib - multibody library) has been developed for the implementation of the control algorithms in the multiaxis board. The kinematic model explained in Sec. 3.4.1 is programmed in this library. All the code for the necessary matrix operations of the previous algorithms have been included also in MBlib library. Additional C or C++ libraries can not be used because of the PowerPC compiler tool.

Programming and management of the control algorithms on the DS1103PPC board is done using the dSpace RTlib libraries and the cited MBlib libraries:

RTlib: The resources of the DS1103PPC board are managed by the dSpace Ind. RTlib library. The functions are written on ANSI-C and control the configuration and the read/write operations on the several resources of the board.
MBlib: The MBlib library holds the necessary matrix functions and the joint primitives of the multibody algorithms that solve the IKP and the FKP problems of the 6-URS platforms. The jacobian matrix of (3.13) is also computed. The library is written in ANSI-C.

For the communication between an external application and the hardware of the DS1103PPC board, the Clib library is needed. This library is proportioned by the board manufacturer. The communication is managed by the DSP slave of the board. The process to establish such communication channel is the next one:

(i) Register a client application on the board DSP.
(ii) Select a board where the real time task is being executed.
(iii) Allocate board resources for the interchange of variables between the dSpace board and the client application.
(iv) Access the real time board variables using the existent functions. (The names of the variables must mate with the ones programmed on the board).
(v) Free the assigned board resources.
(vi) Unregister the application.

Finally some tools for the compilation (Microtec PowerPC C Compiler) and for the program management on the DS1103PPC board (Down1103 program from dSpace Inc.) are needed.

3.5.3 Impedance Mode

Impedance displays are defined as devices that can supply information about its motion at the same time that they can render to the operator a vector of generalized forces. To use a mechanical device as an impedance display, the next requirements must be fulfilled:

- to obtain the position and orientation expressed on the cartesian space, the device must have position sensors and the forward kinematic problem must be solved.
- to render a generalized force vector to the operator, the geometrical jacobian matrix must be evaluated, and the actuators must be controlled by simple open control loop amplifiers.

Fig. 3.5 shows a simplified scheme of the *impedance mode* work-mode (open loop control) programmed on the Magister-P. In this mode, the display is set on open loop mode, and all their control signals are accessible from the outside. This is the widespread control mode in the greater part of the haptic devices,

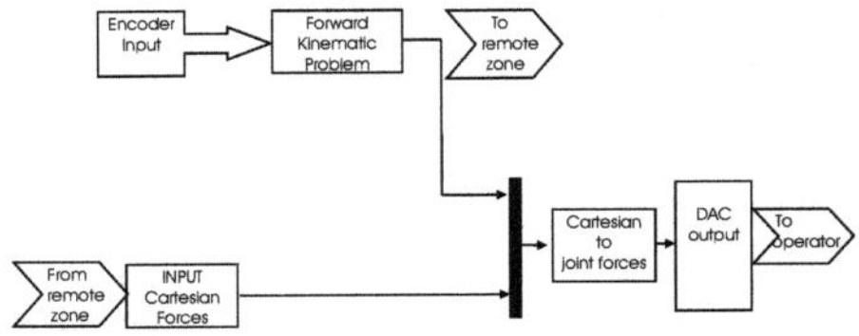

Fig. 3.5. Simplified scheme of the impedance mode (open loop)

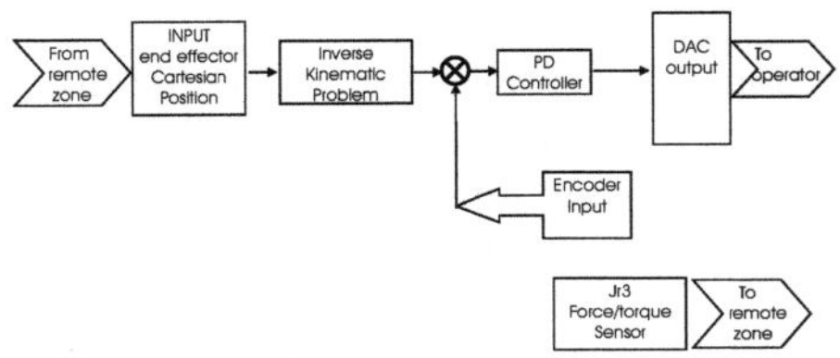

Fig. 3.6. Simplified scheme of the admittance mode (position loop)

and this work-mode allows the implementation of the classic bilateral scheme force-position (force-reflection FR), and others that are derived from this.

3.5.4 Admittance Mode

Admittance displays are defined as devices that can supply information about the generalized force vector exerted by the operator at the same time that they can render to the operator the cartesian position and/or orientation of the remote environment. To use a mechanical device as an admittance display, the next requirements must be fulfilled:

- to obtain the generalized force vector, the device must have force sensors. The most widespread options are a six d.o.f. force sensor on the end effector link or the lecture of the motor intensity and to use the simplified model of the DC motor, where the motor torque is proportional to this intensity. The jacobian matrix converts that torques to generalized forces on the end effector link.
- to render the remote environment positions to the operator, a position loop must be closed on the device. The dynamic model of the display is used to obtain the controllers to establish this loop. The mechanical structure must be stiff enough to render those forces to the operator.

Fig. 3.6 shows a simplified scheme of the *admittance mode* work-mode. This mode allows the implementation of the position-position (Position-reflection PR) bilateral control scheme.

3.6 Experimental Haptic System with a Parallel Device

To test the real performance of the presented device, a dynamic virtual environment simulator has been developed [22]. Among the main characteristics of this

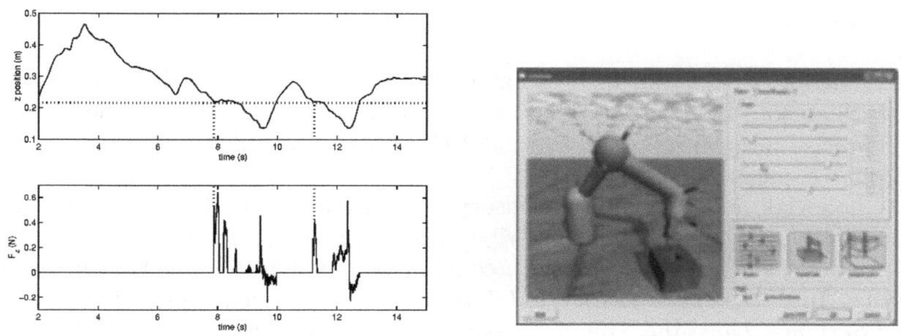

(a) Results (b) Simulator

Fig. 3.7. Peg-in-hole task: position z (top part) and F_z force (bottom part)

virtual simulator are the possibility of modeling a complete dynamic model of a robotic slave and its interaction with other dynamic objects that are in the virtual scene. Another important feature for the experimentation with teleoperated schemes is the capability of controlling the communication time between the local zone and the remote virtual zone, so the virtual zone is controlled by a discrete time interval integration step. This way the communication with the master device can be included on the step simulation. Using the parallel haptic device presented and the dynamic simulator, several experiments have been made. This subsection shows a peg-in-hole insertion. The bilateral scheme used was a classical force-position control. Fig. 3.7(a) shows the results of the simulation. Two separate insertions of the peg where made, (between the 8-10 s and the 11-14 s). The plane marked with the horizontal line in $z = 0.21$ m represents the plane where the hole is allocated, see Fig. 3.7(b). Under this height, the peg is in the hole. At the bottom part of the figure, the Z forces fed back to the operator are plotted. These forces are produced by the friction of the peg with the walls of the hole during the experiment. When the peg is going up, the forces are negative, and when the peg is out of the hole, these forces are null.

3.7 Conclusions

The design and development of an "open control" six degrees of freedom parallel master has been presented, and some performance features have been compared with a well-known serial haptic device. The main conclusion is that the "duality" between parallel and serial morphologies can be exploited to design new haptic devices that have a better performance at some specific tasks.

Besides, the availability of "open control" devices allows the implementation of different bilateral control schemes, being an interesting tool for future developments on teleoperation control.

References

1. R.H. Taylor, H.A. Paul, P. Kazanzides, B.D. Mittelstadt, W. Hanson, J. Zuhars, B. Williamson, B.Musits, E. Glassman, and W.L. Bargar. Taming the bull: Safety in a precise surgical robot. In *Proc. IEEE Int. Conf. on Advanced Robotics (ICAR)*, number 1, pages 865–870, 1991.
2. R. Cholewiak and A. Collins. Sensory and Physiological Bases of Touch In *The Psychology of Touch.* Lawrence Erlbaum Associates, 1991.
3. M. Ueberle and M. Buss. Design, control, and evaluation of a new 6 dof haptic device. In *Proc. of the IEEE/RSJ International Conference on Intelligent Robots and Systems, Laussane*, volume 3, pages 2949–1954, October 2-4 2002.
4. G. Burdea. *Force and touch feedback for virtual reality.* New York Wiley, 1996.
5. R. Kalawsky. *The science of virtual reality and virtual environments.* Addison-Wesley Ltd., 1993.
6. G. Brandt, A. Zimolong, L. Carrat, P. Merloz, H.W.Staudte, K. Lavallee, Radermacher, and G.Rau. Crigos: A compact robot for image guided orthopaedic interventions. In *IEEE Trans. Inform. Technol. Biomed.*, volume 3, pages 252–260, Dec. 1999.
7. L. Birglen, C. M.Gosselin, N.Pouliot, B.Monsarrat, and T. Lalibert. Shade, a new 3-dof haptic device. *IEEE Transactions on Robotics and Automation*, 18 (2):166–175, 2002.
8. P. Arcara, L.Di Stefano, S. Mattoccia, C. Melchiorri, and G. Vassura. Perception of depth information by means of a wire-actuated haptic interface. In *Proc. IEEE Int. Conf. Robotics and Automation, San Francisco*, pages 3443–3448, 2000.
9. Y. Tsumaki, H. Naruse, D.N. Nenchev, M. Uchiyama. Design of a compact 6-dof haptic interface. In *Proc. IEEE Int. Conf. Robotics and Automation*, pages 2580–2585, 1998.
10. H. Iwata. Artificial reality with force-feedback: Development of desktop virtual space with compact master manipulator. *Computer Graphics*, 24 (4):165–170, 1990.
11. H. Ozaki, H. Wang, X. Liu, and F. Gao. The atlas of the payload capability for design of 2-dof planar parallel manipulators. In *Proc. IEEE Inter. Conf. Systems Man and Cybernetics*, pages 1483–1488, 1996.
12. C. Gosselin and J. Angeles. A global performance index for the kinematic optimization of robotic manipulator. *Trans. ASME J. Mech. Des.*, 113:220–226, 1991.
13. J. Yoon and J. Ryu. Design, fabrication and evaluation of a new haptic device using a parallel mechanism. *IEEE/ASME Transactions on Mechatronics*, 6 (3):221–233, 2001.
14. M. Moreyra and B. Hannaford. A practical measure of dynamic response of haptic devices. In *Proc. o the IEEE Int. Conf. on Robotics and Automation*, pages 369–374, 1998.
15. C. Carignan and K. Cleary. Closed-loop force control for haptic simulation of virtual environments. *Haptics-e, The Electronic Journal of Haptics Research*, 2 (2):e–version, 2000.
16. J. E. Colgate and J.M. Brown. Factors affecting the z-width of a haptic display. In *Proc. IEEE Int. Conf. on Robotics and Automation, San Diego, USA*, pages 3205–3210, 1994.
17. J.M. Azorin, J.M. Sabater, L. Paya, and N. Garcia. Kinematics correspondence and scaling issues in virtual telerobotics systems. In TSI Press, editor, *Sixth Biannual World Automation Congress (WAC 2004)*, volume 1 of *CD-Rom Series*, pages 1–10, 2004.

18. L.W. Tsai. *Robot Analysis: the mechanics of serial and parallel manipulators.* Wiley Interscience. John Wiley and Sons, 1999.
19. V. Hayward and O.R. Astley. Performance measures for haptic display. In G. Hirzinger. Springer-Verlag G.Giralt, editor, *Robotic Research: The 7th International Symposium.*, pages 195–207, 1996.
20. J.K. Salisbury and J. Angeles. Articulated hands: force control and kinematic issues. *The International Journal of Robotic Research*, 1 (1):4–17, 1982.
21. J. M. Sabater, J.M. Azorin, R.J. Saltarén, and R. Aracil. Magister-p; a 6-urs parallel haptic device with open control architecture. *Robotica*, 23(1):177–187, 2005.
22. J. M. Sabater, J. M. Azorin, O. Reinoso, R. Neco, and N. Garcia. Dynamic virtual environment to test teleoperated systems with time delay communications. *Journal of Robotic Systems*, 22(4), April 2005.

4

Exoskeletons as Man-Machine Interface Systems for Teleoperation and Interaction in Virtual Environments

Massimo Bergamasco, Antonio Frisoli, and Carlo Alberto Avizzano

Scuola Superiore Sant'Anna
PERCRO
Piazza Martiri della Libertá 33, 56127, Pisa, Italy
{m.bergamasco, a.frisoli, c.avizzano}@sssup.it

Summary. Since the beginning of teleoperation, exoskeleton systems have been considered as man-machine interfaces at the master level. Exoskeletons are robotics systems, that can be put on by the human operator, with a kinematic structure similar to that of operator's upper or lower limbs. They have a twofold functionality: a) to record the movement at the level of human limb, b) to generate forces at the points of attachment with the human limb. The first examples of exoskeletons can be traced back to systems for the arm and the hand specifically designed for replicating grasping and contact forces to the operator. Nowadays exoskeleton systems are successfully employed also as body extenders for human power augmentation and rehabilitation. This chapter describes different types of exoskeleton systems from the point of view of mechanical and control design, and presents different examples of applications from the current state of the art.

4.1 Introduction

Exoskeletons are robotic systems designed to work linked with parts (or the whole) of the human body.

In general robots are designed for a defined workspace where they perform specific tasks autonomously. In such a condition, the issue of the physical interaction between robots and humans is considered in terms of safety. Collision avoidance procedures are usually managed by exploiting data coming from proximity or vision sensors; however, in the case that unwanted physical contact between the human operator and the robot is detected, specific compliance control algorithms can be implemented in order to minimize undesirable effects on the human body.

The design of exoskeleton systems stems from opposite motivations that intend the robotic structure to be always maintained in contact with the human operator's limb. Such a condition is required for several applications that include the use of master robotic arms for teleoperation, active ortheses and technological aids in rehabilitation, robotic extenders, etc. [1].

The condition of physical contact between the exoskeleton and the human body is not related to a single point of contact, like it happens for haptic interfaces

M. Ferre et al. (Eds.): Advances in Telerobotics, STAR 31, pp. 61–76, 2007.
springerlink.com © Springer-Verlag Berlin Heidelberg 2007

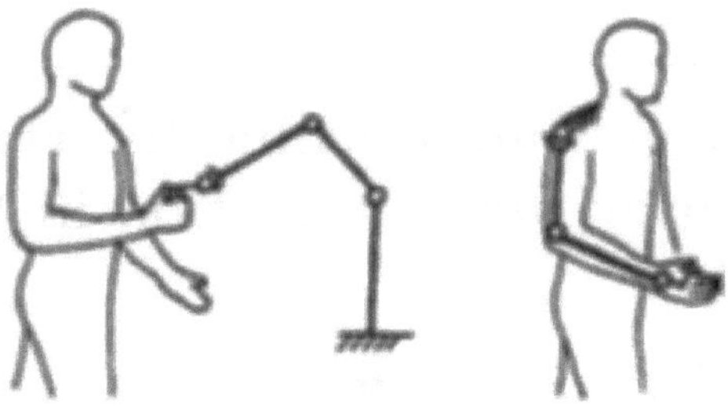

Fig. 4.1. One-point attachement vs. exoskeleton haptic interfaces

presenting a stylus at the end-effector grasped by the human operator (Fig. 4.1). The physical link between the exoskeleton and the human body refers to several points of attachment, usually at least one for each limb: in such a condition the exoskeleton robotic structure presents two possible simultaneous functionalities:

1. Following and tracking the body movements in terms of the complete spatial configuration of the limb;
2. Being able to generate forces and exert them to the human body through the points of attachment.

The strict correspondence of the exoskeleton workspace with the human limb's workspace defines constraints for the kinematics and range of joint motions of the exoskeleton robotic structure.

Another design constraint is represented by the simultaneous presence of the limb's volume and the robotic structure: due to the physical continuity of the body, the mechanical structure of the exoskeleton cannot occupy the same limb's volume and, consequently, it is usually shaped in order to wrap up the limb itself.

The nature of exoskeleton systems of being always maintained in contact with a human limb positions such robotic systems in a more favourable condition with respect to ordinary robots when issues of mobility of the human operator are concerned. Exoskeletons worn by the human operator can be considered as effective tools to allow him/her to extend or to move in workspaces that cannot be easily pre-defined. Mobility issues poses several tough constraints for the design of such systems: structural and actuation/transmission component design as well as weight-compensation control laws must be carefully considered for applications involving the human operator acting in large workspaces.

Since the beginning of the research in teleoperation, robotic exoskeletons have been devised as natural solutions for master arms [2] since they allowed the tracking of the human operator's arm in a direct and immediate way. Experiments on exoskeletons have been performed also at the JPL during 1970s [3]. Sarcos [4] developed a master arm integrating also grasping capabilities for the hand used for the remote control of a robotic arm.

At present, the research on the design of exoskeleton systems is attracting a wide interest in the field of teleoperation for space applications where the requirements for manipulative capabilities of robotic systems are becoming more and more important for near future extra-vehicular activities. Other fields of application in which exoskeletons are acquiring a definitive development niche can be identified as those of rehabilitation [5], ergonomy and vehicle-simulators [6].

Open research issues for exoskeleton systems design are referred to both technology (materials, actuators and transmission, control) as well as to portability and safety. Specific indications are given in Secs. 3-5 where different types of exoskeleton systems are described by following their correspondence with the limbs to which they are attached.

4.2 The Use of Exoskeletons as Haptic Interfaces

Exoskeleton robotic systems are used in the field of teleoperation as master devices to control the movements and operation of robotic systems (slave systems) located in a remote operational environment. In particular, such an option becomes fundamental when the human operator has to control the remote robot to perform manipulative operations in the remote environment, i.e. trying to exploit all movements of the human arm/hand for grasping and fine manipulation tasks. The use of exoskeletons allows to map the joint configuration of his/her hand-arm directly to the joint vector of the remote robotic system [7, 8, 9] while, at the same time, exert contact forces to his/her hand as detected by the robotic end-effector in the remote environment. The research approaches in telerobotics and supervisory control in 1990 exploited the use of 3D graphical representations to simulate the behavior of the robotic system in the operational space. Also in this case the use of exoskeletons was planned to control the motion of the robot [10]. Although exoskeleton-like master devices (as hand controllers) were only shortly employed, the research in telerobotics with haptic feedback capabilities definitely demonstrated that the haptic interface system utilized to interact with virtual environments can be the same of the one utilized to control remote robots in teleoperation conditions. Exoskeletons were employed as haptic interfaces to control both remote robotic arms and virtual hands in Virtual Environments (Fig. 4.2).

The utilization of exoskeleton systems as haptic interfaces implies that such systems could exert controlled forces to the human operator [11]. The study of the replication of external forces by means of so-called force feedback systems lead to the design of different typologies of electromechanical systems, usually conceived for teleoperation applications. From the realization presented in the field of robotics and teleoperation, the following design solutions emerge (Fig. 4.1):

1. Electromechanical systems (e.g. robotic manipulators or purposely designed structures) possessing a kinematics with several DOF. These systems are grasped by the human operator at their end-effector and follow the movement impressed by the human hand; when force controlled, they can exert the desired external forces in correspondence of the human operator's hand. The forces are generated by controlling the torques in each joint of the manipulator according to considerations deriving from the condition of static equilibrium. The mobility of the human operator in the control space is however limited by the constraint of grasping the end effector of the robot. Being the manipulator

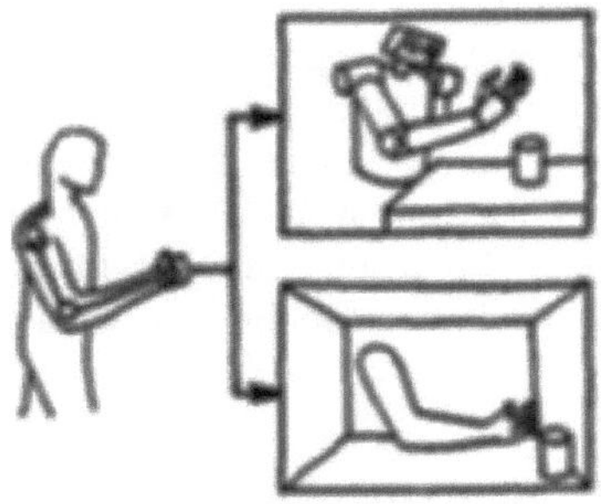

Fig. 4.2. Exoskeleton as devices for remote interaction in telepresence and virtual presence

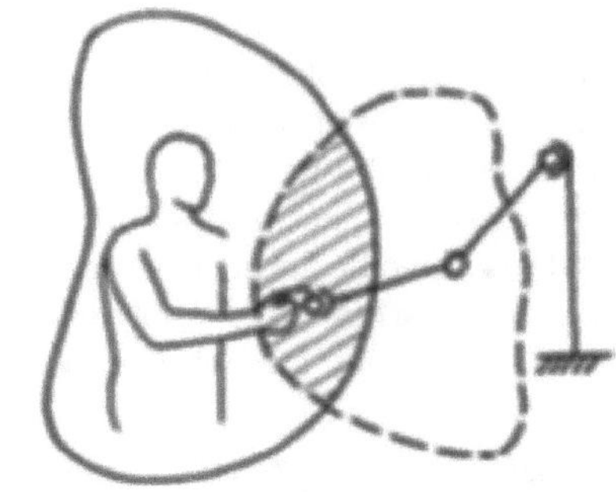

Fig. 4.3. Shared workspace between human and a robotic manipulator

fixed to the ground, the effective workspace for the operator is given by the intersection of the manipulator workspace and the workspace of his/her arm (Fig. 4.3).

By means of this kind of electromechanical systems, the replication of the external force sensation is perfect since forces are really exerted by a system which is external with respect to the operator arm. However, the force replication can occur only at the hand level and not in other parts of the arm; moreover the control of a complete manipulative operation cannot be performed since the finger's mobility is not allowed; this last drawback can be overcome by designing purposely conceived attachments with the dorsal part of the operator's hand and the manipulator end effector;

2. A different solution is obtained by considering arm exoskeleton systems. They consist in actuated and sensorized electromechanical structures, wrapping up the human arm, capable of recording joint movements and, when appropriately controlled, of exerting adequate forces on the arm. This solution introduces a near optimal resultant workspace (very close to that of the human arm) although, due to practical issues of mechanical construction, there are constraints for achieving the complete range of movements for all the human joints. Exoskeleton structures can be directly supported by the shoulders and trunk of the operator, where reaction forces and weight of the system is then distributed (Fig. 4.4).

The utilization of exoskeletons supported by the operator, and not attached to the ground, allows a complete mobility of the human operator inside his/her real control workspace. This is an extremely effective solution in those cases in which the human operator must interact not only with remote/virtual objects, but also, at the same time, with real objects in the control space (e.g. a control panel). A first solution for exoskeleton structures is that presenting only one attachment point with the dorsum of the human operator's hand, and not with other points of the arm or forearm. The resultant force feedback sensation can be considered as perfect in the tract between the hand and the shoulder. On the shoulder and trunk some effects due to reaction torques generated during the application of forces at the distal end are present. A variation with respect to the solution of an exoskeleton with one attachment point at the hand refers to the introduction of a higher number of attachments between the exoskeleton structure and the operator's arm. If attachments are considered also in correspondence of the medium part of the forearm and arm, it is possible to locally replicate desired forces also in these regions.

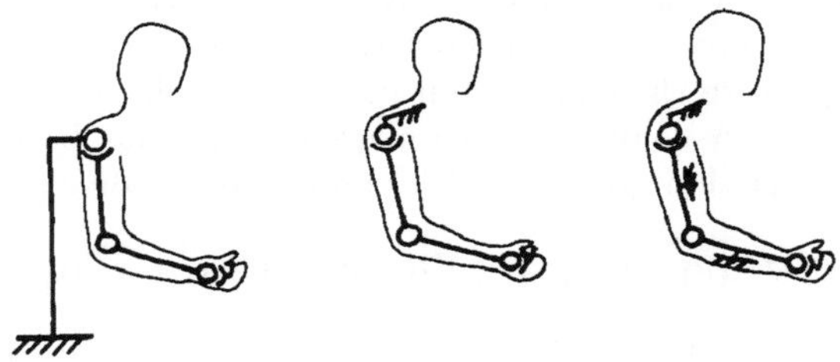

Fig. 4.4. Different schemes of attachment of an exoskeleton device to the human arm

The problem of force replication (generation) by means of an exoskeleton system can be considered identical to the problem of force generation by an ordinary robotic system but with the further constraint of considering that the number of contact points with the human operator can vary according to the number of attachment points between the exoskeleton structure and the human limb.

In the case that the exoskeleton structure is connected to the human arm only at the level of the hand, the system can be considered as an external manipulator with its base link attached to the shoulders and trunk (if dressed) or to the ground. The system can exert applied vector forces, that can be reduced to a resultant wrench, only at the level of the hand. Under this assumption it is possible to replicate external forces by controlling the wrench applied by the exoskeleton to the operator's hand. By considering as an example the kinematic scheme depicted in Fig. 4.5, the variables that can be used to achieve such a control objective are the joint torques.

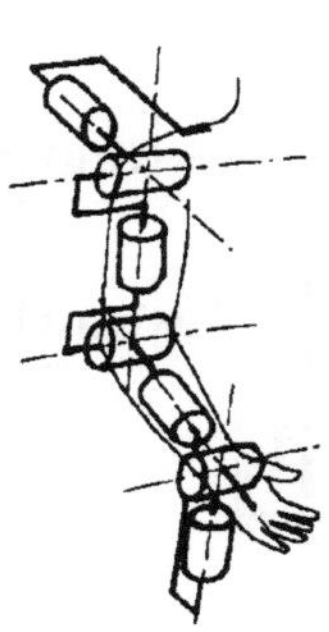

Fig. 4.5. Kinematic scheme of an exoskeleton for the right arm

In quasi-static conditions and with no gravity, the mapping between the applied wrench and the joint torques can be derived by means of the principle of virtual works and is given by the transpose jacobian of the manipulator:

$$\boldsymbol{\tau} = J^T(\mathbf{q})\mathbf{F}, \tag{4.1}$$

where J is the jacobian of the exoskeleton depending on the joint position vector $\mathbf{q}$, $\mathbf{F}$ is the wrench applied on the operator's hand, and $\boldsymbol{\tau}$ is the vector of the joint torques. In dynamic situation and in presence of gravity, the joint torques that must be applied to give a desired wrench $\mathbf{F}_{des}$ must contain additional terms devoted to compensate the exoskeleton inertia, Coriolis and centrifugal effects, friction and gravity. A complete force mapping will thus be dependent on the exoskeleton configuration, joint velocities and accelerations:

$$\boldsymbol{\tau} = M(\mathbf{q})\ddot{\mathbf{q}} + C(\mathbf{q}, \dot{\mathbf{q}}) + D(\mathbf{q}, \dot{\mathbf{q}}) + G(\mathbf{q}) + J^T(\mathbf{q})\mathbf{F}_{des}, \tag{4.2}$$

where M is the inertia matrix of the manipulator, C is the vector of Coriolis and centrifugal terms, D is the vector friction terms, and G is the vector of gravity effects. Notice that the effect of compliance (for instance of the transmission system) is not included in (4.2). The model described by equation (4.2) can be used to build a controller that regulates the wrench $\mathbf{F}$ to a desired reference value. If good performance of the force replication system is required not only in quasi-static conditions, the control must include the compensation of dynamic effects on the exoskeleton. If the requirements are not so strict in terms of bandwidth (say $\omega_0 \leq 3rad/s$), some of the terms of the full dynamic model can be neglected. Quasi-static operation ensures that the effects of Coriolis and centrifugal terms are small and, in the case good backdriveability of joint actuators is ensured by the mechanical design in order to have low values of friction, only the gravity compensation term can be used. The feasibility of the control law will depend on the availability of external wrench and/or joint torque measurements. For the sake of simplicity, in the following the dependency of J and G on joint positions will be omitted. If no force/torque sensors are present, an open loop control law can be devised as follows, as shown in Fig. 4.6:

$$\boldsymbol{\tau} = \hat{G} + J^T\mathbf{F}_{des}, \tag{4.3}$$

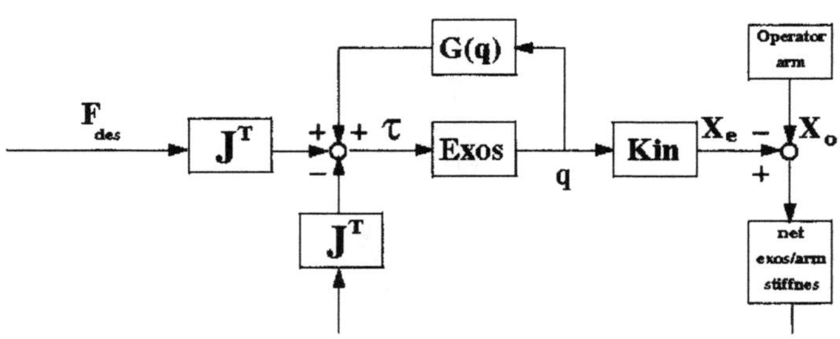

Fig. 4.6. Scheme of the open loop control procedure

where $\hat{G}$ indicates an estimate of G. The open loop scheme cannot be used if non backdriveable drives are present. Furtherly, poor performance is expected due to friction and modeling errors in the estimation of G. If torque sensors are present, the following control law can be devised:

$$\boldsymbol{\tau}_{des} = \hat{G} + J^T \mathbf{F}_{des}, \tag{4.4}$$

where $\boldsymbol{\tau}_{des}$ is a reference value for the joint torque vector. A servo term at the hand level based on the wrench error may be added to $\mathbf{F}_{des}$ in order to improve the tracking of the reference value. A low level joint torque control loop is then used. Joint torque sensing allows to overcome the problems of friction, although the measure of the wrench $\mathbf{F}$ is affected by many errors due to errors in the kinematic model of the exoskeleton, and rough or no modeling at all of the exoskeleton dynamics. If a 6 component force/torque sensor is present, the loop can be closed at the hand level and the following control law can be used, as shown in Fig. 4.7:

$$\boldsymbol{\tau} = \hat{G} + J^T \left(\mathbf{F}_{des} + K(\mathbf{F}_{des} - \mathbf{F}) \right) \tag{4.5}$$

where K is a 6×6 diagonal matrix of constant gains. Closing the loop at the hand level allows a better measurement of $\mathbf{F}$, although the rejection of friction torques cannot be effective and is strongly dependent on the arm configuration $\mathbf{q}$.

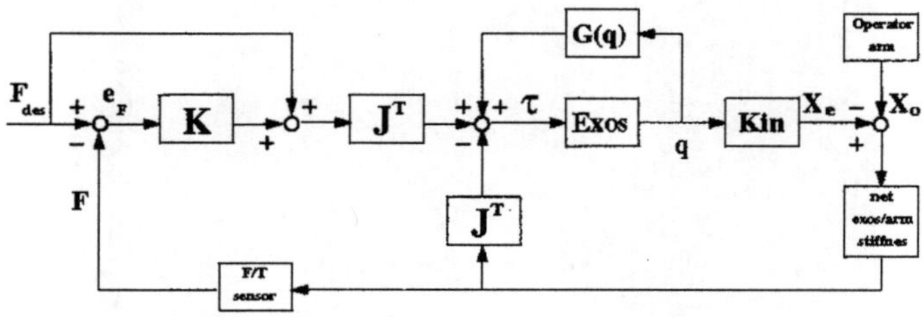

Fig. 4.7. Scheme of the closed loop control procedure

4.3 Arm Exoskeletons

L-EXOS (Light Exoskeleton) [12, 13] is an exoskeleton based haptic interface for the human arm. The L-EXOS has been designed as a wearable haptic interface, capable of providing a controllable force at the center of user's right-hand palm (Fig. 4.8), oriented along any direction of the space. It is a 5 dof robotic device with a serial kinematics, isomorphic to the human arm.

Two configurations of the device have been conceived:

C1 In the configuration C1 (L-EXOS), an handle is mounted on the last link and the system is composed of 5 DOF, of which only 4 actuated. The non actuated DOF is the last one, aligned along the anatomical prono-supination axis of the forearm (Fig. 4.8).

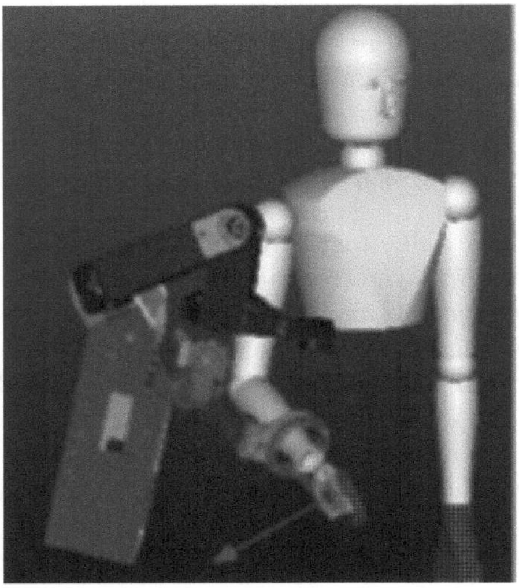

Fig. 4.8. The arm exoskeleton

Fig. 4.9. The arm exoskeleton integrated in the C2 configuration

C2 In the configuration C2, the non-actuated DOF and the handle are replaced with an hand-exoskeleton that can apply generic forces on two fingertips of the hand, preferably thumb and index fingers, as shown in Fig. 4.9.

The configuration C2 is particularly innovative, since it can reach up to three contact points, one located on the palm of the user, and the other two directly on index and thumb. The following section will focus in particular on the mechanical design and the performance of the exoskeleton in its configuration C1.

4.3.1 Kinematics and User Requirements

L-EXOS is characterized by a serial kinematics consisting of five rotational joints (Fig. 4.10) of which the first four are actuated and sensorized, while the last one is only sensorized.

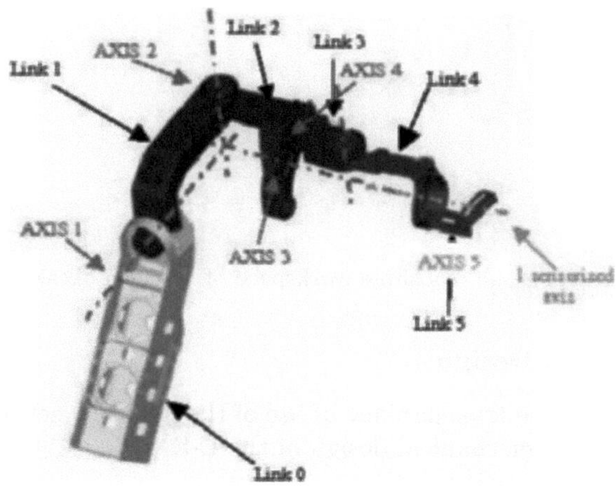

Fig. 4.10. The general kinematics of the L-EXOS

The first three rotational axes are incident and mutually orthogonal (two by two) in order to emulate the kinematics of a spherical joint with the same center of rotation of the human shoulder. The target workspace for the shoulder joint was assumed to be the quadrant of a sphere, as shown in Fig. 4.11. The orientation of the first axis has been optimized to maximize the workspace of the shoulder joint, by avoiding singularities and interferences between the mechanical links and the operator. The optimization process provided also indications for the definition of the shapes of the links. As a result from the kinematic analysis, the final orientation of the first axis (the fixed one) was chosen to be skewed with respect to the horizontal and vertical plane, while the third axis was assumed to be coincident with the ideal axis of the upper arm. This implied that the third joint had to be implemented with a rotational pair aligned with the forearm. The fourth axis was assumed coincident with the elbow joint and the fifth axis with the forearm, in order to allow the prono-supination of the wrist.

The workspace of the elbow coincides with the angle of rotation of the fourth axis. Assuming the position of zero when the forearm and the arm are aligned, the range of motion achieved by the L-EXOS spans approximately from 2.5° to 105°. The wrist has only one non actuated DOF, and its range of motion is of 180°, and precisely ±90° measured when the hand palm is aligned with arm and forearm in the zero position.

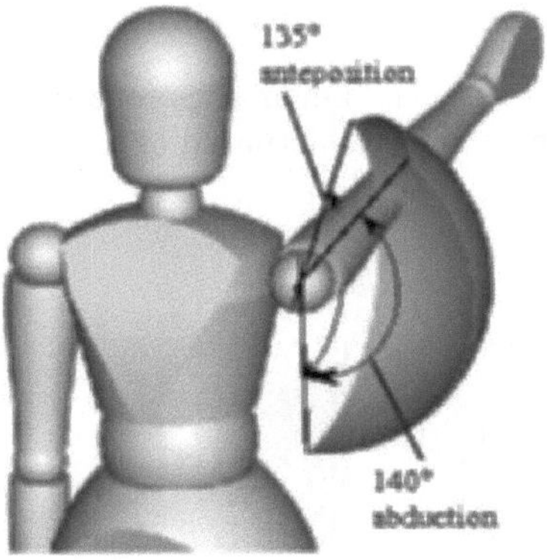

Fig. 4.11. The reachable workspace of the arm exoskeleton

4.3.2 Mechanical Design

In order to improve the transparency of use of the device, a set of guidelines has been adopted for the mechanical design of the L-EXOS:

- Remote placement of motors: this allows to drastically reduce the perceived inertia (increase of interface transparency) and the joint torque required for the gravitational compensation, and so the motors size and the transmissions tension. The remote placement of motor also allows a better weight balancing of the structure once it has been put on;
- Integration of speed reducers at joints: such a choice allows to reduce tendon tension and motor torques by a factor N^2, where N is the reduction ratio at the joint;
- Selection of motors with high torque to weight ratio;
- Use of tendon transmissions: even though such a solution presents low stiffness, it can easily transmit torques to joints placed far apart from motors with zero backlash, low friction and low weight;
- Low transmission ratio: this allows to reduce the contribution of the motors to the perceived inertia at the end-effector and so to lower the perceived transmission friction;
- Low or zero backlash implementation of the joints;
- Use of light materials for the construction of the moving parts;
- Guarantee of a safe, comfortable and ergonomic use of device.

All the motors of the exoskeleton have been located on the fixed frame (Link 0). For each actuated DOF, the torque is delivered from the motor to the corresponding joint by means of steel cables and a reduction gear integrated at the joint axis, as depicted in the scheme of the transmission of axis 2 in Fig. 4.12. Such an arrangement allows to reduce the masses of the moving parts, by reducing the mass of the motors (near 40% of the overall mass of the exoskeleton) and the additional mass of the structural parts, to be reinforced in order to sustain the weight of

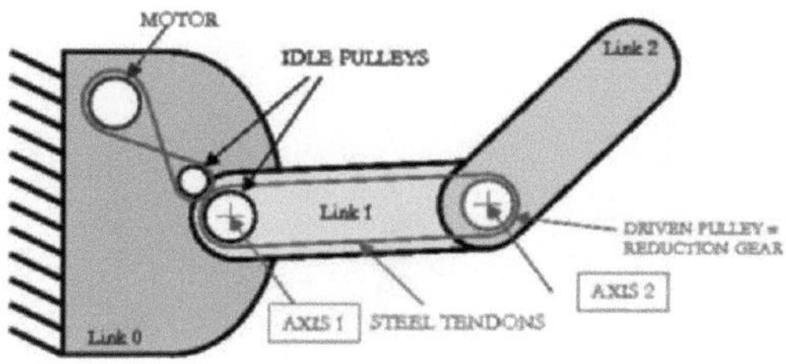

Fig. 4.12. Scheme of the actuation of the joint 2 of the L-Exos

heavier motors. The inertia perceived by the user at the palm is also consequently reduced. The electric actuators offering the best torque to weight and torque to volume ratios have been selected.

As a consequence of grounding the motors, long transmissions were required [14]. Transmissions have been implemented through steel cables that can guarantee low weight and zero backlash, even if presenting low mechanical stiffness. Moreover cable transmissions are more efficient than gear transmissions, thus ensuring a better backdrivability of the system.

To achieve a higher stiffness of the device at the end effector, reduction gears with low reduction ratio were located at the joint axes, thus allowing to reduce the tendon tensions, their elongation and their diameter. The reduction of the tendon diameter led to a consequent saving of mass and volume of all the mechanical parts of the transmission system (pulleys, axles, etc). Thanks to this solution and to the development of expressly conceived speed reducers a mass reduction of about 35% for the transmission system and of 40% for the structural parts has been achieved. The structural components (links) have been designed as thin-wall parts, that can house the mechanical parts of the transmission (pulleys, tendons, axles, spacers, etc) within the links, protecting the inner parts from external interference and the user from potential harm deriving by the tensed steel cables.

Hollow sections, presenting a larger moment of inertia than bulk sections with the same area, have been used to enhance the stiffness of the thin-wall parts. In order to further improve lightness and stiffness, the structural components have been made of carbon fiber. Also aluminum parts were bonded on the carbon fiber structure to realize the connections with other mechanical components.

Carbon fiber parts were manufactured with the vacuum-bag technique, and with dies made of carbon fiber too, due to the low number of manufactured prototypes.

4.3.3 Performance of the L-EXOS

The L-EXOS, shown in Fig. 4.14, can attain very remarkable performance, that can be summarized as follows:

Force	50 N continuous, 100 N peak force;
Backlash	10 mm at the end effector;

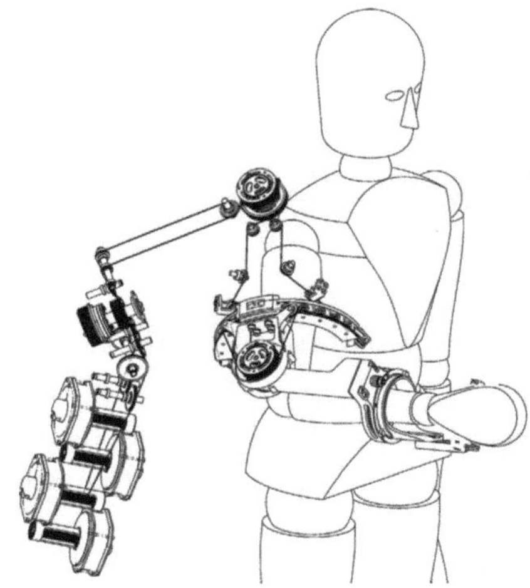

Fig. 4.13. CAD model of the transmission system

Stiffness	Estimated 3 N/mm, measured 2 N/mm;
Workspace	approximately 70% of human arm workspace.

The L-EXOS has a weight of 11 kg, of which approximately 6 kg distributed on the link 0, i.e. the fixed part, and mostly due to the mass of the 4 motor-groups. This means the L-EXOS achieve the desirable very low value of weight/payload ratio close to one (100N vs. 11 Kg). The reported valued of stiffness of 3 N/mm represent the theoretical worst-case condition. The experimental measurements have provided a good confirmation of this value, even if the perceived stiffness seems to be amplified by the backlash introduced by the joint gearheads.

4.4 Hand Exoskeletons

A second type of anthropomorphic haptic interface [15] is an exoskeleton structure conceived to be located on the dorsal part of the operator's hand. It consists of 4 parallel exoskeleton structures covering and connected to the 4 fingers and exerting forces to the phalanges (little finger excluded). Each finger exoskeleton, which kinematic scheme is represented in Fig. 4.15, consists of four links connected by revolute joints disposed as the joints of each finger. For each joint of the finger exoskeleton, the joint axis has been designed in order to approximate the instantaneous position of the flexion-extension axis during operation. At the metacarpo-phalangeal joint a passive abduction-adduction movement has been also integrated.

The actuation system for each finger exoskeleton is based on three DC servomotors and associated tendon transmission systems. Each tendon is pulling on the middle point of each phalanx of the finger in order to execute the extension movement; at each joint, the flexion movement is obtained by a passive torsion spring integrated in the joint axis. The three motors are located on a cantilever structure

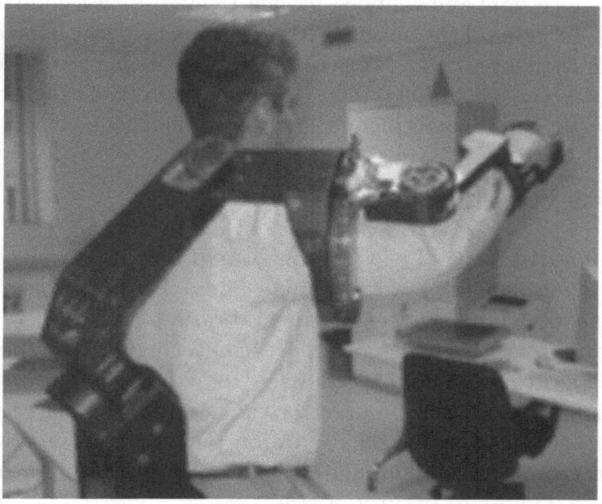

Fig. 4.14. Rear view of the L-Exos in an extended posture, (PERCRO, 2003)

Fig. 4.15. The kinematic scheme of the index exoskeleton

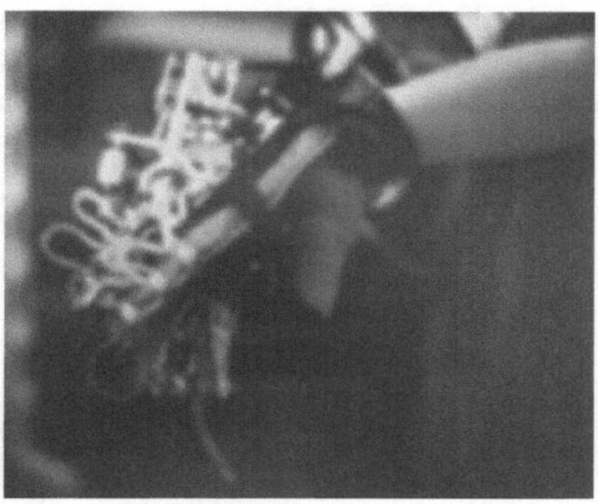

Fig. 4.16. The Hand exoskeleton (PERCRO, 1994)

fixed with the base frame of each finger exoskeleton. Rotation sensors, based on conductive plastics technology, are integrated at each joint, while force sensors, capable of recording the interaction force between the exoskeleton structure and each phalanx, are located directly on the dorsal surface of each phalanx link. A picture of the hand exoskeleton is given in Fig. 4.16. The kinematic structure of the thumb exoskeleton is slightly different to the one of the other fingers. In particular, the cantilever supporting the three motors of the thumb assumes a completely different aspect with respect to the one of the other fingers. One of the critical factors encountered during the design of the system has been that of obtaining a system possessing limited weight and volumes, in such a way to allow a good manoeuvrability of the hand. In terms of the mechanical performances of the hand exoskeleton, a maximum extension force of 0.3 N has been obtained, being the force sensor range of -0.5N – 3.0N. Force resolution is 0.0025N, while the force feedback bandwidth is 0.5 Hz with an angular displacement of 90° for all the 3 DOF.

4.5 Body Extenders

A different innovative type of exoskeleton system is that conceived to follow all limbs (arms, trunk and legs) of the human body. This category of exoskeleton refers to Enhanced Human Power Augmentation research, aimed at developing wearable robotic interfaces which could artificially augment the human strength by means of supporting arm and leg muscles [16]. Such systems are peculiar force feedback interfaces which can be used for the accomplishment of massive effort demanding tasks as well as for rehabilitation purposes: the operator only feels a scaled-down version of a manipulated load or requires only a fraction of the normally required effort to perform a task.

The first attempt towards the creation of an exoskeleton system which could enhance the human power goes back to the early 1960s, when General Electric developed Hardiman [17], a robotic system which could allow the user to carry loads up to 750kg with a force amplification ratio of 25:1. The system was based on a master-slave architecture, with an inner electrical exoskeleton which controlled an outer hydraulic exoskeleton.

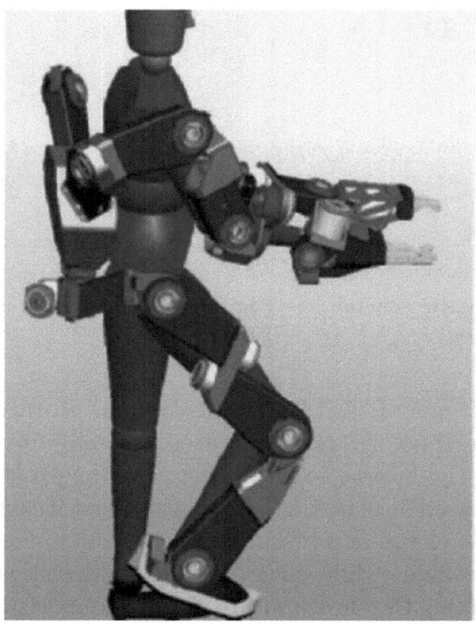

Fig. 4.17. The Full Body Extender design (PERCRO, 2006)

Many prototypes of human power augmentation systems have been developed in the last years. HAL (Hybride Assistive Limb), a lower-limb-strengthening device, has been developed at the Tsukuba University in 2004 [18]. It is controlled by a hybrid predictive approach, based on two modules: a position controller, which operates according to the phase sequence method, and a force controller regulating the actuator torque from the myoelectric signals coming from the muscles involved during locomotion. Power Exoskeleton [19] is another example of lower-limb-strengthening devices. It was developed by Sarcos in 2004 and is driven by piezo-hydraulic actuators. Arm Extender (1996) and BLEEX (Berkeley Lower Extremity EXoskeleton, 2004, [20]), both developed at Berkeley University, are separate upper- and lower-limb-strengthening devices. The control algorithm is based on an inner stabilizing position or velocity controller. A full body exoskeleton system is currently under development at PERCRO laboratory in the framework of a research project funded by the Italian Ministry of Defense. A schematic view of the actual mechanical design of the system is shown in Fig. 4.17.

4.6 Conclusions

This paper has introduced the concept of robotic exoskeletons to be used as haptic interfaces for the control of manipulation tasks in teleoperation conditions or for the interaction with virtual environments. General definitions and functionalities of exoskeleton systems have been given by pointing out the advantages of such devices when there is the need of tracking the spatial configuration of the human arm for specific gesture recognition tasks or rehabilitation procedures. Control schemes for exoskeleton systems have been described. Specific design solutions for arm, hand and full-body exoskeleton systems have been presented. The future utilization of robotic exoskeletons is foreseen in the fields either of teleoperation and robotic rehabilitation and it is tied with future developments and innovation of robotics components and technologies.

Acknowledgements. This work has been performed by the authors in the framework of the ENACTIVE Network of Excellence, contract number : IST-2004-002114-ENACTIVE, and of the SKILLS Integrated Project, contract number: IST-2006-035005-SKILLS, which are 6th Framework Research Programmes of the European Union (EU). The authors would like to thank the EU for the financial support. For more information about the ENACTIVE Network of Excellence and the SKILLS Integrated Project, please visit respectively the web sites: www.enactivenetwork.org and www.skills-ip.eu

References

1. M. Bergamasco. Force replication to the human operator: the development of arm and hand exoskeletons as haptic interfaces. In *Proceedings of 7th International Symposium on Robotics Research*, 1997.
2. AK Bejczy and JK Salisbury Jr. Controlling remote manipulators through kinesthetic coupling. *COMP. MECH. ENG.*, 2(1):48–58, 1983.

3. BM Jau. Anthropomorhic Exoskeleton dual arm/hand telerobot controller. In *Intelligent Robots, 1988., IEEE International Workshop on*, pages 715–718, 1988.
4. A. Nahvi, DD Nelson, JM Hollerbach, and DE Johnson. Haptic manipulation of virtual mechanisms from mechanical CADdesigns. In *Robotics and Automation, 1998. Proceedings. 1998 IEEE International Conference on*, 1, 1998.
5. T. Nef and R. Riener. ARMin-Design of a Novel Arm Rehabilitation Robot. In *Rehabilitation Robotics, 2005. ICORR 2005. 9th International Conference on*, pages 57–60, 2005.
6. A. Frisoli, Marcello Carrozzino, S. Marcheschi, F. Salsedo, and M. Bergamasco. In *Research in Interactive Design Proceedings of Virtual Concept*, chapter Haptic systems for simulation of primary commands of cars. Springer, 2005.
7. M. Bergamasco, B. Allotta, L. Bosio, L. Ferretti, G. Parrini, G.M. Prisco, F. Salsedo, and G. Sartini. An arm exoskeleton system for teleopera-tion and virtual environments applications. In *IEEE Int.Conf.On Robotics and Automation*, pages 1449–1454, 1994.
8. N. Tsagarakis, DG Caldwell, and GA Medrano-Cerda. A 7 DOF pneumatic muscle actuator (pMA) powered exoskeleton. In *Robot and Human Interaction, 1999. RO-MAN'99. 8th IEEE International Workshop on*, pages 327–333, 1999.
9. Y.S. Kim, S. Lee, C. Cho, M. Kim, and C.W. Lee. A new exoskeleton-type masterarm with force reflection based on the torque sensor beam. In *Robotics and Automation, 2001. Proceedings 2001 ICRA. IEEE International Conference on*, 3, 2001.
10. D. McAffee and P. Fiorini. Hand controller design requirements and performances issues in telerobotics. In *Proc. of ICAR, International Conference on Advanced Robotics*, 1991.
11. G.C. Burdea. *Force and touch feedback for virtual reality*. John Wiley & Sons, Inc. New York, NY, USA, 1996.
12. F. Salsedo, A. Dettori, A. Frisoli, F. Rocchi, M. Bergamasco, and M. Franceschini. Exoskeleton interface apparatus. Patent N. WO2004058458.
13. A. Frisoli, F. Rocchi, S. Marcheschi, A. Dettori, F. Salsedo, and M. Bergamasco. A new force-feedback arm exoskeleton for haptic interaction in virtual environments. In *Haptic Interfaces for Virtual Environment and Teleoperator Systems, 2005. WHC 2005. First Joint Eurohaptics Conference and Symposium on*, pages 195–201, 2005.
14. S. Marcheschi, A. Frisoli, CA Avizzano, and M. Bergamasco. A Method for Modeling and Control Complex Tendon Transmissions in Haptic Interfaces. In *Robotics and Automation, 2005. Proceedings of the 2005 IEEE International Conference on*, pages 1773–1778, 2005.
15. F. Barbagli, C. Avizzano, and M. Bergamasco. The Hand Force Feedback: Analysis and Control of a Haptic Device for the Human Hand. In *IEEE International Conference on System Man and Cybernetics*, 2:989–994, 2000.
16. E. Garcia, JM Sater, and J. Main. Exoskeletons for human performance augmentation (EHPA): A program summary. *Journal of Robotics Society of Japan*, 20(8):44–48, 2002.
17. R.S. Mosher. *Handyman to Hardiman*. Society of Automotive Engineers, 1967.
18. H. Kawamoto, S. Kanbe, and Y. Sankai. Power Assist Method for HAL3 Estimating Operator Intention Based on Motion Information. In *Proceedings of IEEE ROMAN*, 2003.
19. http://www.sarcos.com.
20. A. Zoss, H. Kazerooni, and A. Chu. On the mechanical design of the Berkeley Lower Extremity Exoskeleton (BLEEX). In *Intelligent Robots and Systems, 2005.(IROS 2005). 2005 IEEE/RSJ International Conference on*, pages 3465–3472, 2005.

5

Stereoscopic Image Visualization for Telerobotics. Experiments with Active Binocular Cameras

Alexandre Bernardino[1], José Santos-Victor[1], Manuel Ferre[2], and Miguel A. Sanchez-Urán[3]

[1] Instituto Superior Técnico, Instituto de Sistemas e Robótica.
Av. Rovisco Pais, 1
1049-001 Lisbon, Portugal
{alex,jasv}@isr.ist.utl.pt

[2] Universidad Politécnica de Madrid,
Departamento de Automática, Ing. Electrónica e Inf. Industrial
C/. José Gutierrez Abascal 2
28006 Madrid, Spain
mferre@etsii.upm.es

[3] Universidad Politécnica de Madrid,
Departamento de Ingeniería Eléctrica
Ronda de Valencia, 3
28012 Madrid, Spain
mas@ie.upm.es

Summary. The ability to perceive a three dimensional view of a remote workspace is an essential factor for teleoperation performance. In this chapter we present a novel class of 3D video interfaces that actively control the image acquisition system geometry to improve the teleoperation process. We describe a prototype system composed by two main components: (i) a visualization system called 3D-VIP, which receives two video input streams from a binocular camera system and combines them into a single stereoscopic video image displayed on a computer screen, and (ii) an automatic vergence control system that commands the movements of the binocular camera system to minimize the disparity of observed objects, thus facilitating depth perception. Automatic vergence control is based on the estimation of dominant disparity in the stereo image pair. It employs a foveal image representation, whose image resolution decreases toward image periphery, to focus attention on central objects and attenuate the influence of background clutter. Foveal images have smaller sizes than usual cartesian images which allows very fast sample rates and adaptation to working conditions. We present results on real teleoperation experiments which show that active systems compare favorably to conventional ones with respect to working range. Given the increasing availability of high quality cameras and motorized systems, the integration of automatic vergence control with stereoscopic video devices will pave the way for a new advanced generation of human interfaces in teleoperated systems.

M. Ferre et al. (Eds.): Advances in Telerobotics, STAR 31, pp. 77–90, 2007.
springerlink.com

© Springer-Verlag Berlin Heidelberg 2007

5.1 Introduction

Stereoscopic devices are visualization systems whose purpose is to provide users with a three dimensional perception of a remote environment. These systems are constituted by a stereo rig placed at the remote workspace, and a visualization system at the operator's location. Even though the human visual system perceives depth through a wide variety of stimuli, binocular disparity has been the preferred cue employed in stereoscopic devices [1, 3, 4, 5, 6]. Common stereoscopic devices display images independently to each of the operator's eyes, at appropriate rates. The difference between the coordinates of corresponding visual items in right and left eyes (image disparity) constitutes the stimulus for operator's depth perception.

Classically, stereoscopic devices adopt a fixed or manually controlled camera vergence angle. Because proper depth perception in humans is only possible if disparity is within appropriate values, the working area is very limited and may not suffice for certain teleoperation tasks. For instance, in a telemanipulation scenario, where the robot has to interact with objects at different depths, or when it has to move between places, disparities in the image may change significantly[1]. If large disparities are likely to exist, fixed vergence systems are not able to provide proper depth perception in the whole working range. To overcome this problem it is necessary to keep disparities small as the operator moves in the remote work space.

Some automatic vergence control systems have been successfully applied in the domain of autonomous robotics in the last decade, and are exploited in this chapter for teleoperation purposes. Such systems allow the automatic control of disparity within human perception limits by changing the vergence camera angle to match the distance to the object of interest. In such conditions, users would experience a significantly increased working range. In this chapter, a stereoscopic video interface with automatic vergence control is presented and compared to classical fixed configuration systems.

The chapter is organized as follows. In Sec. 5.2, a brief explanation of the visualization process is given. In particular, expressions relating image disparity to object depth and camera system geometry are derived, and the proper working area is calculated as a function of the maximum disparity allowed by the human visual system. Sec. 5.3 is dedicated to describe the automatic vergence control system employed in the experiments. This device controls the cameras' angle to minimize the disparity of objects placed into the observation direction. Sec. 5.4 describes how both systems are integrated. Some experiments have been carried out to illustrate the integrated system operation and performance, both with automatic vergence control and with a fixed vergence configuration. Finally, the conclusions of this chapter are drawn in Sec. 5.5.

[1] On a binocular setup, image disparity depends on camera vergence angle, lens parameters and distance to objects.

5.2 Visualization of Stereoscopic Images on a Computer Screen for Teleoperation Tasks

Most stereoscopic systems used in telerobotics are based on shutter glasses [2, 8], head-mounted displays [7, 9] or polarized images [22]. Systems based on shutter glasses hide user's eyes alternately in synchronization to screen refreshment. The main problem of this kind of interface is image flicking, which degrades users' depth perception. Image flicking is clearly visible when screen refreshment is lower than 60 Hz and disappears when refreshment is higher than 75 Hz. A second type of interfaces is based on polarized images. The user is also required to wear glasses that filter the images received by each eye. Both eyes are continuously receiving images, and there are not refreshment constraints, but the visualization interface becomes more complex since it is necessary to display both images continuously on a special screen. The third type of interface mentioned, the head mounted display, is typically used on virtual reality applications, and is especially designed to immerse users into virtual environments.

In above interfaces, the main depth perception stimulus is provided by the disparity of points in the pair of images. Image disparity can be defined as the horizontal distance between corresponding points in the left and right images. Depending on amount and sign of disparity, users perceive objects at close or further distances. Fig. 5.1 illustrates the cases when disparity is **zero**, **crossed** or **uncrossed**. Zero disparity corresponds to objects that are located at the intersection point of the two camera axes, thus projecting to the same position on both images. Objects behind or in front of the intersection point will have uncrossed (negative) or crossed (positive) disparity. Perceptually, these objects seem to be closer or further from the screen plane, respectively.

Fig. 5.1. Disparity of stereoscopic images displayed on a computer screen

For adequate depth perception in stereoscopic systems both the image display quality and the amount of binocular disparity must meet tight criteria. The visualization system should provide adequate binocular perception not only in terms of individual image quality (color, contrast, brightness) but also in terms of stereo image blending, synchronization, and refreshment rate. Disparities at the objects of interest should be kept under a fair limit so that proper binocular fusion is attained. A telerobotics system should take this into account either to define the range of effective working area or to devise methods to keep the disparity within the human perception limits. In the next section, these aspects will be detailed.

5.2.1 Teleperation Working Area

Though consensus for the value of maximum angular disparity in human perception has not been reached yet, some studies have provided useful information. The most conservative results set the limit for crossed disparity around 27 min-arc and 24 min-arc for uncrossed disparity [10]. Lipton from StereoGraphics Corp., one of the main stereoscopic device manufacturers, suggests a maximum disparity of around 1.5° [11]. Ferre [3] has obtained 2° using the 3D-VIP. In other experiments carried out by Ian Howard [12] limits between 4° – 7° for crossed disparity and between 9° – 12° for uncrossed disparity are given. According to Howard, variability of these findings result from numerous factors affecting the proper image blending, such as scene lighting conditions, contrast between objects or image exposure duration.

To evaluate the effect of human disparity perception limit in the teleoperation working range, disparity is calculated as a function of object to camera distance. Fig. 5.2 shows a situation where cameras are fixating on point A at a distance d, in a symmetric vergence configuration, where α is the angle of camera axes.

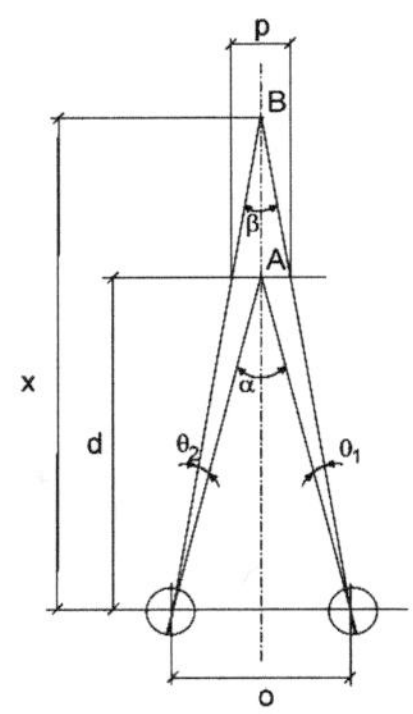

Fig. 5.2. Angular disparity from A and B

Let point B be located in the cyclopean axis at a distance x under an angle β. In the right eye the optical ray from B forms a θ_1 angle and in the left eye it forms a θ_2 angle, with opposite values ($\theta_1 = -\theta_2 = \theta$). If the distance between the eyes is represented by o, then:

$$\tan\frac{\alpha}{2} = \frac{o}{2d} \tag{5.1}$$

$$\tan\frac{\beta}{2} = \frac{o}{2x} \tag{5.2}$$

The angular diference is thus given by:

$$\theta = \frac{\alpha - \beta}{2} = \arctan(\frac{o}{2d}) - \arctan(\frac{o}{2x}) \tag{5.3}$$

Disparity can be measured as an angle (μ-angular disparity) or as a distance (p–linear disparity). Distinction between angular and linear disparity is relevant, since the factor limiting human perception is angular disparity. However, disparity measures on a screen are linear. Of course, relation of both disparities is calculated according to user position. As a consequence, same image disparities can be properly fused by the brain if the user is far enough. The user is usually placed on a fixed position in front of the screen; therefore, a straightforward relation between both disparities comes up. Angular disparity can be computed by:

$$\mu = \theta_1 - \theta_2 = 2\theta \tag{5.4}$$

$$|\mu| = \left|2\left(\arctan\left(\tfrac{o}{2d}\right) - \arctan\left(\tfrac{o}{2x}\right)\right)\right| \tag{5.5}$$

whereas linear disparity (p) is given by:

$$p = 2(x - d)\tan\tfrac{\beta}{2} \tag{5.6}$$

$$|p| = o\left|1 - \tfrac{d}{x}\right| \tag{5.7}$$

Considering fixed camera vergence angles, teleoperation working area will be constrained by the disparity limits allowing a proper stereoscopic image blending. Fig. 5.3 shows the angular disparity evolution of three different curves according to (5.5). In this example, camera axes intersect at 40, 60 and 80 cm. The horizontal line represents the considered disparity limit, in this case 3° for both crossed and uncrossed disparities. Boundaries can be calculated by looking at the curve intersections. For example, the central curve that crosses the axes at 60 cm has a limit between 40 and 120 cm when maximum linear disparity is set to 3 degrees (as the above-marked lines show). Therefore, as illustrated in this example, the disparity limit imposes constraints in the remote working area, which may be too severe for some telerobotics applications. A solution to increase the teleoperation working area is to change the camera axes intersection according to camera to object distance.

Optical parameters have to be considered to calculate the working area were stereoscopic images are fused properly. Above equations consider that human and optical visual parameters are equal. The effect of lens focal distance is equivalent to a scale of the remote working site. Additional factors, such as distance correction, have to be considered when head mounted displays are used for stereoscopic image visualization. An excellent study about this aspect is described in [6, 7].

5.2.2 3-D Video Image Processor

In the human visual system, disparity information is integrated with other visual cues such as textures and occlusions in order to obtain a spatial representation of the scene. Image quality is consequently an important factor in the visual interface, thus requiring adequate camera and display technologies. The 3D-Video

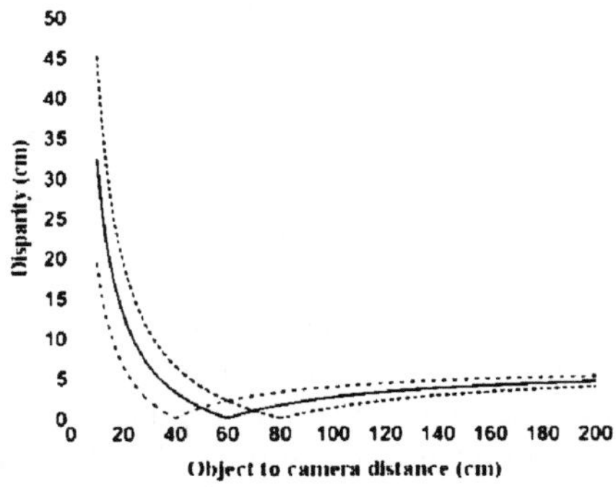

Fig. 5.3. Evolution of angular disparity as object-to-camera distance increases, for different vergence angles

Image Processor (3D-VIP) is a system designed for stereoscopic image visualization on a computer screen using shutter glasses. It has been used successfully in some telerobotics systems and experiments about stereo vision [13, 14]. The 3D-VIP system shows stereoscopic images up to 100Hz refresh rate, which reduces flicking problems. The four main elements of the stereoscopic video system are shown in Fig. 5.4. They are as follows: a pair of video cameras (or a binocular camera), the 3D-VIP device, a computer to display the images and shutter glasses.

3D-VIP is designed as a computer peripheral which displays stereoscopic images on a computer screen window. It has two kinds of video inputs: the VGA video signal coming from the computer, and two PAL video signals coming from two video cameras. Both video signals are digitized and loaded in two separate memories (left frame and right frame). The reconstruction process sends the left and right video images to the VGA output, alternately and synchronously to the computer VGA frequencies and shutter glasses. The stereoscopic video window occupies 30% to 70% out of the screen, depending on the monitor resolution. The screen space left is used for the visualisation of other processes taking place during teleoperation.

To see properly the stereoscopic images, the observer should wear liquid crystal shutter glasses which are synchronized to the display refreshment. The shutter glasses cover alternately each eye, synchronously to the image reconstruction. When the monitor refreshment is lower than 60 Hz flicking problems are strongly noticeable. It is therefore recommended to use monitor vertical frequencies higher than 75 Hz.

5.3 Automatic Vergence Control of Binocular Cameras

As described in previous sections, the working range of telerobotic systems based on stereoscopic devices may increase significantly if the vergence angle is

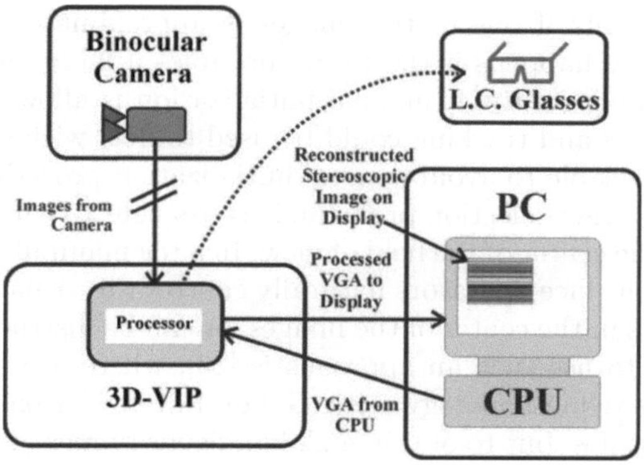

Fig. 5.4. Stereoscopic video system for teleoperation interface

controlled on-line. Some systems have been proposed in the last decade to perform automatic vergence control, but its application has been limited to the domain of autonomous robotics. One of these systems will be exploited here for the purpose of keeping disparity values under human perception limits in remote visualization interfaces. Automatic vergence systems usually analyse the incoming pair of images and compute the average disparity in the object of interest. Then, camera's vergence angle is controlled to reduce disparity as much as possible. Three main problems must be addressed in this process: (i) the target selection problem, i.e. which parts of the image belong to the object of interest; (ii) the computation of target disparities between left and right images; and (iii) how to drive cameras motion according to computed disparities.

Currently existing vergence control systems differ basically on the techniques employed to estimate disparity. Target selection is usually left for user specification, or assume that the target occupies the central part of the images. Camera control is usually done in a negative feedback scheme to reduce disparity until it falls under a threshold.

In our case real-time constrains must also be taken into account: camera vergence angle must respond quickly to environmental changes and computations must be done in real time. Therefore, from the several existing disparity estimation techniques, only a few can be applied to the real-time vergence control problem in telerobotics scenarios.

5.3.1 Disparity Estimation for Vergence Control

One of the first works to present an operational real-time vergence control system is described in [15]. Dominant disparity between the two images is estimated using the *power cepstrum* technique. A similar approach is described in [16], where global disparity is computed by normalized cross correlation between horizontally translated left and right images. The amount of translation that maximizes the correlation is chosen as the estimated global disparity. The previous two methods work reliably if one of the images is approximately a translation of the other. This only happens if the target occupies a large part of the images, which limits its applicability. If manual initialization is allowed, a combination of spatial windowing and tracking could be used to deal with small objects, but in general it is advisable to avoid manual initialization procedures.

To address the target selection problem, it is frequent to consider that objects of interest are at the center of the field of view. In a telemanipulation scenario this is a fair assumption since operators naturally control the visualization system to observe the objects in the center of the images, or can be instructed to do so. One of the first works to use such an approach is [23], where disparity is computed using a phase-based methodology with Gabor Filters. Disparity estimation is done at multiple scales, but to obtain a global disparity measurement large scale filters are applied only at the centre of the images. Since Gabor filters have Gaussian envelopes, objects in the center of the images become more influential in the computation of global disparity.

Another class of methods implements a similar strategy by adopting foveal image representations. Foveal images are obtained subsampling the original images in order to provide higher resolution (more pixels) in the center of the visual field than the periphery. Therefore, assuming that objects of interest occupy the image centre, their disparities will be dominant over peripheral background elements. The minimum object size will depend on the decay of image resolution from central to peripheral regions. For example in [17], a foveated pyramid is used in conjunction with a phase-based method to compute disparity. The work in [18] uses a log-polar representation and correlation measures for computing dominant disparity. In [19] a log-polar representation is also used, and control of vergence is made with optical flow measurements related to image contraction and expansion.

Besides concentrating attention in centered objects, an additional advantage of foveal images is data reduction. Both foveated pyramid and log-polar images represent a reduction of the order of $log(n)$ pixels. In this work the log-polar representation is preferred because it is rotationally invariant (circular symmetry), whereas foveated pyramids retain the square-like symmetry of cartesian images, and measurements become dependent on the object's orientation.

5.3.2 Vergence Control with Log-Polar Images

Most biological visual systems have retinas with a non-uniform distribution of photo-receptors. In particular, humans and other primates have a very high density of photo-receptors in the retina centre (fovea), decreasing significantly toward the periphery. Combined with ocular movements, it is possible to have both a wide field of view and high resolution perception, with limited computational resources.

The log-polar mapping [20] provides a non-uniform resolution geometry which is similar to the distribution of photo-receptors in the human retina. An example of log-polar mapped images is given in Fig. 5.5. Horizontal and vertical coordinates in the log-polar image correspond to radial and angular directions in the

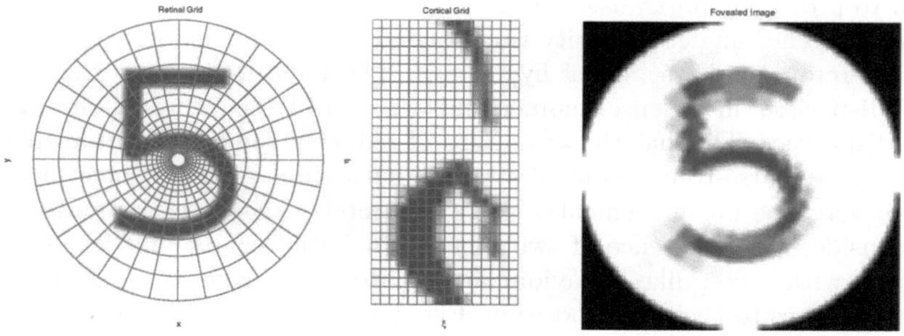

Fig. 5.5. A conventional cartesian image (left), is mapped to log-polar coordinates (middle), and then mapped back to cartesian (right). Notice the reduced resolution in the image periphery.

cartesian image. This image geometry has the advantage of providing a focus of attention in the center of the visual field. This means that the disparity of objects in the image center become dominant over other objects in the periphery, and are implicitly selected as objects of interest. Several properties of the use of log-polar images are described and illustrated experimentally in [18]. In particular, it is shown that with small objects (down to 10% of the full image), global disparity estimation is reliable with log-polar coordinates but unreliable in the cartesian domain (see Fig. 5.6).

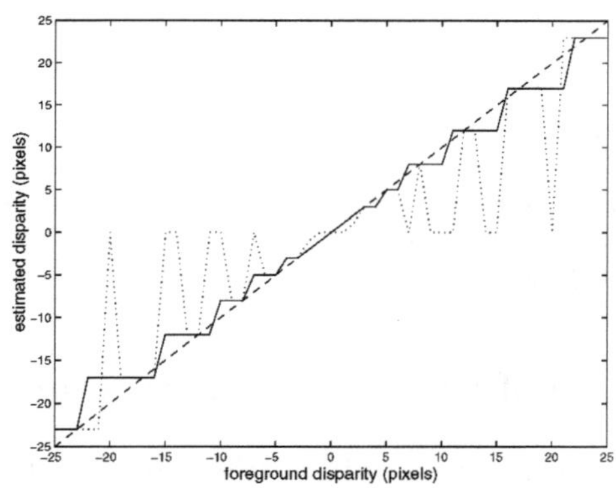

Fig. 5.6. Disparity estimation with cartesian (dotted line) and log-polar images (solid line) with a centered object occupying about 34% of the full image. True disparity is given by the dashed line. The set of disparity channels is $D = \{0, 1, 2, 5, 8, 12, 17, 23\}$.

Having obtained dominant disparity measurement, vergence angle can be controlled by driving the motors with a suitable controller. At each step, motors should be positioned in order to reduce the amount of global disparity present in the stereo pair. In [18], a simple PID control strategy is used. A set of gains is tuned to provide good transient response both for slow and fast changes of the desired vergence angle. Disparity is computed by maximizing image correlation in a discrete and sparse set of hypothesis. The hypothesis cover the range of small disparities in a dense manner and are coarsely spaced over the range of large disparities. This has the advantage of reducing the computation time and results in a coarse-to-fine control strategy with similarities to some models of human vergence eye movements. The "dual-mode" theory of vergence control [21] considers the existence of two phases: an initial coarse and fast transient response with a step-like behaviour and a late more accurate component that brings the eyes to their final position. Fig. 5.7 illustrates the response of a PID vergence angle controller under constant velocity and stepwise motions stimuli. Fig. 5.8 shows the images acquired before and after the vergence process. Initially a large target disparity elicits vergence movements that, after a few time steps, cancel the existing disparity in the target. Notice that, despite the person in the center of the field of view is occupying a small area, it gets the focus of attention implicitly provided by the foveal imaging geometry.

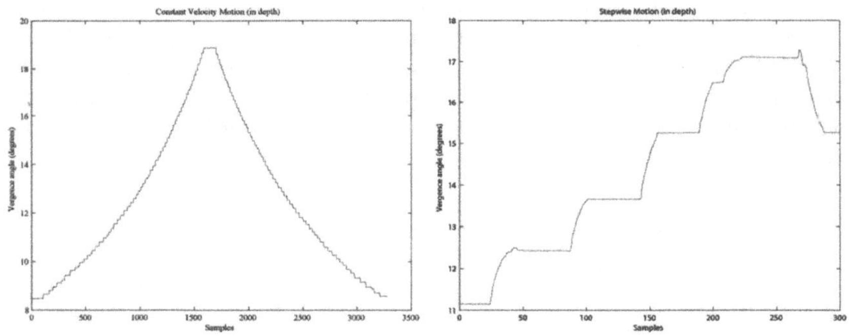

Fig. 5.7. Vergence angle controller responses. Left: when tracking a target moving in depth with constant speed (approaching the head). Right: when the target performs stepwise (discontinuous) motions.

5.4 Performance of the Integrated System

This section describes experiments with a teleoperation setup including the automatic vergence control system described previously. With automatic vergence control the working area is significantly augmented, however, it is necessary to check if camera motions due to vergence control have any negative influence in users precision and comfort during the execution of tasks.

The experiment is focused on checking users depth perception guiding a robot to perform a positioning task. A mobile robot with a controllable stereo vision head has to be positioned with respect to one object. A mobile robot is remotely controlled by the user, which is seeing the stereo images acquired by the binocular

Fig. 5.8. Active vergence experiment. Images from the left and right cameras have been blended for display purposes. The left figure represents the initial state (not verged), where the disparity in the person in front of the setup is noticeable. The right figure shows the system state after convergence, when disparity in the person is cancelled.

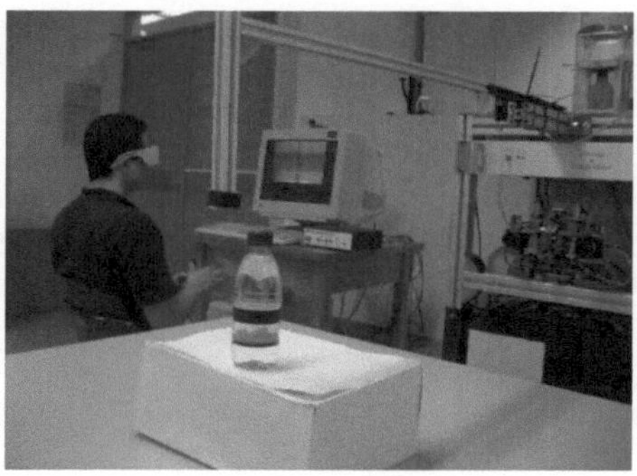

Fig. 5.9. Experiment Setup. Operator is guiding a mobile platform using mono and stereoscopic images.

vision head. The robot is equipped with a pointer that should be placed as close as possible to the target. In this case we consider a small bottle on top of the table. The experimental setup is shown in Fig. 5.9.

The experiment is designed to evaluate users performance in the positioning task with and without vergence control, both in terms of precision and comfort. Fig. 5.10 shows a sequence of displayed stereoscopic images. The experiment using active control is illustrated in images a1 to a5. The stereo head is directed toward the bottle, and the vergence system controls the cameras in order to reduce disparity in the bottle. We can observe that the bottle appears in the same location at both images, which corresponds to a complete cancellation of disparity. The experiment with fixed vergence is shown in images f1 to f5. In this case bottle disparity depends on its distance to the cameras and the bottle

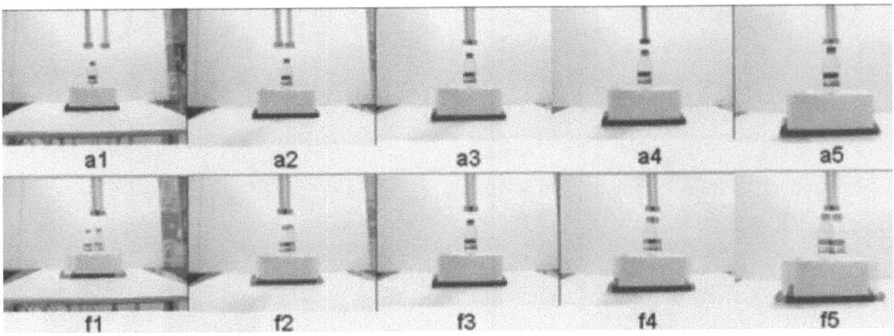

Fig. 5.10. Image sequences. Images a1 to a5: using active control; f1 to f5: without active control.

position is usually different in the two images. Images a4 and f4 correspond to a good positioning of the robot (successful task) because target and bottle have the same disparity, i.e, are at the same distance to the cameras.

Nine subjects were asked to perform the experiment. After a training period, subjects had three trials: one with monocular vision, one with fixed vergence stereo vision, and other with automatic vergence control. The different trials were set in arbitrary order and the subjects were not aware of the experimental conditions. Measures were obtained for the position error between bottle and target, for each user and visualization mode (active vergence control and fixed vergence). Results are presented in Fig. 5.11.

Fig. 5.11. Results of the positioning experiment

Analysis of results indicates that active vergence do not diminish precision in guiding tasks. Inquiring subjects, it was verified that no noticeable discomfort or tiredness arises using active vergence control. Due to significant augmentation of working area, automatic vergence control is a technique worth to explore in situations where large ranges of depths have to be dealt with.

5.5 Conclusions

This chapter discusses the application of stereoscopic devices in teleoperation tasks with extended working ranges. In these cases, conventional systems, with fixed cameras' configuration, are limited due to the limited disparity range of human perception. To address these cases, active camera control is employed in order to minimize disparity in the objects of interest, as the user explores the working space. A system that integrates a stereoscopic visualization interface and an automatic vergence control system is described and tested. Experiments have been carried out to evaluate performance of active vergence in a teleoperation task. Potential disadvantages, such as discomfort and tiredness were not noticed by the subjects. Therefore, automatic vergence control is a promising technique for the design of more flexible and versatile stereoscopic telerobotics systems for operation in large workspaces.

References

1. G. C. Burdea and P. Coiffet. *Virtual Reality Technology. Second edition.* John Wiley & Sons, 2003.
2. R. Aracil, M. Ferre, M. Hernando, E. Pinto and J.M. Sebastian. Telerobotic System for Live-Power Line Maintenance:ROBTET. *Control Engineering Practice*, 10:1271–1281, 2002.
3. M. Ferre, R. Aracil and M. Navas. Stereoscopic Video Images for Telerobotic Applications. *Journal of Robotic Systems*, 22(3):131–146, 1993.
4. W. Barfiel, C. Hendrix and O. Bojorneseth. Comparison of Human Sensory Capabilities with Technical Specifications of Virtual Environment Equipment. *Presence*, 4(4):1–19, 1995.
5. W. Barfiel, C. Hendrix and K.E. Bystrom. Effects of stereopsis and head tracking on performance using desktop virtual environment displays. *Presence*, 8:237–240, 1999.
6. H. Baier, M. Buss, F. Freyberger and G. Schmidt. Interactive Stereo Vision Telepresence for Correct Communication of Spatial Geometry. *Advanced Robotics*, 17(3):219–233, 2003.
7. H. Baier, M. Buss, F. Freyberger and G. Schmidt. Benefits of Combined Active Stereo Vision and Haptic Telepresence. In *Proc. IEEE Int. Conf. on Intelligent Robots and Systems*, pages 702–707, 2000.
8. L. Matthies. Stereo Vision for Planetary Rovers - Stochastic Modeling to near Real-Time Implementation *Int. Journal of Computer Vision*, 8(1), 1992.
9. M. Ueberle, N. Mock and M. Buss. Towards a Hyper-Redundant Haptic Display. In *Proc. of Int. Workshop on Telepresence and Teleaction*, pages 1–22, 2003.
10. Y.Y. Yed and L.D. Silverstein, Limits of fusion and depth judgement in stereoscopic color displays *Human Factors*, 32:45–60, 1990.
11. L. Lipton. *The Crystal eyes handbook* StereoGraphics Corporation, 1991.
12. I.P. Howard and B.J. Rogers. *Binocular vision and stereopsis.* Oxford University Press, Oxford, 1995.
13. M. Ferre and R. Aracil. Interfaces for Telerobotics Studies on Stereoscopic Vision. In *Proc. of the XV IFAC World Congress*, 2002.

14. M. Ferre, R. Aracil, M. Navas and J. A. Escalera. Real time video image processing for teleoperation: Image Blending and stereoscopy. In *Proc. of 9th IEEE Int. Conference on Emerging Technologies and Factory Automation*, pages 539–544, 2003.
15. T. J. Olson and D. Coombs. Real-Time Vergence Control for Binocular Robots *IJCV*, pages 67–89, 1991.
16. K. Pahlavan, T. Uhlin and J. Eklundh. Integrating Primary Ocular Processes. In *Proc. of 2nd ECCV*, pages 526–541, 1992.
17. J. Siebert and D. Wilson. Foveated Vergence and Stereo. In *Proc. of the 3rd Int. Conf. on Visual Search (TICVS)*, 1992.
18. A. Bernardino and J. Santos–Victor. Binocular Visual Tracking: Integration of Perception and Control *IEEE Trans. on Robotics and Automation*, 15(6):137–146, 1999.
19. C. Capurro, F. Panerai and g. Sandini. Dynamic Vergence Using Log-Polar Images *IJCV*, 24(1):79–94, 1997.
20. E. Schwartz. Spatial mapping in the primate sensory projection: Analytic structure and relevance to perception *Biological Cybernetics*, 25:181–194, 1977.
21. J. Semmlow, G. Hung, J. Horng and K. Ciuffreda. Disparity Vergence Eye Movements Exhibit Preprogrammed Motor Control *Vision Research*, 34(10):1335–1343, 1994.
22. G. Hirzinger. Robots in Space-a survey. *Advanced Robotics*, 9(6):625–651, 1995.
23. W.M. Theimer and H.A. Mallot. Phase-Based Binocular Vergence Control and Depth Reconstruction Using Active Vision *CVGIP*, 60:343–358, 1994.

6

Stereoscopic 3-D Acquisition, Processing, and Display for Telerobotic Applications

Fergal Shevlin[1,2], Barry McCullagh[2], David Eadie[2], Manuel Navas-Herreros[1], and Christophe Rabaud[1]

[1] Dyoptyka Ltd.,
7 Westland Court, South Cumberland Street, Dublin 2, Ireland
{fshevlin,crabaud}@dyoptyka.com, mnavas78@yahoo.es

[2] Trinity College
Dept. of Computer Science
Dublin 2, Ireland
{Barry.McCullagh,David.Eadie}@cs.tcd.ie

Summary. We present various solutions developed through our research to problems arising in the stereoscopic 3-D visualisation process for telerobotics applications. We show that *real-time* of processing video imagery is required to rectify geometric distortion that can negatively impact the quality of depth perception; that rectification can be achieved efficiently using both specialised hardware and commodity hardware such as graphic card GPUs; that the solution to the computationally intensive problem of real-time computational depth estimation can be speeded up using commodity graphics card MPEG encoders; and that the problem itself can be simplified through a novel scene illumination and image acquisition strategy. Finally, we show how a display device incorporating an adaptive optics element uses computed depth to display the 3-D scene with appropriate optical distance—thus avoiding a well-known cause of visual discomfort in stereoscopic 3-D visualisation.

6.1 Introduction

In this paper we present some of the benefits of stereoscopic 3-D perception for telerobotics, some of the problems that arise, some solutions to those problems that we have developed through our research—and how these solutions impact image acquisition and display.

The telerobotic application motivating the research described herein is minimally invasive robot-assisted surgery, as exemplified by Intuitive Surgical Inc.'s *da Vinci* system illustrated in Fig. 6.1, however most applications share similar problems and can benefit from the solutions.

In robot-assisted surgery, the surgeon undertakes the procedure watching a binocular stereoscopic 3-D display. Imagery is acquired by miniature stereoscopic 3-D cameras mounted in an endoscope. Hand movements on control devices are relayed in real-time to robotic arms on which laparoscopic instruments are mounted. There are several advantages to this approach, including higher degrees of freedom and a greater working envelope than is the case with conventional

M. Ferre et al. (Eds.): Advances in Telerobotics, STAR 31, pp. 91–105, 2007.
springerlink.com © Springer-Verlag Berlin Heidelberg 2007

minimally-invasive surgery, and the removal of tremor and motion exaggeration inherent in the use of laparoscopic instruments.

Image-guided surgery applications [1, 2] display previously-acquired 3-D graphical data such as Computed-Tomography (CT) scans and Magnetic-Resonance Images (MRIs) over the surgeon's view of the surgical field to facilitate better navigation and decision-making. To ensure correct alignment of the viewed scene with previously-acquired imagery, real-time depth estimates of the scene are required. This necessitates real-time stereoscopic 3-D depth reconstruction from surgical field imagery. Other more sophisticated telerobotic applications can also use real-time depth data, for instance to move instruments back and forth automatically in time with the heart-movements to allow the surgeon to concentrate on the procedure rather than instrument motion [3, 4].

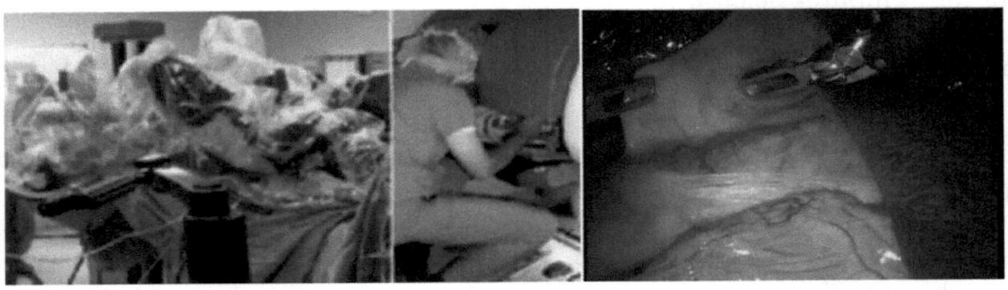

Fig. 6.1. Robot-assisted surgery

Elements of a stereoscopic 3-D system for telerobotics are shown in Fig. 6.2. Essentially, they include a pair of cameras for image acquisition and some sort of image display device(s). In addition to these, we suggest that there are several important benefits to be derived from the inclusion of real-time image processing functionality.

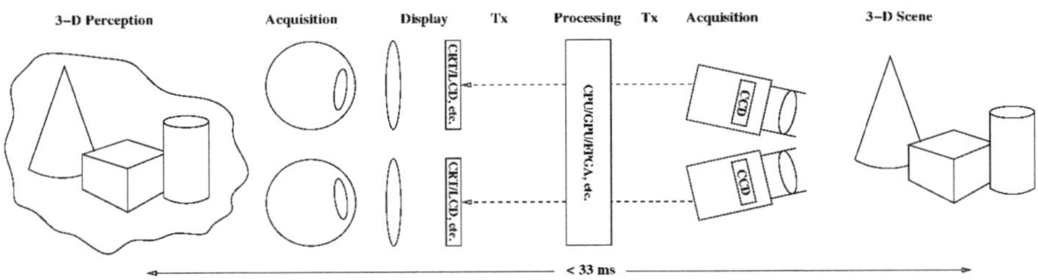

Fig. 6.2. Elements of stereoscopic 3-D acquisition, display, and perception

6.2 Benefits of Stereoscopic 3-D Perception

It is normally accepted that stereoscopic 3-D can improve the speed, the accuracy (reliability, cf. visual acuity), and the precision (repeatability) of depth perception (e.g. [5]). It can also facilitate discrimination of detail from background (cf. visual search). However these advantages are only evident if it is implemented correctly. Otherwise there can arise problems of discomfort which lead to limited usage time and visual and vestibular system after-effects (e.g. [6]). These problems have particular relevance to telerobotic applications such as surgery and hazardous materials handling, where the quality of depth perception is critical and operators may not be able to take breaks to allow their vision systems to recover.

Further emphasising its importance for this application, non-stereoscopic 3-D cues to depth such as shadow and defocus blur are often poor in camera-acquired imagery due to strong illumination, and large depth of field. In addition there is no motion parallax due to head movement with respect to the image display device.

Several studies have sought to quantify the benefits of stereoscopic 3-D in telerobotics applications. Munz et al. [7] measured discrete motion vectors for surgical tasks such as *threading a needle*, *sewing a stitch*, and *tying a knot* performed in stereoscopic 3-D and in binocular 2-D on *daVinci*. The robot end-effectors' individual motion vectors $\mathbf{t}_0 \ldots \mathbf{t}_n$ were measured (depicted in Fig. 6.3).

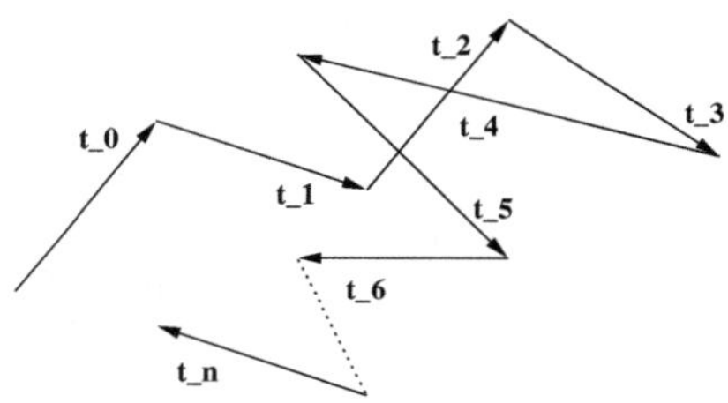

Fig. 6.3. Task trajectory vectors

The results were convincing. The total number of discrete motion per task was lower for 3-D than for 2-D, i.e. the total distance moved per task was reduced, i.e. $\sum_{i=0}^{i=n^{3-D}} |\mathbf{t}_i^{3-D}| < \sum_{i=0}^{i=n^{2-D}} |\mathbf{t}_i^{2-D}|$; speed was increased, i.e. $\sum_{i=0}^{i=n^{3-D}} |\dot{\mathbf{t}}_i^{3-D}| < \sum_{i=0}^{i=n^{2-D}} |\dot{\mathbf{t}}_i^{2-D}|$, and expert-assessed error count $E^{3-D} < E^{2-D}$.

6.3 Acquisition

Obviously there are many issues that need to be considered in the design and operation of a real-time stereoscopic 3-D video acquisition system, e.g. illumination, lens focal length, sensor type, and sensor resolution. We concentrate here on camera convergence, an issue specific to acquisition for stereoscopic 3-D.

An observer adapts their eyes' convergence angle to minimise horizontal disparity at a fixated object, and so achieve a single view with depth perception enhanced by stereoposis. A horizontally-aligned pair of cameras with overlapping fields of view is used to acquire imagery with appropriate horizontal disparity.

The optical axes of the camera lenses can be parallel or convergent. In the convergent case the zone of overlap where stereopsis can be achieved by the observer is much greater than when the axes are parallel. Observer viewing comfort is also greatly increased when camera axes converge to within a few degrees of the eyes' natural convergence angle for the object being fixated (cf. the Bernardino et al. chapter in this book.)

The most straightforward means of making the optical axes convergent is to rotate the cameras towards each other. An alternative method is to translate the sensor with respect to the lens—which requires special mechanisms within the camera. It is possible to simulate sensor translation by using offset subsets of the sensor surfaces, but this can greatly limit the field of view.

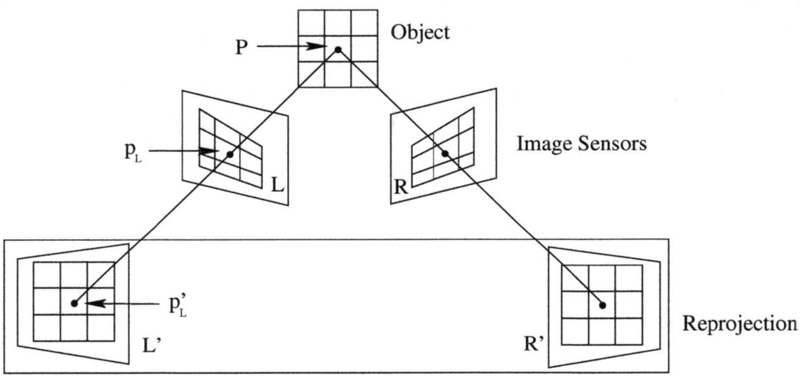

Fig. 6.4. Keystone distortion and its rectification

Fig. 6.4 shows the sensor planes (L and R) of two cameras converged on a planar rectilinear grid. It can be seen that the grid images projected onto the sensors are geometrically distorted such that there is both horizontal and vertical disparity. This is termed *keystone* distortion. It becomes greater as the angle of convergence increases. If these distorted images are presented to an observer in a stereoscopic 3-D display, the planar grid will appear curved. In the more general case of a scene, the entire scene space will appear curved. Hence keystone distortion must be removed in order for the scene to be perceived accurately.

A process of resampling is used to rectify the imagery. This requires that the imagery be processed in *real-time.* The processing speed required for a visualisation system to achieve such performance is not easily defined. We have chosen for our definition the conventional video frame rate of circa 30Hz, i.e. 33ms per frame. This is fast enough to avoid flicker (if displayed appropriately interlaced) and motion lag (the limit for which is circa 15Hz).

6.4 Geometric Correction

Keystone distortion may be removed from the images by *reprojecting* image planes L and R onto a common plane $L' = R'$. With prior knowledge of the cameras' convergence angle and position (i.e. the relative orientation parameters found through camera calibration), a linear projective transform A can be constructed [8]. If necessary, lens distortion can also be removed during this process [9] (if the intrinsic parameters are known). Fig. 6.5 shows the surgical endoscope image from Fig. 6.1 corrected for keystone and lens distortion.

The most practical implementation of reprojection is an inverse mapping strategy, e.g. where each pair of integer pixel homogeneous coordinates p'_L in the left rectified image is mapped by 3×3 transform matrix A to a pair of not necessarily integer coordinates p_L in the left source image, i.e. $p'_L = Ap_L$.

Pixel intensity values for the rectified image are usually found through interpolation of the pixel values in the neighbourhood of the mapped non-integer coordinates. The reprojection requires 6 multiplications, 6 additions and 2 divisions per rectified pixel and bilinear interpolation adds 4 multiplications and 2 additions. A 512×512 stereo image pair requires 6.3M multiplications or divisions and 4.2M additions.

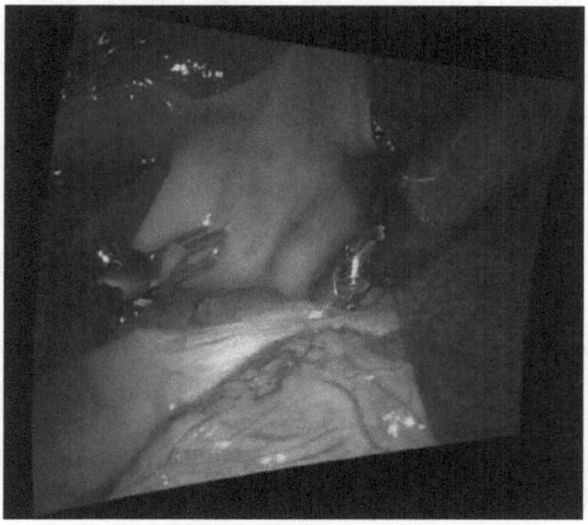

Fig. 6.5. Rectified endoscope image

We have implemented this computation using a general purpose CPU and an FPGA. Execution times were 8ms and 6.5ms respectively. A graphics card GPU implementation was found to take 6.4ms [9]. The CPU was a 2.66GHz Pentium 4 Xeon and used the Intel OpenCV optimised image processing libraries, the FPGA was a Xilinx Virtex II with a fully pipelined 80MHz design, and the GPU was an ATI Radeon 9800. It should be noted that these times were for a single image in isolation rather than as part of a video stream. The difficulty of getting a general purpose CPU to reliably and synchronously process video streams from two cameras should not be underestimated. Both the FPGA and

GPU can exist as parts of reasonably autonomous devices, communicating with the cameras over the a fast PCI Express (or similar) bus, hence they would seem a more realistic proposal for real-time image processing. Fig. 6.5 is the geometrically corrected version of a surgical endoscope image from the same sequence as that shown in Fig. 6.1.

These implementations show that there can be plenty of time left within the circa 33ms period to perform further computation—i.e. without introducing latency into visualisation. This motivated us to investigate real-time computational estimation of depth from the stereoscopic image pair. A relevant benefit of the rectification process described is that vertical parallax between the images is eliminated so that corresponding points between the images are located on the same horizontal row. This reduces the search for matching points from two dimensions to one dimension, reducing the time needed to solve the depth estimation correspondence problem.

6.5 Depth Estimation

The computation of three-dimensional depth maps from pairs of stereoscopic imagery is expensive. This is due primarily to the search required to find corresponding (or homologous) points in the image pairs. Many methods have been proposed to solve this correspondence problem, but very few can achieve real-time video rates (i.e. about 30 frames/sec) required for some robotics and telerobotics applications. It is not unreasonable to claim that no existing method can achieve a good quality solution in real-time for diverse types of imagery.

This paper outlines two novel strategies that are being investigated in our laboratory. The first makes use of MPEG-2 motion vectors to reduce the number of computations required for correlation-based methods. The motion vectors produced by interleaving a stereo pair are used as an initial estimate for a more precise correlation-based search. Independently of this, the vectors produced by encoding each single view are used to identify static regions of the scene avoiding unnecessary disparity recalculation.

The second strategy uses several sources of illumination close to the camera lenses to cast shadows at physical depth discontinuities. These shadows are easily detected in real-time using simple image processing techniques. Knowledge of these discontinuities greatly simplifies the correspondence search because there is no ambiguity between objects' interior texture and their edges.

6.5.1 The Correspondence Problem

Stereo algorithms can be divided into two broad categories: local and global [10]. Global methods produce much better solutions due to the larger number of pixels involved in the decision making process however due to their complex nature are unsuitable for real-time applications, e.g. telerobotic surgery, which must use local methods to achieve real-time frame rates. The increasing performance of general purpose CPUs and single instruction multiple data (SIMD) instructions have allowed real-time stereo correspondence algorithms to be implemented on desktop computers without the need for dedicated special purpose hardware. This performance gain has already led to a 5 fold increase in the performance of the algorithm proposed in [11] on a desktop computer, from 6fps in 1997 to 30fps in 2001.

Traditional correlation-based algorithms (local methods) take a fixed size rectangular window from one image and compare this to a series of windows from the other image, with the location of the highest correlation stored as the disparity value. While these have been used successfully in many systems there are several problems which prevent them from achieving the accuracy of global methods, including the size, position and shape of the window used. The window must be large enough to discriminate between pixels in low texture regions, while being small enough to ensure that it only contains pixels from objects at the same depth and does not cross object borders. Fig. 6.6(a) and 6.6(b) show the original image and ground truth while Figs. 6.6(c) and 6.6(d) show the result of a correlation performed using a small and large window respectively.

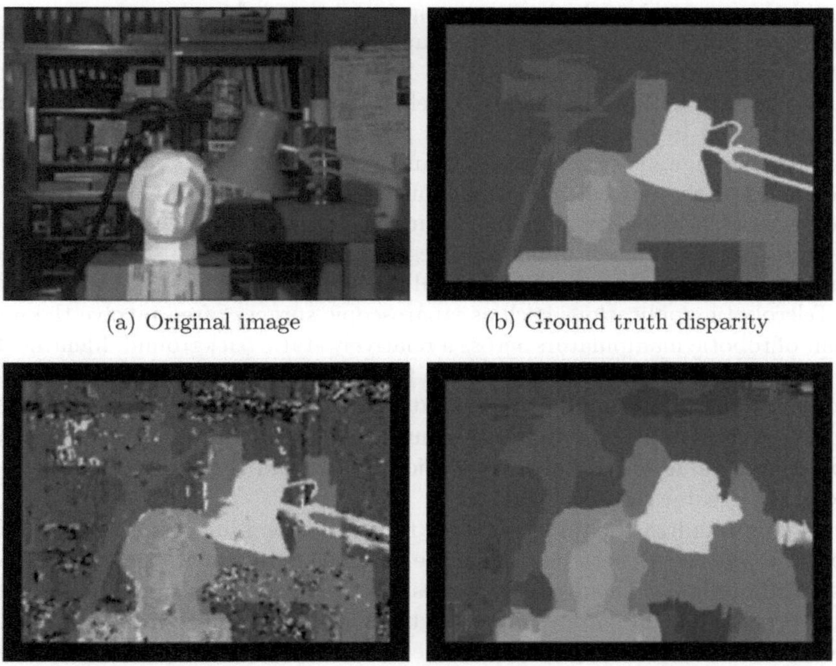

(a) Original image

(b) Ground truth disparity

(c) Disparity, correlation (small window)

(d) Disparity, correlation (large window)

Fig. 6.6. Test data set and correlation results

Not only does the size affect the result but also the position and size of the window relative to object borders. If it crosses an object border there is a high possibility that an occluded region will be included resulting in a poor match for the current pixel of interest. This can also happen if the edge of the window is not the same shape as the object border. To find the optimal window size and position while maintaining real-time capabilities it is necessary to speed up traditional correlation-based techniques so time can be given to placing and shaping the window.

Reducing Computation Time — One way to reduce the computation time required for disparity map calculation is to reduce the number of comparisons required. We propose the use of MPEG-2 motion vectors to achieve this. When encoding an MPEG-2 stream, a series of motion vectors are generated which relate regions (16x16 pixel regions known as macroblocks) between frames. These vectors can reduce the number of comparisons by reducing the search range and identifying static regions of the scene.

Conventional correlation-based methods define a fixed size search range, large enough to encompass objects at every possible horizontal disparity. By establishing an estimate of the actual disparity, the search range can be reduced in size for every pixel reducing the comparisons required. Previously, methods including hierarchical methods [12] and optical flow [13] have been used to narrow the search range for correlation-based approaches. Our investigations show that interleaving a stereo pair as an MPEG-2 stream provides motion vectors which can be used as an initial estimate to refine the search range (spatial encoding). The reduction in computation time is proportional to the reduction in size of the search range. The best position and size of the search range is based on the values of the current and neighbouring motion vectors. This increases the accuracy of the disparity map in some instances by filling 'holes' and forcing the disparity values of small untextured regions, such as specular reflections, to be similar to those of the neighbouring values.

Telerobotic applications such as laparoscopic surgery often involve the movement of robotic manipulators across a relatively static background. Identification of the static areas reduces computation time as their disparities remain the same between frames. Encoding each separate view as an MPEG-2 stream produces motion vectors describing the movement in the scene (temporal encoding). Vectors with a value of zero identify static regions, and the larger the number of these, the greater the speed gain.

As no groundtruth data is available for the surgical imagery used for evaluation, it is only possible to determine the relative error between the proposed and standard methods. The disparity values calculated in untextured regions, where most errors occur in correlation based techniques and where most of the differences are located, are dependent on the search range used. Reducing the search range returns a different disparity value, if both values are incorrect it does not mean one method is less correct than the other. If the proposed method returns a slightly less accurate disparity map this may not impact on the accuracy of the final reconstruction as a small increase in error is not critical when using the least-squared error 3-D data alignment algorithm. The figures quoted here are results obtained when the differences are limited to untextured regions, occlusions and small regions where the traditional method created 'holes' in the disparity map. The gain achieved using the spatial encoding method is approximately 20%, which was predictable since the search range used was 20% smaller, with a difference of less than 10%. The temporal method provides a gain between 15% and 25% with a difference of less than 7%. A gain of 30% with a difference of less than 10% can be achieved by combining the two methods. These results were obtained using a GPU based implementation, which is more than 5 times faster than our CPU based implementation but requires some optimisation to achieve rates similar to those obtained by other stereo implementations e.g. [9, 14, 15].

Increasing Accuracy at Object Edges — The adaptation of window size and shape for different image locations has been proposed to overcome the limitations of a fixed window. One well-known method makes adaptations based on the local variation of intensity and disparity [16]. However, this is a very computationally intensive and the accuracy is strongly influenced by the initial disparity map estimate.

A slightly different approach is to use shiftable windows at depth discontinuities, e.g. [17, 18]. Rather than centring the pixel in the window, several windows are created with the pixel of interest at various locations. The window with the minimum matching error is chosen. The chosen window is more likely to contain pixels of constant depth, as the edge of the window is free to move away from depth discontinuities.

Several different approaches have been presented recently whose aims are to create an optimal window which is free to move away from depth discontinuities. One enlarges the window around the pixel of interest until a grey level edge is found [19]. Another forms an adaptive window by combining the best of several possible windows around the pixel of interest [20]. The more general the shape of the window the better it fits the objects in the scene. Non-rectangular shaped windows have been investigated and found to be very computationally expensive [21].

Illumination Method — Image intensity gradients are not a good indication of physical object boundaries as there may be no detectable intensity change between two objects of similar texture but different disparity, and there are many strong intensity gradients in high texture areas that do not correspond to depth discontinuities. We devised a method that greatly simplifies the correspondence search problem by using unambiguously identified depth discontinuities at physical object boundaries.

Effective and efficient identification of depth discontinuities was proposed by Raskar et al. [22] for the purposes of image enhancement rather than analysis. When illuminated individually in temporal sequence, multiple light sources placed close to the camera cast shadows in different directions along depth discontinuities in the scene. The position of a light source relative to the camera determines which side of an object casts a shadow. The size of the shadow is determined by the distance between the light and the camera, and the difference in depth between the foreground and background objects. We realised they could be used to delimit the size, shape, and position of adaptive windows to ensure they do not cross object borders.

Fig. 6.7(a) shows a simulated shadow image for the scene whose ground truth is known precisely. Fig. 6.7(b) shows the disparity map created using the object boundaries to guide the windows. The map is dense and although not 100% accurate, the areas of low texture are matched correctly and borders resolved accurately. Although this method is actually slower than conventional correlation-based approaches, little or no post-processing is required to correct the results, leading to an overall increase in speed. We have recently discovered that this same fast and accurate method has subsequently been devised independently of ourselves by Feris et al. [23].

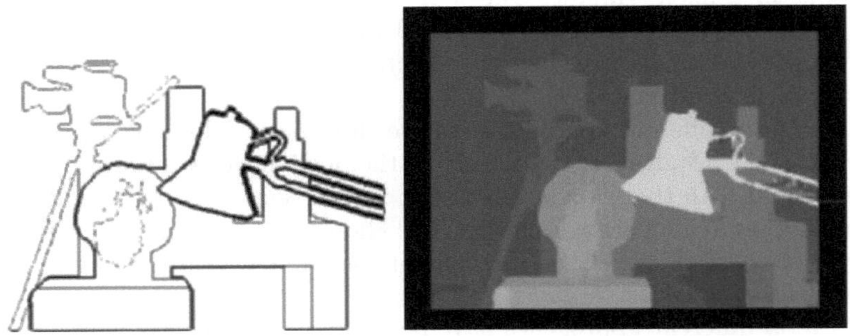

(a) Shadow edges (b) Adaptive window disparity map

Fig. 6.7. Edges detected using shadows and the resulting disparity map

6.6 Display

For the display of stereoscopic 3-D imagery, the most popular, cost-effective, and practical solution is a single high-refresh rate CRT, viewed with LCD shutter glasses to allow temporally interlaced left and right image frames to be viewed by the appropriate eye only, see Fig. 6.8 (left).

However there are some advantages to using fixed-viewpoint binocular eyepieces (see a bench-mounted prototype in Fig. 6.8 (right)): convergence can be

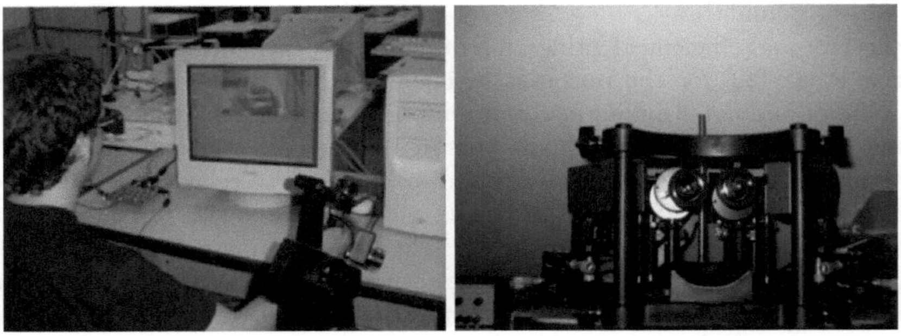

Fig. 6.8. Stereoscopic 3-D display devices

slaved to cameras, they can more economically achieve wider field of view, they can be head-mounted (cf. immersion), and can more easily facilitate variation of *optical distance* to the image.

6.6.1 Optical Distance

The optical distance to an observed point source of light is the radius of curvature of its photon wavefront. When observing a CRT, the optical distance to the various point sources that comprise image pixels is typically the same as the physical distance. When observing through an eyepiece, the optical distance need not match the physical distance. For instance, if the physical distance d_d from the displayed image to the eyepiece is equal to the focal length f of the eyepiece lens, the optical distance to the image $d_o = \infty$. When physical distance $d_d < f$, $d_d \leq d_o < \infty$, see Fig. 6.9.

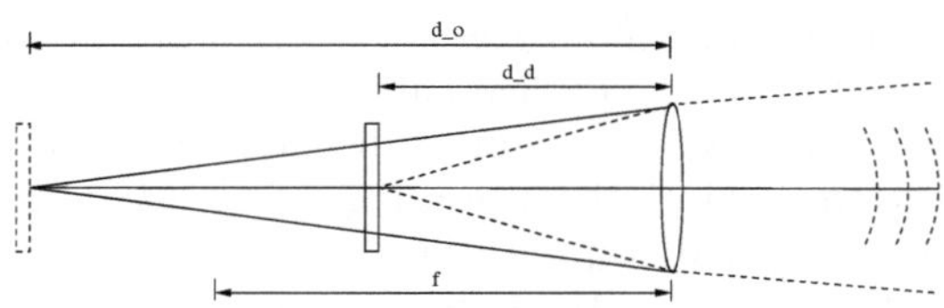

Fig. 6.9. Optical Distance

6.6.2 Accommodation-Convergence Conflict

The observer's convergence system manages the look direction of both eyes such that they converge on the scene point being observed. The function of the observer's accommodative system is to manage the adaptation of the eyes' crystalline lenses to achieve sharp focus of light from the scene point onto the retina. The accommodative system is *coupled* to the convergence system because under natural viewing conditions the convergence distance as dictated by the intersection of look directions is equal to the optical distance to the scene point, i.e. for an angle of convergence α_f, the accommodative system adapts the crystalline lenses to focus light from a distance of $d_f = \frac{1}{2}\text{IPD}/\tan\frac{\alpha_f}{2}$, where IPD is the distance between the eyes' pupils, and α_f is their angle of convergence, see Fig. 6.10.

A problem arises in stereoscopic 3-D viewing if the perceived distance d_f does not match the optical distance d_o (c.f. Sec. 6.6.1). To avoid the retinal image becoming blurred, the natural accommodative response must be *de-coupled* from convergence. De-coupling introduces visual discomfort, loss of visual acuity, and after-effects. It is often cited as the main reason that stereoscopic 3-D displays become intolerable after long periods, e.g. [24].

6.6.3 Multiple Depth Fields

A solution to the problem of accommodation-convergence conflict that has been suggested on several occasions, and evaluated most recently by Akeley et al. [25],

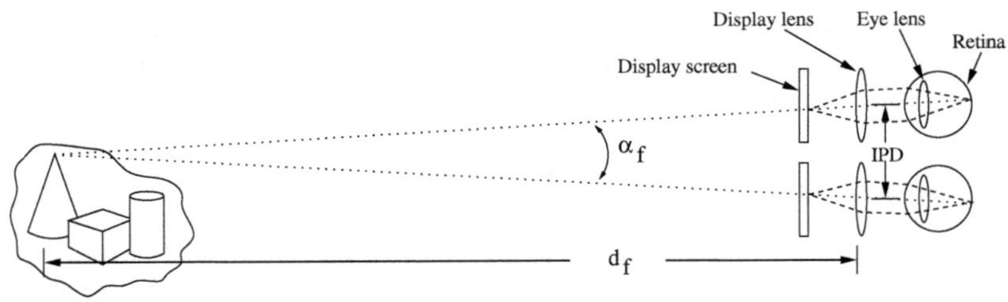

Fig. 6.10. Convergence to perceived depth

is to display different ranges of depth in the stereoscopic 3-D scene at different optical distances so that the observer can accommodate naturally. Implementations of this solution have used a different display device for each depth range, aligned on the same optical path using beamsplitters. This is an impractical solution due to both the number of display devices required and the light loss due to multiple beamsplitters on the optical path.

Our original solution is to use a single display device for each eye, but to dynamically change the optical distance using a fast-switching optical element, as used in *adaptive optics* (AO) applications (e.g. see [26]). A schematic of this *field-sequential depth* concept for three depth ranges d_n, d_m, and d_f is shown in Fig. 6.11. It shows how the AO system changes optical distance d_o to match the depth range being displayed.

A bench-mounted prototype implementation of this system is shown in Fig. 6.12. Its AO element is a deformable mirror 1kHz switching frequency and enough optical power to change d_o from $\frac{1}{3}$m to ∞m. Further details can be found in [27].

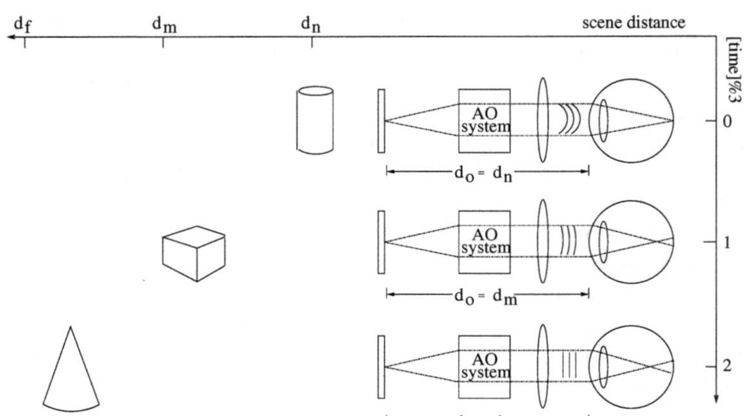

Fig. 6.11. Field-sequential depth concept

Fig. 6.12. Stereoscopic 3-D display prototype using adaptive optics

Such a display device can be used most easily with computer-generated synthetic imagery where the scene depth of each displayed pixel is known (cf. 3-D computer graphics). For use in telerobotic applications, real-time estimation of scene depth for each pixel is required. This has been one of the motivations of our research described in Sec. 6.5.

6.7 Conclusions

In our research we aim to consider the entire range of issues between stereoscopic 3-D acquisition and perception. In the work described herein, we believe that there are at least three original contributions: the use of an MPEG encoder to accelerate the the correspondence search, the use of a novel illumination strategy to simplify the correspondence search, and the use of adaptive optics to modulate optical distance within a stereoscopic 3-D display device.

Acknowledgements. Imagery from the *daVinci* robotic surgical system is courtesy of Intuitive Surgical Inc.

References

1. W.E.L. Grimson, G.J. Ettinger, T. Kapur, M.E. Leventon, W.M. Wells III, and R. Kikinis. Utilizing segmented MRI data in image-guided surgery. *IJPRAI*, 1996.
2. A. State, K. Keller, M. Rosenthal, H. Yang, J. Ackerman, and H. Fuchs. Stereo imagery from the UNC augmented reality system for breast biopsy guidance. *Medicine Meets Virtual Reality (MMVR)*, 2003.

3. M. J. Mack. Minimally invasive and robotic surgery. In *JAMA*, volume 285, pages 568–573, 2001.
4. D. Stoyanov, A. Darzi, and G.Z. Yang. Dense 3-d depth recovery for soft tissue deformation during robotically assisted laparoscopic surgery. In *MICCAI*, pages 41–48, 2004.
5. I.P. Howard and B.J. Rogers. *Seeing in Depth*, volume 2, Depth Perception. I. Porteous, 2002.
6. U.D.A. Müller-Richter et al. Possibilities and limitations of current stereo-endoscopy. *Surgical Endoscopy*, (18):942–947, 2004.
7. Y. Munz et al. The benefits of stereoscopic vision in robotic-assisted performance on bench models. *Surgical Endoscopy*, (18):611–616, 2004.
8. O. Faugeras. *Three dimensional computer vision: A Geometric Viewpoint.* MIT Press, 1993.
9. R. Yang, M. Pollefeys, and S. Li. Improved real-time stereo on commodity graphics hardware. In *CVPR workshop on Real Time 3D Sensors and Their Use*, 2004.
10. M. Brown, D. Burschka, and G. Hager. Advances in computational stereo. *PAMI*, 25(8):993–1008, 2003.
11. K. Konolige. Small vision systems: Hardware and implementation. In *Proc. Eighth Int'l Symp. Robotics Research*, 1997.
12. L. Falkenhagen. Hierarchical block-based disparity estimation considering neighbourhood constraints. In *International workshop on SNHC and 3D Imaging*, September 5-9 1997.
13. J. Mulligan and K. Daniilidis. Predicting disparity windows for real-time stereo. In *ECCV (1)*, pages 220–235, 2000.
14. R. Yang, G. Welch, and G Bishop. Real time consensus based scene reconstruction using commodity graphics hardware. In *Proceedings of Pacific Graphics 2002*, pages 225–234, October 2004.
15. C. Zach, A. Klaus, and K Karner. Accurate dense stereo reconstruction using graphics hardware. In *Eurographics 2003*, pages 227–234, 2003.
16. T. Kanada and M. Okutomi. A stereo matching algorithm with an adaptive window: Theory and experiment. *PAMI*, 16, 1994.
17. A. Fusiello, V. Roberto, and E. Trucco. Efficient stereo with multiple windowing. In *Proc. Computer Vision and Pattern Recognition*, pages 858–863, 1997.
18. A. Bobick and S. Intille. Large occlusion stereo. *International Journal of Computer Vision*, 33(3):181–200, 1999.
19. J. Lotti and G. Giraudon. Correlation algorithm with adaptive window for aerial image in stereo vision. In *Proc. SPIE Image and Signal Processing for Remote Sensing*, 2315, 1994.
20. H. Hirschmuller, P. Innocent, and J. Garibaldi. Real-time correlation-based stereo vision with reduced border errors. *Int'l J. Computer Vision*, 47:229–246, 2002.
21. O. Veksler. Stereo correspondence with compact windows via minimum ratio cycle. *PAMI*, 24:1654–1660, 2002.
22. R. Raskar, K. Tan, R. Feris, J. Yu, and M. Turk. Non-photorealistic camera: Depth edge detection and stylized rendering using multi-flash imaging. In *Proc. ACM SIGGRAPH*, 2004.
23. R. Feris, R. Raskar, L. Chen, K.H. Tan, and M. Turk. Discontinuity preserving stereo with small baseline multi-flash illumination. In *Proceedings of the International Conference on Computer Vision*, Beijing, China, October 15–21 2005. IEEE.

24. J.P. Wann, S. Rushton, and M. Monn-Williams. Natural problems for stereoscopic depth perception in virtual environments. *Vision Research*, 35(19):2731–2736, 1995.
25. K. Akeley, S.J. Watt, A.R. Girshick, and Martin S. Banks. Achieving near-correct focus cues using multiple image planes. In *SIGGRAPH Conference*. ACM, 2004.
26. G. Vdovin, P.M. Sarro, and S. Middelhoek. Technology and applications of micromachined silicon adaptive mirrors. *Optical Engineering*, 36:5509–5513, 1997.
27. F. Shevlin. A fixed-viewpoint volumetric stereoscopic 3-D display using adaptive optics. In *Stereoscopic displays and virtual reality systems XII*. SPIE, 2005.

7

User Voice Assistance Tool for Teleoperation

Oscar Reinoso, César Fernández, and Ramón Ñeco

Universidad Miguel Hernández
Dpto. Ingeniería de Sistemas Industriales
Avda. Universidad. Edif. Torreblanca. 03202 Elche-Alicante, Spain
{o.reinoso,c.fernandez,ramon.neco}@umh.es

Summary. A teleoperation tool that allows to interact with the remote environment in a more comfortable and flexible way is presented in this chapter. Working over a classical teleoperation environment, the goal is to reach a higher level of abstraction in the user commands. The tool allows the operator to interact with the remote environment through natural language recognition. This system is able to interpret and execute the commands formulated by the operator in natural language, according to the elements present in the remote environment. An error feedback module has been designed in order to take into account the on-line correcting information expressed by the operator during the execution of a task in the remote environment. The proposed voice assistance tool has been designed as a module in a novel teleoperation architecture, which allows to integrate multiple assistance tools.

7.1 Introduction

Teleoperation involves cooperation between the operations made by the user in a local environment and the actions exerted by the physical components in the remote environment. In a robotic teleoperated system the operator acts over a robot and the elements in the remote environment to perform the appropriate task [1]. Usually, teleoperation has been used in dangerous or inaccessible environments for the operator such as nuclear industry, reactor maintenance, spatial activities or dismantling tasks [1, 2]. However, the number of teleoperated systems has been significantly increased in the last years due to the possibilities offered by the World Wide Web [3] and the new tools that offer new possibilities to manage the remote environment.

In occasions, the environment is poorly known, and the conditions of perception are difficult. The operation reliability depends thus on all the data sensed around the work area and the way they are reproduced for the operator [2]. Classical telerobotic systems are usually composed of a vision system and a force sensor in the remote environment and a master arm in the local environment in such a way that the operator can manage the robot based on the information reported by these sensors.

However, some of the tasks made by the operator in a teleoperated way require a need of intelligent automation. In teleoperation, this intelligent automation

M. Ferre et al. (Eds.): Advances in Telerobotics, STAR 31, pp. 107–120, 2007.
springerlink.com © Springer-Verlag Berlin Heidelberg 2007

means that the tasks are carried out automatically with a supervisory control by human operators. In this way it is extremely important the assistance to operator's decision. In this chapter, a voice assistance tool that allows to interact with the remote environment in a more comfortable and flexible way is presented.

Natural language programming offers a great possibility to the operator to easily control and manage the remote environment [4, 5, 6]. Voice processing allows to establish a natural dialogue between a human and the robot in the remote environment in such a way that a not-qualified user of the system can manage the robot through a semantics that represents the environment and its relationship with the robot. The tool described in this chapter allows the operator to interact with the remote environment through natural language recognition.

We should mention that, of course, natural language interfaces are not adequate for all types of robotic commands. There are languages or human-friendly interfaces which are not natural but which are better for particular applications. Specifically, in the applications tested using the tool presented in this chapter, it is shown that natural language input can be an efficient technique for high level commands (HLC) and that for low level commands (LLC) [1] other input interfaces less natural are more appropriate.

7.2 Natural Language Interface

One of the simplest, most natural and highly user acceptable interfaces between human and teleoperated robots is surely one where the human operator speaks to the robot. This is the main motivation of the review and work presented in this chapter. One of the main advantages of voice recognition is that a single word or a simple sentence can communicate a complete procedure or a complex data structure. This motivation is confirmed with the fact that in recent years a growing interest exists in robots that are designed specifically to interact with operators who are not roboticists, such that they can respond to commands with voice. In order to be really useful in domestic and industrial environments, the robot should be able to interact in an appropriate way with its environment and with its human operators.

This section describes generalities about a module of natural language processing to teleoperate a robot, and section 7.3 presents a feedback processing and error correction in a real-time robotic teleoperated system. The overall robot control system consists of the following modules: speech recognition module, bilateral control module, vision module, learning module, and the robot controller and interface. The speech recognition module includes a natural processing system that permits understanding of high level natural language commands, and provides a low level program, directly executable by the robot.

Human-robot voice interaction is different from human-computer interaction mainly due to the manipulative power of the robot in a physical environment. The robot has a physical configuration and can operate in the environment during dialogue or as a result of a voice command. However, research on human-computer dialogue management is a good foundation for human-robot dialogue management [7]. In the translation process of a command expressed in natural language to an executable program for the robot, the following characteristics have to be taken into consideration:

- There is no a target language universally-accepted for the high level tasks usually performed by a robot. This means that an appropriate *intermediate language* should be designed [4].
- The robot should have access to static geometric information about its working environment, as well as dynamic information, especially in interfaces for real-time systems.
- The necessary information to process the commands is not as structured as the information needed for other speech interfaces. Therefore, a general *knowledge representation* mechanism should be devised, including all the necessary aspects (information about the context, geometry, the robot's activities, sensor readings, etc.). At the same time, this general mechanism should be adaptive to specific robotic systems.
- The expressions that an operator can speak by using an interface to a robot are very heterogeneous. The translation process is, therefore, more complex since it needs more semantic information.

7.2.1 Related Works

This section describes some works directly related to the topic presented in this chapter. For a more general review (speech recognition details, speech generation) see [8].

Automatic speech recognition and natural language processing have been used as a powerful tool for human-computer communication [9,10]. Some researchers have developed and demonstrated robots with a natural spoken language-based interface in a limited and technically restricted framework. The first complete system was SAM (Speech Activated Manipulator), where the authors implemented a discrete speech recognition system to control a robotic arm [6]. Crangle and Suppes [5,11] devised a theoretical-oriented system to program a mobile robot and a robotic arm. Nevertheless, this system was not linked to a specific speech recognition module. They also gave a first adaptation technique applied to distinguish the different executions of the same verb to different execution circumstances or to different operators, focusing on the specific application of control devices for the disabled.

In the work described in [12] a formal model was designed in order to represent computationally the intentions of the user in dialogue systems. In [13] the authors incorporate a discrete speech recognition system (with some natural language capabilities) to a teleoperated robot which was applied to electrical maintenance tasks. In [14], the author describes new language-processing methods suitable for human-robot interfaces. These methods enable a robot to learn linguistic knowledge from scratch in unsupervised ways, at a low level (speech) processing through statistical optimization. In [15] the author describes a successful natural language interface to a mobile robot working in an office environment.

More recently, in works as [16,17], integration of natural language and gesture understanding was made in order to obtain a more natural interface in space and medical robotic applications. One of the most sophisticated and integrated systems can be found in [18]. It is a communicative humanoid robot with a hand and graphic face, which appears on a small monitor in front of the user. It is capable to perform face-to-face dialog, in real time, with a human user with various hand gestures, facial expressions, body language and meaningful utterances, which can be used to guide the humanoid-like robot.

7.2.2 Language Analysis and Understanding

In order to process natural language, we need to combine our understanding of small textual units to understand larger ones. The main objective of natural language processing theory is to show how these larger units of meaning arise out of the combination of the smaller ones, which is modeled by means of a *grammar*. It is traditional to divide the processing task into syntax and semantics, where syntax describes how the different formal elements of a sentence can be combined and semantics describes how the interpretation is computed. The grammar can be thought as the encoded linguistic knowledge, which is 'static' and separated from the processing components (the analysis algorithms). Basically, the grammar consists of a lexicon (a database of words or groups of words) and rules that syntactically and semantically combine words and phrases into larger phrases and sentences.

More specifically, in natural language processing we can distinguish several processing phases which coincide with the distinct steps in the process of understanding a natural-language command. These phases are the following. (1) Phonological and morphological processing; (2) Syntactic analysis; (3) Semantic analysis; and (4) Pragmatic analysis. Very briefly, each one of these phases is understood as follows.

The phonological and morphological processing is the processing of phonemes into basic units called morphemes, and then the processing of these morphemes into words. The syntactic analysis is the analysis of the order in which words are combined to form commands and, as said before, the syntactic knowledge is represented using formal grammars. The semantic analysis is the phase in which the system obtains the meaning of the individual words and how the whole meaning of a command is built up from the meanings of the words used in it. In the analysis of a verbal command to a teleoperated robot the meanings of the words are all the information we need to make the robot execute the command (including the parameters and the translation into the robot programming language). Pragmatic analysis is the phase in which the system process everything else that affects the use and interpretation of the natural language command in a specific context or situation. In a robotic application, the system uses the pragmatic analysis to find the values of some parameters or to make processing decisions that could not be made just using syntactic or semantic information.

As said before, a grammar describes the sentences that make up a language. It contains a finite number of rules that specify which sentences belong to the language and, at the same time, what is their syntactic structure. This way, we can obtain the underlying structure of the commands to the robot and extract its translation into the robot language. In the tool presented in this chapter we use an extended form of the classical *phrase-structure grammar*, which is defined as a tuple $G = (V, V_T, V_{NT}, P, C)$, where:

- V_T is the finite set of *terminal symbols* which correspond with the words or symbols that can appear in a command.
- V_{NT} is the finite set of *nonterminal symbols* which correspond to grammatical categories in which the language has been structured.
- $V = V_T \cup V_{NT}$ is the set of all the symbols in the grammar.
- P is the set of production rules of the form $p : a \rightarrow b$, which can be read as: from a we may derive b, where a is a combination of nonterminal symbols and b is a combination of nonterminal and/or terminal symbols.

  For example, the production '$p : V \rightarrow drop$' may be read as: the nonterminal symbol V (for verb) may derive or can be rewritten as the terminal symbol 'drop' (a verb), and the production '$p : NP \rightarrow Det\ N$' may be interpreted as: the nonterminal symbol NP (for noun phrase) can be rewritten as the nonterminal *Det* (determinant) followed by the nonterminal N (noun).
- $C \in V_{NT}$ is the start symbol to produce a sentence or command. It is a nonterminal symbol and all the sentences produced by the grammar are derived from C (note that the grammar needs an initial symbol to begin with).

The phrase-structured grammars can be augmented to represent additional information that is important to understand natural-language commands. This information can be incorporated in a grammar by assigning to each symbol (terminal or nonterminal) in the language not just a syntactic category such as verb or adjective but *attributes* that take *values.* This way, we have the so called *augmented* phrase-structured grammars. So for example, the word 'drop' could have the following two basic attributes: *category*, with the value *verb*; and *number*, with the value *singular.* To obtain the translation of a command into the executable instructions to the robot, the word may have additional attributes. For example, for the word 'drop' we can define the attribute *command* with the value *open_gripper(x,y)* where x and y are the coordinates of the point where the operator wants the robot to 'drop' an object. These coordinates depend on the values of the object attributes which appears in the command (for example in the command 'drop the box in the middle of the table', the coordinates x and y are the values of the corresponding attributes of the object 'the middle of the table').

These attribute-value pairs can be represented in matrix form in which the first column contains the attributes and the second one contains the corresponding values. This way, during the parse of a sentence these matrix structures are combined or *unified* using syntactic and semantic analysis algorithms obtaining the command that the robot has to execute.

For more details in grammars, parsing and semantic analysis the reader is referred to texts that deal specifically with these topics (for example, [19,20]).

7.2.3 The Natural Language Tool

The basic objective of the designed natural language interface is to allow an untrained human operator to teleoperate a robotic arm through a semantics that reflects the complex kinematics and dynamics involved in the tasks performed by the robot.

In order to program a robotic arm using natural language, an intermediate language should be defined, including all the actions that the robot can execute. This intermediate language is equivalent to the target languages in database applications. In fact, this intermediate language is the language to which the system will translate the input command, so it can be considered the "target" language for the natural language subsystem. In our architecture, the instructions have been divided in three categories, depending on the type of action to execute. These categories are: (1) Movement instructions; (2) Database access instructions; and (3) Control structures. In the design of the intermediate language, the main objective is to achieve generality. We can consider the intermediate language as a model for the teleoperated robot using natural language.

Once the system has obtained the transcription of the voice command in an ASCII text, the natural language understanding module interprets the sentence and translates it into the intermediate control language. The method consists of programming a grammar or specific transition network [19] for the application and to perform the four analysis phases outlined in section 7.2.2:

- Lexical analysis: the identification of the minimum units from the input command;
- Syntactic analysis: the identification of the syntactic structure of the sentence, obtaining a syntactic tree; and
- Semantic and pragmatic analysis: to obtain the interpretation and final translation of the command.

7.3 Real-Time Control: Error Feedback Using Natural Language

This section presents methods for error feedback processing using natural language. Two different methods have been used such that it is possible to process the feedback from errors produced by a teleoperated robot, using a voice interface with natural language processing capabilities. The solutions proposed in this chapter are motivated by the following questions: (1) How can the feedback be processed during the execution of a teleoperated command, expressed using a voice interface, and (2) how can the system use the information provided by the feedback process so that the robot behaves in the way desired by the operator in successive executions of the task.

We do not consider the automatic recovery from errors [21]. The feedback subsystem using natural language described in this section has two main effects:

(i) The correction in real time of the robot's current action; and
(ii) learning of derived information obtained from the correction such that this information can be used in later executions.

In this section it will be assumed that the system has already a set of tasks that the robot has learned how to execute, or that in some way it has a list of sequential actions associated to each one of the tasks of this set. It is also assumed that most of these tasks depend on a set of parameters, that is, their execution depends on an n-dimensional vector $\mathbf{p}$ of m parameters, $\mathbf{p} = \{p_1, p_2, \ldots, p_m\}$. These parameters can be positions at each step of the execution of the task, speeds, forces, etc.

The value of these parameters is what will be adapted or learned through the feedback. The tool considered in our system can process two types of feedback:

(i) ***Position feedback***, which corresponds to commands to the robot in cases in which some position (final or partial) has not been completely satisfactory. Typical commands that fall in this category are: "*. . . more to the left . . .*", "*. . . much more to the left . . .*", "*. . . a little more to the right . . .*", etc.
(ii) ***Accuracy feedback***, which corresponds with commands that refer to the value of some magnitude (forces, speed, etc.) and they can be expressed using natural language commands such as "*. . . press with more care . . .*", "*. . . not so slowly . . .*", etc.

The automatic adaptation algorithm consists of the following steps:

(i) The operator speaks a command, and then the robot executes this command using m functions f_i $(i = 1, 2, \ldots, m)$ to determine the values of the m parameters $(p_1, p_2, \ldots, p_m)$.
(ii) Depending on the execution, the operator speaks feedback commands.
(iii) The robot processes the feedback commands in two steps:
   a) The robot executes an immediate action according to the feedback command in order to correct in real time the position or accuracy parameter.
   b) The system *adapts* the internal representation of the parameter for future executions of the command.

Depending on the type of functions f_i used in this algorithm, two similar techniques can be distinguished: fuzzy representation techniques and stochastic representation techniques. The main differences between these two techniques reside in the knowledge representation (functions f_i) and the specific algorithm to update this representation, but the main idea is very similar. Secs. 7.3.1 and 7.3.2 describe these two techniques.

7.3.1 Fuzzy Representation Techniques

In the previous algorithm, before feedback takes place on the robot's behavior, the operator should speak a command. When the robot has to decide the value of the parameter vector $\mathbf{p}$ associated with this command, a sampling algorithm is used through the function f_i. An interval $[x_{\min}, x_{\max}]$ is defined for the generation of the values.

Once the robot has chosen the value of the parameter from the current function that represents the knowledge, the user can speak correcting commands associated to that parameter. As reply to a correcting command, the robot executes an immediate action and it adapts the internal representation of the parameter depending on the command. Therefore, the nature of the feedback command determines the adaptation algorithm of the functions f_i. We consider two types of feedback commands: position and accuracy.

Position Feedback

The position feedback in natural language should have an immediate effect on the robot's position, besides modifying the function associated to the position parameters for that task. For example, after a feedback like "*... a little more to the left ...*", the robot will move a longitude toward the left and it will also adjust the coordinates for the next time it executes the task.

The magnitude of the displacement as a consequence of the position feedback depends on the specific feedback parameters that the operator has spoken. A simple way of modelling these displacements is to define n different constants $c_1, c_2, \ldots, c_n$ for each position feedback category, such that:

$$c_1 < c_2 < \ldots < c_n$$

The commands that generate relatively "very small" movements will come defined with the constant c_1, those that generate relatively "small" movements with the constant c_2, those that generate "middle" movements with the constant c_3, "big" movements with the constant c_4, "very big" movements with the constant c_5, and so on. The decision about the value that should have the variable n and the concrete values that should have the n constants c_i depends on the nature of the parameter and on the specific task and they are determined by means of a specific design or in the learning phase of the tasks.

In the fuzzy representation technique, the values of each parameter p_i in the command are represented by a fuzzy membership function in such a way that the specific value of the robot's displacement is multiplied by its dispersion. These membership functions are centered around a real number, with arbitrary functions to both sides of the center, and they are called *Left-Right fuzzy numbers* $(L - R)$, whose membership functions are defined as:

$$f_i(x) = f(x; a, \alpha_L, \alpha_R)_{LR} = \begin{cases} L\left(\frac{a-x}{\alpha_L}\right), & \text{if } x \leq a \\ R\left(\frac{x-a}{\alpha_R}\right), & \text{if } x \geq a \end{cases} \tag{7.1}$$

where a is the center of the fuzzy set, α_L and α_R are positive real numbers which represent the dispersion of the function, and R and L are two functions that satisfy the following conditions:

(i) $R(0) = L(0) = 1$, and
(ii) R and L are non-increasing in the interval $[0, \infty[$.

To decide the change in the robot's position as a consequence of the feedback with voice, as it has been indicated before, the constant c_i is multiplied by the dispersion corresponding to the membership function, given directly by the constants α_L and α_R in the case of $L - R$ fuzzy numbers in (7.1). If triangular or trapezoidal membership functions are used, then the dispersion is given by the distance between the center and the extremes of the triangle or trapezium.

Besides causing a change in the robot's position, this feedback type should cause a change in the generation of the parameter for future executions of the same command expressed in natural language (step 3b in the previous algorithm). In order to achieve this objective, the membership functions should be modified in the following way:

- The center a of the membership function f_i is updated as the last value of the corresponding parameter p_i after the feedback.
- The variation of the dispersion (given by the parameters α_L and α_R) is computed as follows:

$$\alpha_R(i+1) = \sqrt{\alpha_R(i)} \tag{7.2}$$

$$\alpha_L(i+1) = \sqrt{\alpha_L(i)} \tag{7.3}$$

  This correction is made assuming that after several executions and corrections by the user, the dispersion of the function should tend to decrease, since the more corrections carried out, the bigger "reinforcement" of the learning of the parameter. A way of getting this is making the new α_L and α_R values equal to the square root of the previous values. In this way it is possible to decrease the dispersion in the selection of the parameter whose value has been corrected.

Accuracy Feedback

As it has already been mentioned, an accuracy feedback is the speaking of interaction commands in natural language that refer to the value of some magnitude (forces, speed, etc.). These commands are expressed with commands in natural language such as "*...press with more care ...* ", "*...not so slowly ...*", etc. This feedback doesn't cause an immediate change in the position and the robot's current state, but rather it causes a change in the function that is used to generate the magnitude to which refers the command. As in the previous case, the change can be reflected changing the dispersion of the membership function of the fuzzy number directly. If the command refers to the need of increasing the value of a magnitude, then the dispersion should be increased (for expressions like "*...more quickly ...*").

On the other hand, if the command refers to the need of decreasing the value of a magnitude, then the system should decrease the dispersion explicitly (for example, for expressions like "*...with less force ...*". The change that is made to the dispersion should be proportional to the current values of the dispersion, so that the change takes place in a controlled way. A form of getting this is to increase or to decrease the dispersion according to the absolute value of the first derivative of the dispersion with respect to one of the parameters of the membership function (for example, with respect to the center), so that an increase or decrease following the direction of the gradient takes place.

7.3.2 Stochastic Representation Techniques

Another considered possibility to represent the imprecise information in the feedback to the robot using the speech interface is the direct use of probability density functions. For one-dimensional tasks, simple probability distributions of one variable can be used, $f(x)$, where x is the value of the parameter, and f is the probability density function. Using these ideas, a stochastic approach can be designed as an alternative representation of the parameters that may be adjusted. The robot will use as working parameter values of p that belong to an interval around the mean of the distribution, with a dispersion that is given by the variance.

The choice of the function f depends on the robot, the environment and the specific task. In the work described in this paper a well-known function has been used, the beta function or distribution whose probability density is given by [11]:

$$f(x) = \frac{g(\alpha+\beta)}{g(\alpha)g(\beta)} x^{\alpha-1}(1-x)^{\beta-1}, \quad \alpha > 0,\ \beta > 0,\ 0 < x < 1 \qquad (7.4)$$

where g is the gamma distribution, which is defined as $g(x) = \int_0^\infty e^{-t} t^{\alpha-1} dt$.

This function is used in the work presented in this paper to represent probabilistically behaviours of the robot that depend on one or several parameters. Depending on the values of α and β, the function will represent different behaviours. If $\alpha = \beta = 1$, then all the values in the interval $(0, 1)$ are equally probable, representing the idea that the operator doesn't care too much the value of the parameter while it belongs to a specific interval. In the case of position feedback described in the previous sections, this is translated to the fact that it is not relevant the exact point in which the position of a tool is placed, in the range of its longitude (that is, in the range of the function f). For the values $\alpha = \beta = 1$ a uniform distribution in the interval $(0, 1)$ is obtained, that is usually the initial distribution for the task.

7.4 Experimental Results

The work environment of the robotic arm that will be considered in this section consists of a table, a set of pieces and a shelf in which the pieces can be placed (see Fig. 7.1). The operator should be able to communicate with the robot to perform simple assembly tasks. These assembly tasks include simple subtasks such as moving the pieces, sorting, or storing them, etc. We assumed that the robot has already learned how to perform these tasks and the corresponding procedures are stored in a knowledge database.

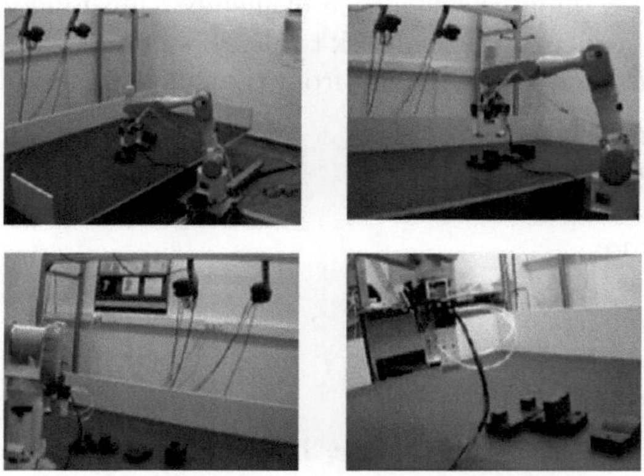

Fig. 7.1. Different images corresponding to the robotic arm working in the remote environment using the natural language interface

The group of techniques described in previous sections has been evaluated with a teleoperated robot as the one described. In Table 7.1 a sequence of values of the parameter **p** is shown. This parameter represents the distance to the origin of coordinates considered in the environment for a command of placement of a piece with position feedback, using possibilistic techniques with $L - R$ fuzzy numbers.

In Table 7.2 results from the same experiment are shown, using the beta probability distribution.

Table 7.1. Position feedback with possibilistic techniques, using $L - R$ fuzzy numbers

p	Position feedback
0.500	*more to the right*
0.725	*much more to the right*
0.911	*a little more to the left*
0.886	*a little more*
0.780	*OK*

Table 7.2. Position feedback with stochastic techniques, using the beta distribution

p	Position feedback
0.500	*more to the right*
0.300	*a little more to the right*
0.401	*a little more to the left*
0.381	*a little more*
0.751	*OK*

In Fig. 7.2 an example the sequence of membership functions is shown, obtained as a consequence of the feedback to the robot using the vector of feedback commands (R_1, R_2, R_8, R_4), obtained from the following list of commands:

- R_1: *more to the right*
- R_2: *much more to the right*
- R_3: *a little more to the right*
- R_4: *a little more*
- R_5: *a little less*
- R_6: *more to the left*
- R_7: *much more to the left*
- R_8: *a little more to the left*

In the experiments five constants c_i have been used for the decision of the robot's displacement from its current position, with the following values: $c_1 = 0.02$, $c_2 = 0.08$, $c_3 = 0.11$, $c_4 = 0.19$, $c_5 = 0.25$. These values are quantified depending on the range of the parameter that is the object of the feedback. In the case of the parameter whose membership functions are shown in Figure 7.2, the values taken by the parameter are in the interval $[0, 1]$. The membership functions can be modeled with the use of fuzzy numbers using identical definitions for the functions L and R, and they are given by the lineal function $L(x) = R(x) = -x+1$.

This function has the two conditions of the definition given previously for the fuzzy numbers ($R(0) = L(0) = 1$) and both are non increasing. All the experiments start with the value for the mean $a = 0.5$. The initial values of the dispersions have been $\alpha_L = \alpha_R = 2$. The initial value of the parameter for the execution of the command is $x = 0.5$ and the initial functions for the representation of the membership functions corresponding to the fuzzy number associated to the parameter are the following: $L\left(\frac{0.5-x}{2}\right) = 0.5x + 0.75$, and $R\left(\frac{x-0.5}{2}\right) = -0.5x + 1.25$.

Fig. 7.2 shows a sequence of probability distribution functions obtained after the processing of feedback position commands, indicating successive displacements toward the right. It can be observed that, besides moving the mean of the distribution, the process also decreases the standard deviation to reinforce the fact that feedback has taken place on the associated parameter, reducing in consequence the effective range in which the sampling algorithm obtains the values. The default initial parameters for the distribution beta are always $\alpha = \beta = 10$ (except for the cases in which explicitly a mean different from 0.5 is indicated). Later on, as a consequence of each feedback command, new values are calculated in such a way that the mean of the distribution approaches to the new obtained value, and such that the standard deviation diminishes approximately according to the constant $1/k$, where $k = 2$ in the lines of Figure 7.2. Note that since the values of α and β are integers, the solution to the equation $\sigma_f(i+1) = (1/k)\sigma_f(i)$ can only be obtained as an integer approximation to obtain the new standard deviation and the new mean values after each feedback iteration.

The two techniques described in this paper allow the processing of the feedback expressed in natural language following the objectives that have been shown

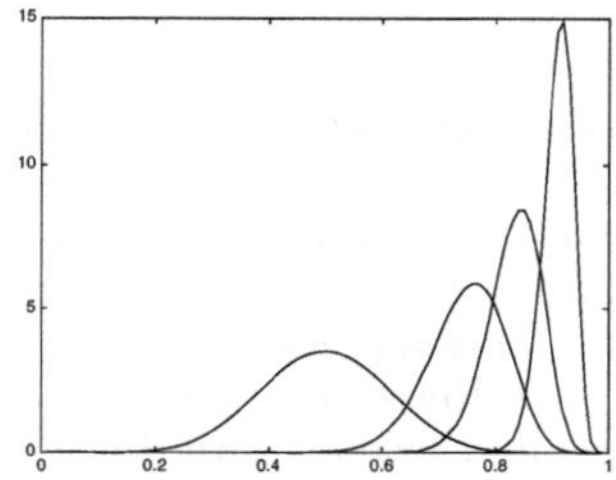

Fig. 7.2. Successive displacements toward the right of the mean of the distribution and decreasing of the standard deviation as a consequence of feedback commands

in this section. However, the fuzzy logic technique presents the advantage of being more intuitive in the definition and interpretation of the obtained representations of the parameters, besides being the most appropriate for a real time system, since the required computational complexity is smaller. The stochastic technique presents the advantage of allowing a more precise feedback of the parameters. Only in the cases in which the task carried out by the operator requires more numeric precision is justified the use of the stochastic technique, keeping in mind the biggest required computational cost.

7.5 Conclusions

The main conclusion of this section is that both representation methods allow the implementation of a feedback system that is very natural to a human operator, specially indicated when programming a robot to perform high-level tasks. It has been also shown that the feedback and correction using natural language commands is adequate for the application to a real-time teleoperated system.

The results obtained with the design of the nucleus of the system that has been shown in this section show that the natural language processing method is the most appropriate for the natural communication with a robot in real time. In particular, the definition of the intermediate language facilitates the design of the syntactic and semantic rules that define the natural language, as well as the augmented transition networks. The lexicalist paradigm used in the design of the lexicon has been the most efficient method for processing in real time. The augmented transition networks are suitable for the initial design of the interface, and for the design of the simple linguistic structures.

The voice assistance tool presented allows the user to interact with the robot in a natural way (using natural language or voice commands), and allows also the adaptation of the system from real-time feedback in natural language. The results obtained with this tool shows that the natural language processing method is the most appropriate for the natural communication with a robot in real time. In particular, the definition of the intermediate language facilitates the design of the syntactic and semantic rules that define the natural language.

References

1. T.B. Sheridan. *Telerobotics, Automation and Human Supervisory Control*. MIT Press, Cambridge, Massachusetts, 1992.
2. M. Decreton. *Teleoperation: Numerical Simulation and Experimental Validation*, chapter Nuclear Teleoperation. Particular Challenges in Decommissioning Applications. Kluwer Academic Publishers, Boston, 1992.
3. J.M. Sabater, J.M. Azorin, O. Reinoso, R. Neco, and N. Garcia. Dynamic virtual environment to test teleoperated systems with time delay communications. *Journal of Robotic Systems*, 22(4):167–181, 2005.
4. R. Neco, O. Reinoso, N. Garcia, and R. Aracil. A structure for natural language programming in teleoperation. In *Sixth International Conference on Control, Automation, Robotics and Vision*, 2000.
5. C. Crangle. Conversational interfaces to robots. *Robotica*, 15(1):117–127, 1997.
6. M.K. Brown, B.M. Buntschuh, and J.G. Wilpon. Sam: A perceptive spoken language understanding robot. *IEEE Trans. on Systems, Man and Cybernetics*, 22(6):1390–1402, 1992.
7. Y. Anzai. Human-robot-computer interaction: a new paradigm of research in robotics. *Int. J. Adv. Robotics*, 8(4):334–358, 1994.
8. R. Prasad, H. Saruwatari, and K. Shikano. Robots that can hear, understand and talk. *Advanced Robotics*, 18(5):533–564, 2004.
9. J. Biing-Hwang and F. Sadaoki. Automatic recognition and understanding of spoken language - a first step toward natural human-machine communication. In IEEE, editor, *Proc. of the IEEE*, volume 88, pages 1142–1165, 2000.
10. M.S. Eid and C. Moghrabi. Natural language interfaces as expert systems for industrial applications. *Computers in Industrial Engineering*, 31(3/4):843–847, 1996.
11. C. Crangle and P. Suppes. *Language and learning for robots*. Center for the Study of Language and Information (CSLI), Stanford, 1994.
12. S. Carberry, J. Chu-Carroll, and L. Lambert. Modeling intention: Issues for spoken language dialogue systems. In *Proc. of the International Symposium on Spoken Dialogue*, pages 13–24, 1996.
13. M. Ferre, J. Macias-Guarasa, R. Aracil, and A. Barrientos. Voice command generation for teleoperated robot systems. In *7th IEEE International Workshop on Robot and Human Communication (RO-MAN'98)*, volume 2, pages 679–685, 1998.
14. N. Iwahashi. Language acquisition through a human robot interface by combining speech, visual, and behavioral information. *Information Sciences*, 156:109–121, 2003.
15. M.C. Torrance. *Natural communication with robots*. MIT, Cambrigde, 1999.
16. E. Alessandri, A. Gasparetto, R. Valencia, and R. Martinez. An application of artificial intelligence to medical robotics. *Journal of Intelligent and Robotic Systems*, 41(4):225–243, 2005.
17. D. Sofge, Bugajska, J.G. Trafton, and D. Perzanowski. Collaborating with humanoid robots in space. *International Journal of Humanoid Robotics*, 2(2):181–201, 2005.
18. K.R. Thorission. A mind model for multi-modal communicative creatures and humanoids. *Int. J. Appl. Artif. Intell.*, 13(4-5):449–486, 1999.
19. J. Allen. *Natural Language Understanding*. Benjaming Cummings Series in Computer Science, 1995.
20. D. Jurafsky and J.H. Martin. *Speech and Language Processing*. Prentice Hall, 2000.
21. H. Aydin, R Melhem, and D. Mosse. Incorporating error recovery into the imprecise computation model. In *Proceedings of the 7th International Conference on Real-Time Computing Systems and Applications*, 1999.

8

Computer Vision Body Modeling for Gesture Based Teleoperation

Manel Frigola, Alberto Rodriguez, Josep Amat, and Alícia Casals

Technical University of Catalonia (UPC)
Automatic Control and Computer Engineering Department
Center of Research on Biomedical Engineering
Edif. U. C/ Pau Gargallo, 5 08028 Barcelona, Spain
{manel.frigola,alberto.rodriguez-garcia,josep.amat,alicia.casals}upc.edu

Summary. Dependable robots and teleoperation, taken in its broadest sense, require natural and friendly human–robot interaction systems. The work presented consists of a methodology for human–robot interaction based on the perception of human intention from vision and force. The vision system interprets human gestures from the integration of a stereovision and a carving system, from which it extracts a model of the human body when a person approaches the robot. The interaction can be performed by contact as well, from the perception of the forces applied to the robot either through a force sensor on the wrist or a sensing skin. The perception of human intention makes possible an intuitive interaction to modify on line the robot trajectory when required.

8.1 Introduction

The introduction of robots in environments close to humans demands for new robot performances and robustness. In such environments robots need to be endowed with high adaptability to changing conditions in order to guarantee safe operation or to deal with uncertainness. Human–robot (H–R) cooperation is performed through teleoperation or other interaction means. In any case, this interaction must offer ease of operation, sociability, etc. The importance of the necessary friendliness, that is, the socially interactive robots requirements as communication, understanding human orders and human intention (face and gestures recognition), learning or imitation are analyzed from different points of view in [1].

With the aim of achieving higher human–robot interaction capabilities and allowing humans to have a bearing on the execution of the robot task different interaction devices are available. Here, interaction is considered in its wide sense, either through local or remote teleoperation, or even through the possibility of changing the robot programmed strategies during their execution.

Apart from using master devices such as robot teach pendant, other manual devices or oral orders, research is going on towards the increase of performance of gesture based systems. For more advanced devices, the measure of forces applied over the robot in conjunction with vision systems can provide very important information for understanding human gestures [2,3,4].

M. Ferre et al. (Eds.): Advances in Telerobotics, STAR 31, pp. 121–137, 2007.
springerlink.com © Springer-Verlag Berlin Heidelberg 2007

This work is oriented to obtain a reliable and precise model of a human body, which operates close to a robot, in order to provide information about human orders, or even their intention, by the interpretation of their movements and gestures. Gesture orders perceived from vision complement other potential information provided by the measure of the forces applied to the robot by the user. This interface provides an intuitive way of controlling a robot so as it can be operated by non expert users. Furthermore, the processing of the sensed data in real time allows a human to modify "on line" the task going on, thus making the adaptation of the robot operation to changing environment conditions possible.

8.2 On-Line Robot Control from Gestures

As more advanced master devices or interaction means are available, a wider range of operations can be performed in cooperation, with a human and the robot. Thus, the concept of teleoperation can be extended to applications having more natural human–robot interaction. Besides the common teleoperation techniques, other kind of interactions during the execution of a programmed task are considered, either to modify it so as to improve the results of its execution or to avoid risky situations.

Direct robot guidance is usually performed using contact devices (i.e. exoskeletons) that have haptic capabilities for teleoperation. Alternatively, non contact techniques based on vision are more flexible and less costly, but they present a lack of force feedback. Recent progresses in robot control show that it is possible to achieve some degree of autonomy alleviating the problem of needing a direct control of the robot movements.

Many emergent teleoperated applications require only some kind of human–robot interaction during the execution of a task, provided that the task has been previously programmed. The required interaction depends on the development of the robot operation. In such cases, it is convenient to introduce more powerful and "natural" tools enabling gesture interaction to operate the robot or to correct its movements, that is, enabling to change the planned strategy.

This kind of requirements appear in robotic fields such as services, assistance, etc., where due to the difficulties in modeling the environment with enough definition, it is necessary to aid the robotic arm to adapt its movements to the real environment or to the user's needs.

Some industrial applications are also susceptible to this requirement when higher flexibility can be useful in response to human supervision. For instance, in polishing tasks, a repetitive primary movement is repeated all along the object surface. If required, a human could change the departing point or correct the main run over the object's surface when required. Also, some maintenance applications are too difficult for full automation. In these cases, it would be extremely useful to program the task in such a way that it could be supervised by humans which can interact with the robot to modify its behavior so as to solve conflict situations produced by unforeseen obstacles or environment constraints.

In the area of entertainment, robots must be open to continuous interaction with humans so as to behave as intelligent mechanisms able to respond to the user's demands. Several human–robot interaction techniques are currently being used to operate a robot or modify its actuation. An example, is presented in [5] where the trajectory of a mobile robot is modified interactively, in 2D, based on B-splines. In [6] a push-pull interaction is simulated via a haptic interface. More broadly, the concepts of human centered robotics and haptic interaction are described in [7], where the path of a mobile arm is modified for obstacle avoidance.

8.2.1 Natural Human–Robot Interaction for On-Line Control

Oral communication, voice, has been for long time the basis of natural language for human–machine interaction. Voice constitutes a remote interaction means with great capabilities, as far as it is possible to understand and interpret different kind of orders. Nevertheless, reliability is not yet enough for many tasks and the interpretation of orders can require several trials. Furthermore, it is difficult to transmit some orders by voice since oral communication is more qualitative than quantitative. An example of such imprecise orders are those corresponding to movements. That is, an order as "follow a given curve in space" can not be clearly defined orally.

Communication by means of gestures emerged later as an alternative to voice or a complementary interaction means. Gestures are especially useful for robot reactive behaviors, as for instance in collision avoidance. The main difficulties for gestures interpretation are due to the complexity in perceiving the position and configuration (posture) of a human body in real conditions, that is, without specific environment adaptation. Current systems achieve limited precision.

Haptics enables human–robot interaction with the robot using a specific device (the master) or the robot itself, a direct interaction. Direct interaction with the robot implies the measure of forces applied over the robot by the user. The robot control unit is involved in the reactive action to the external human guidance. Robots endowed with haptic performances have the capability of perceiving the applied forces and reacting to them according to a predefined behavior. This capability can be used either to modify the robot configuration in function of the applied external forces or to adapt the robot movements to the external interactions produced by physical contact.

This interaction capability can be used in multiple ways: a) classical teleoperation tasks with bilateral control, b) tasks carried out in cooperation, humans and robots, for instance moving an object [7], and c) tasks carried out by one or more robots, but that can be modified on line according to the environment conditions or constraints [3]. The way of implementing such interaction, based on contact and the perception of forces, can rely either on the use of force and torque sensors, or on the capability of the robot control unit for controlling every robot joint, not by position but by force. When an external force and torque sensor is located at the robot wrist a human can interact with the robot applying a force at the end-effector or holding and steering an attached device. But, in this case, the robot can not perceive any force applied in any position above this point, the wrist.

On the contrary, if the force control is performed joint by joint, it is possible to program a given trajectory knowing the efforts caused by the own robot weight and the corresponding forces produced by the robot dynamics. In this way, any action or contact over any point of any moving part of the robot can produce a reaction to stimulus, and therefore, it is possible to plan the adequate reaction strategies.

In the system described the detection of the physical contact is performed by a force and torque sensor on the wrist as well as a sensitive cover. The sensitive cover consists of a set of sensitive surfaces based on four force sensors each, which are located on the four surface supports.

Human–robot interaction performed either through an applied force at the sensor wrist or by means of control based strategies from the force produced at each joint, allows the user to bring the robot to any desired configuration or to reach any position in space. Nevertheless, control with such interaction is limited to fix positions, forces and trajectories, without the possibility of transmitting other kind of orders such as stop, wait, repeat, etc., orders that can be given by means of oral communication. On the contrary, voice commands can not specify a trajectory or modify it adequately.

When remote interaction is more convenient, or is completely necessary, gestures detection and interpretation can be based on magnetic sensors or vision. From these data the body figure is modelled using different techniques, usually fusing several perception techniques.

8.3 Detection and Tracking of Humans in the Working Scenario

Many body detection and tracking systems rely on the use of visual marks attached to the body. For obvious reasons those marks facilitate the process of localization of some body parts in the scene. Marks or beacons can be used; marks are usually passive devices, for instance, reflecting spheres or colored dresses. Beacons are referred as emitting devices, like camera-synchronized light emitting diodes. Both kind of systems are uncomfortable [8], they generally need to be connected through wires and are very sensitive to their placement and surrounding conditions (i.e. when using magnetic trackers in the vicinity of ferromagnetic materials).

The problem of modeling the human body, directly from images, is divided, mainly, into three different and independent parts. The first problem to solve is the extraction of the human figure from 2D images. Background subtraction is one of the most frequently used techniques for that purpose, and the study of its efficient implementation has been tackled by many authors [9,10,11]. Furthermore, quick illumination changes, shadows and slight movements of the background objects constitute additional problems to be solved. Therefore, a posterior region based analysis with specialized techniques like those presented in [12] must be applied, where images are composed by color or texture regions that are tracked along a video stream.

Secondly, the generation of the 3D human model must be extracted from the information obtained from 2D segmentation. Depending on the segmentation technique used, the results will be based on background occlusion or movement detection. In [13] segmented figures are extracted from color analysis. A process for the 3D reconstruction, in a multi camera system is described in [14, 15] and an optimal way to solve the Shape from Silhouette problem is explained in [16], where an octree structure is used for the generation and representation of the reconstructed image.

Finally, from the 3D information extracted in the reconstruction process, a human kinematics model must be adjusted. Recently, many efforts have been devoted to advance in this direction. In [13, 17] a simple model is adjusted by means of local optimization techniques to find the head and extract the main pose of the body. A more accurate body model is adjusted to the segmented body figure in [18, 19, 20].

In order to acquire the operator movements, with high resolution and in real time, a procedure that integrates two different techniques for human body extraction has been designed.

8.4 Recursive Interactive Method for Body Modeling

The method proposed for modeling the human body is based on the combination of two different reconstruction methods, using each of them a different segmentation technique. First, direct triangulation of 2D singular points obtained from movement detection is a method that provides high accuracy, although it is difficult to understand the spatial distribution of the objects just from the information obtained. On the other hand, *Shape from Silhouette* (SFS) constitutes an easy way of combining the information given by N cameras and generating that spatial distribution. However, by using algorithms based on SFS reconstruction, it is computationally impossible to achieve the same accuracy provided by stereovision.

The fusion of both methods aims to obtain better and more reliable results in human modelling by integrating their capabilities and avoiding their drawbacks. Besides, both methods use complementary segmentation techniques: Background subtraction and frame to frame difference. Background subtraction provides a full silhouette of the human, but with possible errors due to background imperfect modelling. On the other hand, frame to frame difference is only useful when the person is moving, but greater accuracy is achievable.

The two reconstruction methods are briefly explained below.

8.4.1 Triangulation of Singular Points Obtained from 2D Movement Detection

The method tries to generate a model of the body with movement information extracted from sequences of 2D images and combining it with stereovision techniques. It is composed of the following steps:

(i) The procedure begins with 2D movement detection. The process segments the image from movement, from frame to frame image comparison using an adaptive acquisition time interval. The time interval is dynamically calculated from an estimation of the human movement speed. The results obtained provide information highly accurate, but with occlusions due to the lack of movement of some parts of the body, as can be seen in Fig. 8.1.

(ii) Secondly, singular points (points with significant local curvature) are extracted from the segmented image.

(iii) Afterwards a matching process is executed in order to locate the 3D position of singular points. Several tests are used to reject incoherent correspondences, based either on stereo or geometrical conditions (test that checks the coherence between the normals associated to the singular points).

(iv) Finally, the head and the arm tip are localized from the list of singular points extracted, based on a multiple hypothesis and fitting test scheme. All possible 3D postures, from those generated by the list of singular points, are tested in order to find the most coherent hypothesis.

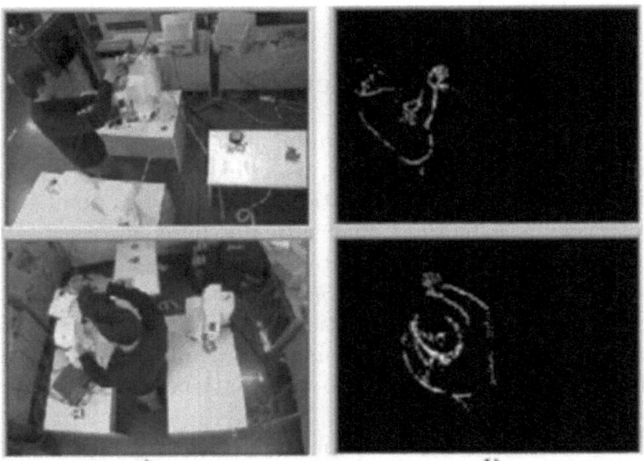

Fig. 8.1. Image segmentation obtained from movement: (**a**) Two simultaneous views of the scene; (**b**) Image difference results

8.4.2 Shape from Silhouette by Carving Reconstruction

SFS constitutes an easy way of combining the information provided by N cameras. The process generates the model of the segmented object by intersecting the volumes induced by the reprojection of the silhouettes obtained in each camera, as shown in Fig. 8.2.

(i) SFS begins with the extraction of the body silhouettes of the body from the images captured by the whole set of distributed cameras. In the present work the algorithm used for the segmentation of the silhouettes is based on background subtraction and consists of a 3-level analysis.
Pixel level: The foreground pixels are segmented from those in the background through direct comparison in the RGB space. The background

Fig. 8.2. Image segmentation obtained from background subtraction, carving reconstruction. (**a**)The segmented silhouettes taken from 4 different points of view; (**b**) The reconstructed 3D image.

model is continuously updated region by region.

Geometric Level: A combination of morphological filters is applied in order to eliminate small holes and dots due to noise and flickers.

Region Level: With a quick implementation of the split and merge algorithm the image is divided into homogeneous colour regions. Once the image is divided into regions it is decided whether each region belongs or not to the final segmented silhouette.

The algorithm is based on the combination of techniques presented in [9, 10, 11], where the study of an efficient and useful implementation of background subtraction is described.

(ii) Afterwards, the process of reprojecting the silhouettes and intersecting them to generate the volume is done by carving. In 1993 Szelinski [16] proposed an octree generation and representation of carving, which led to a quicker, and therefore much more useful, representation of 3D objects. This is the method selected for implementing carving in the present work. However, due to the need to achieve real time generation rates for 3D images, some optimization improvements were required and have been implemented:

Preprocessing of segmented binary Images: The preprocessing consists in the computation of the number of segmented pixels in a square around each pixel. This way, the computation of the segmented pixels inside a region is quicker.

Precomputing and Mapping the Geometrical Camera Model: Depending on the camera model used, the computation of projections and reprojections is really costly. Therefore, if still cameras are used, their geometrical model can be precomputed by mapping its value for each voxel of the camera workspace.

Table 8.1. Carving reconstruction rates for the proposed and the direct algorithms

Carving matrix resolution	Carving matrix generation rate *(direct algorithm)*	Carving matrix generation rate *(proposed algorithm)*	Generation rate for the final integrated model
50 x 50 x 50	1.7 fps	35 fps	22 fps
100 x 100 x 100	0.6 fps	25 fps	15 fps
200 x 200 x 200	0.08 fps	13 fps	7.5 fps

Thanks to this improvements, real time volume generation has been achieved. The results of the proposed algorithm are shown in Table 8.1, that compares the performance, in terms of generation rate, between the direct algorithm (decide for each voxel independently whether it is full or empty) or the proposed one (with optimizations) for obtaining the carving matrix. The last column shows the generation rate of the whole algorithm used for obtaining the integrated model.

(iii) Finally a policylindrical model of the human body is adjusted to the volume obtained from carving using an optimization algorithm. More information about this process can be reviewed in the work presented by Mikic et al. in [20], where a local optimization algorithm is used in conjunction with Kalman estimators in order to adjust a blob model of a human body.

8.4.3 Proposed Fusion Method for Body Modelling

The proposed method for body modelling is based on the implementation and interaction of the two above mentioned methods. Fig. 8.3 shows the diagram of the fusion procedure. The left column corresponds to the first process, the SFS implementation while the right column corresponds to the process of body reconstruction by stereovision. The points where the interaction between both methods is applied, is represented by the crosslink between the two columns. All the steps and crosslink in which the algorithm is divided are explained below.

(i) The images acquired from the multiple cameras placed around the working area are segmented via the two explained techniques: background subtraction and frame to frame image difference.

(ii) In the second stage the reconstruction of the 3D shape of the body takes place from the segmented silhouettes through background occlusion. This is done by applying carving reconstruction.

(iii) A rough body adjustment is applied over the "carving" volume. This primary adjustment is performed by the estimation of the main body axis and possible locations of the head and upper limbs, through global optimizations all over the carving matrix.

(iv) A process of singular points detection in the 2D frame to frame difference images is carried out. The singular points extraction, that consist in the location of tips in the extracted silhouette (assuming that to correspond to the head, arm tips and elbows), is assisted by the information provided by

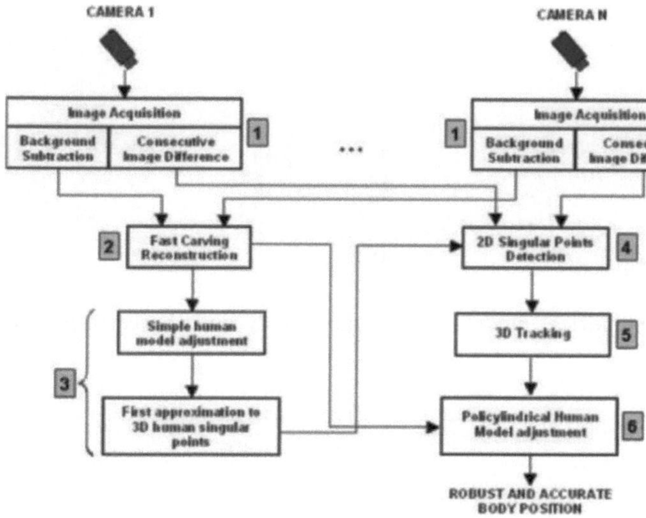

Fig. 8.3. Integration of the stereovision and SFS reconstruction methods

the above mentioned 3D human model fitted with the carving image. With that rough information, the singular points searching area can be limited in order to accelerate their location and avoid possible wrong triangulation results. This information also allows to detect those situations in which any specific body member has moved.

(v) Afterwards, the 3D singular points obtained from triangulation are filtered by a tracking process to improve stability and eliminate possible jitters.

(vi) In a further process, the obtained locations of the tip points are used for the final body adjustment. Some constrains are imposed in order to obtain better results. First of all, the position of the tip points must be coherent with the constraints that characterize a human figure. Also, a connectivity hypothesis along the arms, that has to be compatible with the volumetric data obtained from carving, is considered. After those restrictions, a policylindrical model of the human body is generated by means of local optimization techniques, using the primary data obtained from carving and the processed location of the main points and edges of the body. The model generated is shown in Fig. 8.4.

8.5 Human–Robot Cooperation Strategies

The implemented cooperative control system is based on a high level program that runs in parallel with the execution of the task. That is, a human can interact during the execution of a task changing the trajectories given by the master in teleoperation mode. Human–robot interaction is performed either from the measure of the forces and torques applied at the robot end effector or over the body,

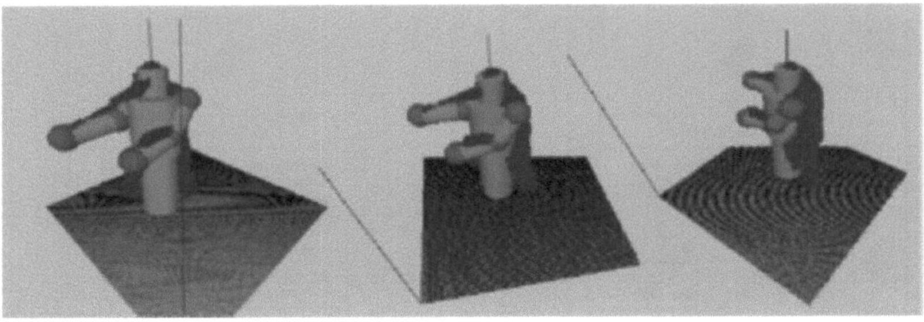

Fig. 8.4. Polycilindrical model of the human body and the corresponding carving volume

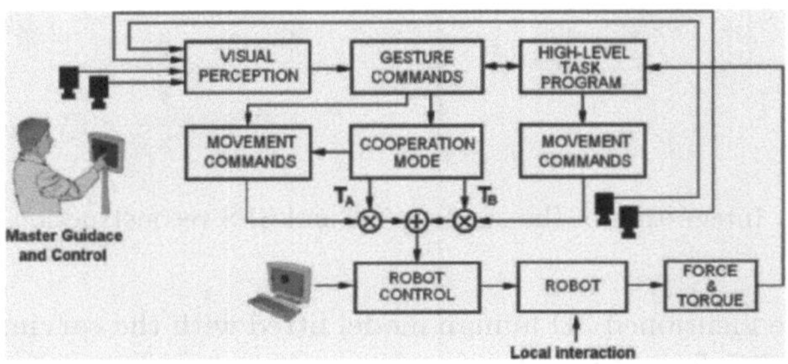

Fig. 8.5. Shared control scheme

or from the visual detection of the movements and interpretation of the gestures of a human operator located in the working area of the teleoperated robot. The structure of the shared control strategy is shown in the schema of Fig. 8.5.

Therefore, the high level task program module receives orders coming from two sides. From one side, it receives the force and torque vectors detected at the end-effector, which can be interpreted as data during the execution of a task or as orders for trajectory modification during the movement from a working point to another. The other orders correspond to the position of the hands of an operator, which is located remotely in front of a control screen or eventually around the teleoperated robot. These continuous positions of the hands are interpreted by means of the gesture commands interpretation module.

The cooperation mode module generates the balanced orders between a direct master–slave control and an assisted master–slave behaviour, giving a weight between 1 and 0 to each control input.

In the first case, the direct master–slave control, the cooperation mode module gives a weight 1 to the direct action T_A and a weight 0 to the assisted action T_B. Instead, when the direct action has to be influenced by the orders given by an external agent, either through the force applied to the end-effector or through the visual detection of this agent by the visual gesture interpretation module, the weighting factors T_A and T_B can swing towards increasing values of T_B, up to the limit that $T_A = 0$ and $T_B = 1$. In this later case, the orders given by the operator close to the robot prevails over those given by the remote master.

The cooperative module operates in two different modes to generate the weights T_A and T_B:

- From explicit commands given by the human that teleoperates the robot. In this mode, the operator can decide the admissible cooperation level from a menu visualized on the control screen. The selection can range from null, scarcely, moderate and intense to totally based on the commands given by an operator close to the robot itself.
- Modifying in an automatic way the weights T_A and T_B in the state of cooperative control. The value of the weights is decided automatically as a function of the analysis of the information received from the force and torque sensor and from the gesture command interpretation module.

For the second mode, an automatic strategy generator has been developed for this experimentation phase. It is based on the "risk" or on the need of a priority intervention. In the first work developed in this direction, the interaction is considered necessary, and thus the value of T_B increases, when the operator touches the robot with higher speed movements. On the contrary, a slow operator movement is considered as a lower priority action, and therefore, the weight of T_B decreases.

In the trials carried out with interaction weights $T_A = 0,5$ and $T_B = 0,5$ the interaction from gestures modifies the trajectory indicated by the operator that performs a master–slave manipulation without interrupting its execution. Instead, when a sudden access over the robot occurs, the present system interprets that this action is motivated by a risky situation, and consequently the weight of T_B increases up to values close to 1.

The interpretation of the degree of interaction priority between the local and the remote control based on the force and torque sensor relies on the intensity and time of application of the effort. Therefore, the action of forces applied produce an interpretation based on proportional–integral criteria, while gestures orders produce actions based on derivative criteria.

When, according to the task and environment conditions it is convenient that the direct orders are influenced by the orders given by an external agent, the weighting factors T_A and T_B can swing towards increasing weights of T_B, up to the limit that T_A takes the value 0 and $T_B = 1$. In this later situation the control orders given by the person that is around the robot working area prevail over those produced at the master control zone. The direct orders given by an external agent can be extracted from the forces that are applied over the robot end-effector, or through the presence of the agent that is tracked by the visual perception modules.

When a task has some physical constraints they have to be considered in the control mode. To clarify this situation, consider the case of a polishing task over a horizontal surface. In this application there is a vertical force restriction ($Fz = \text{cte}$) and additional horizontal position restrictions (X and Y tool robot position are a function of time specified by the high level program task module). Then, all user displacements from the initial position are considered as a force application command in the restricted direction (vertical direction in the example). The other directions are considered as position commands and treated as explained before.

8.6 Potential Applications

The possibility of interpreting the movements or gestures of a human body can be exploited in different ways in teleoperation. Taking into account the cooperation between the human operator (M) and the telerobot (S), four types of control hierarchies are considered: the usual master–slave hierarchy, cooperative work, remote human supervision and remote human supervision with a local interaction. The first kind of control constitutes the classical open loop system, from the functional point of view, Fig. 8.6a. In the second case, the cooperative work admits a local interaction with a second operator (H), Fig. 8.6b. In the third case, the control with the master is weak and limited to the supervision of tasks that are performed mainly in an autonomous way, Fig. 8.6c. The fourth control level considered consists in a supervised cooperative structure, Fig. 8.6d. In this case an autonomous remote system cooperates with a close agent in the execution of the task.

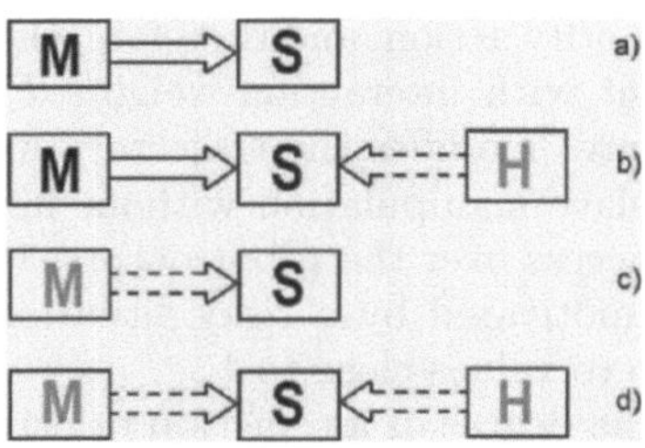

Fig. 8.6. Hierarchical control levels

Three basic tasks are also considered: guiding, operating and manipulation. Looking at the corresponding twelve crossing relationships between cooperation levels and tasks, a suitability Table, Fig. 8.7, shows the evaluation from the experimentation with the developed system.

The use of the developed virtual exoskeleton in the master side provides the user with a more natural interface than those currently used in teleoperation, nevertheless, the limited precision of vision based gesture commands have to be considered. Thus, for applications in a master/slave hierarchy, gesture commands are not very suitable (NS) due to the current lack of precision of vision systems and the absence of force feedback. Further advances in computer vision and artificial intelligence will lead to more precise and reliable systems that would make mimetic behaviour possible. In this kind of hierarchy, the remote robot would translate rough actions of the user to smooth and environment dependant movements, giving place to a very suitable (V.S) gesture based commanding system.

Hierarchy \ Task	Guiding	Operating	Manipulation
Master/Slave	N.S.	N.S.	N.S.
Cooperative	V.S.	S.	P.S.
Supervision	P.S.	P.S.	P.S.
Supervised Cooperative	S.	S.	P.S.

Fig. 8.7. Suitability space for the command mode/kind of task, representation. Non suitable (N.S.), Partially Suitable (P.S.), Suitable (S.), Very Suitable (V.S).

In a cooperative hierarchy, a guiding system based on gesture commands is very suitable (VS) due to its simplicity from the user's point of view. The system can be considered, suitable or partially suitable in operating and manipulation tasks respectively. The reason of the decreasing suitability is the lack of precise and reliable fingers perception, being this limitation more evident in manipulation due to the higher manoeuvrability and force feedback demands. In a supervision hierarchy, the system is considered partially suitable (P.S.) due to the existence of other alternative interfaces such as oral commands or simply push buttons. Finally, in the supervised cooperative mode the above mentioned suitability in the cooperative mode still prevails, but at a lower level due to the rigidity of the previously programmed task.

The system is also analyzed from the point of view of the remote-robot mobility and task complexity. According to the robot mobility performances, three main scenarios can be considered: robots with few degrees of freedom, robotic arms, and complex articulated robotic systems. The term complex articulated system refers here to redundant robots, multi-arm systems or free-floating robots.

As more complex the robot and the task are, the more challenging is the robot control and the human interaction becomes more necessary. Fig. 8.8, gives a view of the kind of scenarios considered, showing in dark grey the kind of applications already explored. The lighter grey areas represent the fuzzy frontiers for gesture interfaces, in which the human interaction is more important, the challenges for the future work.

The system has been experimented in different scenarios. In a first situation, as shown in Fig. 8.9a, a robot presents a collision risk with an object in the scene. In this case, the presence of the hand of an operator produces a modification of the robot trajectory so as to avoid damages in the working environment. A second situation has been tested in an experimental robotized kitchen for the disabled. In this case, an assistant can help a disabled user in the gesture based control of the robot, Fig. 8.9b. Finally, Fig. 8.10 shows a task that requires the manual interaction of an operator that would allow the adjustment of the robot

Task / Ability	Guiding	Operating	Manipulation
Few Degrees of Freedom Robot Systems	Guided tasks over fixed simple structures (Welding, machining..)	Polishing, simple cuts...	Sample picking, rubbish
Robotic Arms	Guided tasks over complex structures (Welding, sealing, gluing...)	Machining Surgery...	Biological samples flexible tissues...
Complex Articulated Robotic Systems	Exploration robots, pipe inspection...	Rescue, oil-ship crash repairing or maintenance...	Bomb deactivation aerospace applications...

Fig. 8.8. Present frontiers of the gesture interface: dark explored, clear unexplored

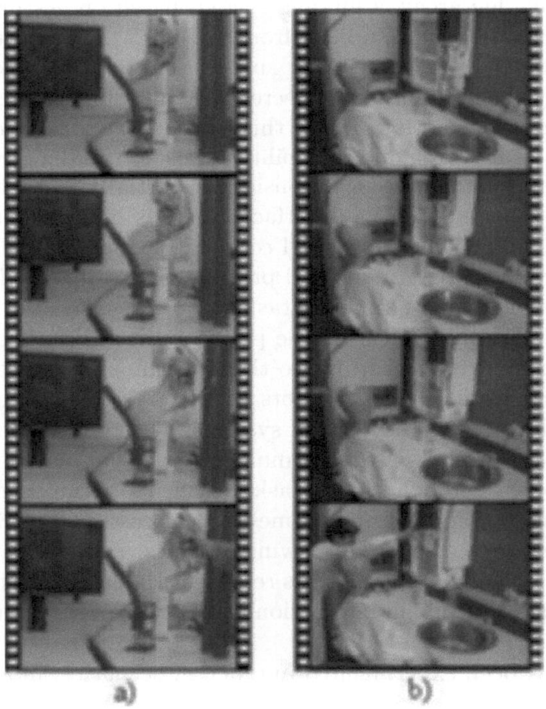

Fig. 8.9. Trajectory modification from visual interpretation of human intention

task to the real conditions. In this case during a polishing task the robot can be steered to a specific point where due to an incidence the task requires a more intense polishing action (i.e. an irregularity in the polished area).

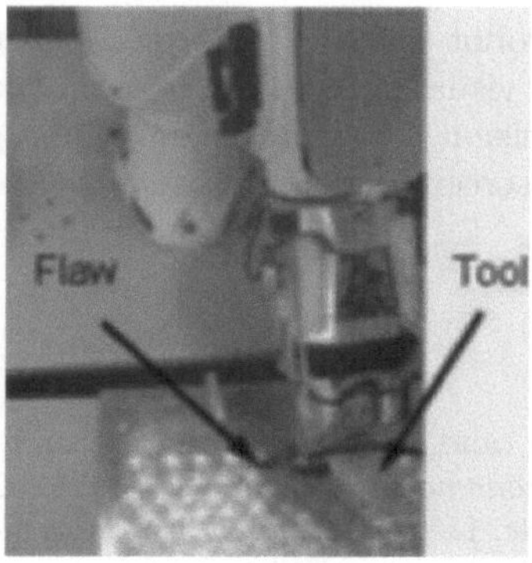

Fig. 8.10. Example of the need of a manual interaction for better finishing a task

In all these cases the possibility of detecting, visually or by touch, the presence of a human and interpret his intention makes possible a better interaction that increases the flexibility and safety in the execution of tasks.

A potential application area of the described gesture based interface for teleoperation could be that of processes in workstations which operation tasks involve multiple robots. In such applications, a previously programmed operation of all the robots guarantees the inexistence of collisions. However, such safe operation mode implies that the rhythm of execution for each robot is adapted to the worse situation so as to assure the synchronism between all the trajectories involved in the execution of the cooperative task. For instance, some machining operations need an undefined execution time depending on the material hardness, irregularities, etc., and a preprogrammed task requires taking the longer period of time possible, while sensor based control would reduce a robot execution time but at the expense of making the coordination between robots more difficult.

On the contrary, using a strategy tolerant to external interactions, the task of each robot can be programmed in a more optimal way, executing their task as a function of their own sensors (efforts, vibrations or dimensions). Operating this way, the operation times are variable but less than those resulting in a preprogrammed application. In this case, the asynchronism between robots that are working very close one to each other (welding car chassis,...) presents collision risks that requires a modification of the robots trajectories. In the mentioned application the interaction is between robots, but the method can be extended to other kind of interactions.

In the works carried out and for each robot run, the capability of absorbing interactions that modify its trajectory has been defined. This degree of tolerance has been established as a function of the distance to the destination, or working points, so as to guarantee the correct execution of the task. In case of conflict between two robots one yields the pass to the one that navigates between fix working points or to the one that is farer to its destination. This procedure yields to execution times significantly lower. In its application to surgery, for instance, the interaction with the robot can be performed through the actions of the surgeon hand, by means of visual perception, or from the force applied over the robot, or when higher precision is required, controlling geometrical constraints graphically defined over a screen, which are converted to virtual obstacles over the working scene.

References

1. T. Fong, I. Nourbakhsh, and K. Dautenhahn. A survey of socially interactive robots. *Robotics and Autonomous Systems*, 42:143–166, 2003.
2. A. Billard and M. Mataric. Learning human arm movements bi imitation: evaluation of a biologically inspired connectionist architecture. *Robotics and Autonomous Systems*, 37:1–16, 2001.
3. M. Frigola, J. Poyatos, A. Casals, and J. Amat. Improving robot programming flexibility through physical human-robot interaction. In *IROS Workshop on Robot Programming by Demonstration*, Las Vegas, USA, 2003.
4. J. Amat, A. Casals, and M. Frigola. Human Body Acquisition and Modeling for Persons - Humanoid Robots Cooperation. In *The Third IARP Workshop on Humanoid and Human Friendly Robotics*, pages 76–78, 2002.
5. J. Hwang, R. Arkin, and D. Kwon. Mobile robots at your fingertip: Bezier curve on-line trajectory generation for supervisory control. In *IROS Workshop on robot programming by demonstration*, Las Vegas, USA, 2003.
6. W. Son, K. Kim, B. Jang, and B. Choi. Interactive dynamic simulation schemes for articulated bodies through haptic interface. *ETRI Journal*, 25:25–33, 2003.
7. O. Khatib, et al. Human-centered robotics and interactive haptic simulation. *The International Journal of Robotics Research*, 2004.
8. T. Molet, R. Boulic, and D. Thalmann. A real-time anatomical converter for human motion capture. In *Proc. 7th Eurographics Workshop on Animation and Simulation* Springer Verlag, Wien, 1996.
9. A. Franois and G. Medioni. Adaptive colour background modelling for real-time segmentation of video streams. Technical Report, University of South California.
10. O. Javed, K. Shafique, and M. Shah. A hierarchical approah to robust background subtraction using color and gradient information. In *Workshop on Motion and Video Computing*, IEEE, pages 22–27, 2002.
11. T. Horprasert, D. Harwood, and L. Davis. A statistical approach for real-time robust background subtraction and shadow detection. In *IEEE ICCV Frame-rate Workshop*, 1999.
12. P. Salembier and F. Marqus. Region-based representations of image and video: Segmentation tools for multimedia services. *IEEE Transactions on circuits and Systems for video technology*, 9:1147–1167, 1999.
13. S. Cheng and M. Trivedi. Human posture estimation using voxel data for "'smart"' airbag systems: Issues and Framework. In *Intelligent Vehicles Symposium*, pages 84–99, 2004.
14. K. Cheung, T. Kanade, and J. Bouguet. A real time system for robust 3D voxel reconstruction of human motions. In *Computer Vision and Pattern Recognition Conference*, 2:714–720, 2000.
15. J. Buades and F. Perales. Voxel matching reconstruction in real image sequences of human avatars. In *1st Ibero-American Symposium in Computer Graphics* Guimaraes, Portugal, 2002.

16. R. Szeliski. Real-time octree generation from rotation objects. Technical Report, Cambridge Research Laboratory. 1990.
17. I. Mikic, M. Trivedi, E. Hunter, and P. Cosman. Articulated body posture estimation from multi-camera voxel data. *Computer Vision and Pattern Recognition*, 2001.
18. F. Caillette and T. Howard. Real-time markerless human body tracking with multi-view 3D voxel reconstruction. In *British Machine Vision Conference*, 2:597–606, 2004.
19. F. Caillette and T. Howard. Real-time markerless human body tracking using colored voxels and 3D blobs. In *ACM International Symposium on Mixed and Augmented Reality*, pages 266–267, 2004.
20. I. Mikic, M. Trivedi, E. Hunter, and P. Cosman. Human body model acquisition and tracking using voxel data. *International Journal of Computer Vision*, 53(3):199-223, 2003.

9

Enhanced Teleoperation Through Virtual Reality Techniques

Abderrahmane Kheddar, Ee-Sian Neo, Riichiro Tadakuma, and Kazuhito Yokoi

AIST/CNRS Joint Japanese-French Robotics Laboratory
Intelligent Systems Research Institute
AIST Central 2, Umezono 1-1-1, 305-305-8568, Tsukuba, Japan
kheddar@ieee.org, {r-tadakuma,rio.neo,kazuhito.yokoi}@aist.go.jp

Summary. Teleoperation effectiveness was almost stagnant during several decades since the discovery of the master-slave structure by Goertz in the 50s; in spite of some technical improvements and computer developments [1]. Recent advances on virtual reality (VR) techniques can be considered as the additional *ingredient* to the total telerobotic renewal, and novel architectures could be developed. Its efficiency is reaching now a considerable level of achievements. It allows a clear improvement in traditional applications such as nuclear or space activities and it also opens many new fields of application, where teleoperation represents a major percentage of robotic development. Build on previous state-of-the-art reports [2, 3, 4], this chapter reviews up-to-date achievements applying virtual reality techniques to teleoperation: we recall some relevant achievements of VR in solving difficult teleoperation problems such as time delay, operator assistance and sharing robot autonomy by combining different supervision strategies or allowing new human-centred teleoperation schemes. The chapter also discusses new robotic applications that have currently appeared which require additional research efforts and call additional investigations on virtual reality techniques. Among them we can find micro- and nano-teleoperation eventually of livings such us cells and DNA molecules, teleoperation of humanoids and animaloids, teleoperation of unmanned air or terrestrial vehicles, hi-fidelity telepresence, multi-operator multi-robots teleoperation, etc. These are exemplified along with their specific challenges.

Virtual reality techniques paved a path in telerobotics thanks to its potential in solving classical problems of early master-slave bilateral coupling schemes. Fortunately, it has been surprisingly noticed that these techniques can offer different solutions in solving a certain problem. For instance, teleoperation time delay has been solved using either predictive displays or teleprogramming; both methods are devised from VR techniques. Moreover, there are several styles and approaches to implement each method. This variety of VR applications makes difficult to infer standard VR-assisted teleoperation architecture. Nevertheless, this difficulty constitutes paradoxically a main strength and cleverness. Next section is therefore focused -without being exhaustive- on the soundest ideas to enhance teleoperation through VR techniques.

M. Ferre et al. (Eds.): Advances in Telerobotics, STAR 31, pp. 139–159, 2007.
springerlink.com © Springer-Verlag Berlin Heidelberg 2007

9.1 Augmented Reality Teleoperators

Augmented reality (AR) techniques have been used in teleoperation in order to search for new solutions to round-trip time-delayed teleoperation. Bilateral force reflection becomes indeed easily unstable in the presence of time-delay. Solutions based on predictive graphic displays break the force feedback loop by enhancing visual feedback and superimposing a virtual representation of the slave robot. This virtual representation is used in a sequential programming/confirming procedure.

When a virtual robot is operated, subsequent motion is displayed in real-time to the operator on top of the video-feedback. Yet, manipulating this virtual robot has no direct effect on the real one: it only acts as a visual predictor. Once the motion is made, replayed as much as needed and confirmed by the operator, the virtual robot controls the real one in a replay scenario where the operator cannot act. Thus, direct manipulation of a virtual robot and a real one are exclusive. Predictive graphic display development was pioneered by Noyes and Sheridan [5]. Few years later predictive display method was improved by Bejczy et al. (1990) [6], see Fig. 9.1.

Fig. 9.1. Predictive display through augmented reality techniques for space teleoperation (Courtesy from Kim, The Jet Propulsion Laboaratory, NASA)

The AR-predictive display method has been recently democratized in many Internet-based teleoperation technologies. Current applications concern: multimedia, tele-surveillance, tele-diagnosis, teleexploration, tele-supervision, etc. In these cases, the principle is to allow human operators to control a remote system which will achieve desired (pre)programmed tasks. Some of these tasks may imply a modification of the environment (physical teleworking) or not (soft teleworking). In all cases, operator needs to be provided with feedback information while sending control commands. Because of time-variability of the delay in Internet, most Internet-based teleoperators offer predictive display functions. They are implemented in different ways. In addition, AR allows further functionalities. For instance, sophisticated mixed rendering can be generated by adequately associating computer synthetic pictures and live-video images and allow enriched feedback such as:

- Symbolic entities for operator assistance or guiding like virtual tools or fixtures; these can also be used in virtual environments [7];
- Additional information such as displaying forbidden areas, robot workspace, specific sensors data, planned motion path, alert data, etc.;
- On-demand predictive simulations of past, ongoing or future tasks;
- Virtual object-models allowing operator to position them on the real one, so that the computer can automatically determine its position into the space using calibration methods;
- Better and comfortable feedback through enhancing and totally removing unnecessary feedback information.

The Fig. 9.2 illustrates an example of a teleworking system in which augmented reality technique is used in superimposing both robot's virtual model

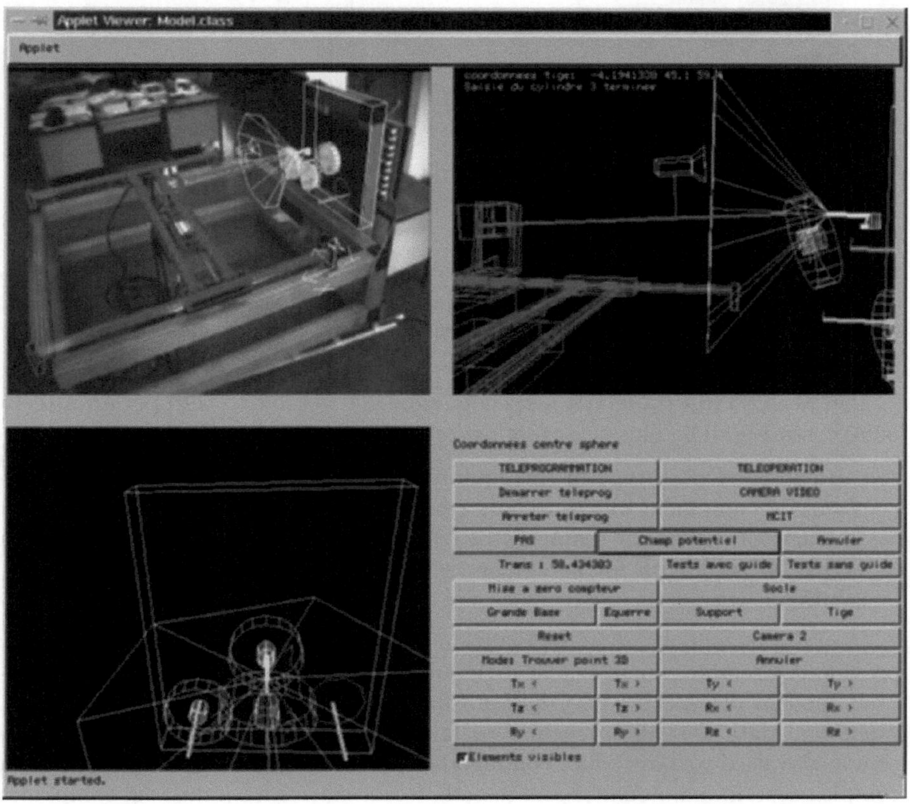

Fig. 9.2. ARITI web based teleoperation interface using augmented reality fixtures for operator assistance

and active virtual guides. Active virtual guides assist the operator or the robot to perform a given task [8, 9]. For example, virtual guide is an attractive potential field, which has been shaped into a cone and attached to the object to be picked. The robot handle (a needle) reaches the spot; the potential field attracts operator actions, and subsequently the needle is moved toward the goal. A tool to generate and dimension various virtual guides is available; different operational control modes have been also implemented.

In most similar teleoperators, control interfaces use Java Applets with different available command buttons allowing any user, connected to Internet, to experience actual remote control of tele-robots. The operator will see the result of her/his actions through a continuous video feedback.

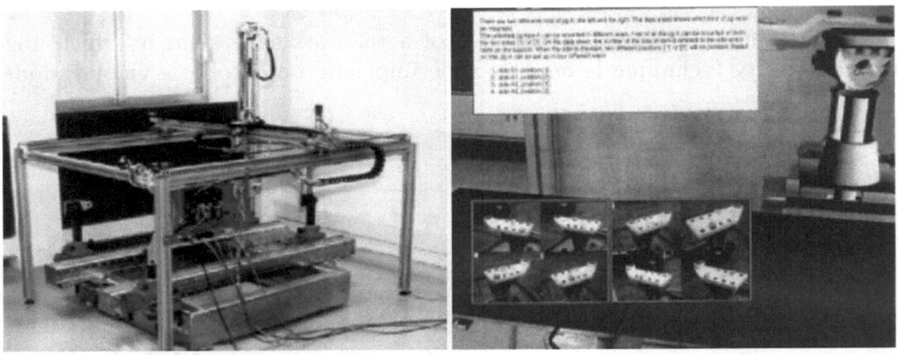

Fig. 9.3. Tele-training/tele-maintenance system: a XYZ table plus yaw pitch camera to explore maintenance equipment (left). Output from the MAESTRO, answer to query: "what are the different jig types" - using augmented reality techniques, (right).

The MAESTRO project is a specific industry web-based teleoperation technology, see Fig. 9.3. This tele-training/tele-maintenance system is made up of:

(i) an extended prototype of the equipment to be maintained, the target customized equipment allows remote control and extensive diagnostic;
(ii) a Cartesian telerobot manipulator embedding a camera and external vision systems, used both for equipment servicing and enhancing user interoperability.

Marking techniques allow registration of virtual models out of video images provided by remote cameras. The virtual model is linked to a database which contains multimedia specific information (operating and maintenance instructions, functional data describing regular or faulty conditions, etc.). Speech recognition is also used to allow voice commands. This development is devoted to train complex system maintenance and installation scenarios for industrial companies which cannot afford on-site training equipment [10].

AR techniques play also a relevant role in enhancing visual feedback. It is carried out by highlighting objects into scenarios with poor visibility or when

Fig. 9.4. Vision feedback with enhanced virtual corridor for unmanned vehicle telepiloting: different environments (courtesy of French DGA)

additional information or sensor data need to be displayed. There are many situations in which such a feedback is needed. For example, in military applications, unmanned rovers and air vehicles evolve in natural or urban environments which are unknown, largely unstructured and hostile. In these cases quick mobility and discretion are two important constraints which prohibit the use of extensive communication and data transmission. Some studies show that unmanned remotely guided vehicles will improve their efficiency if the video feedback were provided with AR interfaces. Here, AR features must be conceived to enable or increase operator perception and apply ongoing task strategies. Teleoperation efficiency is even increased by the possibility to control operators' vision field enrichment, specially superimposed graphic elements on the video feedback. These elements can be known a priori: e.g. virtual navigation corridor (see Fig. 9.4), virtual obstacles processed from numerical terrain or from a on-board unmanned vehicle system (a new obstacle or enemy's vehicle).

Enhancing feedback information is also necessary in teleoperators using virtual environments. The virtual TV metaphor devised in the COSIMIR projective reality concept ([15]) represents an example in which full operator immersion in virtual environment prohibits the use of real feedback. Therefore, the virtual environment needs to be augmented by actual feedback without asking the operator to switch the interface systems, see Fig. 9.5.

Fig. 9.5. Enhancing full virtual immersion with real feedback: TV metaphor in COSIMIR telerobotic system

Most introduced concepts could evolve and extend themselves to other application fields such as minimally invasive surgery, arts... The above mentioned examples, although notably illustrative, are far from being exhaustive regarding AR and modern teleoperators interconnection. Yet, apart from their numerous advantages, some problems using AR predictive display and assistance techniques may appear:

- It is not easy to associate force feedback (even a synthetic one) to AR predictors;
- Virtual objects's occlusion: it requires depth maps which are difficult to obtain;
- Accuracy of video/synthetic objects calibration: it is a well known problem in the vision community and some robust solutions can be found. However, accurate 3D-models of the robot or objects are not always available;
- Stereo-vision feedback introduces specific problems and makes the two previous issues more complex;
- Predictive displays loose efficiency when robot motions are performed in the image plan (normal) direction;
- Predictive displays methods are not efficient in precise manipulations within the range of calibration errors.

As a summary, augmented reality is still a challenging area of research in which considerable advances are being made. Recent contributions in real-time on-line tracking/calibration and enhanced realism superposition have a direct impact on robotics (visual servoying) and more particularly on telerobotics [11,12].

9.2 Full Virtual Environments Teleoperators

Although predictive displays enhance operator performance, they are still based on a move-and-wait strategy. Therefore, if t_{normal} is the 'normal' total time to perform a given task without time delay, its realization using predictive display technology requires:

$$t_{\text{predictive display}} = \sum_{i=1}^{N} n_i t_i + 2N t_{\text{transmission}} + t_{\text{normal}}$$

where N is the number of the necessary subtasks, i labels a given subtask, n_i is the number of predictive trials, t_i represents the mean time for performing one trial of the i-th subtask, and $t_{\text{transmission}}$ is a trip transmission delay. It is necessary to get a feedback before starting a new movement (i.e. matching predictive display and video image).

In fact, predictive display time, $t_{\text{predictive display}}$, is much more important if the lack between force feedback and unsuccessful trials is taken into account. Force display is not able since only a robot model is available. This time is not acceptable in many applications, and a more elegant approach was proposed: *teleprogramming*.

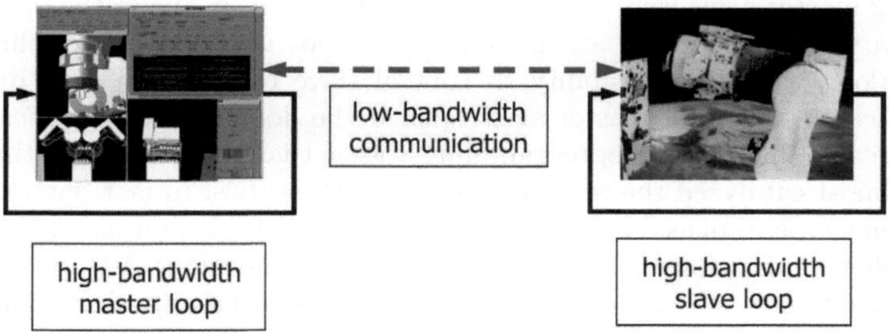

Fig. 9.6. A teleprogramming global view (Images are a courtesy of NASDA ETS7)

As shown in the Fig. 9.6, teleprogramming principle is nearly similar to a predictive teleoperation one but using a full virtual representation of the remote workspace which includes a virtual representation of the robot [13]. Bilateral coupling is achieved by a master arm (or any control device) and a slave robot interacting into the virtual environment (VE). In terms of performance, teleprogramming differs from predictive displays in the following points:

- Using a VE offers additional functionalities which are not available at the direct real feedback. For instance, possibilities to change viewpoints, navigate in scenarios, modulate force feedback, etc.;
- Robot instructions are symbolic (guarded motions) and they integrate model uncertainties;
- Remote location feedback to the master site may concern only the execution status;

If t_{normal} is the execution time of a given task without time delay, total execution using teleprogramming techniques is:

$$t_{\text{teleprogramming}} = \left(1 + 2\frac{1}{n}\frac{t_{\text{transmission}}}{t}\right) t_{\text{normal}}$$

this equation suggests that the executed task by the operator produces n instructions per unit of time on average. Teleprogramming architectures try to obtain $nt \gg t_{\text{transmission}}$.

Many variances were proposed by robotic researches. Hirzinger's team at the German Aerospace Center (DLR) conceived a telesensor-programming mode which is nearly a teaching-by-showing technique as applied to robot off-line programming [14]. In the proposed tele-sensor-programming, complex tasks are split-up into elemental moves keeping a certain constraint frame and a simulated sensor type configuration. This concept has been experimented and validated in many actual space telerobotic applications (ROTEX and ETS7 programs) and, more recently, by the Chinese space robotic program.

Current specific and successful implementations of teleprogramming give green light to many new applications. Teleprogramming is based on responsive VE representation of the remote environment which also means that force reflecting master arms were coupled to the simulation. It allows the operator to control a virtual robot like a controlled real robot via bilateral coupling. This option along with the possibility to control slave robots with very dissimilar kinematical structure of master arms opened the door to many applications of virtual reality haptics. Teleprogramming was actually the true starting point which almost catalyzed the research basis on VR haptics. In fact, virtual robots have been simply replaced by different objects and virtual environment simulates many other applications (e.g. interactive surgery simulation). VR gives a broad perspective to new teleoperation architectures and control paradigms which have been successfully used in real novel fields.

There are many reasons for the need of an entire virtual environment representation in teleoperation:

- Absence of data from remote environment when objects are completely unknown; tasks are not known in advance. This is the case in Mars exploration, or in a rescue mission using telerobotic systems after a disaster;
- Impossibility to relay force feedback information during task execution, for example a blind manipulation. It is the case of nano-scale manipulation by using scanning microscopy as teleoperator;
- A database of predefined skills, procedures to be used in structured or semi-structured environments, it can be slightly adapted for a given mission. In this case the VE plays the role of a teleoperation mission in which preparation, planning, programming and training takes place before execution. It is used for maintenance missions planning of nuclear power plants, or planning/assisting minimally invasive surgery operations;
- Possibility to conceive user-friendly teleoperation interfaces such as in many new teleoperation concepts like the hidden robot one [9] or the projective virtual reality [15], novel computer aided tools, etc.

These added values and settlements are highlighted by an illustrative case study, hereafter introduced. Considering a teleoperation mission where the remote robotic system must perform some tasks under the following conditions:

- Structure and properties of the remote environment are unknown;
- Task cannot be planned in advance; mission is clearly established but not the execution sequence of the basic tasks (we can even consider extreme cases where basic tasks are not known);

These situations are likely to be found in applications such as telerobotic space exploration and telerobotic rescue missions. Experts in rescue or space exploration need to be informed progressively about the environment status in order to elaborate an appropriate sequence plan of tasks. Therefore, a virtual environment is very useful (i) to provide the expert operator with full knowledge of the situation and (ii) to simulate/plan task sequences to achieve the mission target. However, it is necessary to implement task planning and virtual environment simultaneously; in the same spirit as SLAM (simultaneous localization and mapping). This technique is used for robotic research, since the remote environment is not known and no a priori sensors or systems are present to provide part of the representation information. This type of teleoperation mission can be performed following these steps:

(i) Telerobot is located at an initial state in a given environment location;
(ii) Telerobot collects maximum data to make an initial virtual representation of the environment;
(iii) Mission loop begins
- it plans task sequence on the basis of available information and actual VE;
- it simulates and tunes planed tasks (off-line play of the sequence);
- it confirms plan and generate actual command stream;
- it sends command stream to the telerobot;
- telerobot executes commands while collecting and sending sensors data;
- operator's virtual reality program collects feedback sensory data, processes data and:
  a) displays information results to the operator;
  b) enriches and updates remote environment representation;
  c) compares planned actions with the executed ones (playback);

(iv) Mission ends

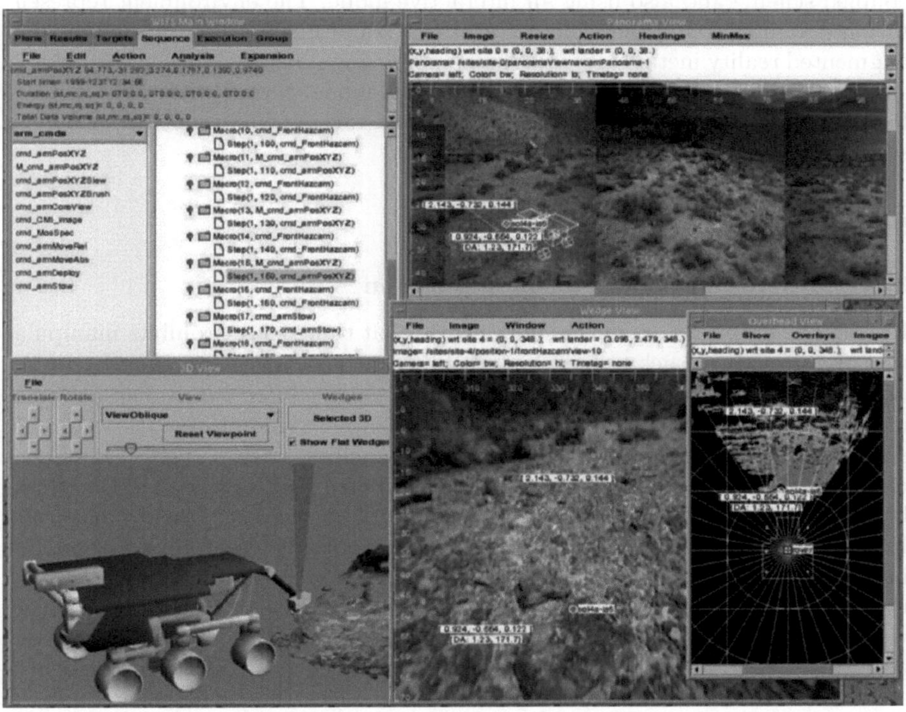

Fig. 9.7. Progressive virtual representation building and mission tasks sequence planning for telerobots (courtesy of the NASA, Sojourner Mars Path Mission)

An illustrative example of this architecture is the Mars Path Mission in which a graphical user interface supplies numerous available rover commands i.e. macro-operations and the corresponding parameters such as: Calibrate Heading with Sun, Capture Images, Drive, GoToWayPoint, etc. Off-line, a 3D terrain model is processed from: (a) previous mission accumulated stereo images obtained from the embedded IMR camera (a stereoscopic imager); and (b) the partial constructed VE from previous missions. Using a simple SpaceBall as an I/O device together with a stereo rendering of the virtual working environment and a virtual model of the Sojourner rover. The operator designs, therefore, plans actions to be achieved by a set of graphic programming metaphors. When the simulation task sequence plan is considered to be satisfactory, then the associated control code is generated and transmitted to the remote robot, see figure 7. This example can be applied to a telerobotic rescue mission and become even more interesting when telerobots of different shapes, functionalities and embedded sensors collaborate together to provide up-to-date information and to build up and up-date virtual representations. We should refer to [16] for the up-to-date developments of the rover sequencing and visualization program (RSVP).

Note that when the communication media allows high bandwidth –time delay being negligible– previous architecture allows also task execution in a bilateral coupled scheme and also using an interactive mode. The environment representation is built in real-time, whereas feedback could made up of combinations of augmented reality metaphors. It is then possible to implement various autonomy-shared control strategies through the switching mode. Nevertheless, this method becomes less efficient in the presence of quickly varying environments such as in some sub-seas exploration/maintenance tasks.

9.3 New Challenges

9.3.1 Micro- and Nano-telemanipulation

Previous telerobotics examples show clearly that time delay prohibits manipulation and direct feedback to be achieved simultaneously. However, there are other applications in which direct manipulation and feedback are not available despite high bandwidth communication media; for example, nano-teleoperation through scanning microscopy. Up-to-date, Scanning Tunnelling Microscopy (STM), Scanning Electron Microscopy (SEM) and Atomic Force Microscopy (AFM) seem to be common tools for scanning and manipulating at the micro- and nano-scale. Each of these techniques has specific limitations and fields of applications. For instance, biological samples such as cells can not be visualized using a SEM since their electrical properties can be altered. From the basic rendering point-of-view, vision feedback through virtualized representation can bring considerable contributions. Field-of-view (or scanning) is, indeed, in most cases restricted to a small area. Distances between objects and lens or to the probe are very short. If we consider AFM example, scanning –which works with seconds or minutes order– does not allow real-time imaging. On the other hand, since the same single probe is used for both functions: scanning and manipulating, they can not be achieved parallelly whatever scanning speed be. Moreover, operations are generally executed within the field-of-view. For those reasons, a 3D virtual environment topology can be built and displayed to the user. Remote features may be augmented with multiple contrastings of colors in order to present data in a comprehensive and easy way.

Telemanipulation at the micro- and the nano-scale worlds raises interesting problems and new challenges that can be efficiently overcome through virtual and augmented reality techniques according to advanced human-machine interfaces. To address challenges of micro- and nano-telemanipulation, problems regarding human interfaces need to be correctly defined. Human operator needs to be *immersed* in a world at a different scale, and do not obey to similar physics. This is one of the most challenging issues. Indeed, remote manipulations are envisaged at scales for which technologies did not mature yet. An automation degree to execute micro- and nano-tasks through off-line programming of micro- and nano-robots is required. These robots also need to be conceived and built. On the other hand, researches on micro- and nano-technologies require experimental tools to provide significant data. Therefore, scientists and engineers have to be provided with flexible investigation tools and methods in different scenarios, which are not necessarily repetitive and/or reproducible at will. As far as micro- and nano-objects are concerned, interactive telemanipulation needs also to be considered –at least, at preliminary steps– as an important functional part of the proposed systems. As stated in previous sections, close field microscopy offers a variety of technologies that can be considered as manipulation or imaging tools (or both). However, the current difficulty lies on the fact that the worlds where tasks are executed are not directly perceivable to human operator. It is not only a matter of time/space scales, but also the very nature of physical signals. Moreover, we can find induced changes in the environment state together with related phenomena which do not obey to the physics that our senses are used to. Hereby the main problem to be tackled is twofold:

(i) How to translate physical signals and phenomena in micro and nanoworlds– eventually measurable through an appropriate technology– into a human-sensory signals and/or language, so that a clear representation of the phenomena could be made by the operator, and
(ii) how to map operator actions in macro world driven by the macro physics, into realizable actions, and subsequently tasks, that obey the micro and nano-world physics. It can be also asked how this mapping could be made in an intuitive, simple and interactive way (i.e. keeping a perception/action loop). It is obvious that a simple interactive bilateral coupling of variables would not have much sense between signals with different physics.

The first point raises the underlying problem of multi-sensory rendering or the question about the appropriate combination of multimodal rendering techniques. This is a difficult problem that requires tedious investigations. Applied methods would benefit from teleoperation technology, virtual and augmented reality techniques, as well as from human-machine interface science. The second point, although related to the first one, raises a highly complex problem: what are the appropriate bilateral mappings that would success making two different physics work coherently?

For example, the lack of direct 3D visual feedback from the micro- and nano-world, on the one hand, and the fragility of the manipulated objects, on the other hand, make force feedback essential, and even an unavoidable component of the macro / micro- and nano-world mapping. But the question about the corresponding parameters/phenomena to feedback forces and their scales is still unanswered.

Fig. 9.8. Examples of nano-telemanipulators using AFM. Left image illustrates the ICA-INPG interactive nano-telemanipulator [17]: the AFM (left to the computer screen) is controlled through a gesture/haptic transducer (held by the right hand), computer screen shows visualization of the nano-interaction forces that occur between the cantilever's tip and the explored surface; forces are amplified and rendered through a haptic transducer; audio substitution of some interaction phenomena is also rendered. Right image illustrates a nano-telemanipulation using one degree of freedom haptic interface for taping operations; left screen shows the obtained scanned image and right screen AFM cantilever's tip (courtesy of the University of Tokyo [18]).

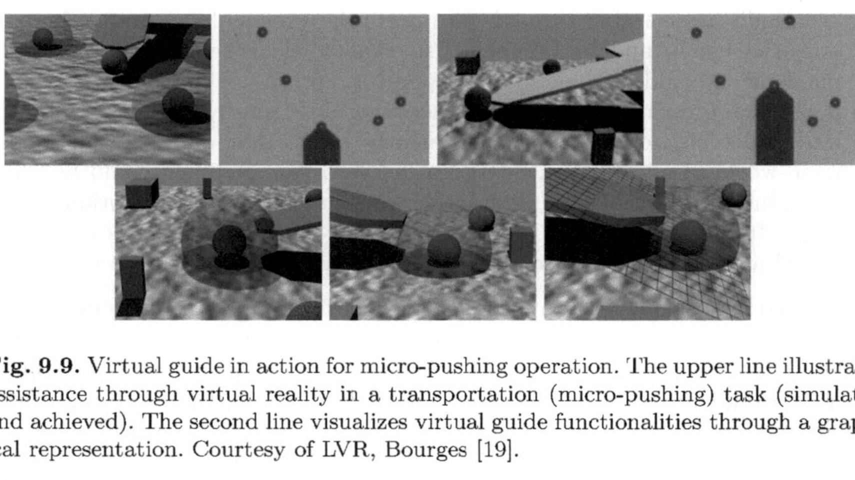

Fig. 9.9. Virtual guide in action for micro-pushing operation. The upper line illustrates assistance through virtual reality in a transportation (micro-pushing) task (simulated and achieved). The second line visualizes virtual guide functionalities through a graphical representation. Courtesy of LVR, Bourges [19].

Several example of tele-nano-manipulation systems have been proposed in the literature, readers may be referred to a recent review of the literature in [20]. As an example, Fig. 9.8 illustrates some nano-telemanipulators using multimodal augmented and virtual realities. Dealing with different physics, scales and signals of non perceivable nature could be tackled by combining and devising perceptual virtual fixtures [7] and active graphic virtual guides [9]. These artefacts allow programming and substituting sensory information in a bilateral way. If these metaphors were well conceived, they will facilitate an intuitive and flexible interaction between macro- and nano-worlds. There are several implementation examples in the literature, Fig. 9.9 illustrates virtual guides in action for a micro-pushing operation.

9.3.2 Humanoids Teleoperation

Humanoids, androids and animaloids are robotic systems with the shape of a human beings or animals respectively. Teleoperating these anthropomorphic robotic systems appear to be not trivial. Lower teleoperation level controls generally slave robot in joint or Cartesian spaces. Due to their multiple degrees of freedom, whole body manipulation requires not only controlling such sub-spaces but also controlling coordination and synchronization. In addition, these systems are not grounded; since motion control should keep a good balance. Besides, some working conditions may induce even more complex coordination [21], whereas simplifications are suspected in some others.

Monitoring humanoid state needs sustained operator attention which is shared with the demanding operator judgment of the remote environment state, so that she or he can perform right actions to complete a given tasks. It is therefore easy to figure out complexity in teleoperating a humanoid system for a single operator without any assistance. Obviously, shared control is a 'must implementation' to release operator from a multi-degree of freedom control and perception complexity. Paradoxically, shared control implementation needs to be flexible to allow redundancy during task execution and switching or combining different control and perceptual subspaces.

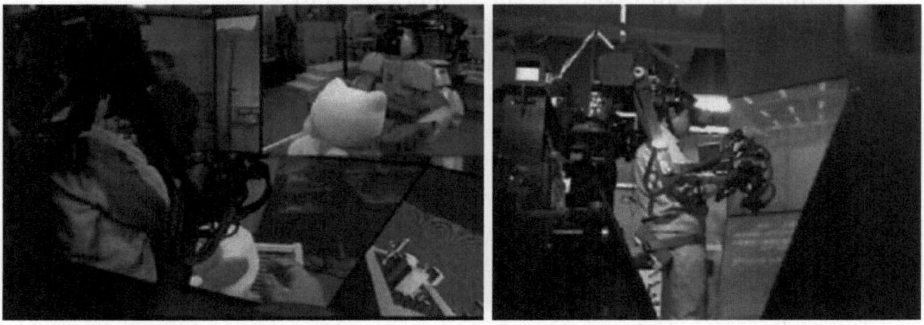

Fig. 9.10. Supercockpit teleoperation system for humanoid HRP-1 (Courtesy Prof. Tachi)

Sensory feedback concerns intrinsic parameters to humanoid such as global behavior, attitude and stability margins feedback, as well as classical parameters inherent to task achievements. Additional features need to be added to help operator perception: virtual fixtures such as visualization of reachable space under humanoid constraints can be very helpful to control parts of the robot during task execution. Virtual reality techniques prove to be a powerful tool for teleoperating such complex systems. Several concepts of isomorphic master 'arms' have been proposed. The idea is to have a reduced shape of the humanoid so that it can be animated like a marionette to directly control its posture/attitude [22]. Virtual reality cockpits with exoskeleton arms have also been proposed by several researchers. Tachi et al. [23] developed the Super Cockpit teleoperation system for the HRP-1 humanoid robot having an exoskeleton master device to teleoperate the upper body movement of the humanoid robot, and voice commands to generate preprogrammed lower body motion. Information feedback uses stereo large panel display technology, see Fig. 9.10. Nishiyama's system uses a modified minimum cockpit. A robot avatar agent and a network-based user interface, see [24]. Systems using exoskeleton or isomorphic masters are able to provide intuitive and flexible manipulations, but they have drawbacks such as difficulty to ensure stable motions and lack of ergonomic design and user-friendliness.

Neo et al. [25] proposed a novel humanoid teleoperation interface concept with a monitoring system using 3D augmented virtual spaces through displays of sensing measurements and embedded recognition algorithms, and an input interface which integrates two joysticks, a mouse and voice command input. The unified interface is made up of the following components:

- An integrated 3D display with environment and robot's perception. Its assistance module makes use of various 3D reconstruction techniques, and virtual reality or augmented reality technologies. The aim is to provide an interface which is not only a display but which also allows an input so that the operator is able to alter the perception process of the robot. Fig. 9.11 shows a simultaneous projection of the simulated humanoid model according to its actual posture obtained from joint sensors, along with the following three levels of visual information:
  - raw image stream taken by the camera system;
  - 3D textured distance information based on stereo vision process;
  - 3D object map based on visual object recognition process;

  In this system, the operator is able to choose the information level she or he wishes to monitor.
- A multi-modal input interface: during task executions, input commands have to be issued to carry out the following basic activities:
  - *selection* by choosing the object of interest, the humanoid body part to be telemanipulated, and the behavior to be taken;
  - Target *positioning and orientation*, the object or the robot part is to be located or oriented;
  - *manipulation*: designing object or humanoid's body part trajectory, direction as well as motion adjustment.

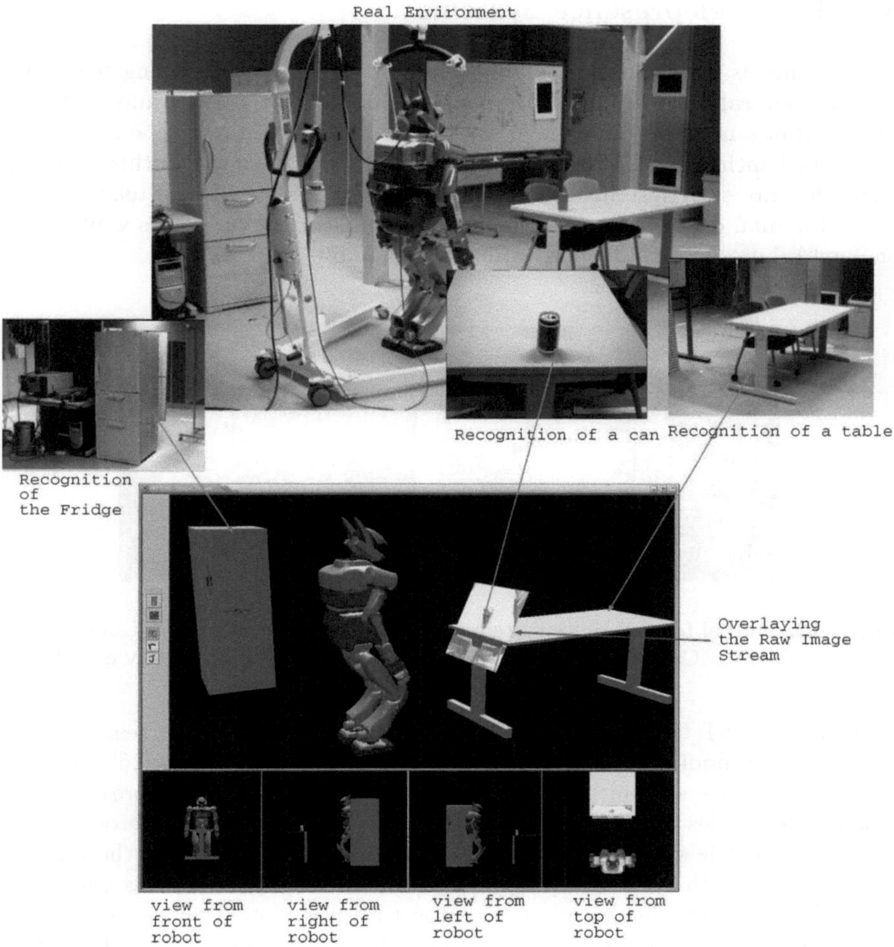

Fig. 9.11. Unified interface for humanoid teleoperation: real environment, image used during object recognition with superimposed recognition results (of a fridge, a can, and a table). An unified interface displays the simulated robot model, the recognized fridge, can and table, along with the raw image stream taken by a camera out of the four embedded humanoids' cameras. By comparing the overlaid raw image stream and the 3D object map created using visual recognition process, the operator is able to confirm the correctness of the humanoid's 'perception' (Courtesy of E-S. Neo and K. Yokoi, ISRI/AIST, Japan).

It is clearly difficult to design an optimal input technique or modality that satisfies all task situations. Combining multiple techniques has been an active research purpose that attempts to bring together the best properties of different input techniques.

9.4 HiFi Telepresence Systems

Telepresence is traditionally a highly sensory and bilateral coupling teleoperation [1]. Several advances have been achieved as to how sensory information is displayed in a modality-immersion way to the human operator. The technology is designed anthropomorphically. The most noticeable advances in this field concerns the way such systems are used not only for remote virtual teleportation (an old human dream; for e.g. touching/manipulating moon stones while being on earth), but also for advanced communication purposes.

Fig. 9.12. Mutual tele-existance system: master operating cockpit (left), and projection slave (right), (Courtesy of R. Tadakuma and S. Tachi, the University of Tokyo)

In the RCAST laboratory at the University of Tokyo, Tachi's team applied humanoids' technologies to enhance their Tele-Existence concepts [26], to a *mutual tele-existence* system [27]. Tele-Existence is an advanced telepresence system which enables an operator to perform tele-manipulation dexterously. The operator can 'tele-exists' in the virtual environment and operates the real environment through a virtual space. One of the main features of this concept is the use of virtual reality to allow operator self-projection. Two self-projections are defined: one of them concerns the 'mapping' of the operator to the virtual robot as if she or he is self-projected to the virtual human in the virtual environment. The second self-projection is the possibility to allow the operator, in an actual operation mode, to switch from a realistic representation to a virtual one so that she or he can plan strategy. Telepresence tasks can be executed thanks to a better understanding of the real environment from its virtual representation. The new mutual existence concept [27] uses an advanced version of seven degrees of freedom anthropomorphic arm with a large display cockpit (same operating configuration as in humanoids' teleoperation techniques). In addition, a humanoid will project operator shape -in a high-fidelity display mode- into the remote location in order to achieve close communication to other people using gesture and interpersonal communication, as shown in Fig. 9.12.

9.4.1 Multi-masters–Multi-slaves Teleoperators

It is well known that collaboration between individuals and between humans and machines can be one of the most stimulating ways to develop new concepts and ideas. Without such collaboration there is less innovation and a lower degree of problem solving. Integration of collaborative robotic systems as active agents operating through various control paradigms within collaborative working environment clusters is a challenging issue for both robotics and collaborative working environments technologies. Hence, it is suspected that robotics will spread smoothly to additional areas of human work, including multi-user/multi-site teleoperation. Telecolloborative working allows new generation of robots to be operated with a multi-user, multi-site teleoperation and telepresence technologies. This operational mode would offer new implications and democratization perspectives in various social domains such as education/edutainment, culture, communication, arts, etc. Such *collaborative working-centred design* of robotic systems requires however the contribution of new technologies such as virtualized environments, multimedia communication, ubiquitous environments, ambient intelligence... which will enable the taking-up and use of robotics in existing and new areas where robotics can support human endeavours.

The idea of multi-operators/multi-robots (MOMR) teleoperation has been proposed in [28] as an extension of their work on single operator multi-robots long distance teleoperation. Note that existing work has also been proposed in a multi-operator single robot teleoperation scheme, e.g. [29]. A MOMR system can be viewed as *distributed resources* of various robots operating at distance, eventually in a shared workspace framework. These robotic resources are supposed to be remotely operated according to classical schemes (coupling, shared autonomy or supervisory) and also according to different taxonomies. For instance, one may imagine a single operator controlling simultaneously a number of robots switching among the available ones to perform a given task. Or, each robot is controlled by a single operator in a collaborative working scheme. Another situation would suggest multiple operators operating a single robot [29]. In the last two cases and considering a general theoretical approach, MOMR teleoperation poses different additional problems that can not be solved by classical single master/slave teleoperation technology. One can imagine a networked architecture where robots are accessible resources that can operated at will by any operator. Kheddar *et al.* [28], enumerated three issues to be tackled:

(i) *interoperability*: it refers to the standardization of control interface API's. It offers the possibility to interchange master device that controls any robot and ensures multi-modal sensory information, which can be fed back to any connected operator in a unified manner. It also means that a connected robot should respond in the appropriate way to desired operator actions whatever master station and location are. This is necessary to ensure MOMR system functionality and open new perspectives in physical teleworking technologies. Related issues may find solution in efforts and institutions devoted to standardization.

(ii) *concurrency*: remote robots can be seen basically as distributed resources to which any operator can access at will; problems such as mutual exclusion, resource protection and handling need to be tackled. Classical tools used in operating computer systems and networking theories can help, as well as those used in handling distribution resources (e.g. access to shared resources). Systems should ensure concurrent robot access exclusion if a given robot can not be operated by more than a single operator, and, more critically, they should guarantee exclusion in tasks and/or objects manipulation among robots (task and context awareness).

(iii) *time-delay*: MOMR teleoperation particularity is that the time delay does not occur only between operator and remote executive robot(s) but also between users operating resources. The latter is crucial when robots are operating within the same shared workspace and controlled, at will, by several distant operators. Solutions do exist for single round-trip delays - single operator / single remote place, using namely VE techniques- which has been thoroughly discussed in previous sections. However, in the presence of multiple users, time delay will compromise their coordination since a certain user is not aware of other user(s)' intentions. This fact raises a crucial problem since not only tasks are harder to be achieved but also solutions to concurrency are difficult to be implemented.

Some ways out of the above mentioned problems that will bring MOMR teleoperation form conceptual theory to actual use can be approached to through a "savant dosage" of: VR techniques (namely human-machine interfaces and distributed VE), recent advances in time-delay control theory (namely in teleoperation), techniques from operating distributed systems and shared resources brought from computer science. For instance, considering interoperability, a dedicated VE architecture could be devised to be used as an intermediary representation for each operator. Indeed, different feedbacks of the same remote workspace can be given to each operator as to the task context, time-delay constraints, available resources... Concurrency is easier when being centralized. Moreover, a dedicated VE offers additional benefits such as the possibility to change users' viewpoint arbitrarily and independently. This is an important issue in favour of VE since this functionality allows reducing the number of on-site cameras for a direct video feedback. VE also provides a large variety of intuitive multi-modal and interactive interfaces that ease operator to plan coordinated tasks using multiple robot resources with the help of distant operators.

The question of concurrency and coordination with different operator requires still solving time-delay issue, namely those occurring among the operators. This problem has been thoroughly studied by Chong and coworkers [30, 31]. To deal with the delay that may occur between operators, and between each operator and robots, predictive strategies are likely to be very efficient. In fact, time delay compromises, for a given operator, appropriate motion control according to other operators' actions. This may make remote robots collide and prohibit coordination among operators. One of the proposed solutions consists in enlarging each robot's geometric envelop with a virtual thickness; it has therefore been called

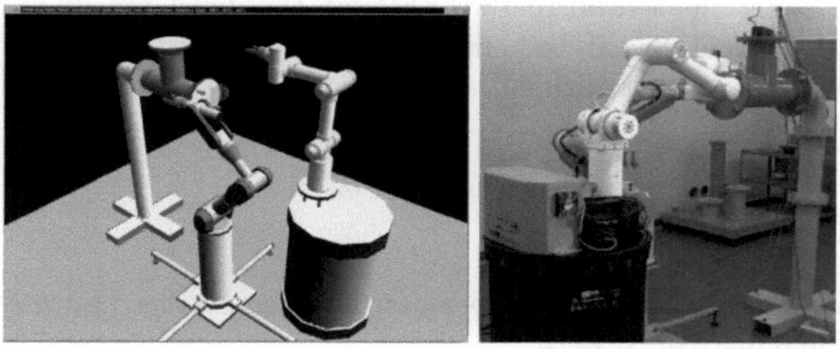

Fig. 9.13. Multi-master and multi-slave teleoperation systems: predictive simulator system in action (courtesy of N.Y. Chong ISRI/AIST)

virtual thickness enlargement (VTE) method. Predictive collision detection is then checked between each robot VTE prior to sending motion. Determination of optimal amount of thickness is the main concern for this method. Current proposed methods need further investigations since in practice not all parts of robots need to be enlarged. Moreover, thickness is a function of minimum distances between robots/objects and each user/workspace round-trip delay estimate. A further predictive solution is called estimated predictor overlay which consists in projecting the wire-frame model of the robot at its predicted position obtained from integrating the actual position with an average velocity over predefined sample periods and combined with an appropriate scaling approach to limit speed around spots of potential collisions. This method combines traditional visual predictors with an estimate of all robot motion that can be refined with more advanced Kalman filtering techniques. Fig. 9.13 shows an actual experimentation of MOMR teleoperation using a VE that implements previously cited solutions.

MOMR teleoperation systems draw potentialities to future teleworking collaborative systems including human/robot teaming. Nevertheless, problems to be solved when considering various application paradigms are not simple. The concurrency of several technologies is necessary and VR techniques are indubitably the key-base on which actual solutions are to be build.

9.5 Conclusion

There is still much to research on the prosperous development of VR techniques for teleoperation. When cleverly devised, intermediary representations contribute in solving several questions related to functionality, human interface technology, etc. VR also offers ingenious metaphors for both operators/users' assistance and novel telerobotic systems such as humanoids and nano-telemanipulators. When remotely controlled systems are complex, there is a need for an augmented reality or virtual environments which may cleverly keep only the functional aspects of task and workspace complexity. It unburdens therefore operator and facilitates the control/supervision of remote robotic systems. In conclusion, VR techniques have still to bring out a total renewal through the development of new operating concepts. Currently, searched efficiency could be a reality allowing, not only a clear improvement in traditional applications, but also opening new fields of interest.

References

1. T.B. Sheridan. *Telerobotics, automation and human supervisory control.* MIT Press, Cambridge, USA, 1992.
2. A. Kheddar and P. Coiffet. *Téléopération et réalité virtuelle.* Hermès-Lavoisier, 2002.
3. P. Coiffet and A. Kheddar. *Téléopération et Télérobotique.* Hermès-Lavoisier, 2002.
4. A. Kheddar, R. Chellali, and P. Coiffet. *Hand Book of Virtual Environments, K.M. Stanney (Eds)*, Chapter 49, Virtual environment-assisted teleoperation, pages 959–1026. Lawrence Erlbaum Associates, 2002.
5. M.V. Noyes and T.B. Sheridan. A novel predictor for telemanipulation through a time delay. In *Annual Conference Manual Control*, Moffett Field, CA, 1984.
6. A.K. Bejczy, W.S. Kim, and S. Venema. The phantom robot: predictive displays for teleoperation with time delay. In *IEEE Int Conf on Robotics and Automation*, 1990.
7. L.B. Rosenberg. Virtual Fixtures: Perceptual Tools for Telerobotic Manipulation. In *IEEE Virtual Reality*, pages 76–82, 1993.
8. R.J. Anderson. Teleoperation with virtual force feedback. *Control and Information Sciences 200: Experimental Robotics III*, pages 366–375, 1994.
9. A. Kheddar. Teleoperation based on the hidden robot concept. *IEEE Transaction on Systems Man and Cybernetics*, 31(1):1–13, 2001.
10. P. Bisson and V. Conan. The MAESTRO project: an augmented reality environment for telemaintenance. In *6ème Journées de travail du GT-RV*, Issy-les-mounlinaux, 1998.
11. P. Fuchs, F. Nashashibi, and D. Maman. Assistance for telepresence by stereovision-based augmented reality and interactivity in 3D space. *PRESENCE: Teleoperators and Virtual Environments*, 11(5):525–535, 2002.
12. T. Burkert, J. Leupold, and G. Passig. A photorealistic predictive display. *PRESENCE: Teleoperators and Virtual Environments*, 13(1):22–43, 2004.
13. R.P. Paul, T. Lindsay, C.P. Sayers, and M. Stein. Time-delay insensitive, virtual-force reflecting, teleoperation. In *i-SAIRAS*, pages 55–67, Toulouse, 1992.
14. G. Hirzinger, B. Brunner, J. Dietrich, and J. Heindl. ROTEX– the first remotly controlled robot in space. In *IEEE Int Conf on Robotics and Automation*, pages 2604–2611, 1994.
15. E. Freund and J. Roßmann. Projective virtual reality as a basis for on-line control of complex systems–not only–over the Internet. *Journal of Robotic Systems*, 22(3):147–155, 2005.
16. J. Wright, F. Hartman, B. Cooper, S. Maxwell, J. Yen, and J. Morrison. Driving on Mars with RSVP. *IEEE Robotics and Automation Magazine*, 13(2):37–45, June 2006.
17. A. Luciani, D. Urmaa, S. Marlière, and J. Chevrier. PRESENCE: the sense of believability of inaccessible worlds. *Computers & Graphics*, 28:509-517, 2004.
18. M. Sitti, B. Aruk, H. Shintani, and H. Hashimoto. Scaled teleoperation system for nano-scale interaction and manipulation. *Advanced Robotics*, 17(3):275–291, 2003.

19. M. Ammi, A. Ferreira, and J-G. Fontaine. Virtualized reality interfaces for telemicromanipulation. In *IEEE Int Conf on Robotics and Automation*, New Orleans, 2004.
20. A. Ferreira and C. Mavroidis. Virtual reality and haptics for nano robotics: A review study. *IEEE Robotics and Automation Magazine*, 13(3):78–92, 2006.
21. K. Yokoi, K. Nakashima, M. Kobayashi, H. Mihune, H. Hasunuma, Y. Yanagihara, T. Ueno, T. Gokyuu, and K. Endou. A tele-operated humanoid operator. *International Journal of Robotics Research*, 25(5–6):593–602, May-June 2006.
22. S. Kagami, J.J. Kuffner, K. Nishiwaki, T. Sugihara, T. Michikata, T. Aoyama, M. Inaba, and H. Inoue. Design and implementation of remotely operation interface for humanoid robot. In *IEEE Int Conf on Robotics and Automation*, pages 401–406, Seoul, Korea, 2001.
23. S. Tachi, K. Komoriya, K. Sawada, T. Nishiyama, T. Itoko, M. Kobayashi, and K. Inoue. Development of telexistence cockpit for humanoid robot control. In *Int. Symposium on Robotics*, pages 1483–1488, 2001.
24. T. Nishiyama, H. Hoshino, K. Sawada, Y. Tokunaga, H. Shinomiya, M. Yoneda, I. Takeuchi, Y. Ichige, S. Hattori, and A. Takanishi. Development of user interface for humanoid service robot system. *IEEE Int Conf on Robotics and Automation*, pages 2979–2984, 2003.
25. E-S. Neo, O. Stasse, Y. Kawai, T. Sakaguchi, and K. Yokoi. A unified on-line operation interface for humanoid robots in a partially-unknown environment. In *IEEE Int Conf on Robotics and Automation*, 2006.
26. S. Tachi. Real-time remote robotics- towards networked telexistence. *IEEE Computer Graphics & Applications*, 18(6):6–9, 1998.
27. R. Tadakuma, Y. Asahara, H. Kajimoto, N. Kamakami, and S. Tachi. Development of Anthropomorphic Multi-D.O.F. Master-Slave Arm for Mutual Telexistence. *IEEE Transactions on Visualization and Computer Graphics*, 11(6):626–636, 2005.
28. A. Kheddar, P. Coiffet, T. Kotoku, and K. Tanie. Multi-robots teleoperation–analysis and prognosis. In *IEEE International workshop on Robotic and Human Communication, ROMAN*, pages 166–171, Sendai, Japan, Sept. 29-Oct. 1 1997.
29. K. Goldberg, B. Chen, R. Solomon, S. Bui, B. Farzin, J. Heitler, D. Poon, and G. Smith. Collaborative teleoperation via Internet. In *IEEE Int. Conference on Robotics and Automation*, pages 2019–2024, San Francisco, CA, April 2000.
30. N.-Y. Chong, S. Kawabata, K. Ohba, T. Kotoku, K. Komoriya, K. Takase, and K. Tanie. Multioperator teleoperation of multirobot systems with time delay: Part I–aids for collision-free control. *PRESENCE: Teleoperators and Virtual Environments*, 11(3):277–291, 2002.
31. N.-Y. Chong, T. Kotoku, K. Ohba, H. Sasaki, K. Komoriya, and K. Tanie. Multioperator teleoperation of multirobot systems with time delay: Part II–testbed description. *PRESENCE: Teleoperators and Virtual Environments*, 11(3):292–303, 2002.

Part II

Control

10

Bilateral Control Architectures for Telerobotics

Sandra Hirche[1], Manuel Ferre[2], Jordi Barrio[2], Claudio Melchiorri[3], and Martin Buss[4]

[1] Tokyo Institute of Technology
Fujita Lab, Dept. of Mechanical and Control Engineering
152-8552 Tokyo, Japan
S.Hirche@ieee.org

[2] Universidad Politécnica de Madrid
Dpto. Automática, Ing. Electrónica e Inf. Industrial
C/. José Gutierrez Abascal, 2. 28006 - Madrid, Spain
{mferre,jbarrio}@etsii.upm.es

[3] University of Bologna
Laboratory of Automation and Robotics, DEIS
Viale Risorgimento 2, 40136 Bologna, Italy
cmelchiorri@deis.unibo.it

[4] Technische Universität München
Institute of Automatic Control Engineering
D-80290 Munich, Germany
M.Buss@ieee.org

Summary. This chapter introduces and surveys bilateral (master-slave) control methods and system architectures for telerobotics known in the literature. The discussion takes a focus most relevant to subsequent chapters, see [1, 2, 3] for complementary and more complete surveys. Firstly, an introductory example of a control-oriented block diagram formulation for a typical two-channel force-velocity architecture is presented, where the human system interface (master) is force controlled and the teleoperator (slave) is velocity controlled. Then a network theoretic view using n-ports is taken and the basic concept of passivity is explained meaning that an interconnection of passive subsystems results in an overall passive system. Transparency is defined, meaning that the human operator should ideally feel as if directly acting in the remote environment (is not able to feel the technical systems/communication network at all). The well known and for ideal transparency required four-channel architecture is presented. The important discussion of time delay in the communication network, which impairs or might threaten stability properties of the closed loop teleoperation system, is then discussed. The most successful approach using the scattering (wave variable) transformation approach is presented, which guarantees that the communication two-port is lossless (passive) for arbitrary constant time delay. Finally, we mention the key challenges in the design of bilateral teleoperation systems and give a short preview of the subsequent chapters addressing some of these challenges.

M. Ferre et al. (Eds.): Advances in Telerobotics, STAR 31, pp. 163–176, 2007.
springerlink.com © Springer-Verlag Berlin Heidelberg 2007

10.1 Introduction

The most common way to classify control architectures in telerobotic systems is according to their coupling characteristics between the human and the remote environment or the respective degree of autonomy in the slave system. The coupling can be realized through the direct exchange of motion and force signals between master and slave device like in bilateral control, or on symbolic basis like in supervisory control architectures. While bilateral systems imply a very low degree of autonomy of the slave system, supervisory control requires at least semi-autonoumous behavior. For an overview on the different types of control architectures the reader is referred to [1, 2, 3] and references therein.

In bilateral control, the human operator synchronously manipulates and perceives the resulting reaction force through direct feedback. Therefore the human operator applies a force to the master device, the human system interface (HSI), resulting in a displacement. While the teleoperator (slave robot) follows the motion of the HSI (master), the force resulting from the interaction with the environment is displayed to the human operator via this interface. Bilateral control architectures provide the most natural way of interaction with the remote environment. Direct force feedback increases the sense of being present in the remote environment and thereby improves the ability to perform complex manipulative tasks [4]. Interest in this type of telerobotic control architectures started with Goertz' early works [5] and has since then attracted a large number of researchers, see e.g. [3] and references therein. The advantages of bilateral systems over other alternatives[1] have been highlighted in a large number of experiments. Some of these results are reported in [6, 7, 8, 9].

A large variety of bilateral control architectures has been developed in the past two decades. The choice depends on the intended application; the kind of devices used and communication issues play a role. It goes beyond the scope of this introductory chapter to provide complete overview over all available approaches. The focus is here on the fundamentals of bilateral control. Stability and transparency are the major goals in every bilateral control design that are briefly reviewed in this chapter. Some prominent control architectures without and with time delay between HSI (master) and teleoperator (slave) are discussed.

This chapter provides an introduction into bilateral control architectures. Fundamental issues are discussed in Sec. 10.2. As an example for bilateral control without time delay the four-channel architecture – a generalization to many existing approaches – is introduced in Sec. 10.3. Bilateral control with time delay is considered in Sec. 10.4 where the wave (scattering) variable transformation is presented. The discussion of control challenges and an outlook to the subsequent chapters of part II of this book follows in Sec. 10.5.

[1] Alternatives with non-autonomous slave systems are unilateral systems where only motion information is transmitted from the master to the slave without force feedback information, and systems based on sensorial substitution, i.e. where the environment interaction force is represented through a different perceptual modality, e.g. through vision or audio.

10.2 Fundamentals

The design of bilateral architectures includes the design of the local control measures and the coupling control. The local control loops ensure the proper display of the reaction force (motion) at the HSI (master) side and the tracking of the commanded motion (force) at the teleoperator (slave) side. Coupling control basically determines which quantity is transmitted. The interaction between human and the telerobotic system together with the control measures is illustrated in the following example of the simple force-velocity control architecture.

10.2.1 Control Architecture Example: Force-Velocity Control

The term *force-velocity* refers to the local controlled quantities[2]: The force signal is transmitted from the teleoperator (slave) to the HSI (master) with local force (F) control, the velocity signal is transmitted to the teleoperator with local velocity (V) control. The basic control architecture is illustrated in Fig. 10.1. The human operator manipulates the HSI (master), a multi-degree-of-freedom (DoF) robot, and perceives the force information from the remote environment via this interface. He/she moves his/her arm holding on to the HSI (master) with the velocity $v_m = v_h$ and applies the force f_h consisting of a reaction force induced by the arm impedance (dynamics) Z_h and a voluntarily applied force f_h^*. The mechanical impedance[3] Z relates the reaction force to the velocity $Z : v \rightarrow f$. The HSI (master) velocity is transmitted via the communication channel and acts as reference input v_s^d for the local velocity control loop of the teleoperator, also a multi-degree-of-freedom (DoF) robot, possibly with different kinematics than the HSI (master). The local velocity controller C_s ensures the tracking of this reference velocity. As a result, the teleoperator (slave) interacts with the remote environment with velocity which in continuing contact is equal to the environment velocity $v_s = v_e$. The environment reacts according to its impedance (dynamics) Z_e with a reaction force f_e, which is measured by a force/torque sensor. The environment force information is transmitted via the communication channel to the HSI acting as reference input f_h^d to the local force control loop with the controller C_m. The local control measures in this force-velocity control architecture consist of the HSI (master) force control and the teleoperator (slave) velocity control, the coupling is performed via the exchange of HSI (master) velocity and environment force signals.

Remark 1. The key challenges of bilateral teleoperation control systems are *stability* and *transparency*. Stability is essential as a bounded response of the system is an essential requirement for the human operator to be able to perform teleoperation tasks, and for safety in particular. Time delay in the communication network is to be considered when designing stable control schemes. Transparency, as stated in Definition 1 below and further discussed in chapter 12, means that

[2] Here, we choose the convention *force-velocity* F-V architecture to refer to the local control mode in the order HSI (master) then teleoperator (slave).

[3] In some literature references researchers use "generalized impedance" with the force being a reaction to a position.

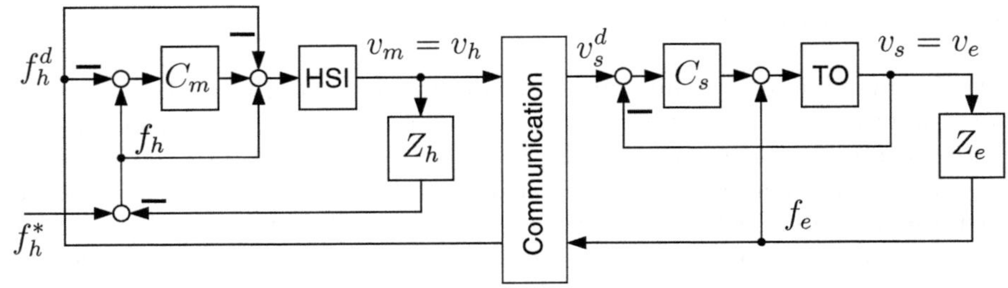

Fig. 10.1. Force-Velocity (F-V) control architecture

HSI, teleoperator velocities, forces be the same $v_h = v_e$, $f_h = f_e$ or that the impedance felt by the human is equal to the environment impedance.

10.2.2 Modeling by n-Ports

For the analysis and control synthesis the modeling of the bilateral teleoperation system as interconnection of two-ports, see Fig. 10.2 and e.g. [10], is convenient. This view enables to draw on results from network theory discussing variables as *effort* and *flow*. Here, we choose the analogy that effort, flow correspond to mechanical variables force, velocity, respectively; duality allows the opposite correspondence. Note that by analogy between electrical and mechanical systems that there is a correspondence of effort, flow to voltage, current, respectively.

The n-port view of a bilateral teleoperation system is shown in Fig. 10.2. The teleoperator interface is used to transmit the environment impedance Z_e to the operator site, where it is felt by the human as the transmitted impedance Z_t. The applied force f_h and the velocity v_h of the human operator's hand are related by this impedance as $f_h = Z_t(v_h)$. The teleoperator interface includes the HSI (master), the communication network, and the teleoperator (slave) with the respective local and coupling control measures. Obviously, the human operator feels exactly the mechanical properties of the environment at the HSI (master) if the transmitted impedance is the same as the environment impedance, i.e. if the teleoperator interface of Fig. 10.2 is totally transparent as if it were not existent.

10.2.3 Transparency

Transparency of the telerobotic system is the major goal in bilateral control architecture design.

Definition 1 (Transparency). *The telerobotic system is transparent if the human operator feels as if directly interacting with the (remote) task [10].*

Formally, transparency is achieved if the transmitted and the environment impedances match [11] as also indicated above

$$Z_t = Z_e \ , \tag{10.1}$$

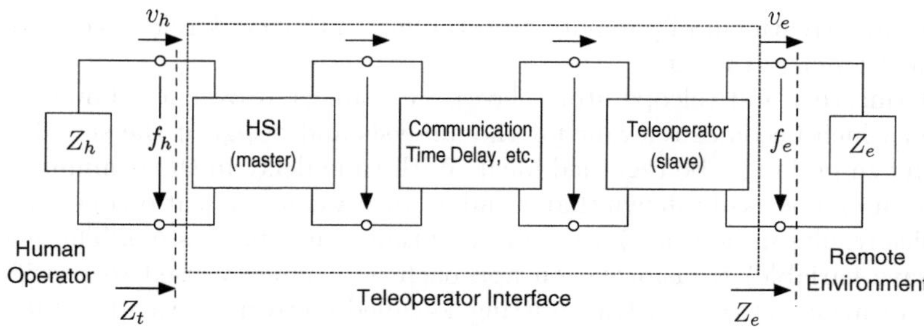

Fig. 10.2. Bilateral telerobotic system as interconnection of n-ports

or alternatively if HSI (master) and teleoperator (slave) movements are equal and the force displayed to the human operator is exactly the reaction force from the environment [12]

$$x_h = x_e \quad \text{and} \quad \mathrm{f_h} = \mathrm{f_e}. \tag{10.2}$$

Transparency in this sense is in practice not achievable as the device dynamics comprising inertia and friction cannot completely cancelled by control. Communication effects, especially time delay, severely degrade the achievable transparency. The development of quantitative measures is part of the transparency analysis. For a more detailed transparency discussion please refer to chapter 12 of this book.

10.2.4 Stability by Passivity

The interconnected n-ports "human operator", "HSI (master)", "teleoperator (slave)" and "remote environment" exchange energy. Stability of the overall interconnected system is the basic prerequisite and primary objective for the control design. Unstable systems are not operable and represent a severe hazard to the human operator and the environment. Stability basically requires the limitation of the system energy and therewith the boundedness of all system variables. The largely unknown models of human and environment, both part of the global control loop, represent major challenges for the synthesis of stable bilateral control architectures. The powerful concept of passivity has become the major tool to deal with these challenges and with unreliabilities in the communication channel. Passivity (stability) of the overall system is infered based on the known fact that interconnected passive subsystems (n-ports) result in overall passivity.

Proposition 1 (Passivity Argument). *Consider that the subsystems HSI, teleoperator are passive (can be made passive by suitable controller design). Assume no time delay in the communication network. Most mechanical environments are passive. If the human operator behaves in a passive manner (cooperatively, which is to be expected), then the overall teleoperation system is a connection of passive subsystems (n-ports) and therefore passive.*

For a formal treatment of passivity and stability in this book refer to chapter 15, and the references therein.

HSI (master) and teleoperator (slave) are connected over a communication channel which becomes relevant for the analysis and design if the signal transmission can no longer be regarded ideal. With time delay in the communication network it can be easily shown that standard direct communication of effort/flow variables results in non-passivity and potentially unstable system. The scattering (wave variable) transformation approach is commonly used to passify the communication network by transmitting so-called scattering (wave) variables. A detailed time delay discussion follows in Sec. 10.4. Other communication effects like time-varying transmission delay, the loss of data e.g. in Internet-like communication networks further compromise stability and transparency if not treated by appropriate control measures. Control approaches using the passivity concept have been successful proposed for such systems. It goes beyond the scope of this introductory chapter to provide a full overview on available methods. The reader is referred to the overview works on control architectures [1, 3, 13, 14, 15], and [16, 17] for a comparison of control schemes.

10.2.5 Classification of Bilateral Control Schemes

The variety of bilateral control architectures in the literature address the different stability and transparency issues arising from the dynamics of the devices and communication effects. Classification schemes for bilateral control architectures consider for example the kind of signals that are transmitted from the HSI (master) to teleoperator (slave) and vice versa. The force-velocity architecture introduced in Sec. 10.2.1 is one example. Another classification scheme considers the number of quantities communicated, i.e. the number of virtual channels used for the interconnection between HSI (master) and teleoperator (slave). Two-, three-, and four-channel architectures have been considered so far in the literature. The force-velocity control architecture is a typical example for a two-channel architecture.

10.3 Four-Channel Control Architecture

The four-channel architecture [11] depicted in Fig. 10.3 captures in its generality a large variety of bilateral control schemes, including the two-channel architectures like the force-velocity architecture presented in Fig. 10.1. In its original form the HSI (master) and teleoperator (slave) velocities and forces are transmitted. The notations are the same as in Sec. 10.2.1; additionally here the external environment force f_e^* is considered. The HSI (master) dynamics is represented by the impedance Z_m, the teleoperator (slave) dynamics by the impedance Z_s. The local control is given by the HSI (master) controller C_m and the teleoperator (slave) controller C_s. The four blocks C_1, C_2, C_3, C_4 denote the four communication channels including coupling control for velocity forward, force backward, force forward, velocity backward, respectively. Note that all blocks in

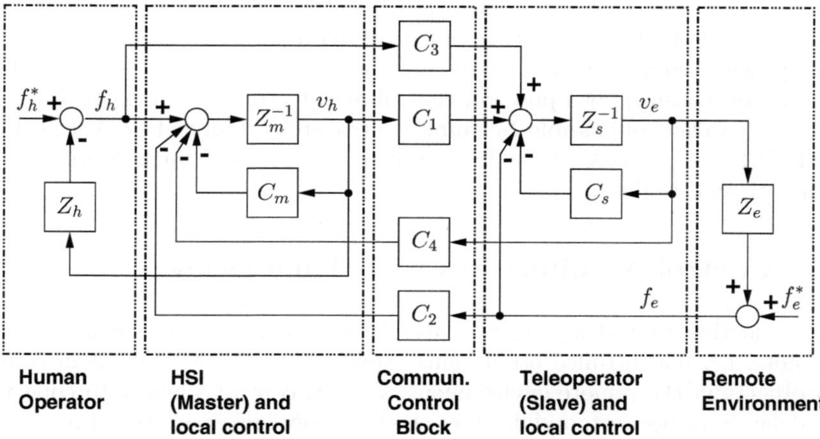

Fig. 10.3. Block diagram of a four-channel bilateral teleoperation system

Fig. 10.3 can be nonlinear and may involve dynamics. The C-blocks are to be designed. Therefore the dynamics of the overall system is for the linearized time-invariant case (and with $f_e^* = 0$) is conveniently characterized by the general hybrid matrix formulation in the Laplace domain as

$$\begin{bmatrix} f_h(s) \\ -v_e(s) \end{bmatrix} = \begin{bmatrix} h_{11}(s) & h_{12}(s) \\ h_{21}(s) & h_{22}(s) \end{bmatrix} \begin{bmatrix} v_h(s) \\ f_e(s) \end{bmatrix} = H(s) \begin{bmatrix} v_h(s) \\ f_e(s) \end{bmatrix} \tag{10.3}$$

The parameters of the hybrid matrix for the four-channel architecture in Fig. 10.3 compute to

$$h_{11} = (Z_m + C_m)D(Z_s + C_s - C_3C_4) + C_4 \tag{10.4}$$
$$h_{12} = -(Z_m + C_m)D(I - C_3C_2) - C_2 \tag{10.5}$$
$$h_{21} = D(Z_s + C_s - C_3C_4) \tag{10.6}$$
$$h_{22} = -D(I - C_3C_2) \ , \tag{10.7}$$

where $D = (C_1 + C_3Z_m + C_3C_m)^{-1}$.

Remark 2. It has been shown that ideal transparency requires the four-channel architecture [11]. Specifically, the choice

$$C_1 = Z_{cs} \tag{10.8}$$
$$C_2 = I \tag{10.9}$$
$$C_3 = I \tag{10.10}$$
$$C_4 = -Z_{cm} \ . \tag{10.11}$$

results in ideal transparency [11, 18, 19]: $h_{11} = h_{22} = 0$, $h_{12} = -h_{21} = I$. However, this is technologically difficult because the HSI, teleoperator dynamics need to be cancelled by their exact inverse dynamic models. More insights about the block parameters h_{ij} along with stability arguments are given in [11].

Remark 3. The force-velocity control architecture discussed in Sec. 10.2.1 is a special case of the four-channel architecture with $C_3 = C_4 = 0$, i.e. only two of the possible communication paths are used. Abbreviate this as F-V control architecture, consider also position control architectures abbreviated by P, we have a large variety of possible 2-channel control architectures: F-V, V-F, F-F, V-V, F-P, P-F, P-P, P-V, V-P, etc. In chapter 11 the P-P and the F-P architecture are investigated in detail.

10.4 Control Architectures with Time Delay

Communication time delays in a bilateral telerobotic system usually generate instability, i.e. uncontrolled oscillations expecially when the slave is in contact with objects in the remote environment. Various control schemes to cope with time delay have been proposed. Most classical approaches are based on a small gain argument, some use predictive techniques, environment models, and knowledge about the time delay. Most of the experimentally successful approaches are based on the scattering (wave variable) transformation [20, 21, 22].

10.4.1 The Scattering/ Wave Variable Transformation

The basic idea is to transform the effort and flow variables by a linear transformation (coordinate rotation) into wave variables. As a result the communication 2-port becomes lossless (passive) for arbitrarily large *constant* time delay. One superior strength of this approach is that the time delay value does not have to be known.

Definition 2 (Scattering / Wave Variables). *The wave variables u/w are defined based on the effort/flow variables f/v as follows:*

$$\begin{aligned} u_m &= \tfrac{1}{\sqrt{2b}}(f_m + bv_m) \qquad & u_s &= \tfrac{1}{\sqrt{2b}}(f_s + bv_s) \\ w_m &= \tfrac{1}{\sqrt{2b}}(f_m - bv_m) \qquad & w_s &= \tfrac{1}{\sqrt{2b}}(f_s - bv_s) \end{aligned} \tag{10.12}$$

where $b > 0$ is the characteristic impedance, *also called wave impedance, of the communication line.*

Instead of the effort and flow variables, the wave variables are transmitted over the communication channel with time delay, i.e.

$$u_s(t) = u_m(t-T) \qquad w_m(t) = w_s(t-T), \tag{10.13}$$

where T denotes the time delay in the forward and backward communication channel.

In Defintion 2, f_m, f_s denote effort variables, v_m, v_s flow variables. On the teleoperator (slave) site v_s or f_s can be interpreted as the command signal depending on the choice of control architecure (local slave control mode); accordingly f_m or

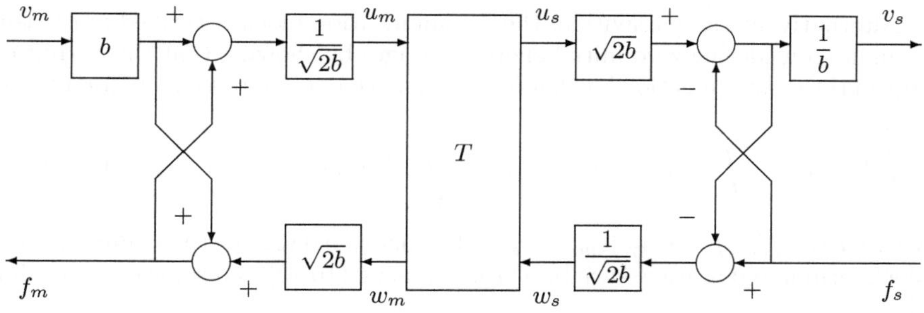

Fig. 10.4. Transmission line based on passivity

v_m are the feedback signals on the HSI site. Specifically for the force-velocity control architecture we have $v_s := v_s^d = v_e^d$, $f_m := f_h^d$, cf. Sec. 10.2.1. The modified communication block containing the time delaying communication channel and the scattering / wave variable transformation blocks on each side is illustrated in Fig. 10.4 for the example of force-velocity control.

10.4.2 Passivity Based Control with Constant Time Delay

The scattering / wave variable transformation maps the time delaying two-port on a long, energetically lossless transmission line in an electrodynamical sense. As a result the communication block is passified for arbitrarily large *constant* delays. The interconnection of the lossless communication block with the passively controlled teleoperator/HSI and the assumed passive human operator/environment is a passive overall system, cf. Definition 1. The overall system is stable for arbitrarily large time delay [20, 21, 22].

The passivity of the communication block is conveniently analyzed via the scattering operator S defined via the hybrid matrix (10.3) in the Laplace domain as

$$S(s) = (H(s) - I)(H(s) + I)^{-1}.$$

It is well known that the system described by $H(s)$ is passive if and only if the largest singular value of the scattering operator is less or equal to one, i.e.

$$||S|| = \sup_{\omega} ||S(j\omega)|| \leq 1 \ .$$

For the communication block with constant time delay $||S(j\omega)|| = 1$ holds for all ω and for arbitrarily large constant time delay implying that it is *lossless*. Contrarily, without the scattering / wave variable transformation, i.e. with the direct transmission of the effort and flow variables

$$f_m(t) = f_s(t - T) \qquad v_s(t) = v_m(t - T),$$

the scattering operator is shown to be unbounded. Hence without the transformation the communication channel with time delay is not passive and may cause instability. For a detailed analysis refer to [20].

Alternatively, the passivity of the communication block is verified in the time domain (considering zero initial energy storage) by showing that the output energy of the communication block is less or equal than the input energy for all times

$$\int_0^t P(\tau)\,\mathrm{d}\tau = \int_0^t P_{in}(\tau) - P_{out}(\tau)\,\mathrm{d}\tau = \int_0^t f^T(\tau)v(\tau)\,\mathrm{d}\tau \geq 0 \quad \forall t \geq 0\ , \tag{10.14}$$

where $f := [f_m,\ f_s]$, $v := [v_m,\ -v_s]$. The energy balance of the communication block formulated in wave variables (10.12), using (10.13) and some computations yield

$$\begin{aligned} \int_0^t P(\tau)\,\mathrm{d}\tau &= \frac{1}{2}\int_0^t u_m^T u_m - u_s^T u_s + w_s^T w_s - w_m^T w_m\,\mathrm{d}\tau \\ &= \frac{1}{2}\int_{t-T}^t u_m^T u_m + w_s^T w_s\,\mathrm{d}\tau \geq 0, \end{aligned} \tag{10.15}$$

where additionally it is assumed that initially at $t = 0$ no energy was stored in the communication channel, i.e. $u_m(t) = w_s(t) = 0$ for all $t \in [-T, 0]$. The energy balance of the communication block is positive for all possible time delays; it does not generate energy, i.e. is passive. In fact, the communication block is *lossless* as the energy is only temporarily stored for the time T of transit.

Impedance Matching in Two-Port Models

It is widely known from the theory of electric networks that it is necessary to match impedance at the end of an electrical line. Likewise, the application of scattering variables in teleoperation implies paying attention to *wave reflections* at both sides of the telemanipulator, which can be avoided by proper impedance matching. Additional scaling factors are inserted to achieve impedance matching, see Fig. 10.5 (cf. Fig. 10.4) and [21]. Formally, we have

$$\begin{cases} v_m'(t) &= v_m(t) - \frac{1}{b}f_s(t) \\ f_s'(t) &= f_s(t) + bv_s(t) \end{cases} \tag{10.16}$$

where v' and f_e' are the new input variables for the communication channel at both sides of the network after matching the impedance b and $1/b$.

Another dynamic impedance matching approach uses dynamic filters (instead of the static gains b, $1/b$) to passify non-passive environments [23].

10.5 Control Design Challenges

The control design for telerobotic systems faces multiple interdisciplinary challenges in the intersection of the fields of robot control, human-machine interaction, and communication systems. Stability and transparency are the main issues to be addressed. Some of the major challenges are summarized in the following.

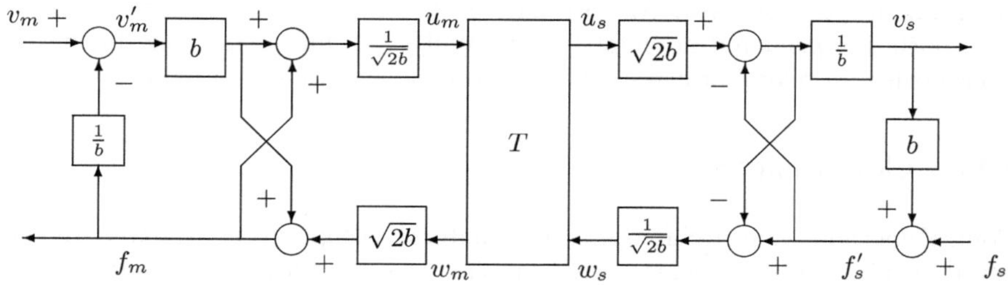

Fig. 10.5. Transmission line based on passivity with impedance matching

Packet Switched Communication Networks — *The* major challenge in control design for bilateral teleoperation systems is due to communication network characteristics such as: communication time delay, time-varying time delay, and disruption/loss of communication. Most of the known approaches addressing those aspects are based on the passivity paradigm using the scattering / wave variable transformation, see Sec. 10.4. However, the scattering / wave variable transformation is not sufficient to guarantee stability in the presence of time-varying delay and loss of data. Extensions targeting the time-varying delay problem are proposed in [24, 25, 26, 27]. The problem of data loss is addressed in chapter 15 and also in [28, 29, 30, 31, 32]. Transparency with time-varying delay and packet loss is investigated in [33].

Human Sensorimotor (Perception-Action) Issues — The human is the 'supervising' component in the telepresence system. His/her subjective feeling of presence and interaction with the remote environment is essential for the evaluation of transparency. However, generally a trade-off between (robust) stability and transparency must be made in the control design. Ideal transparency in the sense of definition 1 is not achievable in real telerobotic systems [11, 34]. Human sensorimotor (perception and action) characteristics can be used to find this trade-off in a human perception/action appropriate sense. Accordingly, the exploitation of human perception models in system design is considered beneficial. A perception oriented transparency analysis of constant time delay is performed in chapter 12, the influence of other communication related effects and the corresponding control architectures are investigated from a human perception point of view in [33, 35, 36, 37].

Some of these challenges are addressed in the subsequent chapters of this part II of the book. The experimental results for different two-channel architectures without time delay and for the wave (scattering) variable approach with time delay in chapter 11 provide a deeper understanding of the theoretical results. Time delay and other communication issues and their influence on stability on transparency are the dominating topics in chapters 12-16. In chapter 13 a very promising approach to identify the mechanical properties of remote environments is discussed. An alternative modeling approach using Hamiltonian ports is considered in chapter 14 to analyze scaling issues (e.g. for micro-telerobotics) in telerobotic systems with time delay and loss of data.

10.6 Conclusions

Control methods and architectures for bilateral teleoperation systems are surveyed in this chapter presenting an examplary two-channel force-velocity control architecture, the four-channel architecture (or at least three channels with appropriate local control [18]) required for ideal transparency, and the powerful scattering (wave variable) transformation to cope with constant time delay was introduced. Some key challenges for the design of teleoperation control systems are mentioned and a preview of subsequent chapters in part II of this book is attempted. Probably one of the most exciting but also difficult challenge in the area are interdisciplinary investigations between system (information) theory and psychology (psychophysics). All readers are welcome to be inspired by this and the following chapters and to take up some exciting challenges.

References

1. M. Buss and G. Schmidt. Control Problems in Multi-Modal Telepresence Systems. In P.M. Frank, editor, *Advances in Control: Highlights of the 5th European Control Conference ECC'99 in Karlsruhe, Germany*, pages 65–101. Springer, 1999.
2. S.E. Salcudean. Control for Teleoperation and Haptic Interfaces. In B. Siciliano and K.P. Valavanis, editors, *Lecture Notes in Control and Information Sciences 230: Control Problems in Robotics and Automation*, pages 51–66. Springer, London, 1998.
3. P.F. Hokayem and M.W. Spong. Bilateral Teleoperation: An Historical Survey. *automatica*, 42(12):2035–2057, 2006.
4. T.B. Sheridan. *Telerobotics, Automation, and Human Supervisory Control.* MIT Press, Cambridge, Massachusetts, 1992.
5. R. Goertz. Manipulator Systems Development at ANL. In *Proceedings of the 12th Conference on Remote Systems Technology, ANS*, pages 117–136, 1964.
6. A.K. Bejczy and J.K. Salisbury. Controlling remote manipulators through kinesthetic coupling. *ASME Computers in Mechanical Engineering*, 2(1):48–60, 1983.
7. B. Hannaford, L. Wood, D. McAffee, and H. Zak. Performance evaluation of a six axis generalized force reflecting teleoperator. *IEEE Transactions on Systems, Man, and Cybernetics*, 21:620–633, 1991.
8. M. Massimino and T.B. Sheridan. Sensory substitution for force feedback in teleoperation. *Presence*, 2(4):344–352, 1993.
9. M. Ferre, R. Aracil, J.M. Bogado, and R.J. Saltaren. Improving force feedback perception using low bandwidth teleoperation devices. In *Proceedings of EuroHaptics Conference EH'2004*, 2004.
10. G.J. Raju, G.C. Verghese, and T.B. Sheridan. Design Issues in 2-Port Network Models of Bilateral Remote Teleoperation. In *Proceedings of the IEEE International Conference on Robotics and Automation*, pages 1317–1321, Scottsdale (AZ), US, 1989.

11. D.A. Lawrence. Stability and Transparency in Bilateral Teleoperation. *IEEE Transactions on Robotics and Automation*, 9(5):624–637, October 1993.
12. Y. Yokokohji and T. Yoshikawa. Bilateral Control of Master-Slave Manipulators for Ideal Kinesthetic Coupling Formulation and Experiment. *IEEE Transactions on Robotics and Automation*, 10(5):605–619, 1994.
13. S.E. Salcudean. Control for Teleoperation and Haptic Interfaces. In B. Siciliano and K.P. Valavanis, editors, *Lecture Notes in Control and Information Sciences 230: Control Problems in Robotics and Automation*, pages 51–66. Springer, London, 1998.
14. A. Eusebi and C. Melchiorri. Force Reflecting Telemanipulators with Time Delay: Stability Analysis and Control Design. *IEEE Transactions on Robotics and Automation*, 14(4):635–640, 1998.
15. T.L. Brooks. Telerobotic response requirements. In *Proceedings of the IEEE International Conference on Systems, Man, and Cybernetics*, pages 113–120, Los Angeles (CA), US, 1990.
16. P. Arcara. *Control of Haptic and Robotic Telemanipulation Systems.* PhD thesis, Universita degli Studi di Bologna, 2001.
17. P. Arcara and C. Melchiorri. Control Schemes for Teleoperation with Time Delay: A Comparative Study. *Robotics and Autonomous Systems*, 38(1):49–64, 2002.
18. K. Hashtrudi-Zaad and S.E. Salcudean. On the Use of Local Force Feedback for Transparent Teleoperation. In *Proceedings of the IEEE International Conference on Robotics and Automation*, pages 1863–1869, Detroit (MI), US, 1999.
19. K. Hashtrudi-Zaad and S.E. Salcudean. Analysis of Control Architectures for Teleoperation Systems with Impedance/Admittance Master and Slave Manipulators. *International Journal of Robotics Research*, 20(6):419–445, 2001.
20. R.J. Anderson and M.W. Spong. Bilateral Control of Teleoperators with Time Delay. *IEEE Transactions on Automatic Control*, 34(5):494–501, 1989.
21. G. Niemeyer and J.-J.E. Slotine. Stable Adaptive Teleoperation. *IEEE Journal of Oceanic Engineering*, 16(1):152–162, January 1991.
22. G. Niemeyer. *Using Wave Variables in Time Delayed Force Reflecting Teleoperation.* PhD thesis, MIT, Department of Aeronautics and Astronautics, September 1996.
23. S. Hirche and M. Buss. Passive Position Controlled Telepresence System with Time Delay. In *Proceedings of the American Control Conference*, pages 168–173, Denver (CO), US, 2003.
24. K. Kosuge, H. Murayama, and K. Takeo. Bilateral Feedback Control of Telemanipulators via Computer Network. In *Proceedings of the IEEE/RSJ International Conference on Intelligent Robots and Systems IROS*, pages 1380–1385, Osaka, Japan, 1996.
25. Y. Yokokohji, T. Imaida, and T. Yoshikawa. Bilateral Control with Energy Balance Monitoring under Time-Varying Communication Delay. In *Proceedings of the IEEE International Conference on Robotics and Automation*, pages 2684–2689, San Francisco (CA), US, 2000.
26. R. Lozano, N. Chopra, and M. Spong. Passivation of Force Reflecting Bilateral Teleoperators with Time Varying Delay. In *Proceedings of the 8. Mechatronics Forum*, pages 954–962, Enschede, Netherlands, 2002.
27. S. Stramigioli. About the Use of Port Concepts for Passive Geometric Telemanipulation with Time Varying Delays. In *Proceedings of the 8. Mechatronics Forum*, pages 944–953, Enschede, The Netherlands, 2002.

28. G. Niemeyer and J. E. Slotine. Towards Force-Reflecting Teleoperation Over the Internet. In *Proceedings of the IEEE International Conference on Robotics and Automation*, pages 1909–1915, Leuven, Belgium, 1998.
29. Y. Yokokohji, T. Tsujioka, and T. Yoshikawa. Bilateral Control with Time-Varying Delay including Communication Blackout. In *Proceedings of the 10th Symposium on Haptic Interfaces for Virtual Environment and Teleoperator Systems*, Orlando (FL), US, 2002.
30. C. Secchi, S. Stramigioli, and C. Fantuzzi. Dealing with Unreliabilities in Digital Passive Geometric Telemanipulation. In *Proceedings of the IEEE/RSJ International Conference on Intelligent Robots and Systems IROS*, Las Vegas (NV), US, 2003.
31. B. Berestesky, N. Chopra, and M. W. Spong. Discrete Time Passivity in Bilateral Teleoperation over the Internet. In *Proceedings of the IEEE International Conference on Robotics and Automation ICRA'04*, pages 4557–4564, New Orleans, US, 2004.
32. S. Hirche and M. Buss. Packet Loss Effects in Passive Telepresence Systems. In *Proceedings of the 43rd IEEE Conference on Decision and Control*, pages 4010–4015, Paradise Island, Bahamas, 2004.
33. S. Hirche. *Haptic Telepresence in Packet Switched Communication Networks.* PhD thesis, Technische Universität München, Institute of Automatic Control Engineering, July 2005.
34. K. Hashtrudi-Zaad and S.E. Salcudean. Analysis and Evaluation of Stability and Performance Robustness for Teleoperation Control Architectures. In *Proceedings of the IEEE International Conference on Robotics and Automation*, pages 3107–3113, San Francisco (CA), US, 2000.
35. S. Hirche and M. Buss. Insights on Human Adapted Control of Networked Telepresence and Teleaction systems. *International Journal of Assistive Robotics and Mechtronics*, 7(1):20–31, March 2006.
36. S. Hirche, P. Hinterseer, E. Steinbach, and M. Buss. Transparent Data Reduction in Networked Telepresence and Teleaction Systems Part I: Communication without Time Delay. *PRESENCE: Teleoperators and Virtual Environments*, 2007, to appear.
37. S. Hirche and M. Buss. Transparent Data Reduction in Networked Telepresence and Teleaction Systems Part II: Time-Delayed Communication. *PRESENCE: Teleoperators and Virtual Environments*, 2007, to appear.
38. J.C. Willems. Dissipative dynamical systems- part i: General theory. *Arch. Rational Mechanics and Analysis*, 45:321–251, 1972.
39. C.A. Desoer and M. Vidyasagar. *Feedback Systems: Input-Output Properties.* Academic Press, 1975.
40. A. van der Schaft. *L_2 Gain and Passivity Techniques in Nonlinear Control.* Springer, 2000.
41. M. Krstic, I. Kanellakopoulos, and P. Kokotovic. *Nonlinear and Adaptative Control Design.* John Wiley and Sons Inc., New York, 1995.

11

Experimental Results on Bilateral Control Using an Industrial Telemanipulator

Manuel Ferre[1], Jordi Barrio[1], Claudio Melchiorri[2], Juan M. Bogado[3], Pedro L. Castedo[1], and Juan M. Ibarra[4]

[1] Universidad Politécnica de Madrid
Dpto. Automática, Ing. Electrónica e Inf. Idustrial
C/. José Gutierrez Abascal, 2. 28006 - Madrid, Spain
m.ferre@upm.es, jbarrio@etsii.upm.es, luis.castedo@upm.es

[2] University of Bologna
Department of Electronics, Computer Science and Systems
Via Risorgimento 2, 40136 Bologna, Italy
cmelchiorri@deis.unibo.it

[3] Universidad Simón Bolivar
Dpto. de Electronica y Circuitos
Edif. Fisica y Electronica I. 1080A - Miranda, Venezuela
jbogado@usb.ve

[4] Instituto Politécnico Nacional. Centro de Investigacin y de Estudios Avanzados (CINVESTAV). 07300, Mexico DF, México
jibarra@ctrl.cinvestav.mx

Summary. The goal of this chapter is to experimentally study the behaviour of bilateral controllers. Previous chapter introduced a general framework for designing bilateral systems. In this chapter, performance of some of the most common control schemes is analyzed. In case of negligible time delay, experiments have been carried out by implementing position-position and force-position controllers. Force-velocity control scheme has been used in presence of communication time delays. Experiments have been run with a constant delay of 100 ms in the communication channel. In this case, master-slave interconnection has been described by means of the corresponding scattering variables. Experimental results of the different control schemes have been obtained using an industrial telemanipulator of 6 degrees of freedom manufactured by Kraft Telerobotics.

11.1 Introduction

This chapter is focused on interpreting the performance of some bilateral controllers. Theoretical results of chapter ten have been used in order to understand how the telemanipulator controllers work. However other effects appear when experiments are carried out. The most important are gravitational and inertial effects which imply that upward and downward trajectories possess different profiles.

The environment where the experiment was performed has been a hard object with very little elasticity. It is equivalent to a quite rigid spring. In block diagrams of the different bilateral controllers, the environment has been modelled like a

M. Ferre et al. (Eds.): Advances in Telerobotics, STAR 31, pp. 177–190, 2007.
springerlink.com © Springer-Verlag Berlin Heidelberg 2007

spring that simplifies the analysis of bilateral controllers, and equally obtaining valid conclusions. The environments are also modelled by an impedance, which represents a more complex combination of spring, inertia and a damping. Details about environment modelling can be found in [1, 2].

The outline of this chapter is described as follows. Sec. 2 analyses the performance of position-position and force-position control schemes without communication time delays. Next section explains cases of time delays in the communication channel. In such case, experiments have been carried out by using a force-velocity controller with a time delay of 100 ms in the communication channel. This channel is modelled using the corresponding wave variables. Finally, the last section summarizes the main results obtained in this work. The test bed used for these experiments is based on the Grips telemanipulator manufactured by Kraft Telerobotics. It is shown in an appendix located at the end of this chapter.

11.2 Analysis of Position-Position and Force-Position Control Schemes

Position-position and force-position are classical control schemes used in telerobotics. They have been chosen in experiments of communication channel without time delay due to their simplicity and robustness. Next sections describe the corresponding block diagrams and the results obtained for each controller. Differences between the two controllers are highlighted in order to better understand their performance. In the following diagrams, it is assumed that the master and the slave controllers are of the proportional type (i.e described by the static gains Kpm and Kps), so as to simplify calculations and better reach conclusions. Moreover, experiments have been carried out using PD controllers, as they improve the dynamic behaviour of both devices.

11.2.1 Performance of Position-Position Control Scheme

The simplest bilateral controller to be implemented is position-position scheme since no force sensors are required. It is only required to know the master and the slave joint position values. This scheme has been applied since the development of the first nuclear industry[3, 4]. Fig. 11.1 shows the position-position control scheme. In this block diagram, the master and slave dynamics are described by $M(s)$ and $S(s)$ respectively. The controllers are represented by the gains K_{pm} and K_{ps}. Finally, K_e represents the remote environment. Therefore, interaction forces between slave and environmente are proportional to slave position variation.

In Fig. 11.1, f_h represents the force input applied by the operator to the master. If this force is not compensated by the force reflected from the master f_m, then the master moves. As a consequence of such movement, new position references will be given to the slave control loop. In this manner, the operator guides the slave and becomes aware of the exerted forces. Position trajectory of each device will be used as reference for the other devices. Therefore, the position error e_p is calculated as the difference between master and slave positions $(x_m - x_s)$.

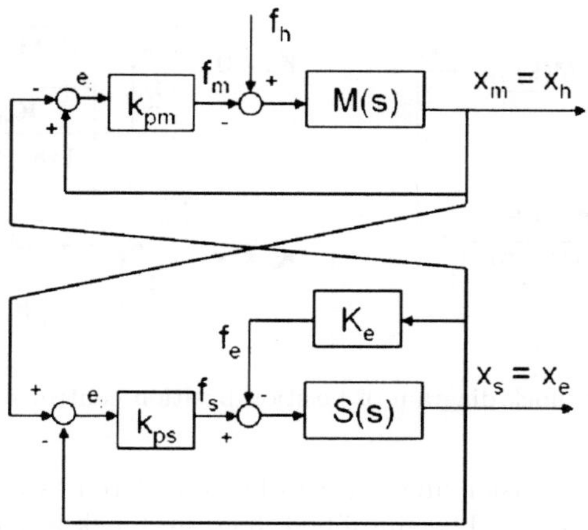

Fig. 11.1. Position-position control scheme

The force reflected to the operator (f_m) is proportional to the position error according to the master controller gain (K_{pm}). In a similar way, the force exerted by the slave (f_s) is also proportional to the position error according to the slave controller gain (K_{ps}). Therefore, position error between both devices is displayed on the operator as a resistance against his movements. Mathematically, the position-position bilateral control scheme is described as:

$$f_m(s) = k_{pm}\ e_p(s) \tag{11.1}$$

$$e_p(s) = x_m(s) - x_s(s) \tag{11.2}$$

$$x_m(s) = M(s)\ (f_h(s) - f_m(s)) \tag{11.3}$$

$$x_m(s) = x_h(s) \tag{11.4}$$

$$x_s(s) = S(s)\ (f_s(s) - f_e(s)) \tag{11.5}$$

$$f_s(s) = k_{ps}\ e_p(s) \tag{11.6}$$

$$f_e(s) = Z_e(s)\ x_s(s) \tag{11.7}$$

$$x_s(s) = x_e(s)^1 \tag{11.8}$$

It would be recommendable to transform the block scheme shown in Fig. 11.1 in order to obtain the transfer function $H(s) = f_m(s)/x_m(s)$. It defines the relation between master's force and position signals. $H(s)$ represents how the master reflects the remote environment to the operator. This transfer function relates positions generated by the operator to the forces feed backed from the master.

[1] It implies that slave is in contact with the environment.

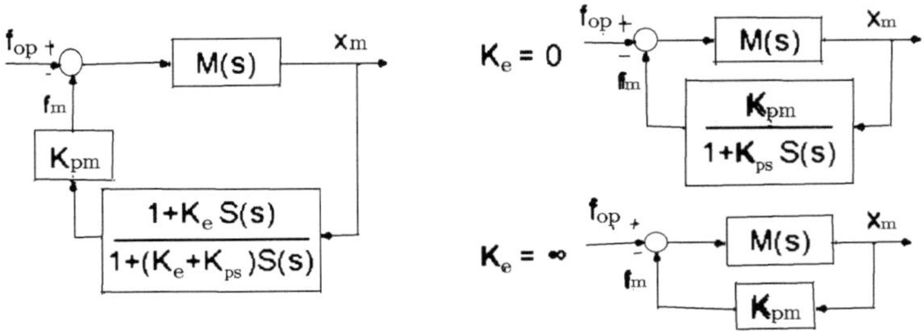

Fig. 11.2. Simplified block diagram of position-position control seen from the master

In other words, this transfer function can be considered as a sort of impedance transmitted to operator [2]. The simplified diagram block is shown in Fig. 11.2.

$$H(s) = \frac{f_m(s)}{x_m(s)} = K_{pm} \frac{1 + S(s)K_e)}{1 + S(s)\,(K_e + K_{ps})} \tag{11.9}$$

An interesting way to assess the influence of each parameter in $H(s)$ is to analyse extreme cases, i.e. when the environment impedance is null or infinite. When impedance is null ($K_e = 0$), then $H(s)$ is given by $\frac{K_{pm}}{1+S(s)K_{ps}}$; i.e. operator will perceive some kind of strength although no force is being exerted by the slave. The second option is an infinite value of K_e, which represents the case under which the slave is in contact with an ideally rigid environment, and no deformation takes place apart from the exerted force. In this case, $H(s)$ is equal to K_{pm}. It implies that position-position control scheme has a maximum impedance to be perceived by the operator and this value is given by K_{pm}.

Experiments Using Position-Position Control

Forces reflected on the operator are proportional to the position error, which represents the error between the master and slave positions. In this experiment, actuation on the slave drives the electrohydraulic servo-valves, which is also proportional to the position error. Fig. 11.3 shows wrist-pitch joint evolution during a contact. Position of the master x_m, slave x_s and force reflected f_m to operator are shown. Data about this articulation are shown because they carry much information such as effect of gravity and sensitivity of exerted forces due to the proximity to the robot wrist. These experiments have been carried out by a telemanipulator from Kraft Telerobotics as described in the appendix.

As Fig. 11.3 shows the behaviour of position-position control. There are three different steps in this experiment:

- Downward trajectories. In this step master and slave positions are similar. It is advisable to highlight that a slight force is reflected to the operator. This

[2] Note that physically an “impedance” relates an “effort” (e.g. a force) to a “flow” (e.g. a velocity, not a position).

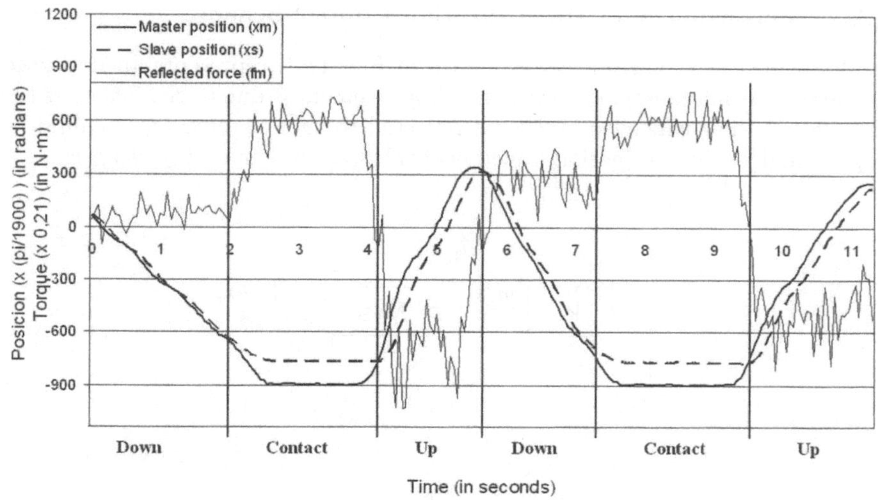

Fig. 11.3. Position-position control experiments (K_e:rigid environment, $K_{pm} = 2.75$, $K_{ps} = 6.5$)

force is due to the drag effect of the position-position control. However, this force is reduced due to an equal direction of gravity. Inertia forces are also present but they affect less than drag effect.

- Contact steps. In this case, the slave exerts higher forces on the environment as position error increases. Moreover, operator receives higher force in this case. The contact with a rigid object is thereby perceived as an elastic contact.
- Upward trajectories. The effect of gravity on this joint can be clearly identified in this step. As the master position increases, position error increases faster than downward trajectories. This phenomenon is due to the effect of gravity, which means that the WP joint moves against the gravity when master position increases. Therefore, the system responds more slowly, which means that position error is greater.

As conclusions, little sensibility being offered is a major inconvenience of this type of bilateral control. It is due to the drag effect that the master has on the slave, since all movements imply a position error; likewise, such error is also reflected on the master. In this experiment, forces reflected to operator in upward trajectories are similar to the contact forces since gravity and drag effect are added. Moreover, only elastic objects can be reflected to operator since collision with rigid objects are transformed to elastic contacts. Therefore, position-position controller offer a poor operator perception. However, main advantages are robustness in the control and the easiness of implementation, since position signals are the only ones transmitted and no force sensors are needed.

11.2.2 Performance of Force-Position Control Scheme

The force-position control scheme, shown in Fig. 11.4, represents an improved version of position-position control. The improvement is due to the fact that the drag effect is now cancelled and force reflected to the operator (f_m) should be proportional to the interaction force between slave and environment (f_e).

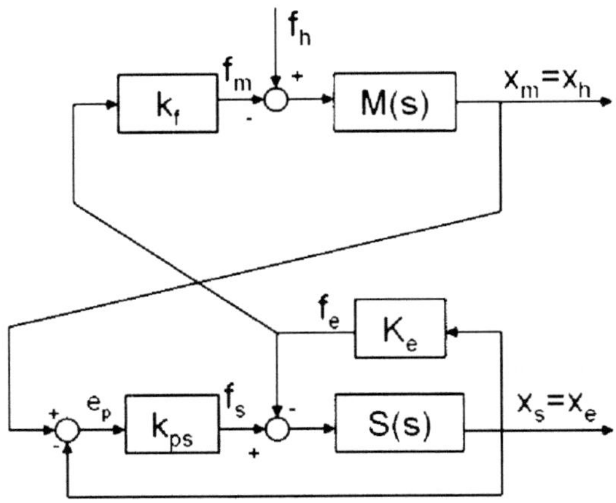

Fig. 11.4. Force-position control scheme

This kind of control has been implemented in many bilateral systems, such as those described in [5, 6, 7, 8]. The telemanipulation system with force-position control is described by:

$$f_m(s) = k_f \; f_e(s) \tag{11.10}$$

$$e_p(s) = x_m(s) - x_s(s) \tag{11.11}$$

$$x_m(s) = M(s) \; (f_h(s) - f_m(s)) \tag{11.12}$$

$$x_m(s) = x_h(s) \tag{11.13}$$

$$x_s(s) = S(s) \; (f_s(s) - f_e(s)) \tag{11.14}$$

$$f_s(s) = k_{ps} \; e_p(s) \tag{11.15}$$

$$f_e(s) = K_e \; x_s(s) \tag{11.16}$$

$$x_s(s) = x_e(s)^3 \tag{11.17}$$

Compared to the position-position scheme, only (11.10) differs. This equation represents the forces reflected by the master. Fig. 11.5 reports a block diagram showing the relation between forces applied by the operator f_h and the master

[3] It implies that slave is in contact with the environment

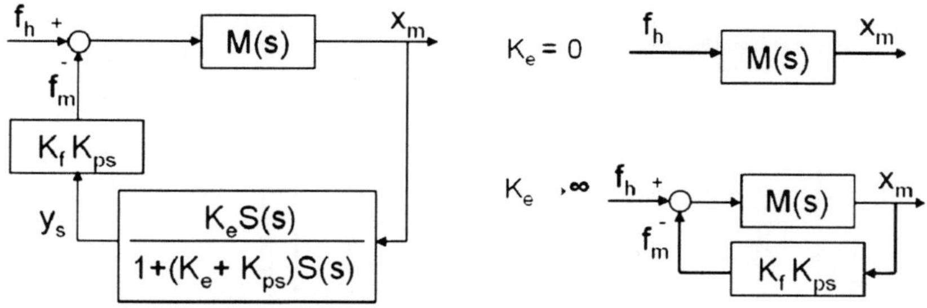

Fig. 11.5. Simplified block diagram of force-position control seen from the master

position $x_m(s)$. The value of $H(s)$ for this control scheme is described in the following equation.

$$H(s) = \frac{f_m(s)}{x_m(s)} = K_f \frac{K_e S(s) K_{ps}}{1 + S(s)(K_e + K_{ps})} \tag{11.18}$$

Note that this expression of $H(s)$ is significantly different with respect to equation (11.9). In this case, K_e is a factor in the numerator, it implies that low values of K_e also give low values of $H(s)$. This represents an improvement if compared to the position-position value of $H(s)$, since soft contacts can be properly reflected. Moreover, when environment contact is missing, $H(s)$ value is null. That is, applied force by the operator is only used for moving the master device. For high K_e values, maximum $H(s)$ value is equal to the product of $K_f K_{ps}$. Therefore, maximum $H(s)$ value is limited again.

Experiments Using Force-Position Control

The application of such control scheme requires knowing forces applied by the slave. Kraft hydraulic telemanipulator comes with sensors that measure the difference of pressure in the hydraulic servo-valve. It is as an average measurement of the torque applied on the corresponding joint.

Fig. 11.6 shows an experiment implementing force-position control when a collision happens. Master position is the reference in the slave control loop; therefore, actuation on their articulation is proportional to the position error. This graph shows the following steps:

- Downward trajectories. A small force is reflected to the operator in this case. This disturbance is due to the effect of gravity and inertia. Drag effect does not exist for this controller; therefore the operator perception is improved.
- Contact step has two stages; first stage is elastic, where master and slave positions are very similar and force sensor increases almost linearly. Second stage; the contact stops being elastic, the slave and the master are stopped and in such case, the torque reflected becomes constant. Object is rigid but telerobot joints have some elasticity. When the object becomes rigid, the

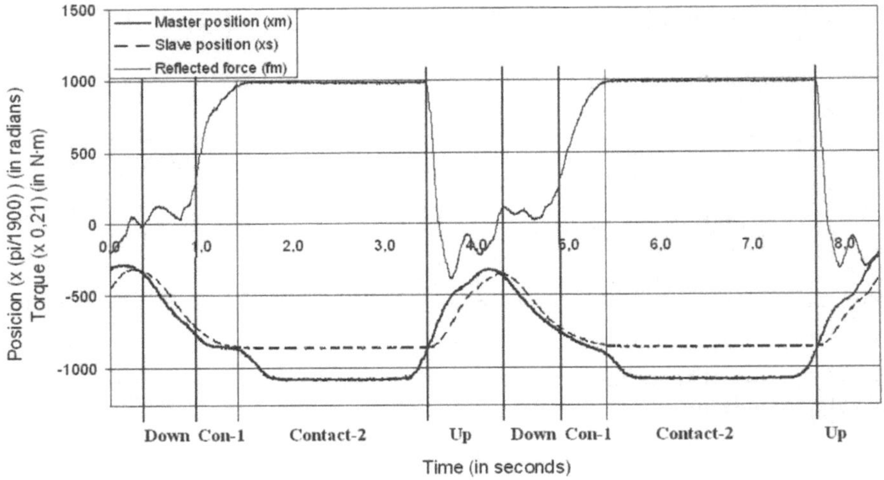

Fig. 11.6. Force-position control experiments. (K_e:rigid environment, $K_f K_{ps} = 6.875$)

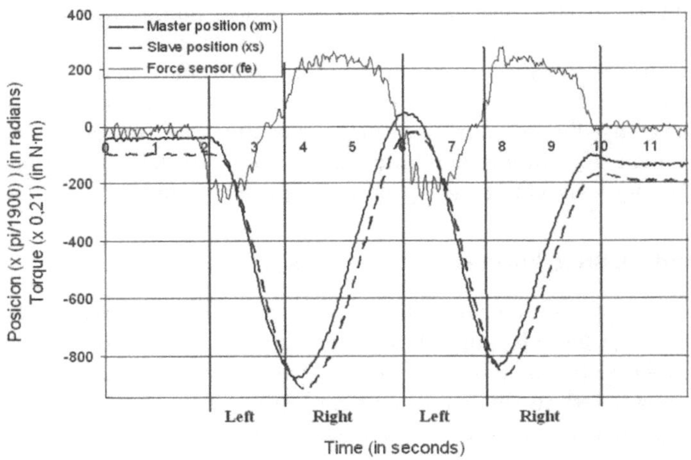

Fig. 11.7. Reverse behaviour of force-position control. ($K_e = 0$, $K_f K_{ps} = 25$)

slave then stops. Nevertheless, the operator perceives the object elastic, which allows the operator to move the master.

- Upward trajectories. In this case gravity appears again and has a higher effect since the movement is against the gravity.

As conclusion, force-position control has had a better performance than position-position since reflected forces are closer to the forces exerted by the slave robot. However, gravity and inertial slightly deteriorate operator perception.

Fig. 11.7 shows the bilateral system behaviour after exerting a force on the SA slave articulation. Once force is exerted on the slave, it is reflected on the master-arm, which then provokes a torque in the corresponding articulation that consequently moves the joint. The mentioned movement of the master provokes simultaneous movements on the slave, which is a reverse flow of the system. It means that applying torque on the slave the master is moved. Great sensitivity of this behaviour is shown by the force-position control in its force reflection control in spite of having big difference in the scale of devices. This reverse behaviour of the system is also possible with algorithm of position-position control. If the system is dimensioned for small devices to move bigger devices, scale of devices thereby play an important role. On the other hand, forces are reproduced in a reversed scale.

11.3 Experiments Implementing Force-Velocity Control with Communication Time Delays

As described in the previous chapter, passive techniques have been successfully applied in making telemanipulators stable upon presence of communication time delays [9, 10, 11]. Fig. 10.5 describes a block diagram that includes all transformations applied to the communication line variables in order to guarantee a passive channel. Velocities (v) and forces(f) are the variables that characterizes a bilateral telerobot. v_m and f_{md} refered to the master side of the communication channel and v_{ds} and f_s to the slave side. b is a parameter that define the remote environment impedance and the wave variables are u_m, u_s, w_m and w_s. Wave variables are used for describing the transfer of energy between the master and the slave.

Diagram 10.5 defines force reflected by the master (f_{md}) and slave velocity (v_{sd}) are calculated according to the values of (i) master velocity (v_m), (ii) interaction force between the slave and the environment (f_s), (iii) communication time delay (t) and (iv) line impedance (b). This diagram permits to obtain (11.19) among the above variables.

$$\begin{cases} f_{hd}(s) = \frac{b}{2}v_m(s) + \frac{1}{2}e^{-sT}f_s(s) \\ v_{sd}(s) = \frac{1}{2}e^{-sT}v_h(s) - \frac{1}{2b}f_s(s) \end{cases} \tag{11.19}$$

11.3.1 Experiment in Presence of Communication Delays

In order to obtain a successful application of this control algorithm, it is important to comply with some requirements regarding communications between the master and the slave. Some requirements such as guarantee a constant delay and avoiding loss of data have to be complied. Fig. 11.8 shows the results obtained from the teleoperation of the Kraft system with a communication time delay of 100ms. No passivity has been applied in this case. As it can be observed in the figure, the system becomes unstable when the slave collides with rigid objects.

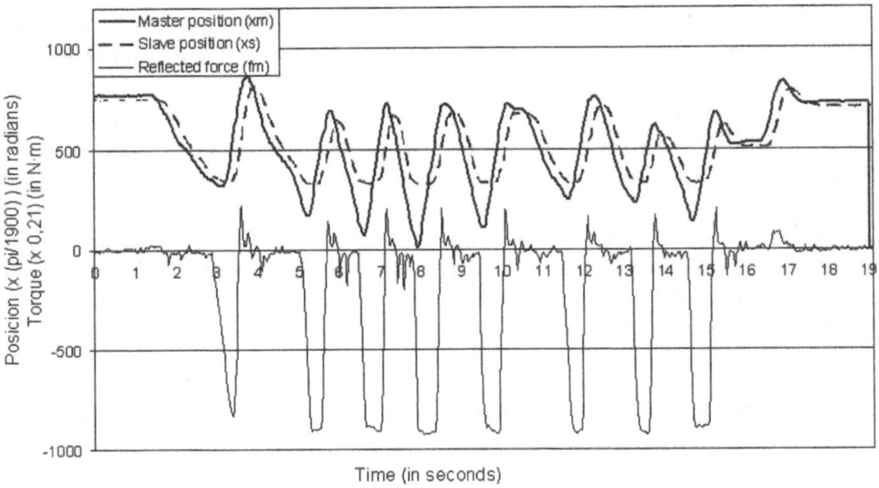

Fig. 11.8. Teleoperator master-slave system with a 100ms delay, without a controller based on passity

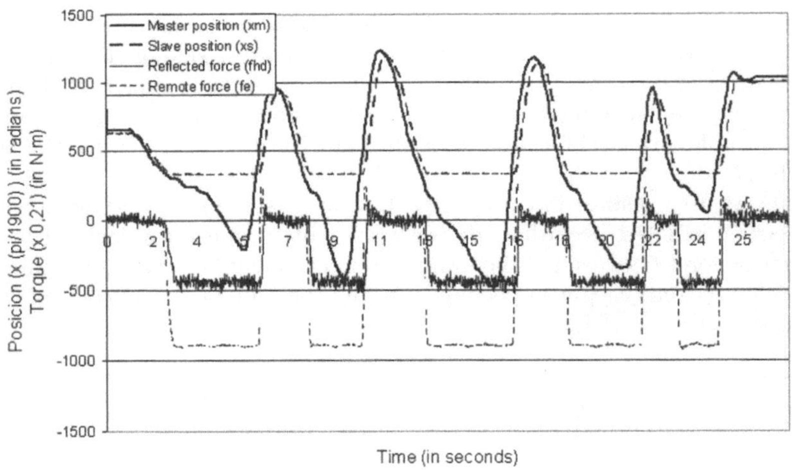

Fig. 11.9. Teleoperator master-slave system with a 100ms delay, with a controller based on passity

Upon collision of the slave, the master reflects a delayed force to the operator which does not correspond to the current master movement. It implies an excessive force reaction reflected on the operator. Consequently, the master and slave have oscillatory movements around contact points. Due to the delay, if the theory of passivity is not applied when the slave collides with a rigid object, an excessive force is reflected on the master, which makes the control of the bilateral system difficult upon collision.

Fig. 11.9 shows the behaviour of the system that already includes a controller based on passivity. In such case, how the system behaves in a stable manner upon collision with rigid objects in spite of a 100 ms delay will be analyzed. In Fig. 11.9, the delay between the position of the master and slave is known on the one hand, and on the other hand, it can be observed that upon collision with a rigid object, force reflected by the master is limited, independently to the position that has been reached by the master. This is fundamental due to the regulator that corrects forces reflected on the master, and also by the correction of the slave position, (11.19). These corrections allow the limitation of applied forces upon contact with the environment. They are done by the correction of slave velocity (v_{ed}) and also by the reflected force on the operator(f_{hd}). Such behaviour makes the system stable.

Stability of the system clearly shows the fact that the telerobot can be maintained balanced in spite of the communications delay. They can be clearly seen in collisions where a permanent contact is done without making the system unstable. If passivity was not applied, such contact to the operator would repel, as shown in the Fig. 11.8. It is why bilateral system is maintained stable even if an error clearly exists in the position between the master and the slave. Such error disappears when the slave does not collide with the environment.

11.4 Conclusions

Bilateral controllers with or without communications time delays have been experimentally analyzed in this chapter. First, position-position and force-position controllers have been tested. Position-position scheme allows designing simple controllers. However, operator gets a poor perception of the remote interaction forces since reflected forces include drag and gravity effects. Moreover, all objects are perceived by the operator as elastic. Therefore, the operator obtains a very poor perception of remote exerted forces. Force-position control represents a significant improvement with respect to position-position. Main advantage of this controller is that remote forces are directly reflected on the operator as the drag effect is cancelled. Therefore, the system becomes more sensible to the forces exerted by the slave. However, the inertia and gravity of master and slave devices deteriorate the operator perception since they are superimposed to the remote environment forces.

Finally, force-velocity controllers in presence of communication delays have been implemented and experimentally tested. Passivity theory has been applied in implementing bilateral systems with time delay. In the experiments, passive techniques have demonstrated their usefulness in guaranteeing stability of bilateral systems in presence of significant communication time delay (100 ms). Nevertheless, significant difference between master and slave positions can be observed when the slave collides with an object.

References

1. R. Volpe and P. Khosla. A theoretical and experimental investigation of impact control of manipulators. *Int. Journal of Robotics Research*, 12 (4):351-365, 1993.
2. F. Janabi-Sharifi. Collision: Modelling , simulation and identification of robotic manipulators interacting with environments. *Journal of Intelligent and Robotic Systems*, 13 (1):1–44, 1995.
3. T. Raimondi and L. Galbiati. Manipulators mascot iv used in jet and propects of enhancement. In EURATOM, editor, *Teleoperation: Numerical Simulation and Experimental Validation. Edited by M.C. Becquet*, pages 139–161, 1992.
4. B. Schott, K. Hendrich, K. Dring, and H. Bressem. First experience with the electrical master-slave manipulator emsm3. In *ANS 7th Topical Meeting on Robotics and Remote Systems*, 1997.
5. A.K. Bejczy and J.K. Salisbury. Controlling remote manipulators through kinesthetic coupling. *ASME Computers in Mechanical Engineering*, 2 (1):48–60, 1983.
6. B. Hannaford, L. Wood, D. McAffee, and H. Zak. Performance evaluation of a six axis generalized force reflecting teleoperator. *IEEE Transactions on Systems, Man, and Cybernetics*, 21:620–633, 1991.
7. R.W. Daniel and P.R. McAree. Fundamental limits of performance for force reflecting teleoperation. *Int. Journal of Robotics Research*, 17(8):811–830, 1998.
8. P.J. Fischer. *A Novel Input Device for Force Controlof Robots.* PhD thesis, University of Oxford, 1993.
9. R.J. Anderson and M.W. Spong. Bilateral control for teleoperators with time delay. *IEEE Transactions on Automatic Control*, 34 (5):494–501, 1989.
10. G. Niemeyer and J.E. Slotine. Stable adaptive teleoperation. *IEEE Journal of Oceanographic Engineering*, 16 (1):152–162, 1991.
11. P. Arcara and C. Melchiorri. Control schemes for teleoperation with time delay: a comparative study. *Robotics and Autonomous Systems*, 38, 2002.
12. R. Aracil, M. Ferre, M. Hernando, E. Pinto, and J.M. Sebastian. Telerobotic system for live-power line maintenance:robtet. *Control Engineering Practice*, 10:1271–1281, 2002.
13. Texas Instruments. *TMS320F2810 and TMS320F2812, Digital Signal Processors.* Data Manual, Texas Instrument. Literature Number: SPRS174J, 2003.

Appendix: Test Bed for Teleoperation

A test bed has been developed in order to experimentally evaluate bilateral control algorithms. This test bed is made up of the Grips telemanipulator from Kraft Telerobotics and a control system which was specifically designed for controlling theses devices. This control device is based on a DSP from Texas Instruments. Both components are explained, with their major relevance being highlighted in the following sections.

Telerobotics Description

Grips telemanipulator from Kraft Telerobotics has been used in a number of teleoperation applications such as underwater tasks and maintenance of electrical lines [12]. A major advantage of this equipment is that it allows evaluating control

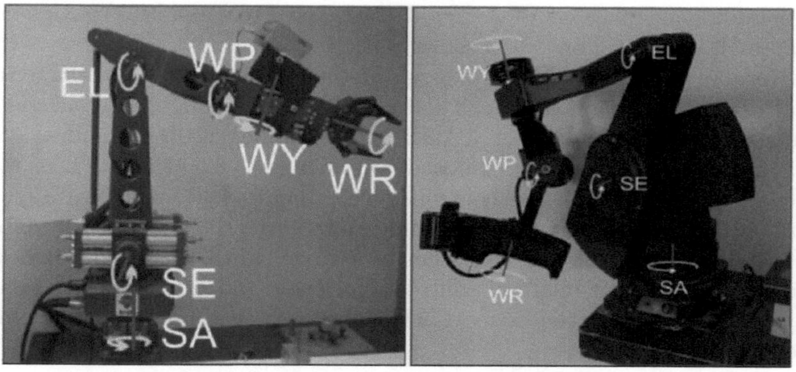

Fig. 11.10. Slave robot (left) and master-arm (right) by Kraft Telerobotics

techniques on a commercial system and therefore being more accustomed to habitual environments than prototypes made for such tests. Fig. 11.10 shows both devices master and slave.

The slave is a hydraulic robot of 6 DoF, which is shown in the left picture of Fig. 11.10. It is robust device and has a high load capacity up to 50 kg. Its reach is a semi-sphere with 1.2 meters of radius. In order to control this device, the information about the position of every join and an average measurement of the applied torque in the first five joints are read. A potentiometer for measuring the angular position of each joint is used in the six DoF. Pressure sensors in five joins allow calculating the applied torque in each articulation. Fig. 11.10 shows the six DoF of the slave manipulator. Abbreviations are as follows: SA-Shoulder Azimuth, SE - Shoulder Elevation, EL - Elbow Elevation, WY - Wrist Yaw, WP - Wrist Pitch, WR - Wrist Rotation.

The master arm is light and has an anthropomorphic design which adapts comfortably to the operators hand and forearm, thereby position master references are generated by the hand movements. They are sent to the slave controller in order to describe the proper trajectory. Master-arm has electrical motors in its first five joins that allow reflecting torques on the corresponding master DoF. The angular position of every joint is measured by a potentiometer. This device is shown in the right picture of Fig. 11.10. As shown in the figure, it exists a direct correspondence between the articulations of the master and slave, master joints have the same names as slave joints. It is important to mention that there is a slight difference between the kinematics of the master and the slave. It consists of changes in the 4th and 5th DoF of both devices. This change eases movement in the operator's wrist and does not affect its propioception guiding the slave.

A control board has been specifically designed using a DSP Texas Instruments (TMS320F2810) [13]. This device has been used in implementing all schemes of control for the master and slave. This controller is in charged of doing all corresponding calculations to the regulators, converters and scale of signals and manages all input and output communication.

Performance of this controller depends on the DSP used. Its principal technical characteristics are as follows; a calculating power of 150 MIPS, 256 KB memory, Communication ports SPI, CAN, USB and a number of digital and analogical inputs and outputs. Six control loops of every device to 8.5 kHz have been closed. However, all resources of the controller were required to concentrate on such task. In order to do all exterior communications correctly, lowering speed for the processing of control loop to 2 kHz has been done, which has been sufficient for the implementation of bilateral control algorithms and transmission of data in real time. With such speed, a good behaviour is reached in the control of every device separately.

12

Human Perceived Transparency with Time Delay

Sandra Hirche[1] and Martin Buss[2]

[1] Tokyo Institute of Technology
Fujita Lab, Dept. of Mechanical and Control Engineering
152-8552 Tokyo, Japan*
S.Hirche@ieee.org

[2] Technische Universität München
Institute of Automatic Control Engineering
D-80290 Munich, Germany
M.Buss@ieee.org

Summary. Transparency – in the sense that the technical systems and communication network should not be felt by the human – is one of the key issues in telerobotics control design. The communication characteristics is one of the crucial factors for the achievable transparency level in bilateral telerobotic control architectures. Especially time delay – resulting from the communication network between the operator site and the tele-robot – deteriorates the realistic (transparent) perception of the remote environment. The quantitative analysis of the time delay influences on transparency is the major goal of this chapter. Technical measures along with human haptic perception characteristics play a key role when evaluating transparency. The guiding questions for this chapter are: how does constant time delay modify the mechanical properties displayed to the human, and can the human perceive this distortion or not. The mass, spring, and damper characteristics as displayed to the human are derived as function of the time delay and the environment parameters. Known psychophysical facts are applied to analyze and interpret the results from a human perception point of view. The results are validated in simulations, experiments, and human user studies.

12.1 Introduction

The communication between the operator side with the multi-modal human system interface and the remote side with the telerobot typically takes place over a communication network as depicted in Fig. 12.1. As a result the motion and force signals arrive delayed at the corresponding receiver side. In terrestial telerobotics application the transmission time delay is typically in the range of some milliseconds up to several hundred milliseconds depending on distance and communication infrastructure; in space application the data transmission may easily take several seconds. Time delay in the haptic feedback loop represents one of the key challenges in control design with respect to stability and transparency. Without appropriate control measures even small time delay may destablize the telerobotic system [1] resulting in a severe hazard to the safety of the human and the remote environment. The wave variable (scattering)

* This research has been performed while the first author was with the Institute of Automatic Control Engineering, Technische Universität München, D-80290 Munich, Germany.

M. Ferre et al. (Eds.): Advances in Telerobotics, STAR 31, pp. 191–209, 2007.
springerlink.com © Springer-Verlag Berlin Heidelberg 2007

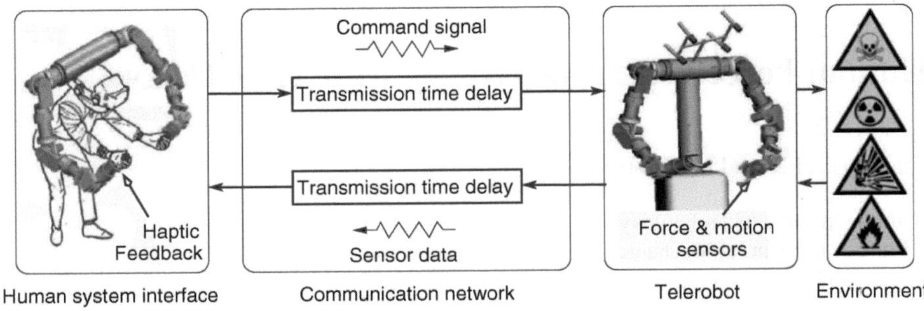

Fig. 12.1. Telerobotic system with transmission time delay

transformation introduced in Sec. 10.4 represents one of the most prominent control approaches to stabilize the telerobotic system with time delay, see also [2, 3]. Stability is guaranteed for arbitrarily large constant time delay. Nevertheless, the time delay value has a significant influence on transparency.

Transparency – in the sense that the technical systems and communication network should not be felt by the human, i.e. the operator should feel as if directly being present and active in the remote environment – is aside from stability one of the key issues in telerobotic systems. Ideally, the human operator feels as if directly interacting with the (remote) task [4]. This is expressed in transparency criteria requiring that mechanical properties of the environment are exactly transmitted to the human operator, motion and forces at the master and the slave device should be equal [5, 6]. Naturally, there is a trade-off between transparency and robust stability in all control schemes, i.e. ideal transparency is not achievable in real systems [5, 7, 8]. Transparency further deteriorates in telerobotic systems with time delay as the bandwidth of the closed loop system has to be severely reduced in order to achieve stability [9]. If the time delay between operator action and the corresponding haptic feedback is too large the telerobotic systems becomes inoperable. Nevertheless, direct haptic feedback is still beneficial for task completion even with a significant time delay of up to one second as experimentally validated in [10, 11]. Other typical communication effects like time-varying delay and the loss of data additionally influence stability and transparency, but are beyond the scope of this chapter. For control architectures considering time-varying delay refer to [12, 13, 14, 15, 16, 17] and for packet loss to [18, 19, 20, 21]. A transparency analysis considering communication effects can be found in [22].

Main focus in this chapter is on *how* constant time delay and the tuning of the control influence the human perception of the remote environment. The *feel* of the human is essential, hence, the consideration of human factors is an important issue for transparency evaluation. The transparency measures known from the literature, such as the maneuverability index [6], impedance error norms [23], or the Z-width [24] are well-suited for the comparison of control schemes in technical terms. However, the interpretation considering human factors is difficult. One of the main contributions of this chapter is a time delay transparency analysis from a human haptic perception point of view using methods and results from psychophysics. Humans may not discriminate arbitrarily small differences in a physical quantity, expressed by the *just noticeable difference*

(JND). Numerous psychophysical studies on the JND of mechanical impedance parameters, such as stiffness [25, 26], inertia [27], and viscosity [27, 28], exist. These results are applied in the transparency evaluation on the basis of a comparison between the mechanical parameters of the impedance transmitted to the human operator and of the real environment impedance. Exemplarily, the widely used wave (scattering) variable control approach, see Sec. 12.4 and [2, 3] for details, is considered in the analysis. For a comparison of other control schemes using a similar approach refer to [29]. Using a low frequency approximative analysis based on a Padé first order model for the time delay it is shown that environment mechanical parameters are distorted by communication time delay and the wave (scattering) variable approach, i.e. a) in free space motion communication time delay introduces artificial inertia; b) stiff environments are displayed softer; c) displayable stiffness is upper bounded; d) environment stiffness discrimination is limited; e) as a result there is a detection threshold for relative changes in time delay. The results are analyzed from a human perception point of view using psychophysical insights of JNDs for mechanical parameters; design issues are discussed. The results are validated in experiments and human user studies. The remainder of this chapter is organized as follows: in Section 2 transparency measures and psychophysical aspects are reviewed. Section 3 provides the objective transparency analysis followed by a perception oriented analysis in Section 4 and design issues in Section 5. Experimental validation results are presented in Section 5.

12.2 Background

In the following the state of the art in transparency analysis and the psychophysical results, related to this work are briefly reviewed. Further the assumptions for the subsequent analysis are clarified.

12.2.1 Transparency

The telerobotic system design goal is that the human cannot distinguish between direct interaction with an environment and teleoperated interaction with an remote environment. Then the system is called *transparent*. Ideally the human feels as if directly performing the task in the (remote) environment [4]. According to [6] a system is transparent if the positions and forces at master and slave device are equal, i.e. $x_h = x_e$ and $f_h = f_e$. The derived transparency measure, called maneuverability, comprises two values based on the integral norm in the frequency domain for the transfer functions in operator force to position error and operator force to force error, respectively. Small values indicate a good level of transparency. Alternatively, in [5] for transparency the equality of the impedance transmitted to the human operator and the environment impedance is required

$$Z_t = Z_e, \tag{12.1}$$

see Fig. 12.2 for a visualization of these impedances. Measures derived from this criterion are integral impedance error norms in the frequency domain as applied in [23] and the Z-width [24], which expresses the dynamic range of the impedance transmitted to the operator. It is quantified for the two extreme values of environment impedance $Z_e = 0$

(free space motion) and $Z_e \to \infty$ (infinitely stiff wall). Accordingly, transparency is good if $Z_t \to 0$ and $Z_t \to \infty$, respectively. All mentioned transparency measures are well-suited for the qualitative comparison of control schemes, however, a direct relation of the obtained values to psychophysical findings and such the interpretation using human haptic perception characteristics is difficult.

12.2.2 Human Haptic Perception

The level of transparency, according to the transparency definition, can be interpreted as the human perceived performance of the telerobotic control system. Hence the *feel* of the human is essential for transparency evaluation. To measure this *feel* depending on physical stimuli is the goal of psychophysics, which is the branch of psychology concerned with the quantitative relation between physical stimuli and the sensations and perceptions evoked by these stimuli, see [30] for a comprehensive introduction.

It is well known that the human cannot discriminate arbitrarily small differences in a physical quantity, expressed by the *just noticeable difference* (JND), the smallest difference in a sensory input that is perceivable by a human being. For many sensory modalities the JND is an increasing function of the base level of input. For most force-related physical properties the ratio of the two is roughly constant over a large range [31] and can therefore be represented by Weber's law [32]

$$\frac{\Delta I}{I} = c, \tag{12.2}$$

where I is the original intensity of stimulation, ΔI is the addition to it required for the difference to be perceivable, and c is the Weber fraction. Discrimination thresholds for mechanical parameters such as stiffness, inertia, and viscosity are typically given as the percentual change with respect to the original intensity of stimulation. Accordingly, we will refer to the percentual discrimination threshold as JND in this chapter. Some of the most relevant JNDs for haptic telerobotic systems are summarized in Table 12.1.

Remark 1. It is well known that the experimental conditions have a significant influence on the results gained in psychophysical experiments. This explains the in some cases wide variation for JNDs. Additionally, some parameters, as e.g. inertia, are suspected of not [33] following Weber's law given by (12.2). As the results are empirically obtained, they generally represent a statistical quantity, i.e. individual differences exist.

Table 12.1. Perceptual discrimination thresholds (JND) for haptics related properties

Physical property	JND [%]	Experimental conditions
Stiffness	23 ± 3 [25]	arm/forearm, cross-limb-matching
	8 [26]	pinch-fingers, work/maximum force applied
Viscosity	34 ± 5 [28]	arm/forearm, cross-limb-matching, > 20 Ns/m
	13.6 ± 3 [27]	pinch-fingers, at 120 Ns/m
Inertia	21 ± 3.5 [27]	pinch-fingers, at 12 kg

12.2.3 Assumptions

For discussion simplicity the following assumptions are made for the subsequent transparency analysis:

A1 A telerobotic system with one degree of freedom (1DoF) is assumed, with the master and the slave device being kinematically similar. An extension of the proposed approach to the more general case is straightforward if the cartesian directions in the master and the slave device are dynamically decoupled which can be achieved by appropriate local control.

A2 The slave device is assumed to be controlled such that its own dynamics is negligible, i.e. the slave device velocity is equal to the desired slave device velocity $v_s = v_e = v_s^d$, see Fig. 12.1. The same assumption is made for the master device, such that the force displayed to the human is equal to the desired master force $f_h = f_h^d$. The extension to the case with non-negligible master and slave dynamics is discussed later in this chapter for the 1DoF case. The extension to multi-DoF devices with non-negligible dynamics is straightforward in the case of decoupled cartesian directions which can be achieved by appropriate local control.

A3 Assume that the environment impedance Z_e can be approximated by a linear time-invariant (LTI) system, which is valid for most considered environments. The environment impedance can then be represented by the transfer function $Z_e(s) = f_e(s)/v_e(s)$ with $s = \sigma + j\omega$ denoting the Laplace variable.

A4 The time delays T_1, T_2 in the forward and backward path, respectively are assumed to be constant and arbitrarily large.

A5 The wave (scattering) variable control approach is applied to stabilize the overall telerobotic system in the presence of constant time delay, see Sec. 12.4 and [2, 3] for details.

12.3 Transparency Analysis

The existence of JND results for mechanical impedance parameters such as stiffness, damping, and inertia encourage a transparency evaluation based on the comparison of the parameters of the impedance displayed to the human and the real environment impedance. Another advantage is that the analysis can conveniently be performed in the frequency domain. The goal is to derive the displayed stiffness, damping and inertia depending on the time delay value. Therefore the displayed impedance Z_t is expressed as function of the round-trip time delay $T = T_1 + T_2$ and the environment impedance Z_e. Straightforward manipulation of the equations for the wave (scattering) variable transformation (10.12)with $f_m = f_h$, $v_m = v_h$ using assumptions A1-A5 yields

$$Z_t(s) = b\frac{1 + R(s)\, e^{-sT}}{1 - R(s)\, e^{-sT}} \quad \text{with} \quad R(s) = \frac{Z_e(s) - b}{Z_e(s) + b}, \tag{12.3}$$

where $b > 0$ represents the wave impedance, the parameter of the wave (scattering) variable transformation. Note that for zero time delay $T = 0$ the displayed impedance is equal to the environment impedance, meaning ideal transparency in the sense of (12.1). For non-zero time delay the displayed and environment impedance differ in general.

Due to the delay element the transfer function has an infinite number of poles and zeros rendering the interpretation of the displayed impedance as simple mass-spring-damper system difficult. Therefore, the transfer function from (12.3) is approximated by a lower order system.

12.3.1 Analytical Low Frequency Approximation

The approximation of the displayed impedance transfer function is derived employing the commonly used Padé series of finite order to approximate the delay transfer functions e^{-sT} in (12.3). The order of the displayed impedance approximation depends on the order N of the Padé approximation. A Padé approximation of order N is valid for frequencies $\omega < N/(3T)$. In order to simplify the analysis the time delay element is approximated here by a first order, i.e. $N = 1$, Padé series

$$e^{-sT} \approx \frac{1 - \frac{T}{2}s}{1 + \frac{T}{2}s}. \tag{12.4}$$

This comes at the cost that for large round-trip time delay the approximation validity range does not fully cover the frequency range of human proprioreceptive and kinesthetic perception (up to approximately 60 Hz). Inserting (12.4) in (12.3) yields the approximated displayed impedance

$$Z_t(s) \approx Z_t^{\text{app}}(s) = b\frac{2Z_e(s) + bTs}{2b + TsZ_e(s)} \tag{12.5}$$

In accordance to the limited frequency range of approximation validity for further analysis this transfer function is split into a low frequency component $Z_{t,lf}^{\text{app}}$ and a high frequency component F_{hf}

$$Z_t^{\text{app}}(s) = Z_{t,lf}^{\text{app}}(s)F_{hf}(s) \tag{12.6}$$

with the high frequency component having approximately unity gain at lower frequencies. The component $Z_{t,lf}^{\text{app}}$ represents a good approximation of the low frequency behavior of the displayed impedance. The mechanical parameters of the approximated displayed impedance $Z_{t,lf}^{\text{app}}$ can be derived analytically as a function of the round-trip time delay T, the wave impedance b, and the environment impedance Z_e, which is exemplarily carried out in detail for the prototypical cases *free space motion* and *contact with a stiff wall.*

12.3.2 Analysis for Prototypical Environment Impedances

Free Space Motion

In free space motion, no environment force is exerted on the telerobot $f_e = 0$, i.e. the environment impedance is $Z_e = 0$. The exact displayed impedance (12.3) is

$$Z_t(s) = b\frac{1 - e^{-sT}}{1 + e^{-sT}}. \tag{12.7}$$

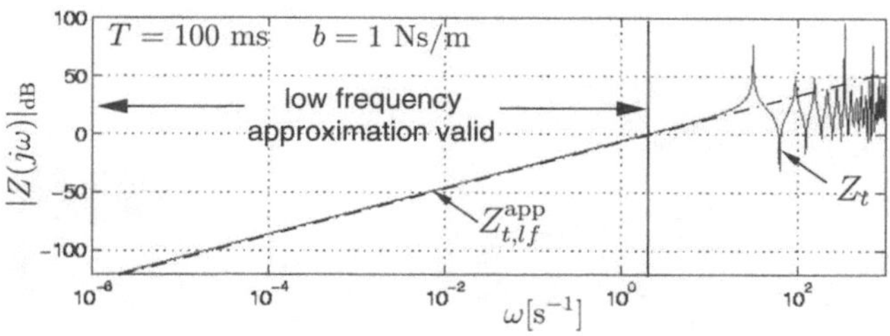

Fig. 12.2. Amplitude/frequency characteristics of the exact and the approximated displayed impedance in *free space motion*

Inserting the environment impedance into (12.5) gives the approximation of the displayed impedance valid for low frequencies

$$Z_t^{app}(s) = m_t s \frac{1}{1 + \frac{T}{2}s}. \tag{12.8}$$

with

$$m_t = \frac{bT}{2}. \tag{12.9}$$

The lefthand part sm_t in (12.8) represents the dominant low frequency component $Z_{t,lf}^{app}$ from (12.6). The righthand factor is the high frequency component F_{hf} satisfying $|F_{hf}(0)|= 1$. The similarity of the exact and the approximated displayed impedance for low frequencies can also be observed from their frequency responses for a simulated example, see Fig. 12.2. The displayed impedance is an inertia with the mass m_t (12.9). A similar result is presented for the static case in [34], its validity is extended here to a low frequencies.

Contact with a Stiff Wall Environment

In contact with a stiff wall, a force proportional to the wall penetration depth with the stiffness k_e acts on the teleoperator; the environment impedance is described by the transfer function $Z_e = k_e/s$. The exact displayed impedance (12.3) is

$$Z_t(s) = b\frac{k_e + bs + (k_e - bs)e^{-sT}}{k_e + bs - (k_e - bs)e^{-sT}}. \tag{12.10}$$

The approximation (12.5) of the displayed impedance for low frequency is analogously computed to the *free space motion* case

$$Z_t^{app}(s) = \frac{k_t}{s}\left(1 + \frac{bT}{2k_e}s^2\right) \tag{12.11}$$

with

$$\frac{1}{k_t} = \frac{1}{k_e} + \frac{T}{2b}. \tag{12.12}$$

The lefthand factor $Z_{t,lf}^{app} = k_t/s$ is the low frequency component from (12.6). The right hand factor in (12.11) exhibits high pass behavior satisfying $|F_{hf}(0)| = 1$. A simulation

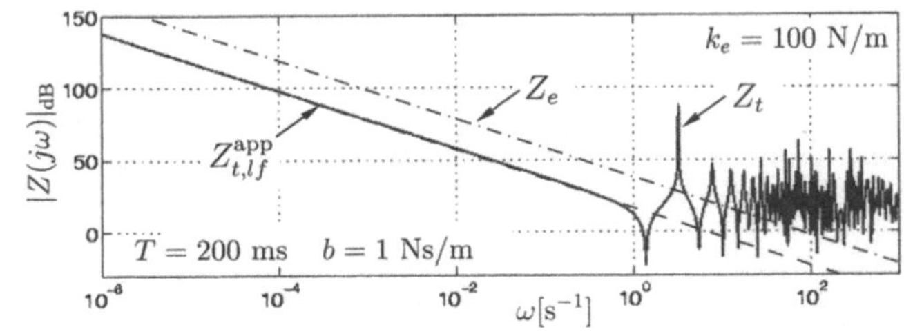

Fig. 12.3. Amplitude/frequency characteristics of the exact and the approximated displayed impedance in *contact with a stiff wall*

example in Fig. 12.3 shows the frequency responses for the exact and the approximated displayed impedance, which are similar at low frequencies. The displayed impedance in *contact with a stiff wall* exhibits a springlike behavior at low frequencies, however, with a lower stiffness k_t than the environment stiffness k_e, see also Fig. 12.3. As observable from (12.12), the communication subsystem including the wave (scattering) variable transformation can be interpreted as a rod with a stiffness coefficient $2b/T$ in mechanical series connection with the environment.

12.4 Perception Oriented Time Delay Transparency Analysis

On the basis of the above analysis and taking into account relative human perception limitations we obtain the following insights along with design guidelines for bilateral telerobotic systems.

12.4.1 Communication Induced Inertia Perception

In *free space motion* an inertia is displayed to the human operator even though no inertia is contained in the environment. The inertia characteristics is induced by the wave (scattering) variable transformation and the communication delay. With increasing round-trip time delay T and wave impedance b the displayed inertia m_t proportionally grows (12.9) as shown in a simulation example in Fig. 12.4. Given a time delay $T > 0$, free space motion is transparent in the sense of (12.1), i.e. $m_t = 0$, only if $b = 0$ which is unfeasible in terms of the tuning requirement $b > 0$. Considering human perception, an inertia is not perceivable if it is below the absolute human perception threshold Δm for inertia.

Example 1. Lets assume free space motion of the slave device, a communication round-trip delay $T = 200$ ms, typical for the communication over the Internet, and the wave impedance tuned to $b = 1$ Ns/m. Then the operator feels an inertia $m_t = 0.1$ kg. If the wave impedance is chosen to be $b = 1000$ Ns/m then the displayed mass is already increased to $m_t = 100$ kg.

12.4.2 Communication Induced Stiffness Reduction

If the environment exhibits spring characteristics a substantially reduced stiffness is displayed to the human. The environment feels softer than it really is. The displayed

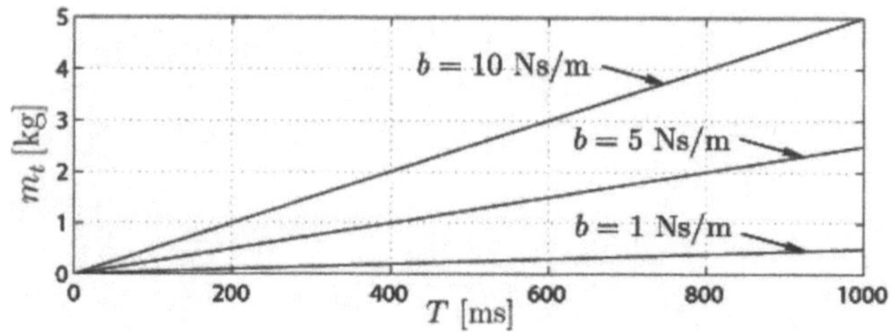

Fig. 12.4. Displayed inertia m_t in *free space motion* depending on round-trip time delay T and wave impedance b

stiffness coefficient (12.12) nonlinearly depends on the communication time delay as shown in Fig. 12.5 for different environment stiffness values. Ideal transparency in the sense of (12.1), i.e. $k_t = k_e$, is not achievable for non-zero time delay. Considering the human haptic perception limits, however, a transparency degradation should not be perceivable if the displayed stiffness is within the JND range of the environment stiffness $k_t > (1 - \text{JND}_k)k_e$ with $0 < \text{JND}_k < 1$ the stiffness JND. Accordingly, a stiff environment appears transparent to the human even for non-zero round-trip time delay as long as it satisfies

$$T < \frac{\text{JND}_k}{1 - \text{JND}_k} \frac{2b}{k_e} \tag{12.13}$$

which follows from inserting (12.12) in the previous equation. Note, that increasing environment stiffness reduces the allowable time delay margin. In contrast, a high value of the wave impedance b increases the delay margin, and reduces the impact of the time delay on stiffness reduction as observable from (12.12), i.e. increases the transparency of stiff environments. This, however, contradicts the design rule for free space motion. Good transparency in free space motion and for arbitrary stiff environments is not achievable at the same time.

Example 2. Consider a stiff wall with $k_e = 30000$ N/m, and the wave impedance tuned to $b = 1$ Ns/m. Already a very small round-trip delay of $T = 1$ ms substantially decreases the displayed stiffness to $k_t = 1875$ N/m, a reduction by 94%. At a delay of $T = 200$ ms the operator perceives only a stiffness of $k_t = 10$ N/m, hence 0.03% of the environment stiffness. Contacting a soft environment with $k_e = 10$ N/m, see Fig. 12.5, the displayed stiffness at $T = 1$ ms is still $k_t = 9.95$ N/m, at $T = 200$ ms still $k_t = 5$ N/m. Increasing the wave impedance for the hard wall to $b = 1000$ Ns/m at $T = 200$ ms the displayed stiffness is $k_t = 7500$N/m. Note the increased inertia in free space from the previous example with these values.

12.4.3 Communication Induced Stiffness Bound

The displayed stiffness (12.12) cannot exceed

$$k_{t,\max} = \lim_{k_e \to \infty} k_t = \frac{2b}{T}. \tag{12.14}$$

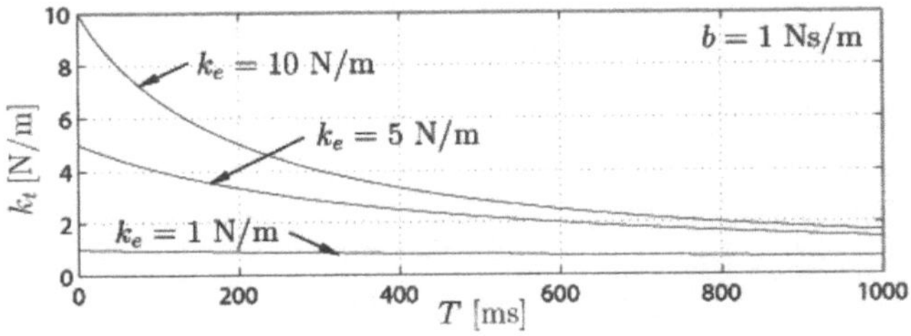

Fig. 12.5. Displayed stiffness k_t in *contact with a stiff wall* depending on round-trip time delay T and environment stiffness k_e

This result is also indicated by the asymptotic behavior of the displayed stiffness for increasing environment stiffness shown for a simulation example in Fig. 12.6. Considering the psychophysical fact that the human feels a wall to be rigid for $k_t \geq 24200$ N/m [35] it becomes clear that only for a very small time delay and a very large wave impedance b a rigid wall can be realistically displayed with this control architecture. For large time delay the stiffness, especially in case of hard walls, is not transparent. Appropriate tuning (high values) of the wave impedance b increases the transparency in terms of the maximum displayable stiffness.

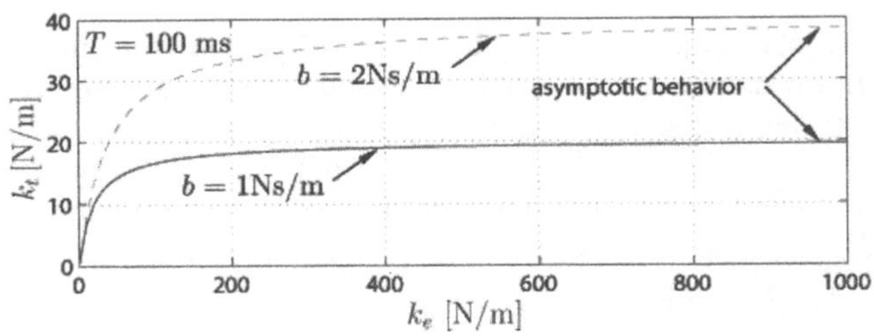

Fig. 12.6. Displayed stiffness k_t depending on environment stiffness k_e and wave impedance b

Example 3. Assuming a communication delay $T = 200$ ms with a wave impedance tuned to $b = 1$ Ns/m the maximum displayable stiffness is only $k_{h,max} = 10$ N/m. Any stiff environment feels very soft.

12.4.4 Bounded Displayable Stiffness Difference

In some tasks not only the absolute value of the displayed stiffness is important but also the possibility to distinguish between various stiff environments. This is especially important for e.g. tele-surgery applications, where different characteristics have to be distinguished. As indicated by the asymptotic behavior of the displayed stiffness in Fig. 12.6 at higher values of the environment stiffness, a stiffness difference in the environment results in a smaller difference in the displayed stiffness. However, a difference between a reference value k_e^0 and a value k_e of the environment stiffness is perceivable by the human only if the corresponding percentual difference in the displayed stiffness

$$\delta k_t = |k_t - k_t^0|/k_t^0 \tag{12.15}$$

is larger than the stiffness JND

$$\delta k_t = \frac{2b\delta k_e}{2b + Tk_e} \geq \text{JND}_k \tag{12.16}$$

with the percentual difference in the environment stiffness δk_e defined analogously to (12.15) and the displayed reference stiffness $k_t^0 = k_t(k_e^0)$ according to (12.12). The percentual difference δk_t of the displayed stiffness and the environment stiffness δk_e is equal only for the marginal cases of zero delay $T = 0$ or infinite wave impedance $b \to \infty$. At high delay and high environment stiffness, a large difference in the environment stiffness may result in a non-perceivable difference of the displayed stiffness. According to (12.16) the appropriate tuning (high values) of the wave impedance b increases the transparency in terms of the range of environment stiffness where a difference is perceivable by the human.

Example 4. Let us assume a communication delay is $T = 200$ ms and the wave impedance tuned to $b = 1$ Ns/m. If the environment stiffness coefficient is $k_e > 40$ N/m than a difference to *any* larger environment stiffness is not perceivable under the 23%-JND assumption.

12.4.5 Just Noticeable Difference for Time Delay

So far the distortion induced by the absolute value of the time delay has been investigated. This section discusses when a relative increase of the time delay can be perceived by the human operator. Assuming that the delay difference is haptically perceived only by the difference in the mechanical properties of the displayed impedance, the just noticeable difference for time delay can be derived from the results from Sec. 12.3.2 and the well known JND's for mechanical properties. This result is interesting with respect to the design of control architectures for telerobotic sytems over the Internet coping with time-varying delay, where data buffering strategies, as e.g. in [19], introduce additional delay. If the additional delay results in a distortion below the human perception threshold then no change in transparency should be perceived.

The inertia $m_t^0 = m_t(T^0)$, see (12.9), and the stiffness $k_t^0 = k_t(T^0)$, see (12.12), represent the displayed mechanical properties at the reference time delay T^0. An additional time delay $\Delta T = T - T^0 > 0$ results in a further increased displayed inertia in *free space motion* and further reduced displayed stiffness in *contact with a stiff wall.* The further distortion due to the time delay difference is just noticeable by the human if the corresponding percentual difference of the displayed mechanical property is equal to the JND

$$\text{(a)} \quad \delta m_t(T^0, \Delta T) = \text{JND}_m \qquad \text{(b)} \quad \delta k_t(T^0, \Delta T) = \text{JND}_k, \tag{12.17}$$

where δm_t denotes the percentual difference of the displayed inertia defined similar as δk_t in (12.15), and JND_m the inertia JND. The just noticeable time delay difference is computed straightforwardly using (12.9) and (12.12) in (12.17)(a) and (b), respectively. In *free space motion* a time delay difference is expected to be just perceivable by the human if

$$\frac{\Delta T}{T^0} = \text{JND}_m. \tag{12.18}$$

Obviously, the just noticeable time delay difference in free space motion follows a linear law similar to Weber's law, see Sec. 12.2.2. In *contact with a stiff wall* a time delay difference is expected to be just perceivable by the human if

$$\frac{\Delta T}{T^0} = \frac{\text{JND}_k}{\text{JND}_k + 1}\Big(\frac{2b}{k_e T^0} + 1\Big). \tag{12.19}$$

In contrast to the free space motion case, the just noticeable time delay depends on the environment stiffness k_e, the wave impedance b and the reference time delay T^0. In both cases, the absolute just noticeable difference ΔT increases with the reference time delay. Accordingly, any additional time delay should be avoided in the haptic telerobotic system, especially if the reference time delay is small. At high reference time delay an additional delay may not further perceivably degrade transparency. In consequence, if buffering strategies in telerobotic systems with time varying delay induce an additional time delay below the just noticeable difference, then human perceived transparency is not further degraded. For further results on the influence of communication effects on transparency refer to [22].

Remark 2. For the transparency analysis in this chapter the environment is assumed to be constant. Dynamic transitions between different environments, e.g. from free space motion to contact with a stiff wall, also have an influence on the perceived transparency as indicated in [36]. The analysis requires different techniques and is beyond the scope of this chapter.

Remark 3. The displayed impedance parameters in (12.9) and (12.12) are derived for the wave (scattering) variable approach. Consequently, all results in (12.14), (12.16), (12.18), and (12.19) are valid only for this specific control architecture. A perception oriented transparency analysis of other control architectures is straightforward by using the corresponding expressions for the parameters of the displayed impedance, e.g. from [29].

12.5 Perception Oriented Design Aspects in Real Systems

The tuning of the wave impedance b has a high impact on the transparency of the communication subsystem as observed in the previous sections. The transparency criterion (12.1) requires for free space motion $b \to 0$ as observable from (12.9). In contact with a stiff environment, time delay has no influence on transparency if $b \to \infty$, see (12.12), (12.14), (12.16). These are contradicting design rules, that can be relaxed by considering human haptic perception and a real telerobotic system as shown in the following.

In real telerobotic systems with limited control input and robustly designed controllers the dynamics of the master and the slave device is generally not negligible.

As a result, even without time delay transparency in the sense of (12.1) is not achievable. In the following, master and slave dynamics refers to the locally controlled device dynamics.

In free space motion, at least the inertia m_m induced by the master dynamics is displayed to the human. If the wave impedance is chosen

$$b < \frac{2}{T}\mathrm{JND}_m m_m, \tag{12.20}$$

then the displayed overall inertia, the sum of the master and the communication induced inertia, is within the JND range of the master inertia $m_t < (1 + \mathrm{JND}_m)m_m$ as straightforward derivable from (12.9). No additional communication induced transparency degradation should be perceivable by the human then. The original design requirement $b \to 0$ is relaxed.

In order to avoid contact instability or oscillations, the slave device is typically compliance controlled. The resulting stiffness $k_{s/e}$ of the slave device together with the environment computes from the environment stiffness k_e and the stiffness k_s of the compliance controlled slave device according to the serial connection of springs $k_{s/e}^{-1} = k_e^{-1} + k_s^{-1}$. If the wave impedance is chosen to be

$$b > \frac{T}{2}(\mathrm{JND}_k^{-1} - 1)k_{s/e}, \tag{12.21}$$

then the communication induced reduction is within the JND range of the combined slave device/environment stiffness $k_t > (1 - \mathrm{JND}_k)k_{s/e}$. The upper bound of the slave device/environment stiffness $\sup_{k_e} k_{s/e} = \lim_{k_e \to \infty} k_{s/e} = k_s$, i.e. the slave device compliance, determines the lower bound of a transparently designed wave impedance b (12.21). Clearly, the original transparency requirements for the communication subsystem design $b \to \infty$ are relaxed by (12.21).

Example 5. The haptic input device ViSHaRD10 (see chapter 2) displays without time delay an inertia of at least $m_m = 8$ kg. With a for the Internet realistic round-trip time delay of $T = 160$ ms and a inertia JND assumption of $\mathrm{JND}_m = 21\%$ [27] for communication transparent design in free space motion the wave impedance should $b < 21$ Ns/m (12.20). In contact with a stiff wall, assuming a slave device compliance of $k_s = 900$ N/m as in [37] and a stiffness JND of 23% [25], the wave impedance should be $b > 241$ Ns/m (12.21). The gap between the design requirements for the wave impedance b for free space motion and contact with a stiff wall derived from the strict transparency criterion (12.1) become smaller by considering human perception aspects and the real telerobotic system.

12.6 Experiments

In the first experiment the dependency of the displayed impedance parameters on the round-trip time delay obtained in Sec. 12.3 by (12.9), (12.12) is validated. In the second one, a human user study is conducted on how a relative increase of time delay further degrades the human perceived transparency. In both experiments, the prototypical cases of *free space motion* and *contact with a stiff wall* (stiffness coefficient $k_e = 12.5$ kN/m) are investigated.

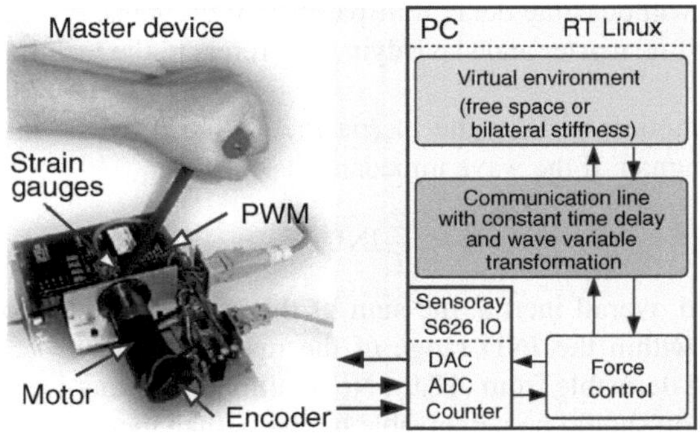

Fig. 12.7. Experimental system architecture with a 1DoF telerobotic system

12.6.1 Experimental Setup

The experimental setup, see Fig. 12.7, consists of a single degree-of-freedom force feedback paddle connected to a PC. The paddle DC motor torque is controlled by the PWM amplifier. The force applied to the paddle lever is measured by a strain gauge bridge, the position of the lever by an optic pulse incremental encoder. A virtual environment is used instead of a real slave device/environment in order to seperately consider the prototypical environment scenarios, especially in the human user studies. The virtual environment, the control loops, the model of the communication subsystem with different constant delay and the wave (scattering) variable transformation with a wave impedance $b = 125$ Ns/m are composed of MATLAB/SIMULINK blocksets; standalone realtime code for RT Linux is automatically generated from that. All experiments were performed with a sample time interval $T_A = 0.001$s.

12.6.2 Objective Measurements

The displayed inertia m_t in free space motion and the displayed stiffness k_h in contact with the wall are determined depending on the round-trip time delay that is varied within the interval $T \in [5, 400]$ ms. The parameters m_t and k_t are determined by a least squares identification from the measured slave position and slave force signals. The results for the displayed inertia in free space motion are shown in Fig. 12.8 (a) [1], and for the displayed stiffness in contact in Fig. 12.8 (b). The theoretically obtained dependencies of these parameters on the round-trip time delay given by (12.9) and (12.12) are convincingly validated. The slightly reduced stiffness and the higher inertia in the experiments result from the limited bandwidth of the conservatively tuned force control loop at the slave.

12.6.3 Human User Study

The hypothesis to be validated is that at low reference time delay the relative time delay increase detection threshold is smaller than at high reference time delay, see Sec. 12.4.5.

[1] The inertia results for $T < 100$ ms are missing because of numerical unreliabilities in the least squares estimation.

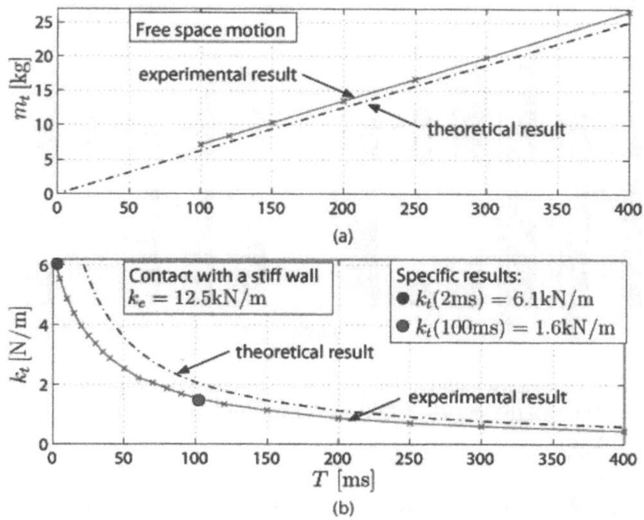

Fig. 12.8. Experimentally obtained displayed inertia m_t (a) and stiffness k_t (b) depending on round-trip delay T compared to theoretical results

Four experiments with 7 subjects (aged 20–30, 3 female, 4 male) were performed for two different reference round-trip time delays $T^0 = 2\,\text{ms}$ and $T^0 = 100\,\text{ms}$ for each of the considered prototypical cases *free space motion* and *contact with a wall* using the same parameters as in the foregoing experiment. The subjects were told to operate with their preferred hand. They were equipped with earphones to mask the sound the device motors generate. No visual feedback of the virtual environment was provided. The subjects were not refunded. During a familiarization phase subjects were told to feel operation for the reference round-trip delay configuration. As soon as they felt familiar with the system the measurement phase began.

12.6.4 Procedure

In order to determine the detection thresholds for the time delay difference the three interval forced choice (3IFC) paradigm has been applied, which is a common experimental tool in psychophysics to determine detection thresholds in human haptic perception [30]. The main feature is that the subjects are presented three consecutive time intervals, 20s duration each, two with the reference value T^0 of the time delay, one - randomly chosen which - with a different time delay value T. The subject has to tell which of the intervals felt different. Starting from a non-perceivable delay difference ΔT this value is increased after every incorrect answer until three consecutive correct answers on the same value ΔT are given. No feedback on the correctness of the answer was given. Three of these passes are performed, the mean value over the passes is considered the subject specific discrimination threshold. The experiment started with a delay of $T_{2\text{ms}} = 3\,\text{ms}$ ($T_{100\text{ms}} = 103\,\text{ms}$), i.e. a delay difference

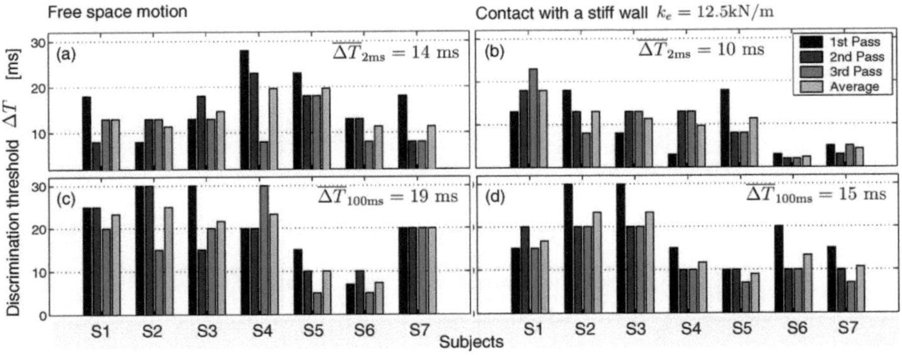

Fig. 12.9. Results of human user study: discrimination thresholds for time delay ΔT at reference time delays $T_0 = 2$ ms (a), (b) and $T_0 = 100$ ms (c), (d) for free space motion (a), (c) and contact with a stiff wall (b), (d)

Table 12.2. Results of human user study: Average detected delay differences and corresponding percentual parameter differences for different reference time delays T_0

	Free space $T_0 = 2$ms	Free space $T_0 = 100$ms	Contact $T_0 = 2$ms	Contact $T_0 = 100$ms
$\overline{\Delta T}$ [ms]	14	19	10	15
$\lvert m_t^0 - m_t\rvert$ [kg]	n.a.[1]	0.048		
δm_t [%]	n.a.[1]	17		
$\lvert k_t^0 - k_t\rvert$ [kN]			1.2	0.16
δk_t [%]			20	10

of $\Delta T_{2ms} = 1$ ms ($\Delta T_{100ms} = 3$ ms), where values $(\cdot)_{2ms}$ indicate a reference round-trip delay of $T_0 = 2$ ms, and accordingly $(\cdot)_{100ms}$ refers to $T_0 = 100$ ms.

Results

The results for all four experiments are shown in Fig. 12.9, where $\overline{\Delta T}$ denotes the average over all subjects, see also the first row in Table 12.2. As expected from the theoretical results in Sec. 12.4.5, in both scenarios, the average detected delay difference is smaller for low reference time delay $\overline{\Delta T}_{2ms} < \overline{\Delta T}_{100ms}$. The Student's test is performed giving a statement about the statistical significance of the discrimination thresholds difference $(\Delta T_{2ms} - \Delta T_{100ms})$. Even with this rather small number of subjects, for *contact with a stiff wall* the mean discrimination threshold for low reference delay is statistically significant (95%) smaller than for high reference delay. For *free space motion* it is not significant (90%) in a statistical sense: As expected time delay is more crucial for the transparency degradation in very stiff environments. For a cross check the percentual differences (12.15) of the displayed inertia δm_t and stiffness δk_t corresponding to these just noticeable time delay differences are computed using the results from the previous experiment. For example, the average discrimination threshold $\overline{\Delta T}_{2ms} = 10$ ms for *contact with a stiff wall* corresponds to a percentual difference in the displayed stiffness of $\delta k_{h,2\text{ms}} = 20\,\%$. The percentual differences, see Table 12.2[2], are all in the range of the JNDs reported in literature (for stiffness: 8% [26], $(23 \pm 3)\%$ [25], for inertia $(21 \pm 3.5)\%$ [27]), see also Table 12.1 for comparison.

In summary, the results indicate, that the discrimination threshold for time delay is lower for low round-trip delay as predicted in Sec. 12.4.5. Accordingly, any additional time delay should be avoided if the communication delay is very low. For high communication delay some additional delay does not further degrade human perceived transparency.

12.7 Conclusions

The consideration of human factors is important for the design and evaluation of telerobotic systems. In this chapter a method for the transparency analysis of haptic telerobotic systems is presented with the goal to quantify the degradation induced by communication time delay from a human perception point of view. Therefore the effect of constant time delay and of the wave (scattering) variable control approach on the mechanical properties displayed to the human is analyzed. The interpretation of the results using known psychophysical facts reveals important insights with implications for the control design and the range of tele-applications depending on the communication time delay: a) in free space motion time delay introduces artificial inertia; b) stiff environments are perceived softer; c) displayable stiffness is upper bounded; d) environment stiffness discrimination is limited; e) there is a detection threshold for relative changes in time delay. Nevertheless, ideal transparency requiring the displayed impedance to be exactly equal to the environment impedance is not necessary for the telerobotic system to be perceived transparent by the human. The consideration of human haptic perception limits leads to relaxed design requirements still guaranteeing human perceived transparency. Psychophysically motivated design guidelines for the wave impedance as well as upper time delay bounds for human perceived transparency are derived in this chapter. The just noticeable difference for time delay gives implications for the design of telerobotic systems with time-varying delay. The obtained results are validated in experiments and human user studies.

The approach in this chapter constitutes a first step towards a human perception oriented analysis of the time delay on the transparency. Exciting challenges incorporate the systematic human perception oriented transparency analysis of general communication unreliabilities, and a rigorous human perception oriented design of multimodal, multi-DoF telerobotic systems.

Acknowledgements

This work is supported in part by the DFG Collaborative Research Center SFB453 and the Technische Universität München.

[2] As the inertia for $T_0 = 2\,\text{ms}$ could not be identified from the sensor signals in the foregoing experiment, the corresponding percentual difference in inertia not available (n.a.) in Table 12.2.

References

1. W.R. Ferrell. Remote Manipulation with Transmission Delay. *IEEE Transactions on Human Factors*, 6:24–32, 1965.
2. R.J. Anderson and M.W. Spong. Bilateral Control of Teleoperators with Time Delay. *IEEE Transactions on Automatic Control*, 34(5):494–501, 1989.
3. G. Niemeyer and J.-J.E. Slotine. Stable Adaptive Teleoperation. *IEEE Journal of Oceanic Engineering*, 16(1):152–162, January 1991.
4. G.J. Raju, G.C. Verghese, and T.B. Sheridan. Design Issues in 2-Port Network Models of Bilateral Remote Teleoperation. In *Proceedings of the IEEE International Conference on Robotics and Automation*, pages 1317–1321, Scottsdale (AZ), US, 1989.
5. D.A. Lawrence. Stability and Transparency in Bilateral Teleoperation. *IEEE Transactions on Robotics and Automation*, 9(5):624–637, October 1993.
6. Y. Yokokohji and T. Yoshikawa. Bilateral Control of Master-Slave Manipulators for Ideal Kinesthetic Coupling Formulation and Experiment. *IEEE Transactions on Robotics and Automation*, 10(5):605–619, 1994.
7. B. Hannaford. Stability and Performance Tradeoffs in Bi-Lateral Telemanipulation. In *Proceedings of the IEEE International Conference on Robotics and Automation*, pages 1764–1767, Scottsdale (AZ), US, 1989.
8. K. Hashtrudi-Zaad and S.E. Salcudean. Analysis and Evaluation of Stability and Performance Robustness for Teleoperation Control Architectures. In *Proc. of the IEEE Int. Conf. on Robotics and Automation*, pages 3107–3113, San Francisco (CA), US, 2000.
9. J. Vertut, A. Micaelli, P. Marchal, and J. Guittet. Short Transmission Delay on a Force Reflective Bilateral Manipulator. In *Proceedings of the 4th Rom-An-Sy*, pages 269–274, Zaborow, Poland, 1981.
10. C.A. Lawn and B. Hannaford. Performance Testing of Passive Communication and Control in Teleoperation with Time Delay. In *Proceedings of the IEEE International Conference on Robotics and Automation*, pages 776–783, Atlanta (GA), US, 1993.
11. G. Hirzinger. ROTEX–The First Robot in Space. In *Proceedings of the ICAR International Conference on Advanced Robotics*, pages 9–33, Tokyo, Japan, 1993.
12. G. Niemeyer and J. E. Slotine. Towards Force-Reflecting Teleoperation Over the Internet. In *Proceedings of the IEEE International Conference on Robotics and Automation*, pages 1909–1915, Leuven, Belgium, 1998.
13. R. Lozano, N. Chopra, and M. Spong. Passivation of Force Reflecting Bilateral Teleoperators with Time Varying Delay. In *Proceedings of the 8. Mechatronics Forum*, pages 954–962, Enschede, Netherlands, 2002.
14. N. Chopra, M.W. Spong, S. Hirche, and M. Buss. Bilateral Teleoperation over Internet: the Time Varying Delay Problem. In *Proceedings of the American Control Conference*, pages 155–160, Denver (CO), US, 2003.
15. Y. Yokokohji, T. Imaida, and T. Yoshikawa. Bilateral Control with Energy Balance Monitoring under Time-Varying Communication Delay. In *Proceedings of the IEEE International Conference on Robotics and Automation*, pages 2684–2689, San Francisco (CA), US, 2000.
16. S. Munir and W.J. Book. Internet Based Teleoperation using Wave Variable with Prediction. *ASME/IEEE Transactions on Mechatronics*, 7(2):124–133, 2002.
17. S. Stramigioli. About the Use of Port Concepts for Passive Geometric Telemanipulation with Time Varying Delays. In *Proceedings of the 8. Mechatronics Forum*, pages 944–953, Enschede, The Netherlands, 2002.
18. Y. Yokokohji, T. Tsujioka, and T. Yoshikawa. Bilateral Control with Time-Varying Delay including Communication Blackout. In *Proceedings of the 10th Symposium on Haptic Interfaces for Virtual Environment and Teleoperator Systems*, Orlando (FL), US, 2002.

19. B. Berestesky, N. Chopra, and M. W. Spong. Discrete Time Passivity in Bilateral Teleoperation over the Internet. In *Proceedings of the IEEE International Conference on Robotics and Automation ICRA'04*, pages 4557–4564, New Orleans, US, 2004.
20. S. Hirche and M. Buss. Packet Loss Effects in Passive Telepresence Systems. In *Proceedings of the 43rd IEEE Conference on Decision and Control*, pages 4010–4015, Paradise Island, Bahamas, 2004.
21. C. Secchi, S. Stramigioli, and C. Fantuzzi. Dealing with Unreliabilities in Digital Passive Geometric Telemanipulation. In *Proceedings of the IEEE/RSJ International Conference on Intelligent Robots and Systems IROS*, Las Vegas (NV), US, 2003.
22. S. Hirche. *Haptic Telepresence in Packet Switched Communication Networks*. PhD thesis, Technische Universität München, Institute of Automatic Control Engineering, July 2005.
23. X. Wang, P.X. Liu, D. Wang, B. Chebbi, and M. Meng. Design of Bilateral Teleoperators for Soft Environments with Adaptive Environmental Impedance Estimation. In *Proceedings of the IEEE International Conference on Robotics and Automation*, pages 1139–1144, Barcelona, Spain, 2005.
24. K. Hashtrudi-Zaad and S.E. Salcudean. Analysis of Control Architectures for Teleoperation Systems with Impedance/Admittance Master and Slave Manipulators. *International Journal of Robotics Research*, 20(6):419–445, 2001.
25. L.A. Jones and I.W. Hunter. A Perceptual Analysis of Stiffness. *Experimental Brain Research*, 79:150–156, 1990.
26. H.Z. Tan, N.I. Durlach, G.L. Beauregard, and M.A. Srinivasan. Manual Discrimination of Compliance Using Active Pinch Grasp: The Role of Force and Work Cues. *Perception and Psychophysics*, 57:495–510, 1995.
27. G.L. Beauregard and M.A. Srinivasan. The Manual Resolution of Viscosity and Mass. *ASME Dynamic Systems and Control Division*, 1:657–662, 1995.
28. L.A. Jones and I.W. Hunter. A Perceptual Analysis of Viscosity. *Experimental Brain Research*, 94(2):343–351, 1993.
29. P. Arcara and C. Melchiorri. Control Schemes for Teleoperation with Time Delay: A Comparative Study. *Robotics and Autonomous Systems*, 38(1):49–64, 2002.
30. G.A. Gescheider. *Psychophysics: The Fundamentals*. Lawrence Erlbaum and Associates, 3rd Edition, Hillsdale, 1997.
31. G.C. Burdea. *Force and Touch Feedback for Virtual Reality*. John Wiley, 1996.
32. E. H. Weber. *Die Lehre vom Tastsinn und Gemeingefühl, auf Versuche gegründet*. Vieweg, 1851.
33. L. A. Jones and I. W. Hunter. Human Operator Perception of Mechanical Variables and Their Effects on Tracking Performance. *ASME Advances in Robotics*, 42:49–53, 1992.
34. G. Niemeyer and J.-J.E. Slotine. Telemanipulation with Time Delays. *International Journal of Robotic Research*, 23(9):873–890, September 2004.
35. H.Z. Tan, M.A. Srinivasan, B. Eberman, and B. Cheng. Human Factors for the Design of Force-Reflecting Haptic Interfaces. *ASME Dynamic Systems and Control Division*, 1:353–359, 1994.
36. N.A. Tanner and G. Niemeyer. Improving Perception in Time Delayed Telerobotics. *International Journal of Robotics Research*, 2005.
37. B. Stanczyk and M. Buss. Experimental comparison of interaction control methods for a redundant telemanipulator. In *Proceedings of the International Symposium on Methods and Models in Automation and Robotics MMAR'2005*, pages 677–682, Poland, 2005.

13

Environment Estimation in Teleoperation Systems

Luigi Biagiotti and Claudio Melchiorri

University of Bologna
Department of Electronics, Computer Science and Systems
Via Risorgimento 2, 40136 Bologna, Italy
{lbiagiotti,cmelchiorri}@deis.unibo.it

Summary. One of the main objectives of control algorithms for teleoperation systems is to have a master device mimicking the response of the remote environment, while the slave device is requested to behave as the human operator. In general, the remote environment is compliant, with a quite different behavior with respect to perfectly rigid surfaces (e.g. in surgery or human-centered applications). In these cases, the knowledge of the dynamical properties of the remote environment can be used in order to improve the transparency of the overall system. A number of analytical and computational models have been proposed in literature in order to describe the behavior of compliant materials but, for sake of simplicity, design and simulation of controllers for robotic telemanipulation are still tied to classical linear spring-damper models. On the other hand, previous experimental activities with soft materials and human tissues have demonstrated that they are characterized by dynamical effects (relaxation and creep phenomena), which cannot be taken into account by means of linear, low-order models. In this Chapter, we study the suitability of a class of nonlinear contact models to describe and emulate compliant visco-elastic environments. Their parameters, estimated on-line, can then be used to command a suitable behavior to the master device in order to render a better contact sensation to the user.

13.1 Introduction

On-line estimation of the environment dynamics plays an important role in the field of master/slave bilateral teleoperation systems, used by human operators to interact with remote environments. It is known [1] that the fidelity of force feedback, usually defined *transparency*, is affected by controllers and time delays in data transmission. This sensation is important e.g. in surgical applications, where the surgeon should be able to recognize human tissues by touching them and to identify their mechanical properties. Therefore, on-line estimation of mechanical properties of *soft* tissues can be used to (partially) recover transparency and telepresence sensation, as suggested by Colgate [2] and Buss [3], thus making force feedback surgical systems more effective.

The first step towards the implementation of on-line impedance estimation of objects interacting with a robotic device is the choice of suitable contact models [4], able to describe contact dynamics, i.e. the relation between contact forces and deformations of contacting bodies. Focusing on one-dimensional contacts, in this chapter it is shown that linear models, commonly used to describe the interaction with stiff environments, cannot be applied to describe a unilateral contact with soft materials, where viscous

M. Ferre et al. (Eds.): Advances in Telerobotics, STAR 31, pp. 211–231, 2007.
springerlink.com © Springer-Verlag Berlin Heidelberg 2007

effects are relevant. More generally, linear models are not suitable to emulate the behavior of compliant interfaces, when large deformations are imposed. For these reasons, more sophisticated nonlinear models, suitable for realtime computation and online estimation of the parameters, are introduced. In particular, the *Hunt-Crossley* model [5] and the *Quasi-linear* model [6, 7] are considered.

This chapter is organized as follows: in Sec. 13.2 a general architecture for telemanipulation systems including an online impedance estimator is introduced. Then, Sec. 13.3 provides an overview on contact models, from the linear ones to Hunt-Crossley and quasi-linear models. Sec. 13.4 describes the techniques for on-line recursive estimation of the parameters of the proposed models. These algorithms have been applied to different materials in order to provide experimental validation (Sec. 13.5) to the theoretical considerations. Conclusions are reported in the final Sec. 13.6.

13.2 Telemanipulation Systems with Environment Impedance Estimation

Estimation of remote impedance has been introduced in different ways in bilateral telemanipulation schemes in order to improve their performance. For example, in [8] the estimation of the environment parameters (stiffness and damping coefficients) is used to achieve a good force tracking at the slave side. However, the main goal of control architectures involving the identification of the environment dynamics, is the achievement of "transparency" [9], that is an ideal kinesthetic coupling between the human operator and the remote environment. In other terms, the impedance perceived by the operator must be equal to the environmental impedance. This is a basic requirement of any ideal telemanipulation system, but, in particular, it is of great importance for medical/surgical applications, in which the possibility of perceiving the impedance(/stiffness) of the remote environment is a fundamental factor for the success of the task performed.

In order to enhance the transparency of the overall system, the typical solution consists in including the environment properties in the master controller (and, possibly, the human operator characteristics in the slave controller). This approach, called *impedance reflection* originally proposed by Hannaford in [10], is conceptually reported in the scheme of Fig. 13.1, which represents the most general form of this approach, including the reflection of the environment impedance and of the operator impedance as well, and the exchange of both position and force information from master to slave and viceversa. In many cases the teleoperation schemes are simplified versions of the general approach, in which, for example, only the remote impedance is reflected to the human operator to improve the perception of the environment and, in the other direction, only the master position is transmitted to the slave controller [11, 12]. In other cases the impedance of the environment, see the impedance shaping telemanipulation architecture of Colgate [2], is applied to the operator with proper modifications which allow to obtain particular impression or perception, e.g. the kinematic similarity between the actual impedance and the apparent impedance in tasks with different scales.

In principle, the telemanipulation schemes with impedance reflection can be assimilated to haptic systems. In an haptic system the operator interacts, by means of the master devices, with a model-based virtual environment. For this reason, the models

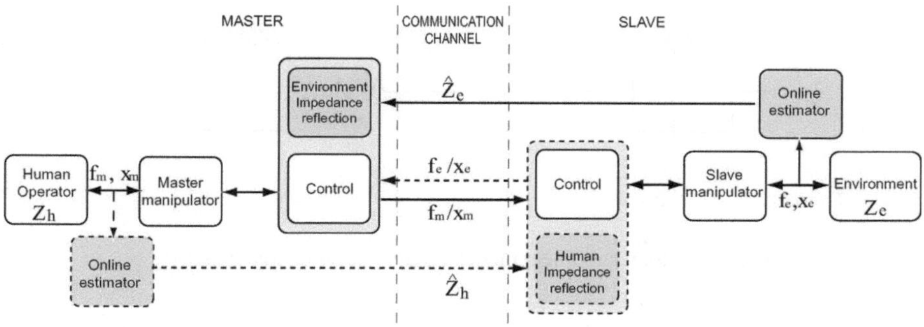

Fig. 13.1. Conceptual scheme of the impedance reflecting teleoperation

proposed in the next sections can be adopted, without any difference, in this field. In teleoperation, the user (partially or totally) interacts with an emulated environment, but in this case the model is built online, with the characteristic parameters of the contacted environment to be properly estimated. Therefore, a very crucial element in the overall system is the estimator, which must guarantee at the same time a fast convergence rate and the necessary approximation of the identified parameters to true values. For this reason, standard estimators are based on linear systems, whose algorithms are well-settled and computationally efficient. On the other hand, it is worth to notice that recently in teleoperation there is a strong interest towards the interaction with soft environments (e.g. for surgery and medical palpation) which involves large deformations and nonlinear phenomena. Therefore, in order to obtain very realistic responses, the linear contact models usually adopted should be revised and improved by considering nonlinear terms. At the same time, the identification of the parameters should remain feasible from a computational point of view.

13.3 Contact Models

In the following, some contact models that can be used to describe the interaction between robot and environment are discussed. Only one-dimensional models are described in order to present the main ideas and problems. Extension to multi-dimensional models is a current research activity.

13.3.1 Linear Contact Models

In many cases, within certain ranges of the involved variables, the linear theory is sufficient to describe the mechanical impedance of an object. In this case, a number of mechanical models, composed by linear springs and linear viscous dampers arranged in different configurations (e.g. Maxwell and Kelvin models), can be adopted. These models are widely used because of their simplicity and their clear physical interpretation [13], [4]. In this case, a general representation of the force-displacement relation is given by ordinary differential equations with constant coefficients, which depend on the

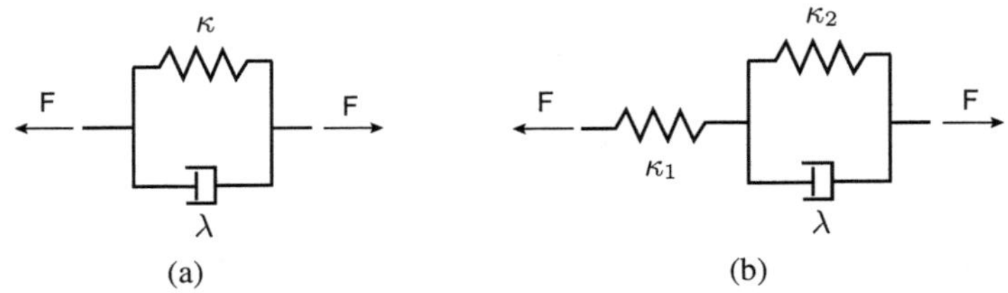

Fig. 13.2. Standard linear models: *Kelvin-Voigt* (a) and *Zener* (b)

model structure as well as on the considered material. The simplest example of linear model is the *Kelvin-Voigt* contact model (see Fig. 13.2.a), composed by the mechanical parallel of a linear spring and a damper.

If F is the force exerted by the material on a probe during contact, the linear model is expressed by:

$$F(t) = \begin{cases} \kappa\delta(t) + \lambda\dot{\delta}(t) & \delta \geq 0 \\ 0 & \delta < 0 \end{cases} \tag{13.1}$$

where $\dot{\delta}$ is the penetration velocity of the probe and k and λ are the elastic and viscous parameters of the contact. When the visco-elastic behavior of the material/object under analysis is remarkable, more complex models are available. A standard linear viscoelastic solid model (*Zener model*) is shown in Fig. 13.2.b. Such a model is frequently adopted to represent behaviors in which instantaneous and delayed elasticities arise. Force and displacements are related by the following differential equation:

$$F + \frac{\lambda}{\kappa_1 + \kappa_2}\dot{F} = \frac{\kappa_1}{\kappa_1 + \kappa_2} \cdot (\kappa_2\delta + \lambda\dot{\delta}) \qquad \delta \geq 0 \tag{13.2}$$

where the meaning of the parameters κ_1, κ_2 and λ is shown in Fig. 13.2.b.

13.3.2 Nonlinear Contact Models

Lumped Parameters Systems with Non-linear Coefficient: The Hunt-Crossley Case

When the linear hypothesis is not applicable, non linear models must be considered. For the sake of simplicity, many authors developed dynamic models of contacting systems drawing inspiration from classical linear models and introducing nonlinearities in the definition of the spring stiffness and of the dashpot viscosities [14, 15].

A noteworthy nonlinear model has been proposed by Hunt and Crossley [16] and adopted by several authors [17, 18, 19] to describe the dynamic behavior of the interface between a robot and the environment when a contact occurs. It consists of a Kelvin model, (13.1), where the stiffness of the spring and the viscosity of the damper are expressed by a power-law of the displacement. The differential form of this model is:

$$F(t) = \kappa\delta^n + \lambda\delta^q \cdot \dot{\delta} \tag{13.3}$$

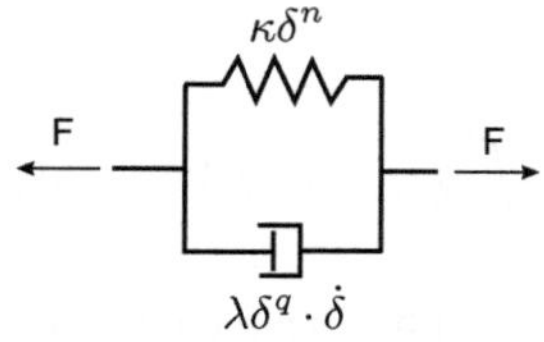

Fig. 13.3. Hunt-Crossley model

where usually it is assumed $q = n$, with the exponent $n \in \mathbb{R}$ usually close to the unity, that takes into account the geometry of contact surfaces. Notice that when $n = 3/2$ the elastic term of (13.3) exactly matches the force resulting from the Hertzian theory for spheres contacting in static conditions [6].

An Integral Form for Non-linear Systems: The Quasi-linear Model

Models like that of Hunt and Crossley are based on constitutive equations expressed in differential form. A different approach consists of considering directly the force response due to a change step in displacement, which is expressed by the *relaxation function*

$$F_0(t) = \Psi(\delta_0, t) \tag{13.4}$$

and superimposing each contribution of a displacement history by applying a suitable superposition principle. In order to simplify the treatment, following the hypothesis originally formulated by Fung [20], the relaxation function is assumed to be of the form:

$$\Psi(\delta, t) = F^{(e)}(\delta) \cdot g(t) \qquad \text{with} \qquad g(0) = 1 \tag{13.5}$$

where $F^{(e)}(\delta)$ is the *elastic response*, that is the amplitude of the force generated instantaneously by a displacement δ, while $g(t)$ is the *reduced relaxation function*, describing the time-dependant behavior of the material.

In order to find the force response to a generic displacement $\delta(t)$, one can model the input function as a sequence of steps. The force produced by an infinitesimal displacement $d\delta(\tau)$, superposed in a state of displacement δ at an instant of time τ is, for $t > \tau$:

$$dF(t) = \frac{\partial\Psi[\delta(\tau), t - \tau]}{\partial\delta} d\delta(\tau) \tag{13.6}$$

and, considering the Fung's hypothesis, expressed in (13.5), one obtains

$$dF(t) = g(t - \tau)\frac{dF^{(e)}[\delta(\tau)]}{d\delta} d\delta(\tau) \tag{13.7}$$

By applying a modified superposition principle, discussed in [7, 21, 22], the total force at the instant t is the sum of the contribution of all the past changes, i.e.

$$F(t) = \int_{-\infty}^{t} g(t - \tau)\frac{dF^{(e)}[\delta(\tau)]}{d\delta}\frac{d\delta(\tau)}{d\tau} d\tau \tag{13.8}$$

That equation can be rewritten as

$$F(t) = \int_0^t g(t-\tau)\, K^{(e)}[\delta(\tau)]\, \dot{\delta}(\tau) d\tau \tag{13.9}$$

where the lower limit of the integral has been changed assuming that the motion starts at time $t = 0$ and $F^{(e)} = 0$, $\delta = 0$ for $t < 0$. The term $K^{(e)}(\delta) = \frac{dF^{(e)}[\delta]}{d\delta}$ is the elastic stiffness and $\dot{\delta}(\tau)$ is the rate of displacement.

The expressions of the reduced relaxation function and of the elastic response in (13.9) determine the model of the contact. According to the Fung's hypotheses, expressed by (13.5), the relaxation function $g(t)$ is a decreasing, continuous function of the time, normalized to 1 at $t = 0$. It is usually assumed to be composed by a linear combination (with the coefficient c_i depending on the material) of exponential functions

$$g(t) = \sum_{i=0}^{r} c_i e^{-\nu_i \cdot t} \quad \text{with} \quad \sum_{i=0}^{r} c_i = 1 \tag{13.10}$$

whose exponents ν_i identify the rate of the relaxation phenomena. The number r and the value of such parameters depend on the behavior of the system under analysis, while $\nu_0 = 0$. From this definition of the reduced relaxation function $g(t)$ it descends that $F^{(e)}(\delta)$ can be approximated by the force response in a loading experiment with a sufficiently high rate of displacement.

The nonlinear elastic response can be modelled by different analytical expressions. Two of the most significant models adopted in the literature are:

$$K^{(e)}(\delta) = m \cdot e^{b\delta} \tag{13.11}$$

$$K^{(e)}(\delta) = p \cdot \delta^{q} \tag{13.12}$$

where (m, b) and (p, q) are parameters dependent on the material and the geometry taken into account. The expression of $F^{(e)}(\delta)$ descends directly from the integration of (13.11) or (13.12), with respect to δ:

$$F^{(e)}(\delta) = \frac{m}{b}(e^{b\delta} - 1) \tag{13.13}$$

$$F^{(e)}(\delta) = \frac{p}{q+1}\delta^{q+1} \tag{13.14}$$

The quasi-linear model has a simple and meaningful interpretation. (13.9) describes in the time domain the signal $K^{(e)}(\delta)\, \dot{\delta}$ filtered by a linear system represented by its transfer function:

$$G(s) = \mathcal{L}\{g(t)\} = \frac{c_0}{s} + \frac{c_1}{s+\nu_1} + \ldots + \frac{c_i}{s+\nu_i} + \ldots + \frac{c_r}{s+\nu_r} \tag{13.15}$$

The number r of characteristic modes of the filter $G_F(s)$ should be carefully chosen on the basis of the viscoelasticity of the surface in contact. Strongly viscoelastic

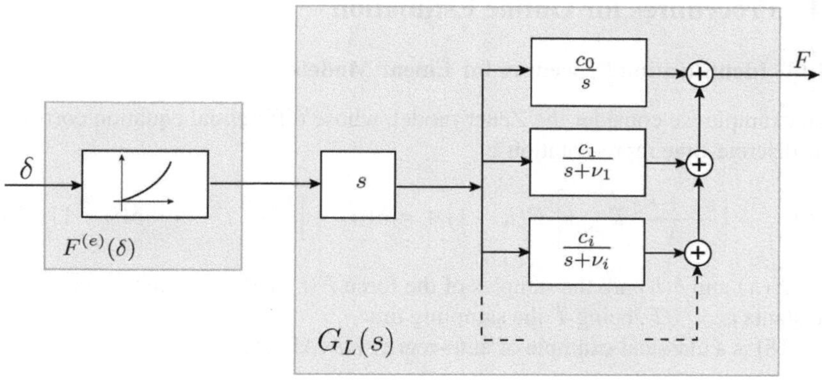

Fig. 13.4. Block diagram of the quasi-linear model: static nonlinear part and dynamic linear function

materials require a number r quite large, while quasi-elastic materials can be modelled by assuming $r = 1$. The input signal of the linear filter can be rewritten as:

$$K^{(e)}(\delta)\,\dot{\delta} = \frac{dF^{(e)}(\delta)}{d\delta}\dot{\delta} = \frac{dF^{(e)}(\delta(t))}{dt} \tag{13.16}$$

Being the derivative a linear operator, from (13.9) and (13.16) it comes out that it exists a linear relation between the elastic response $F^{(e)}(\delta)$ and the force produced by the system. Such a relationship is given by the transfer function

$$\begin{aligned} G_L(s) &= s\left(\frac{c_0}{s} + \frac{c_1}{s+\nu_1} + \frac{c_2}{s+\nu_2} + \ldots + \frac{c_r}{s+\nu_r}\right) \\ &= \frac{\left(\sum_{i=0}^{r} c_i\right)\, s^r + \ldots + c_0\left(\prod_{i=0}^{r}\nu_i\right)}{s^r + \ldots + \left(\prod_{i=0}^{r}\nu_i\right)} \end{aligned} \tag{13.17}$$

where, for both the numerator and the denominator, only the coefficients of terms of maximum and minimum degree are explicitly reported. It is worth to notice that the relative degree of the function $G_L(s)$ is zero. Therefore, when a step (of amplitude $\widehat{F^e}$) is applied to the system, its response will be discontinuous and will start from $\widehat{F^e}$ (note that in (13.17) $\sum_{i=0}^{r} c_i = 1$).

In Fig. 13.4 the complete model of a viscoelastic material is reported; it is composed by two elements, connected in cascade:

- a nonlinear static block which has the displacement δ as input and provides the instantaneous elastic response $F^{(e)}$;
- a linear dynamic block $G_L(s)$, which takes into account the (typically slow) dynamic behavior of the material.

Such a model is usually referred to as Hammerstein model.

13.4 Procedures for Online Estimation

13.4.1 Identification Procedure for Linear Models

As an example we consider the Zener model, whose differential equation corresponds to the discrete-time representation[1]:

$$F(h) = -\left[\frac{\kappa_1+\kappa_2}{\lambda}T-1\right]F(h-1)+\kappa_1\delta(h)+\left[\frac{\kappa_1\kappa_2}{\lambda}T-\kappa_1\right]\delta(h-1) \quad (13.18)$$

where $F(h)$ and $\delta(h)$ are the samples of the force $F(t)$ and of the displacement $\delta(t)$ at the instants $t_h = hT$, being T the sampling time.

(13.18) is a classical example of auto-regressive (ARX) model,

$$y(h) = \theta^T\varphi(h) \quad (13.19)$$

where $y(h) = F(h)$, $\varphi(h) = [-F(h-1)\ \ \delta(h)\ \ \delta(h-1)]^T$ and $\theta = [\theta_1\ \theta_2\ \theta_3]^T = \left[\left(\frac{\kappa_1+\kappa_2}{\lambda}T-1\right)\ \ \kappa_1\ \ \left(\frac{\kappa_1\kappa_2}{\lambda}T-\kappa_1\right)\right]^T$. If one assumes a model of this kind, the online estimation of the unknown parameters $\hat{\theta}$ is straightforward. A standard recursive algorithm [23], derived from classical least-square methods employed for off-line identification, can be adopted with the aim of minimizing the cost function:

$$V_N(\theta) := \frac{1}{N}\sum_{h=1}^{N}\alpha_h\varepsilon^2(h) \quad (13.20)$$

$$\varepsilon(h) := y(h) - \theta^T\varphi(h) \quad (13.21)$$

where N is the number of measures acquired from the beginning of the estimation process, and α_h are proper weights. In particular, since the aim is to obtain an estimation of possibly time-varying parameters, it has been assumed

$$\alpha_h = \beta^{N-h} \quad (13.22)$$

where $\beta < 1$ is the forgetting factor. With this expression of the weights, it is clear that older samples have less importance than recent ones. The expression of the recursive estimators is:

$$\begin{aligned}\hat{\theta}(h) &= \hat{\theta}(h-1) + Q(h)\left[y(h) - \varphi(h)^T\hat{\theta}(h-1)\right]\\ Q(h) &= P(h-1)\varphi(h)\left[\beta + \varphi(h)^T P(h-1)\varphi(h)\right]^{-1}\\ P(h) &= \tfrac{1}{\beta}\left[\mathbb{I} - Q(h)\right]P(h-1)\end{aligned} \quad (13.23)$$

At each sampling time, the *recursive least square algorithm (RLS)* computes new values of the unknowns and update its status on the basis of new data, but to start the estimation process the initial conditions on $\hat{\theta}(0)$ and $P(0)$ are necessary. A common choice is to take $P(0) = c \cdot I$, where c is a proper constant scalar, and $\hat{\theta}(0) = 0$, but it can see that the importance of these initial values decays with the time, as the number of new data

[1] The Euler approximation $\frac{dy(t)}{dt} \approx \frac{y(h)-y(h-1)}{T}$ has been adopted in this case.

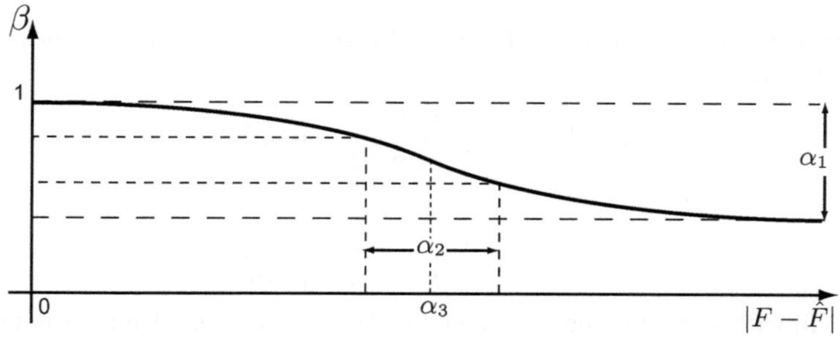

Fig. 13.5. Forgetting factor adaptation depending on error. Parameters α_1, α_2, and α_3 must be adjusted on the basis of the noise level.

grows. Also for this reason, a critical point in the definition of the estimator is the choice of the forgetting factor β, which must guarantee a proper tradeoff between tracking ability and noise sensitivity. The need to perceive variations in the touched materials suggests to compute the forgetting factor β dynamically, according to the magnitude of the error $|F(h) - \hat{F(h)}|$ between the estimated and the measured force [19]. In particular, when the error is small, the forgetting factor should be close to one, while it has to be decreased for large errors, so that the weight related to older samples decays. In particular, as reported in Fig. 13.5, β is computed as

$$\beta = 1 - \alpha_1 \left(\frac{1}{\pi} \arctan \left(\alpha_2 |F(h) - \hat{F}(h)| - \alpha_3 \right) + \frac{1}{2} \right) \tag{13.24}$$

where $1 - \alpha_1$ is the forgetting factor value for large errors, α_3 has the meaning of a threshold between the small error and the large error condition, while the amplitude of the transition region is governed by α_2.

13.4.2 Identification Method for Hunt-Crossley Parameters

Also in the case of the Hunt-Crossley model, a linear regression algorithm is applied. Since (13.3) is non linear with respect to the exponent n, the main idea is to separate the estimation of k and λ from the estimation of n. In this way, we can write two recursive least-squares estimators Γ_1 and Γ_2 [23] interconnected via feedback, as shown in Fig. 13.6.

In particular, Γ_1 estimates k and λ minimizing the cost function $V_N(k, \lambda)$, which has the same structure of (13.20) with the error defined as follows:

$$\varepsilon_1(h) := F(h) - [k + \lambda \dot{\delta}(h)]\delta^n(h) \tag{13.25}$$

Therefore, by assuming the vector of estimates at time $t = hT$ $\hat{\theta}_1(h) = [\hat{k}(h), \hat{\lambda}(h)]^T$, the vector of input signals $\varphi_1(h) = [\delta^n(h), \delta^n(h)\dot{\delta}(h)]^T$ and the system output $y_1(h) = F(h)$, the estimator Γ_1 is implemented in the standard recursive form (13.23), already used for linear models.

On the other side, an expression of (13.25) that is linear with respect to the parameter n can be obtained as

$$\ln \varepsilon_2(h) = \ln \frac{F(h)}{k + \lambda\dot{\delta}(h)} - n \ln \delta(t) \tag{13.26}$$

with

$$\varepsilon_2(h) := 1 + \frac{\varepsilon_1(h)}{[k + \lambda\dot{\delta}(h)]\delta^n(h)} \tag{13.27}$$

If ε_1 is small with respect to the force computed according to the Hunt-Crossley model described by (13.3), it is possible to write the following series expansion of (13.26):

$$\ln \varepsilon_2 = \ln \left(1 + \frac{\varepsilon_1}{[k + \lambda\dot{\delta}]\delta^n}\right) \simeq \frac{\varepsilon_1}{[k + \lambda\dot{\delta}]\delta^n} \tag{13.28}$$

Therefore, if the previous assumption holds and $\langle\varepsilon_1\rangle$ is independent on the force computed according to (13.3), $\langle\ln(\varepsilon_2(h))\rangle$ can be considered as a stochastic process with zero mean and it is possible to estimate n by means of a recursive least-square procedure minimizing the cost function $W_N(n)$:

$$W_N(n) := \frac{1}{N}\sum_{h=1}^{N} \ln^2(\varepsilon_2(h)) \tag{13.29}$$

and the implementation of the estimator Γ_2 is again analogous to (13.23), with $\hat{\theta}_2(h) = \hat{n}(h)$, $\varphi_2(h) = \ln \delta(h)$ and $y_2(h) = \ln(F(h)/(k + \lambda\dot{\delta}(h)))$.

Because of the feedback interconnection (see Fig. 13.6), the value of n used by Γ_1 as well as the value of these parameters used by Γ_2 is not the "true" value but the estimated one. For this reason, beside measurement noise and model error, each estimator behaves as an additional source of noise that could compromise the convergence of the algorithm. These considerations lead to rewrite estimation errors (13.25) and (13.26) as:

$$\varepsilon_1'(h) = F(h) - \left(k + \lambda\dot{\delta}(h)\right)\delta^{\hat{n}}(h) \tag{13.30}$$

$$\ln \varepsilon_2'(h) = \ln \frac{F(h)}{\hat{k} + \hat{\lambda}\dot{\delta}(h)} - n \ln \delta(h) \tag{13.31}$$

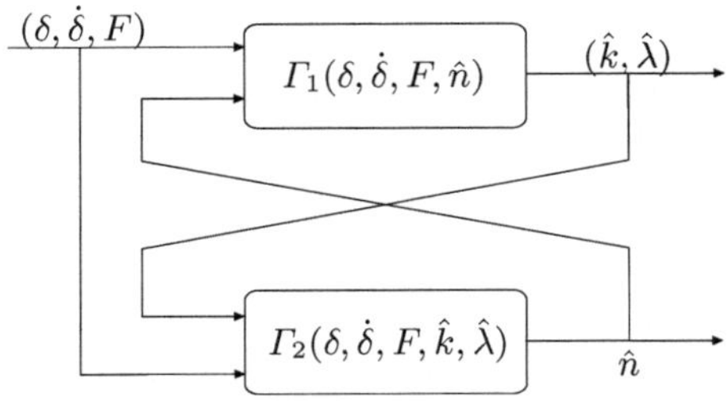

Fig. 13.6. On-line parameters estimator for the Hunt-Crossley model

where the effect introduced by the use of estimates instead of true parameters values is considered. Therefore, provided that elementary estimators would converge, the convergence of their feedback interconnection is obtained if additional disturbances do not bias the stochastic processes $\langle \varepsilon_1'(h) \rangle$, $\langle \ln(\varepsilon_2'(h)) \rangle$ expressing residuals between measured and estimated forces for $(\hat{k}, \hat{\lambda})$ and $\hat{n}$ respectively.

In particular, (13.30) can be rewritten as:

$$\varepsilon_1'(h) = \varepsilon_1(h) + \left(k + \lambda\dot{\delta}(h)\right) \delta^n(h) \left(1 - \delta^{\delta n}\right) \tag{13.32}$$

and, if the estimation error δn is small, the following approximation holds:

$$1 - \delta^{\delta n} \simeq \delta n \ln \delta(h) \tag{13.33}$$

Therefore, $\langle \varepsilon_1'(h) \rangle$ is a zero mean stochastic process if:

$$E\left[\delta n\right] E\left[\left(k + \lambda\dot{\delta}(h)\right) \delta^n(h)\right] = 0 \tag{13.34}$$

and this condition is satisfied if the estimator Γ_2 converges independently on Γ_1.

Since the relation between ε_2 and ε_2' is given by:

$$\begin{aligned} \ln \varepsilon_2'(h) &= \ln \varepsilon_2(h) - \ln \frac{\hat{k} + \hat{\lambda}\dot{\delta}(h)}{k + \lambda\dot{\delta}(h)} \\ &= \ln \varepsilon_2(h) - \ln \left(1 + \frac{\delta k + \delta\lambda\dot{\delta}(h)}{k + \lambda\dot{\delta}(h)}\right) \end{aligned} \tag{13.35}$$

and by computing the series expansion, also recalling (13.28), we obtain:

$$\ln \varepsilon_2' \simeq \ln \varepsilon_2 - \frac{\delta k + \delta\lambda\dot{\delta}}{k + \lambda\dot{\delta}} \simeq \frac{\varepsilon_1 - \delta^n(\delta k + \delta\lambda\dot{\delta})}{\delta^n(k + \lambda\dot{\delta})} \tag{13.36}$$

Hence, the additional noise due to feedback interconnection has to be dominated by $\varepsilon_1(h)$ so that $\langle \ln(\varepsilon_2'(h)) \rangle$ is still a zero mean process:

$$\|\delta k + \delta\lambda\dot{\delta}\| \ll \left\|\frac{\varepsilon_1}{\delta^n}\right\| \tag{13.37}$$

This condition is not too restrictive, because at the beginning of the estimation process, when the estimation errors δk and $\delta\lambda$ can be considerable, the penetration δ is small. Moreover $\dot{\delta}$ is also small, since the Hunt-Crossley model is valid for a limited range of impact velocities. Therefore, the use of $(\hat{k}, \hat{\lambda})$ within the Γ_2 estimator does not alter its convergence properties and the overall feedback estimator provides unbiased estimates $\hat{k}$, $\hat{\lambda}$, $\hat{n}$ of parameters of the Hunt-Crossley model.

13.4.3 Estimation of Quasi-linear Model Parameters

In order to estimate the characteristic parameters of the quasi-linear model it is necessary to have a sampled-data system. This can be obtained by simply using a digital filter in lieu of $G_L(s)$ (see Fig. 13.4), whose general form is

$$G_L(z^{-1}) = z^{-k} \frac{b_0 + b_1 z^{-1} + \ldots + b_l z^{-l}}{1 + a_1 z^{-1} + \ldots + a_r z^{-r}} \tag{13.38}$$

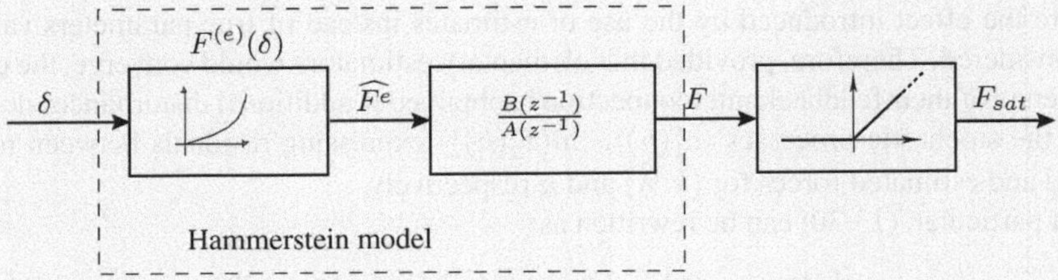

Fig. 13.7. Hammerstein model with saturated output of a contact with a viscoelastic material

Therefore, from the scheme of Fig. 13.4 the model reported in Fig. 13.7 is obtained, where the continuous transfer function $G_L(s)$ has been replaced by the digital filter $G_L(z^{-1})$[2]. Note that, from the considerations reported in sec. 13.3.2, it follows that:

- no pure delay exists in the system between the input (of the linear systems) $\widehat{F}^e$ and the output F ($k = 0$);
- the response depends on the input at the same instant; therefore the orders of the numerator of $G_L(z^{-1})$ and that of the denominator must be equal ($l = r$).

Finally, it is worth to notice that in the general case the model, composed by a cascade of a nonlinear and of a linear part, includes a redundancy in the parameters definition. As a matter of fact, the gain of overall system results from the product of the gain of nonlinear characteristic and of that of the linear part. In our case, this ambiguity has been solved by assuming that the coefficient b_0 in (13.38) is equal to 1. In this way, the initial value of F to a unit step input

$$F(0^+) = \lim_{z \to +\infty} G_L(z^{-1}) \frac{1}{1 - z^{-1}} = b_0 = 1$$

is the same of that produced by $G_L(s)$

$$\lim_{s \to +\infty} s(G_L(s)\frac{1}{s}) = \lim_{s \to +\infty} sG(s) = g(0^+) = 1$$

and, accordingly, the (physical) meaning of the elastic response $F^{(e)}$ remain unchanged.

The scheme of Fig. 13.7 reproduces the well-known Hammerstein model, frequently adopted to model and identify nonlinear systems [24]. This structure is quite interesting, since it is composed by a no-memory nonlinear gain and a dynamic linear block, resulting particularly suitable for identification [25, 26].

In particular, by exploiting the model structure with the two temporally distinct elements (the static part - elastic response $F^{(e)}$ - produces an instantaneous response, while the dynamic one - reduced relaxation function - reacts with a slower dynamic and in the first time instants does not influence the elastic response, see Fig. 13.9), which reflects

[2] The saturation on the model output avoids that the force response becomes negative, but does not interfere with the identification process.

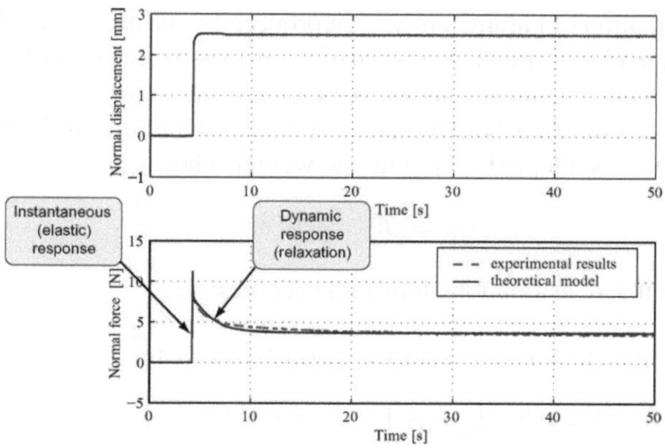

Fig. 13.8. Dynamic response of polyurethane gel (dashed) and estimated model (solid)

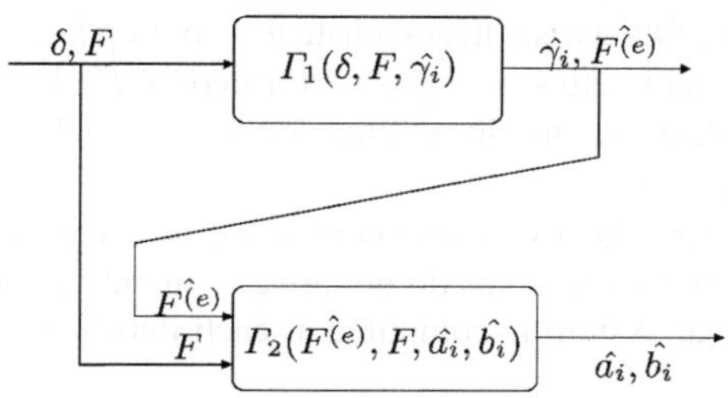

Fig. 13.9. Online parameters estimator for the Hammerstein model

the typical behavior of soft materials during an interaction, it is possible to conceive a two phases estimation algorithm, see Fig. 13.9:

- during the contact, provided that the velocity of the probe is fast enough (in principle an ideal step deformation should be used), the parameters of $F^{(e)}$ are estimated, by neglecting the dynamic terms;
- by using the output of the elastic response, the coefficients of the linear filter can be identified.

For the elastic part, the expression based on the power law is the most convenient[3]

$$F^{(e)}(\delta) = \gamma_1 \delta^{\gamma_2} \tag{13.39}$$

since it can be easily transformed in a linear form, like in the case of Hunt-Crossley model:

$$\ln(F^{(e)}) = \ln(\gamma_1) + \gamma_2 \ln(\delta) \tag{13.40}$$

[3] Compare with (13.14), where $\gamma_1 = \frac{p}{q+1}$, $\gamma_2 = q + 1$.

Therefore, the estimation of parameters γ_i, can proceed by means of the *RLS* algorithm as in (13.23), with $\hat{\theta}_1(h) = [\ln(\hat{\gamma_1}(h)), \hat{\gamma_2}(h)]^T$, $\varphi_1(h) = [1, \ln(\delta(h))]^T$ and $y_1(h) = \ln(F(h))$. In particular, to compensate for the fact that the minimization is performed with respect to the norm of the squared error of $\ln(F^{(e)})$ and not of $F^{(e)}$, a weighted least-square algorithm is adopted [27], with the weights chosen as

$$\alpha_h = \beta^{N-h} e^{y_1(h)} \tag{13.41}$$

In this case, the recursive estimation algorithm becomes:

$$\begin{aligned} \hat{\theta}_1(h) &= \hat{\theta}_1(h-1) + Q(h)\left[y_1(h) - \varphi_1(h)^T \hat{\theta}_1(h-1)\right] \\ Q(h) &= P(h-1)\varphi_1(h)\left[\left(\frac{\beta}{e^{y_1(h)}}\right) + \varphi_1(h)^T P(h-1)\varphi_1(h)\right]^{-1} \\ P(h) &= \frac{1}{\beta}\left[\mathbb{I} - Q(h)\right] P(h-1) \end{aligned} \tag{13.42}$$

It is worth to notice that only the data, which are acquired when the speed of the contacting probe (and therefore the speed of the deformation $\dot{\delta}$) is bigger than a given threshold $\dot{\delta}^\star$ (and $F \approx F^{(e)}$) [28]. Otherwise, the estimate θ_1 is not updated, and only the estimation of the linear filter parameters $[\hat{a}_i, \hat{b}_i]$ is performed by Γ_2. In this case, the standard *RLS* method is applied, being the linear filter a classical *ARX* system, with the input $\hat{F^{(e)}}(h)$ and the output $F(h)$[4].

Note, that in this case the two estimators work in a cascade configuration (see Fig. 13.9), and convergence problems do not arise. The only interaction is that the the input signal of the linear system is computed at each iteration on the basis of the the updated values of γ_i.

13.5 Experimental Results

In order to verify the above estimation algorithm, a laboratory setup has been implemented. The setup consists of a linear electric motor equipped with a position sensor and a load cell, Fig. 13.10.

The measures required by the estimation algorithm have been obtained by imposing a motion profile to a linear motor in contact with different materials. The contact force F is measured by means of the load cell, the penetration δ is measured by comparing the current motor position with the position measured at the time of impact, and finally the penetration velocity $\dot{\delta}$ is obtained from x by means of a state variable filter. The linear motor and the load cell have been connected, by means of a AD/DA board, to a standard PC running control and estimation algorithms in a mixed MATLAB/RTLinux environment. In the experimental activity, the sampling time has been set to $T = 1$ msec. Several materials have been used and here, in particular, the results obtained with a thin layer of plastic material, characterized by a stiff behavior, and a soft gel, whose viscoelasticity is relevant, are described.

[4] Being the parameter $b_0 = 1$ the estimation (13.23) for the linear filter is performed with $\hat{\theta}_2(h) = [-\hat{a}_1, -\hat{a}_2, \ldots, -\hat{a}_r, \hat{b}_1, \hat{b}_2, \ldots, \hat{b}_r]^T$, $\varphi_2(h) = [F(h-1), F(h-2), \ldots, F(h-r), \hat{F^{(e)}}(h-1), \hat{F^{(e)}}(h-2), \ldots, \hat{F^{(e)}}(h-r)]^T$ and $y_2(h) = F(h) - \hat{F^{(e)}}(h)$.

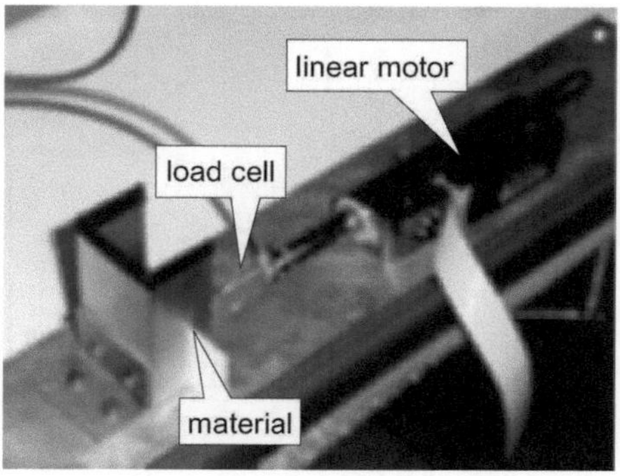

Fig. 13.10. Experimental setup

13.5.1 Hunt-Crossley Parameters Estimation

Stiff Material

Experimental results related to the thin layer of plastic materials are presented in Fig. 13.11 and in Fig. 13.12. In particular, the hysteresis loop reported in Fig. 11(a) shows that energy dissipation is low and the behavior of the material depends essentially on the elastic coefficient k. Moreover, hysteresis loop and the related power exchange, reported in Fig. 11(b), show good correspondence between the experimental curves and the estimated one. Finally, Fig. 13.12 shows the parameters estimates. In particular, the following values are obtained: $k \simeq 1.15e4$, $\lambda \simeq 70$, $n \simeq 1.2$. According to what previously discussed about properties of the Hunt-Crossley model, the exponent n, that takes into account geometry of contact surfaces, is about one. We notice that k and n are quite stable around their final value, while λ presents a more oscillatory behavior, because of

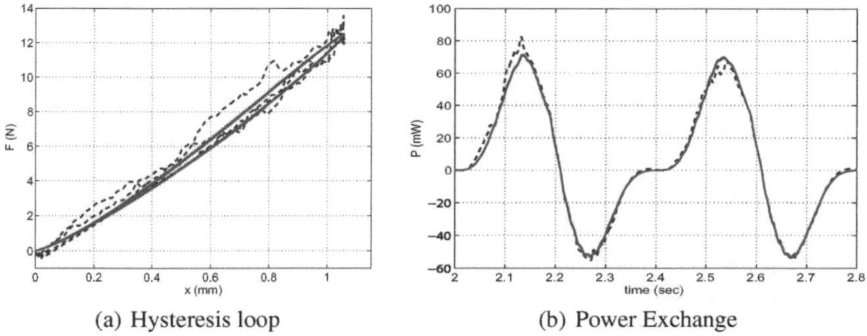

(a) Hysteresis loop (b) Power Exchange

Fig. 13.11. Hunt-Crossley model: measured (dashed) and estimated (solid) hysteresis loop and power exchange for a thin layer of plastic materials

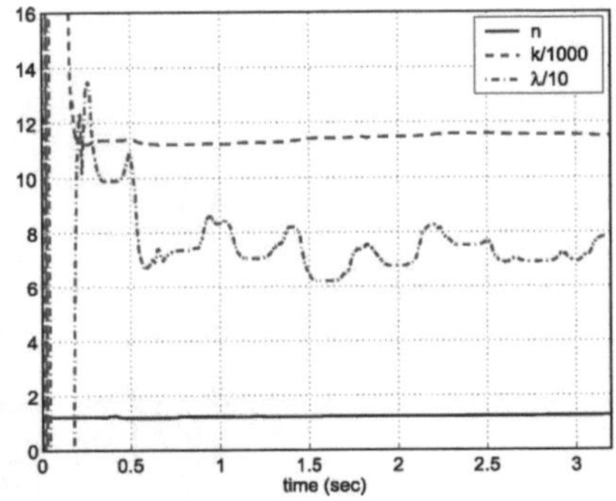

Fig. 13.12. Hunt-Crossley model: estimation of n (solid), k (dash) and λ (dash-dot) for a thin layer of plastic material

numerical differentiation used to compute $\dot{x}$ which makes this parameter more sensitive to measurement noise.

Soft Material

For compliant materials, the advantages of the Hunt-Crossley model are more evident. The application of a linear regression algorithm to estimate parameters K and B of the Kelvin-Voigt model for a soft gel provides the results of Fig. 13.13, with $K \simeq 2.26e3$ N/m and $B \simeq 123$ Ns/m. In particular, drawbacks related to nonzero estimated force when $x = 0$ are evident, as well as inconsistencies in power exchange between the probe device and the soft gel. On the contrary, hysteresis loop estimated by means of the Hunt-Crossley model is more similar to the measured one, Fig. 13.13(a), as well the estimated power exchange, shown in Fig. 13.14(b), that does not exhibit positive "spikes" present in Fig. 13.13(b).

Finally, Fig. 13.15 shows the convergence of parameters to their final values $k \simeq 1.35e3$, $\lambda \simeq 36$, $n \simeq 1.35$. Notice that in this case the value of n is greater than that

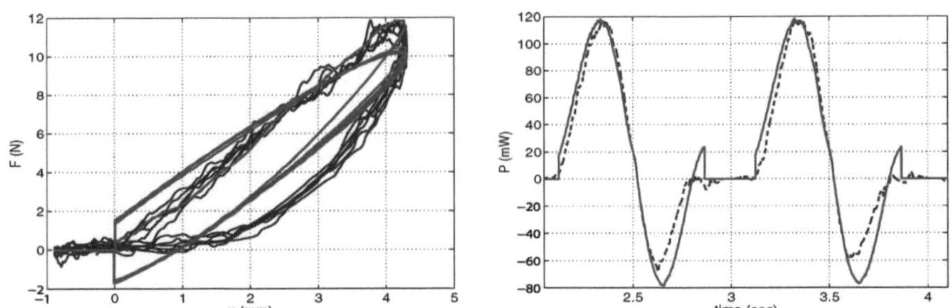

(a) Hysteresis loop (b) Power Exchange

Fig. 13.13. Linear model: measured (dash) and estimated (solid) hysteresis loop and power exchange for soft gel

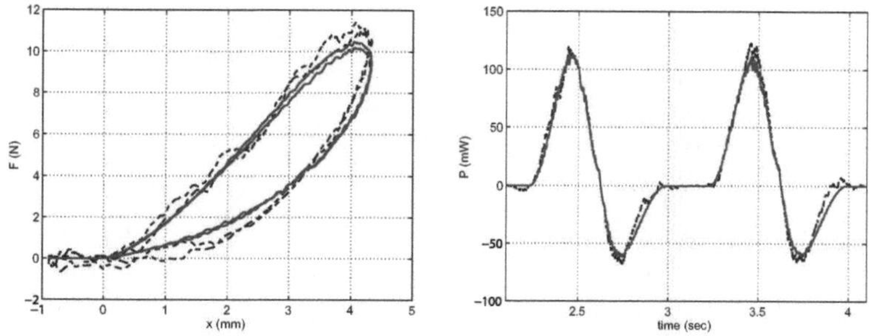

(a) Hysteresis loop (b) Power Exchange

Fig. 13.14. Hunt-Crossley model: measured (dashed) and estimated (solid) hysteresis loop and power exchange for soft gel

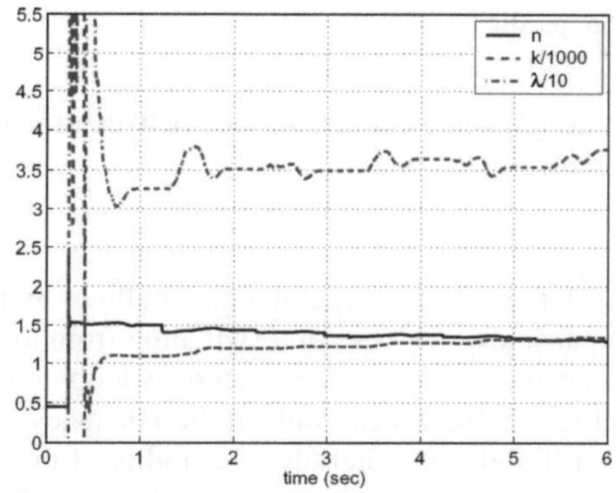

Fig. 13.15. Hunt-Crossley model: estimation of n (solid), k (dash) and λ (dash-dot) for soft gel

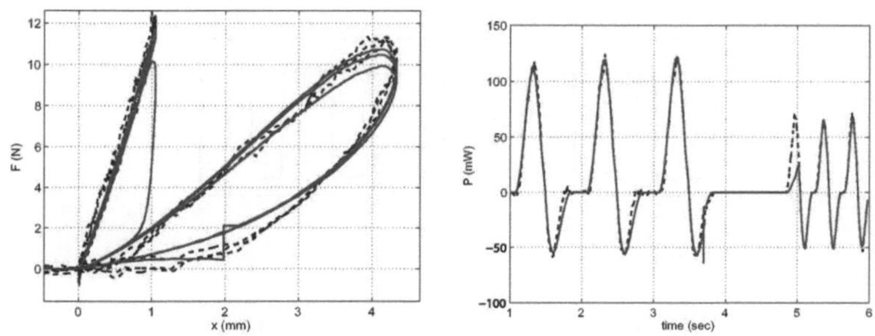

(a) Hysteresis loop (b) Power Exchange

Fig. 13.16. Hunt-Crossley model: measured (dashed) and estimated (solid) hysteresis loop and power exchange when soft gel is substituted by the thin plastic layer

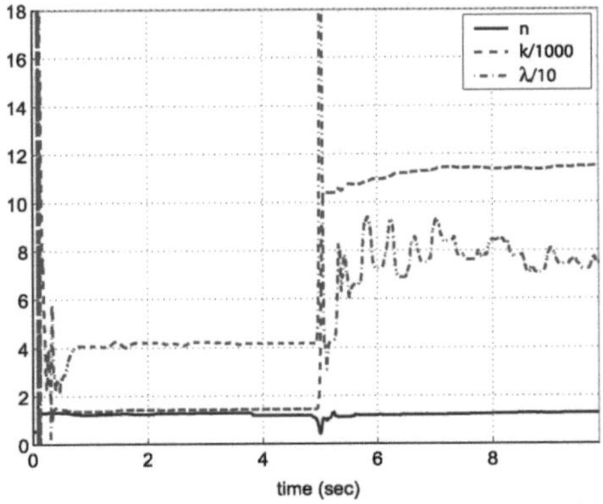

Fig. 13.17. Hunt-Crossley model: estimation of n (solid), k (dash) and λ (dash-dot) when soft gel is substituted by the thin plastic layer

obtained for the stiff material, since contact surface between the probe device and the soft gel is slightly different.

Change of Material

As mentioned in Sec. 13.4, the use of a forgetting factor allows to improve the detection of material changes. In particular, the case of a switching from the soft gel to the layer of plastic materials has been considered. The hysteresis loop and the power exchange diagram are shown in Fig. 13.16, which confirm the adequacy of the Hunt-Crossley model to describe both stiff and soft materials. The ability of the estimation algorithm to detect a change in the touched material is shown in Fig. 13.17. Indeed, after a short transient, the convergence to parameters of the new material is achieved and previous estimates do not affect final values.

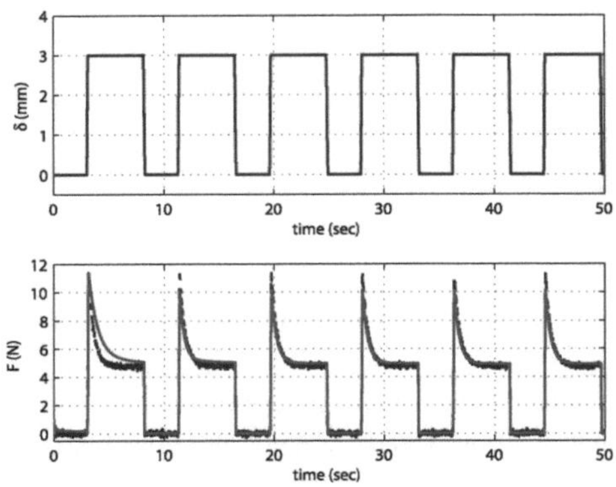

Fig. 13.18. Contact with a layer of polyurethane gel: measured data (dashed) and estimated model (solid)

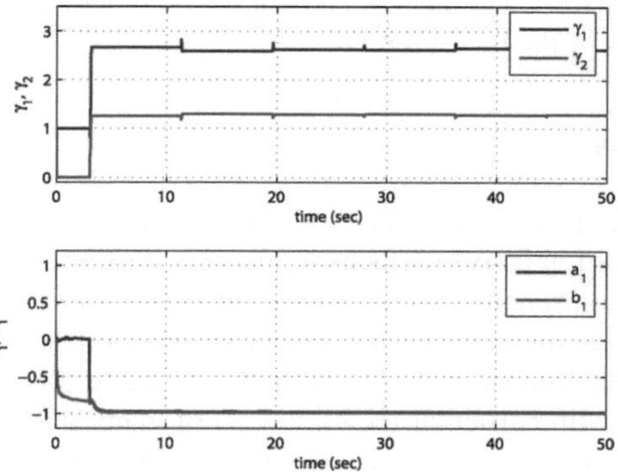

Fig. 13.19. Contact with a layer of polyurethane gel: parameters estimation

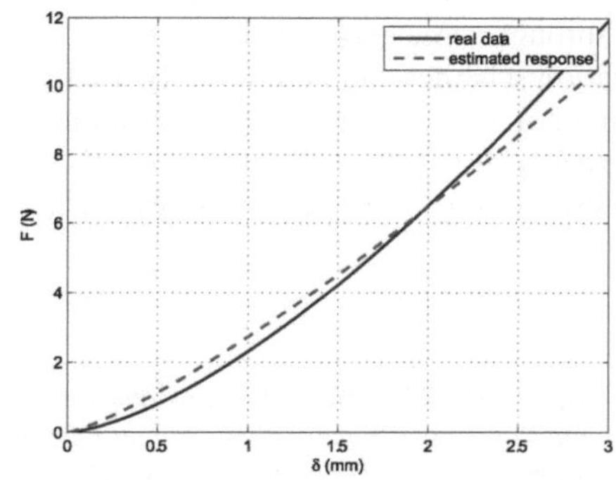

Fig. 13.20. Elastic response of polyurethane: real (solid) and estimated (dashed) response

13.5.2 Hammerstein Model Estimation

The Hammerstein model is suitable to describe the interaction with both elastic and viscoelastic materials. In particular, it is quite convenient for modelling the behavior of those objects which are characterized by a very slow dynamics (e.g. biological tissues). In Fig. 13.18 the estimation process described in Sec. 13.4.3 is performed, by considering the contact with a layer of polyurethane gel. Finally in Fig. 13.20, the comparison of the elastic response, as it has been estimated at the end of the identification process, with the real one is reported. Also in this case, the approximation can be considered satisfactory. In this case, the deformation has been applied according to a step (or, to be precise, as a feasible approximation of the step with a rising ramp performed at $30mm/s$) and there are not speed limitations during the contact (differently from Hunt-Crossley model, whose validity is limited under a certain velocity). As shown in Fig. 13.19, the estimator converges rather quickly to a stable parameters configuration[5]

[5] In this case, for the linear filter, it has been assumed $r = 1$.

(for both the elastic response and the reduced relaxation function), which guarantees an optimal approximation of the measurements.

13.6 Conclusions

In this chapter, an overview of modelling and estimation techniques for nonlinear contact dynamics with hard and soft surfaces is given, with the purpose to use such results in telemanipulation schemes to improve their performance, and in particular the transparency towards the human operator. The proposed models are the Hunt-Crossley and the quasi-linear/Hammerstein model, both able to capture the nonlinearities characterizing the behavior (in particular the elasticity) of many materials during physical interactions. The former has been extensively used to describe the interaction of robot manipulators with the environment, the latter is very suitable to model behavior of visco-elastic materials (e.g. biological tissues), characterized by very slow dynamic modes.

For both models some suitable identification algorithms have been developed, which allows to estimate online linear and non-linear terms. Finally, the proposed estimation techniques are validated through some experimental tests.

This kind of identification schemes can be profitably used for real-time impedance estimation in telemanipulation architectures. The extension to 3D models, for general applications with 6- degrees of freedom manipulators interacting with a remote environment, is therefore necessary and a topic of current research.

References

1. P. Arcara and C. Melchiorri. *MISTRAL: Methodologies and Integration of Subsystems and Technologies for Anthropic Robots and Locomotion*, chapter Comparison and Improvement of Control Schemes for Robotic Teleoperation Systems with Time Delay. Springer-Verlag, 2003.
2. J.E. Colgate. Robust impedance shaping telemanipulation. *IEEE Transactions on Robotics and Automation*, 9(4), 1993.
3. S. Hirche and M. Buss. Passive position controlled telepresence systems with time delay. In *American Control Conference*, volume 1, Denver, USA 2003.
4. D. Erickson, M. Weber, and I. Sharf. Contact stiffness and damping estimation for robotic systems. *Int. Jour. of Robotic Research*, vol. 22, 2003.
5. K.H. Hunt and F.R.E. Crossley. Coefficient of restitution interpreted as damping in vibroimpact. *ASME Journal of Applied Mechanics*, pages 440–445, 1975.
6. K. L. Johnson. *Contact Mechanics*. Prentice Hall International Editions, 1995.
7. W.N.Findley, J.S. Lai, and K. Onaran. *Creep and Relaxation of Nonlinear Viscoelastic Materials*. North-Holland Publishing Company, 1976.
8. S. Misra abd A. M. Okamura. Environment parameter estimation during bilateral telemanipulation. In *Symposium on Virtual Interfaces for Virtual Environment and Teleoperator Systems*, Alexandria, Virginia, USA, 2006.
9. Y. Yokokohji and T. Yoshikawa. Bilateral control of master-slave manipulators for ideal kinesthetic coupling. In *IEEE International Conference on Robotics and Automation*, Nice, France, 1992.
10. B. Hannaford. A design framework for teleoperators with kinesthetic feedback. *IEEE Transactions on Robotics and Automation*, 5(4), 1989.

11. G. De Gersem, H. Van Brussel, and J. Vander Sloten. Enhanced haptic sensitivity for soft tissues using teleoperation with shaped impedance reflection. In *World Haptics Conference*, Pisa, Italy, 2005.
12. K. Hashtrudi-Zaad and S.E. Salcudean. Adaptive transparent impedance reflecting teleoperation. In *IEEE International Conference on Robotics and Automation*, Minneapolis, Minnesota, 1996.
13. K. B. Shimoga and A. A. Goldenberg. Soft robotic fingertips - part ii: A comparison of construction materials. *The Int. Journal of Robotic Research*, 15(4), 1996.
14. H. C. Güler, N. Berme, and S. R. Simon. A viscoelastic sphere model for the representation of plantar soft tissue during simulations. *Jour. of Biomechanics*, 31:847–853, 1998.
15. M. Shibata and S. Hirai. Stability and graspability analysis in grasping task taking fingertip dynamics into consideration. In *Proc. of IEEE Int. Conf. on Intelligent Robots and Systems, IROS'04*, Sendai, Japan, 2004.
16. K.H. Hunt and F.R.E. Crossley. Coefficient of restitution interpreted as damping in vibroimpact. *ASME Journal of Appl. Mech.*, 1975.
17. D.W. Marhefka and D.E. Orin. A compliant contact model with nonlinear damping for simulation of robotic systems. *IEEE Trans. on Systems, Man and Cybernetics*, 29(6), 1999.
18. P.R. Kraus and V. Kumar. Compliant contact models for rigid body collisions. In *Proc. of IEEE Int. Conf. on Robotics and Automation, ICRA'97*, 1997.
19. N. Diolaiti, C. Melchiorri, and S. Stramigioli. Contact impedance estimation for robotic systems. *IEEE Trans. on Robotics and Automation*, 21(5), 2005.
20. Y.C. Fung. *Biomechanics: Mechanical Properties of Living Tissues*. Springer-Verlag, 1993.
21. W. N. Findley and J. S. Y. Lay. A modified superposition principle applied to creep of nonlinear viscoelastic material under abrupt changes in state of combined stress. *Trans. of the Society of Rheology*, vol. 11(3):361–380, 1967.
22. A. C. Pipkin and T. G. Rogers. A non-linear integral representation for viscoelastic behaviour. *Jour. of Mech. Phys. Solids*, vol. 16, 1967.
23. L. Ljung and T. Södertström. *Theory and Practice of Recursive Identification*. The MIT Press, Cambridge, Massachusetts, 1983.
24. K.J. Hunt, M. Munih, N. Donaldson, and F.M.D. Barr. Investigation of hammertein hypothesis in the modelling of electrically stimulated muscle. *IEEE Transactions on Biomedical Enginering*, 45(8), 1998.
25. E. Eskinat, SH. Johnson, and W.L. Luyben. Use of hammerstein models in identification of nonlinear systems. *AIChE Journal*, 37(2), 1991.
26. K.S. Narendra and P.G. Gallman. An iterative method for the identification of the nonlinear systems using the hammerstein model. *IEEE Transactions on Automatic Control*, 12, 1966.
27. I. Kao and F. Yang. Stiffness and contact mechanics of soft fingers in grasping and manipualtion. *IEEE Trans. on Robotics and Automation*, 20(1), 2004.
28. L. Biagiotti, C. Melchiorri, P. Tiezzi, and G. Vassura. Modelling and identification of soft pads for robotic hands. In *Proc. of IEEE Int. Conf. on Intelligent Robots and Systems, IROS'05*, 2005.

14

Power Scaling in Port-Hamiltonian Telemanipulation over Packet Switched Networks

Cristian Secchi[1], Stefano Stramigioli[2], and Cesare Fantuzzi[1]

[1] University of Modena and Reggio Emilia
DISMI
Viale Allegri 13, 42100 Reggio Emilia, Italy
{secchi.cristian,fantuzzi.cesare}@unimore.it

[2] University of Twente
EEMCS
P.O.Box 217, 7500 AE Enschede, NL
S.Stramigioli@ieee.org

Summary. In several important applications involving bilateral telemanipulation systems (e.g. macro-micro teleoperation, telesurgery) master and slave act at different power scales. In this contribution, we illustrate how to embed power scaling into port-Hamiltonian based telemanipulation schemes over packet switched networks. We propose a discrete scattering based communication strategy to scale the power exchanged between master and slave sides and a way to handle lost packets that allow to get power scaling while preserving a stable behavior of the system independently of any communication delay and of any possible loss of packets.

14.1 Introduction

Passivity is a very suitable tool to stabilize a telemanipulator; in fact, implementing each part of a telerobotic system as a passive system and interconnecting them in a power preserving way it is possible to achieve an intrinsically passive system which is consequently characterized by a stable behavior.

In [1] scattering theory has been exploited to build a communication channel that is passive independently of any constant transmission delay. In [2] the concept of wave variables, which allow to give a clear energetic interpretation of the scattering based communication strategy proposed in [1], is introduced. Several passivity based strategies have been proposed to control master and slave robots. In particular, in [3], the port-Hamiltonian framework ([4]) has been applied to telemanipulation: both master and slave robots, which can be modeled as port-Hamiltonian systems, are controlled by means of intrinsically passive port-Hamiltonian impedance controllers (IPC, [5]) which guarantee a stable interaction with any, possibly unknown, passive environment. In this way both master and slave sides can be modeled as passive port-Hamiltonian systems that are then

M. Ferre et al. (Eds.): Advances in Telerobotics, STAR 31, pp. 233–256, 2007.
springerlink.com © Springer-Verlag Berlin Heidelberg 2007

interconnected by means of a scattering based communication channel that allows a passive exchange of information independently of any transmission delay. The resulting telemanipulation system is intrinsically passive and, therefore, characterized by a stable behavior. In [6, 7] the discrete nature of the controllers is considered and techniques presented in [8] are used to obtain passive discrete port-Hamiltonian controllers and to interconnect them in a power preserving way to the continuous plants. Furthermore, discrete scattering is defined and used to implement a packet switching transmission line that has been proven to be passive also when communication delay is variable and some packets are lost. For an overview of the various techniques proposed in the literature see e.g. [9].

In several tasks involving bilateral telemanipulators, such as telesurgery and telemanipulation of huge robotic arms for extra-vehicular activity in space applications, master and slave act at different power scales and therefore, it is necessary to scale velocities and forces that are exchanged in order to take into account for this difference. Ideally, the system should transmit to the human operator a scaled version of the environment impedance such that the dynamic character of the environment remains undistorted.

Several researchers addressed this problem and several control algorithms for achieving velocity/force scaling have been proposed. In [10, 11] linear teleoperators have been considered; in [10] loop-shaping compensators have been introduced in a position/force architecture to increase transparency of the overall telemanipulation scheme. In [11] H_∞ control and μ-synthesis have been used to implement power scaling over the Internet. In [12] robots characterized by a constant inertia tensor have been considered and a control law based on the cancellation of dynamics has been suggested. In [13] the telemanipulator is decomposed in a shape and locked system and a passive feed-forward action is used to implement a scaled coordination between master and slave robots.

Scaling forces and velocities means scaling the power exchanged between master and slave side. Thus, it is natural to assign the issue of power scaling to the medium through which master and slave side exchange power, namely the communication channel. In [14, 15] linear telemanipulators are considered and the problem of scaling is tackled from this point of view. In [14] it is shown that a scaling of the exchanged power can be performed without affecting the passivity of the overall scheme and in [15] this concept is further analyzed and a necessary condition for the stability in case of multichannel amplification is provided. However, in both cases, negligible communication delay has been assumed. The aim of this contribution is to provide a communication strategy that allows power scaling over a packet switched communication channel in a digital port-Hamiltonian based telemanipulation scheme [6, 7]. We will firstly prove that, analogously to what has been done in [14] for linear telemanipulators, in case of negligible communication delay, power scaling can be achieved without compromising the stable behavior of the telemanipulation scheme. Secondly, we will propose a novel scattering based communication strategy that allows to scale the power exchanged also in case of non negligible communication delay. Finally we will consider the problem of loss of packets in the communication; we will show that if lost packets are not properly handled the system can become unstable and then we will propose a strategy that preserves the stable behavior of the overall scheme independently of the loss of packets.

In Sec. 14.2 some background on port-Hamiltonian based telemanipulation is given and in Sec. 14.3 the problem of power scaling in case of negligible communication delay is addressed. In Sec. 14.4 it is shown how to embed power scaling in digital port-Hamiltonian based telemanipulation in case of non negligible transmission delay. In Sec. 14.5 we address the problem of packets loss and in Sec. 14.6 we provide some simulations in order to validate the proposed scaling strategy. Finally in Sec. 14.7 some concluding remarks are reported and in Appendix A the mathematical proofs of the obtained results are given.

14.2 Background

14.2.1 Port-Hamiltonian Systems

We will now try to give an intuitive description of port-Hamiltonian systems; more formal descriptions can be found in [4]. Port-Hamiltonian systems are probably the most powerful representation which allow to model many physical systems using generalizations of concepts widely used in network theory.

A fundamental concept in engineering sciences is the notion of an *open system*, that is a system having a direct interface with its environment. The concept of an open system is directly linked to the notion of a *network*, where open systems are coupled to each other through their interfaces. Complementary to the network modeling of complex systems is the *design* and *control* of systems with a required functionality by coupling open system components.

Port-Hamiltonian systems are a perfect language to tackle network modeling and control of complex *physical* systems, using an integrated system approach allowing to deal with physical components stemming from different physical domains [16, 17] (electrical, mechanical, thermodynamical, ..), both in the lumped-parameter and in the distributed-parameter case. In order to describe and to manipulate these dynamical models in a systematic way it is mandatory to develop a coordinate-free, *geometric* framework for their mathematical formulation, especially because of the intrinsic and strong nonlinearities in their system behavior. In the port-Hamiltonian methodology, physical components are formulated as generalized Hamiltonian systems, coupled to each other through power ports. The resulting complex physical system is then geometrically described as a Hamiltonian system with respect to the geometric object of a Dirac structure. Apart from the great advantages for simulation and analysis this Hamiltonian framework immediately provides a powerful starting point for design and control of multi-domain technological systems.

But what is a power-port? A power port is the means by which interaction can take place between parts of a physical system or, in certain situations, between a physical system and a controller which has been designed following this philosophy.

A network structure defines then a certain relation on the "to be connected ports" which should be power continuous (does not destroy or generate energy) and which will describe the power flows in the system. Beside the network structure, there will be elements which will store energy like a spring, a mass, a capacitor or an inductor or which will transform energy irreversibly to heat like a resistor or friction.

With these components we are able to describe a big variety of physical (sub)-systems stemming out from different domains, and this gives rise to a real systematic analysis, control and design of Mechatronics systems and it is perfectly suitable in all those situations where multi-port analysis is useful like in telemanipulation.

Mathematically a power port is the pair of two entities which, if properly combined, will express power flowing between the subsystems the port connects. Consider for example the interconnection of a resistor with a capacitor in parallel. This interconnection can be described by considering the pair (v, i) of voltage and current common to the two elements. Instantaneously, the power flowing from the capacitor to the resistor will be equal to $P = vi$. The product of these two variable should be always power and for this reason they are called *power conjugate variables.* Other power conjugate variables are force and velocities, pressure and flow-rate, temperature and entropy flow. These variables are called efforts and flows and this nomenclature is the usual one used also in bond-graphs [18].

This can be generalized to much more complicated physical entities which are not scalars but do have a tensorial (geometrical) structure like vector forces and velocities or even more generally like twists and wrenches in multi-body mechanics [5]. In this case the power flow will still be expressible as the pairing of two dual[1] variables. After a proper coordinate choice, this product turns out to be the usual scalar product of 2 vectors:

$$P : V \times V^* \to \mathbb{R} \; ; \; (e, f) \mapsto e^T f.$$

Once we have the concept of power ports available, which will be the interface between subcomponents, we can look at a physical system as it would be a collection of parts like springs, dampers, resistors, flying wheels and others, connected through ports to a network structure which represents the energetic interconnections. This network structure is mathematically represented by what is called a Dirac structure.

If for example we consider 2 capacitors, a resistor and an inductor, there are many ways how we can interconnect them, and the resulting behavior, due to the different interconnections, will be completely different; the elements are the same, but the network structure is different.

Once we have a proper definition of the interfaces (power ports) and a description of the network topological structure (Dirac structure), we clearly need the components to be connected like springs, dampers etc. Properly speaking, it is correct to talk about ideal elements instead of components since a "component spring" could be modeled as an interconnection of an "ideal element" spring with some extra parasitic effect like "ideal damping" and "ideal inertial" properties.

[1] Duality is a well defined mathematical concept which allows in an intrinsic way to associate to two dual variables a scalar.

Fig. 14.1. The electrical (left) and electro-mechanical part (right)

For this reasons, we will classify ideal elements as *pure storages* of energy like potential or kinetic energy, electrical or magnetic energy and *irreversible transducers* like resistors.

We will now introduce as an example the modeling of a DC motor in the port-Hamiltonian setting. The example is chosen to be very simple on purpose, in order to make the procedure clear to the reader.

In the electrical network reported on the left of Fig. 14.1, there are 4 ports which will be used: a port connected to a voltage source (e_s, f_s), a port connected to a an inductor which is a storage element (e_i, f_i), a port connected to a resistor (e_r, f_r) and an interconnection port which will be used to interconnect the system to something else (e_c, f_c).

We can represent the Dirac structure in many different ways, but the most straight forward in this case will be by means of a skew-symmetric matrix. For example, the skew-symmetric matrix representing the interconnection between the parts of the electrical network can be deduced using Kirchhoff's laws and it is given by:

$$\begin{pmatrix} f_s \\ e_i \\ f_r \\ f_c \end{pmatrix} = \begin{pmatrix} 0 & -1 & 0 & 0 \\ 1 & 0 & -1 & -1 \\ 0 & 1 & 0 & 0 \\ 0 & 1 & 0 & 0 \end{pmatrix} \begin{pmatrix} e_s \\ f_i \\ e_r \\ f_c \end{pmatrix} \tag{14.1}$$

The skew symmetry ensures power continuity required for the Tellegen's theorem:

$$e_s^T f_s + e_i^T f_i + e_r^T f_r + e_c^T f_c = 0$$

Consider the electrical network on the left of Fig. 14.1. It is a series or a source, an inductor, a resistor and an open port which can be used to interconnect the system to the rest of the world.

The inductor is a storage of magnetic energy. The energy stored in the inductor is a function of the flux which is the proper energy variable[2]. If we consider a linear inductor, the energy function would be:

$$H(\lambda) = \frac{1}{2L}\lambda^2 \tag{14.2}$$

[2] If we consider the energy function, the inductor is a function of the flux linkage and not of the current. Properly speaking, the energy expressed as function of current is called co-energy.

The generalized effort corresponding to each storage element is the partial derivative of the energy function to the energy variable which in this case is the flux:

$$e = i = \frac{\partial H}{\partial \lambda} = \frac{\lambda}{L} \tag{14.3}$$

$$f = v = \dot{\lambda} \tag{14.4}$$

The resistor will satisfy Ohms law: $e_r = Rf_r$. It could be shown that the system's equation could be easily written in the following form:

$$\begin{aligned} \dot{x} &= (J(x) - R(x))\frac{\partial H}{\partial x} + g(x)u \\ y &= g^T(x)\frac{\partial H}{\partial x} \end{aligned} \tag{14.5}$$

where $x = \lambda$ is the physical state of the subsystem $H(x) = \frac{1}{2L}\lambda^2$ is the energy stored in the sub-system, $J(x)$ is in general a skew symmetric energy representing the network structure, $R(x)$ is a semi-positive matrix representing dissipation and $g(x)$ is an input matrix representing the interconnection with external ports or power supply. In details we have for our case:

$$J(x) = 0 \qquad R(x) = R \qquad g(x) = (1\ 1) \qquad u = \begin{pmatrix} u_s \\ u_i \end{pmatrix}$$

where u_i indicates the voltage source value and (u_i, i_i) the interconnection port which will be connected to the mechanical part.

It would be possible to follow the same procedure for the mechanical part represented in Fig. 14.1. and get to an equation of exactly the same form as (14.5). In this case would be

$$x = p \qquad J(x) = 0 \qquad R(x) = b \qquad g(x) = K \qquad H(p) = \frac{1}{2I}p^2$$

and where b is the damping coefficient and K the motor constant relating current to torque and angular velocity to the e.m.f.

It could be also seen that the interconnection of the electrical part and the electro-mechanical one would result once again in equations of the same form as (14.5) with $H(\lambda, p) = \frac{1}{2L}\lambda^2 + \frac{1}{2m}p^2$ and:

$$x = \begin{pmatrix} \lambda \\ p \end{pmatrix} \qquad J(x) = \begin{pmatrix} 0 & -1 \\ 1 & 0 \end{pmatrix} \qquad R(x) = \begin{pmatrix} R & 0 \\ 0 & b \end{pmatrix} \qquad g(x) = 0$$

Using this trivial example, it has been shown that each physical subsystem is characterized by a network structure $J(x)$, an energy storage function $H(x)$, some extra terms representing dissipation $R(x)$ and external interconnection $g(x)$. Furthermore, it has also been shown that the interconnection of two subsystems still results in a system of the same form whose energy is the sum of the energy of the subsystems.

Something which is very important to note is that an interconnection of systems with ports does result in the real physical behavior of the interconnected parts. This is NOT the case in general if physical parts are represented by signal transfer functions, like it can be seen by a cascade interconnection of electrical filters models.

In a slightly more formal and geometrical terms, a port-Hamiltonian system is composed of a state manifold $\mathcal{X}$, a lower bounded energy function $H : \mathcal{X} \to \mathbb{R}$ corresponding to the internal energy, a network structure $D(x) = -D^T(x)$ whose graph has the mathematical structure of a Dirac structure ([4]), which is in general a state dependent power continuous interconnection structure, and an interconnection port represented by a pair of power variables $(e, f) \in V^* \times V$ which is geometrically characterized by dual vector elements which are called effort and flow. This port is used to interact energetically with the system: the power supplied through a port is equal to $e(f)$ or, using coordinates, to $e^T f$. We can furthermore split the interaction port in more sub-ports, each of which can be used to model different power flows. We will indicate with the subscript I the power ports by means of which the system interacts with the rest of the world, with the subscript C the power ports associated with the storage of energy and with the subscript R the power ports relative to power dissipation.

The dissipation in the system can be modeled using as characteristic equations $e_R = R(x) f_R$ with $R(x)$ a symmetric and positive semi-definite matrix. If we furthermore set $\dot{x} = f_C$ and $e_C = \frac{\partial H}{\partial x}$, we obtain the following power balance due to the skew-symmetry of $D(x)$:

$$\dot{H}(t) + f_R^T(t) R(x) f_R(t) = e_I^T(t) f_I(t) \tag{14.6}$$

which clearly says that the supplied power $e_I^T f_I$ is either stored or dissipated and that, therefore, a port-Hamiltonian system is passive.

A very broad class of physical systems, both linear and non linear, can be modeled within the port-Hamiltonian framework which can therefore be used to model telemanipulation systems endowed also with nonlinear robots. In the mechanical domain, the efforts and the flows are given by generalized forces, or *wrenches*, and by generalized velocities, or *twists*, respectively and, therefore, a power port is given by a wrench-twist pair (w, t) [19, 20].

In [7, 8] it has been shown how to discretise a continuous port-Hamiltonian system preserving passivity in its discrete counterpart and a strategy to interconnect in a power preserving way discrete and continuous port-Hamiltonian systems has been illustrated. In discretized port-Hamiltonian systems the power balance reported in (14.6) holds in its discrete counterpart.

14.2.2 Port-Hamiltonian Based Bilateral Telemanipulation

It is possible to use port-Hamiltonian systems for building an intrinsically passive sampled data telemanipulation system which is represented in fig. 14.2 using a bond-graph notation [18]. The double band on the bond represents a discrete exchange of energy.

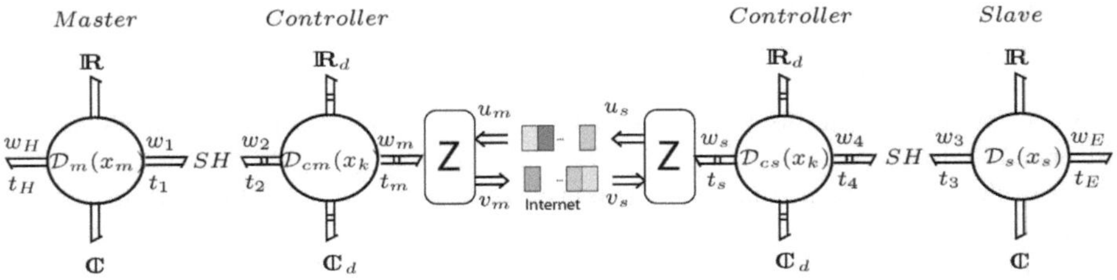

Fig. 14.2. The Passive Sample Data Telemanipulation Scheme

We can see that the continuous plant (i.e. a port-Hamiltonian system representing either master or slave) is interconnected in a passive way (by means of the Sample & Hold algorithm SH proposed in [7, 8]) to the discrete passive impedance controller (namely to a passive discretized port-Hamiltonian system which represents a virtual mechanical impedance). Master and slave communicate by means of a discrete communication channel. Each discrete power port (either the master or the slave one) of the discrete communication channel is characterized by a wrench $w(k)$ and by a twist $t(k)$. The energy flowing into master and slave sides from the communication channel in one sample period is given by $T_s w_m^T(k) t_m(k)$ and $T_s w_s^T(k) t_s(k)$ respectively, where T_s is the sample period. We can always make the following decomposition of the power flow into an incoming power wave and an outgoing power wave in such a way that:

$$\begin{cases} w_m^T(k)t_m(k) = \frac{1}{2}\|u_m(k)\|^2 - \frac{1}{2}\|v_m(k)\|^2 \\ w_s^T(k)t_s(k) = \frac{1}{2}\|v_s(k)\|^2 - \frac{1}{2}\|u_s(k)\|^2 \end{cases} \tag{14.7}$$

where

$$\begin{cases} u_m(k) = \frac{1}{\sqrt{2}}N^{-1}(w(k) + Bt(k)) \\ v_m(k) = \frac{1}{\sqrt{2}}N^{-1}(w(k) - Bt(k)) \end{cases} \qquad \begin{cases} u_s(k) = \frac{1}{\sqrt{2}}N^{-1}(w(k) - Bt(k)) \\ v_s(k) = \frac{1}{\sqrt{2}}N^{-1}(w(k) + Bt(k)) \end{cases} \tag{14.8}$$

and $B = NN > 0$ is the characteristic impedance associated with the wave variables [3, 2]. Integrating in a discrete sense, we get that the energy that is flowing from the communication channel to the master and slave side respectively during one sample period is given by:

$$T_s w_m^T t_m = \frac{T_s}{2}\|u_m\|^2 - \frac{T_s}{2}\|v_m\|^2$$

$$T_s w_s^T t_s = \frac{T_s}{2}\|v_s\|^2 - \frac{T_s}{2}\|u_s\|^2$$

The discrete wave based communication channel can be used to model a packet switched communication channel. In fact, we can interpret $\frac{T_s}{2}\|u_m\|^2$ ($\frac{T_s}{2}\|v_s\|^2$)

and $\frac{T_s}{2}\|v_m\|^2$ ($\frac{T}{2}\|v_m\|^2$) respectively as an incoming and an outgoing energy packet. At each sample time the system will read the incoming energy packet and the wrench $w(k)$ and will calculate the outgoing energy packet and the twist $t(k)$. The communication strategy usually adopted for non scaled bilateral telemanipulation is given by:

$$\begin{cases} v_s(k) = v_m(k-T) \\ u_m(k) = u_s(k-T) \end{cases} \tag{14.9}$$

where the indexes m and s stand for master and slave and $T = dT_s$ is the communication delay. If the communication is implemented through (14.9), only a non scaled exchange of energy between master and slave is allowed: the energy injected into the communication channel at one side is brought, after a delay T, to the other side as it is and, therefore, no scaling is possible. Both master and slave sides are passive systems and the communication channel is a lossless system. Thus, the overall telemanipulation system consists of the power preserving interconnection of passive systems and, consequently, it is passive and characterized by a stable behavior both in free motion and in case of interaction.

In some packet switched networks (e.g. Internet) some packets can be lost during the communication because of some traffic problems or some troubles on the servers each packet has to cross. When a packet is lost, nothing is received at the master (or slave) side and, therefore, the controller has to replace the undelivered information with something else.

The so called *Hold the Last Sample (HLS)* strategy is quite used in web-based software application for dealing with loss of information and it consists of replacing the unreceived packet by the value of the last received packet. Suppose that a packet is lost in the communication between master and slave side and that nothing is received at time k. Using the HLS strategy we would have that the packet $v_m(k-T)$ is replaced with the packet $v_m(k-T-1)$. If

$$\frac{T_s}{2}\|v_m(k-T)\|^2 < \frac{T_s}{2}\|v_m(k-T-1)\|^2$$

we have that the HLS strategy induces a production of extra energy within the communication channel which, therefore, becomes a non passive system. Thus, this strategy, widely used for software application, is not suitable for bilateral telemanipulation since it destroys the passivity of the overall telemanipulation system leading to a potentially unstable behavior.

Another possible strategy consists of replacing the undelivered packet with a null packet and it has been proposed in [6, 7]. Suppose again that a packet is lost in the communication between master and slave side and that nothing is received at time k. Using this strategy we would have that the packet $v_m(k-T)$ is replaced with the packet 0. We always have that

$$0 \leq \frac{T_s}{2}\|v_m(k-T)\|^2$$

and, therefore, we always replace a lost packet with a packet with 0 energy content or, in other words, we dissipate the energy content of the lost packet. This makes the communication channel a strictly passive system and therefore, the intrinsic passivity of the overall telemanipulation system is preserved. For a more formal treatment of the problem of packets loss in bilateral telemanipulation, the interested reader is addressed to [7, 21].

Thus, using the communication strategy proposed in (14.9) together with the algorithm developed in [6, 7] for dealing with lost packets, it is possible to obtain an intrinsically passive bilateral telemanipulation system despite of the packet switching nature of the communication channel and of the possible loss of packets and, therefore, it is suitable for being used in port-Hamiltonian based telemanipulation.

14.3 Power Scaling in Case of Negligible Delay

In this section we consider port-Hamiltonian based bilateral telemanipulators where the communication channel is characterized by a negligible delay. We will prove, generalizing what has been proven in [14] for linear teleoperators, that it is possible to implement power scaling while preserving a stable behavior of the overall scheme.

Consider the telemanipulation scheme reported in fig. 14.2. The master (slave) robot interacts with the human operator (environment) through the continuous power port $(w_H(t), t_H(t))$ $((w_E(t), t_E(t)))$. Furthermore, $(w_m(k), t_m(k))$ and $(w_s(k), t_s(k))$ are the discrete power ports through which master and slave side are interconnected to the communication channel respectively. If the communication delay is negligible, master and slave controllers can be connected directly through power variables and, in non scaled discrete port-Hamiltonian based telemanipulation, master and slave sides are interconnected through a discrete common effort interconnection [6, 7] which, in the mechanical domain, is described by:

$$\begin{cases} w_s(k) = w_m(k) \\ t_s(k) = -t_m(k) \end{cases} \tag{14.10}$$

In this way $P_s(k) = w_s^T(k)t_s(k) = -w_m^T(k)t_m(k) = -P_m(k)$, that is the power supplied to the slave side is exactly that extracted from the master side, namely no power scaling takes place. In order to allow a scaling in the interconnection between master and slave side the following *discrete power scaled common effort interconnection* has to be considered

$$\begin{cases} w_s(k) = \alpha w_m(k) \\ t_s(k) = -\beta t_m(k) \end{cases} \qquad \alpha, \beta \in \mathbb{R}^+ \tag{14.11}$$

where α and β are the scaling factors which can be different since it can be desirable to differently scale the motions and the exchanged forces. In this case we have that the power variables at the slave side are scaled with respect to those at the master side and that

$$P_s(k) = w_s^T(k)t_s(k) = -\alpha\beta w_m^T(k)t_m(k) = P_m(k) \tag{14.12}$$

namely the power supplied to the slave is equal to the power extracted from the master side scaled by a factor $\alpha\beta$. In the following, we will use the following notation for discrete derivative and discrete integral:

$$dg(k) = \frac{g(k+1) - g(k)}{T_s} \qquad I_h^k = \sum_{i=h}^{k-1} g(i)T_s \tag{14.13}$$

where $g(\cdot)$ is a generic sequence. Furthermore, to lighten the notation, we will not explicitly indicate the dependence of the energy function on the state. The following result, whose proof is reported in Appendix A.1, holds:

Proposition 1. *If the interconnection between master and slave is made through a discrete power scaled common effort interconnection reported in (14.11), then the overall systems is dissipative.*

Since the nature of the digital geometric telemanipulation system is mixed (continuous and discrete), both continuous and discrete terms appear in the power balance expressing the energetic behavior of the system. Setting

$$\dot{\hat{H}}_m(t,k) = \dot{H}_m(t) + dH_{cm}(k) \qquad \dot{\hat{H}}_s(t,k) = \dot{H}_s(t) + dH_{cs}(k)$$

and

$$\hat{P}_{dm}(t,k) = P_{dm}(t) + P_{dcm}(k) \qquad \hat{P}_{ds}(t,k) = P_{ds}(t) + P_{dcs}(k)$$

we have that, using the results obtained in Appendix A.1, the following power balance holds:

$$e_H^T(t)f_H(t) = \dot{\hat{H}}_m(t,k) + \hat{P}_{dm}(t,k) + \frac{1}{\alpha\beta}(\dot{\hat{H}}_s(t,k) + \hat{P}_{ds}(t,k) - e_E^T(t)f_E(t)) \tag{14.14}$$

We can see that, using the communication strategy reported in (14.11), the human operator perceives the impedance of the master side and a scaled version of the slave side and of the environment impedance. Thus, the effect of the power scaled common effort interconnection is to transmit a scaled version of the slave side to the master side as ideally should be, as reported in [10].

Remark 1. The scaled interconnection between master and slave prevents from proving passivity of the overall system and allows only proving dissipativity. Dissipativity comes from the fact that it is necessary to scale the energetic behavior of the slave side in order to give to the operator a realistic feeling and thus, the power balance of the overall system has to contain a scaled version of the power exchanged with the environment instead of the "real" power. Nevertheless the fact that the telemanipulation system is dissipative with respect to the supply rate $w_H^T(t)t_H(t) + (1/\alpha\beta)w_E^T(t)t_E(t)$ is still sufficient for guaranteeing that the overall system is characterized by a stable behavior when interacting with any passive environment. In fact, as it happens in non scaled telemanipulation, $-w_E^T(t)t_E(t) = \dot{H}_e(t)+P_{de}(t)$, where $H_e(t)$ and $P_{de}(t)$ are the lower bounded storage function and the non negative dissipation function characterizing the passive environment. Replacing this power balance in (14.14) we can see that the overall system is passive and, therefore, characterized by a stable behavior. ◁

The scaling of the power variables gives a consequent scaling of the power exchanged between master and slave as expressed by (14.12). The power extracted from the master side is scaled by a factor $\alpha\beta$ and is supplied to the slave side. Thus, while common effort interconnection is power preserving, namely it simply allows a transfer of power, the power scaled common effort interconnection in general changes the amount of power that is transferred. Therefore, the scaled interconnection is NOT a passive element but, nevertheless, as shown in Proposition 1, it can be safely used in telemanipulation. This happens because the effect of the scaling is to "mask" the slave side and to make it appear to the human operator as if it acted at the same power scale of the master side. However, this amplification/attenuation of the power transmitted does NOT modify the kind of dynamic behavior of the slave side which keeps on being passive. As a result we have two systems characterized by a passive behavior but acting at different power scales and this leads to a behavior which is passive with respect to the sum of the power injected at one side and the *scaled* power injected at the other side.

14.4 Power Scaling in Case of Non Negligible Communication Delay

In case of non negligible communication delay, the interconnection between master and slave sides cannot be implemented using directly power variables since, as it has been proven in [2] for non scaled telemanipulation, in this way the communication channel would become a non passive system and it would destabilize the overall telemanipulator. In non scaled telemanipulation, the energetic interconnection between master and slave sides is implemented through wave variables and, in this way, passivity of the overall system is preserved independently of any constant communication delay. In this section we want to exploit wave variables to implement the power scaled common effort interconnection over a delayed transmission line while preserving a stable behavior of the overall system.

In order to find the wave based communication strategy for power scaling interconnections, we proceed as in the case of power preserving interconnections. First we find the equivalent expression, in terms of wave variables, of the power scaled common effort interconnection reported in (14.11); secondly we analyze the behavior of the overall telemanipulation scheme when the energetic interconnection between master and slave is made through the wave based power scaled common effort interconnection and the communication channel is characterized by a non negligible delay.

Consider the power scaled common effort interconnection described by (14.11). Given a power port $(w(k), t(k))$, the following relations can be easily derived from (14.8):

$$\begin{cases} w_m(k) = \frac{N}{\sqrt{2}}(u_m(k) + v_m(k)) \\ t_m(k) = \frac{1}{\sqrt{2}}N^{-1}(u_m(k) - v_m(k)) \end{cases} \qquad \begin{cases} w_s(k) = \frac{N}{\sqrt{2}}(v_s(k) + u_s(k)) \\ t_s(k) = \frac{1}{\sqrt{2}}N^{-1}(v_s(k) - u_s(k)) \end{cases} \tag{14.15}$$

Using (14.15) with (14.11) we have that the wave based expression of the discrete power scaled common effort interconnection is given by:

$$\begin{cases} v_s(k) + u_s(k) = \alpha(u_m(k) + v_m(k)) \\ v_s(k) - u_s(k) = -\beta(u_m(k) - v_m(k)) \end{cases} \tag{14.16}$$

where the pedices m and s indicate master and slave side respectively. After some simple computations we have that (14.16) can be rewritten as:

$$\begin{cases} v_s(k) = \frac{2\alpha\beta}{\alpha+\beta}v_m(k) + \frac{\alpha-\beta}{\alpha+\beta}u_s(k) \\ u_m(k) = \frac{\beta-\alpha}{\alpha+\beta}v_m(k) + \frac{2}{\alpha+\beta}u_s(k) \end{cases} \tag{14.17}$$

Consider now that a communication delay T is present in the interconnection between master and slave. The wave based communication strategy can be obtained by (14.17) and is given by:

$$\begin{cases} v_s(k) = \frac{2\alpha\beta}{\alpha+\beta}v_m(k-T) + \frac{\alpha-\beta}{\alpha+\beta}u_s(k-T) \\ u_m(k) = \frac{\beta-\alpha}{\alpha+\beta}v_m(k-T) + \frac{2}{\alpha+\beta}u_s(k-T) \end{cases} \tag{14.18}$$

Remark 2. Consider the master side and the incoming wave u_m. The term $u_s(k-T)$ arrives from the communication channel while the term $v_m(k-T)$ has to be added at the master side; this can be done by implementing a buffering strategy that stores a copy of the outgoing wave v_m for a time interval of length T. In case the communication delay is unknown, the delay can be recovered by attaching a time stamp to the transmitted wave. Similar considerations hold for the slave side ◁

The following result, whose proof is reported in Appendix A.2, holds:

Proposition 2. *If the interconnection between master and slave sides in the scheme reported in Fig. 14.2 is made through the discrete delayed power scaled common effort strategy reported in (14.18), then the overall telemanipulation system is dissipative.*

Remark 3. In case of $\alpha = \beta$ we have that the communication strategy is simplified and it results

$$\begin{cases} v_s(k) = \alpha v_m(k-T) \\ u_m(k) = \frac{1}{\alpha} u_s(k-T) \end{cases} \tag{14.19}$$

In this case no buffering is needed for the implementation of the interconnection. ◁

Remark 4. In case $\alpha = \beta = 1$, namely in case of no scaling, we recover the communication strategy that is usually exploited in port-Hamiltonian based telemanipulation, namely

$$\begin{cases} v_s(k) = v_m(k-T) \\ u_m(t) = u_s(k-T) \end{cases} \tag{14.20}$$

Thus, the communication strategy proposed in (14.18) represents a generalization what is commonly used in passivity based bilateral telemanipulation. ◁

Using the results obtained in Appendix A.2, we have that the following power balance holds:

$$e_H^T(t) f_H(t) = \dot{\hat{H}}_m(t,k) + \hat{P}_{dm}(t,k) + \\ + \frac{1}{\alpha\beta}(dH_{ch}(k) + \dot{\hat{H}}_s(t,k) + \hat{P}_{ds}(t,k) - e_E^T(t) f_E(t)) \tag{14.21}$$

Thus, analogously to what happens in case of negligible delay, the human operator feels the impedance of the master side and a scaled version of the slave side. The proposed interconnection strategy transforms, analogously to what happens in non scaled telemanipulation, the communication channel in a subsystem characterized by a lossless dynamics which is perceived by the human operator as part of the slave side. In the delayed case as well, the presence of scaling prevents from proving passivity of the overall telemanipulation scheme but the considerations made in Remark 1 are still valid.

When implementing a delayed interconnection by means of wave variables, the communication channel acts like a buffer along which power packets travel before reaching either master or slave side. In non scaled telemanipulation, the master side exchange the outgoing power packet, v_m, with the slave side while in scaled telemanipulation, as it can be seen from (14.29) in Appendix A.2 , the master side exchanges a scaled version of the outgoing power packet. This means that while in non scaled telemanipulation power is simply exchanged, in scaled telemanipulation the power extracted from the master side is scaled by a factor $\alpha\beta$ and then supplied, through the communication channel, to the slave side.

14.5 The Problem of Packet Loss

When using packet switched transmission lines to interconnect master and slave sides, it is possible that some packets are lost in the communication. In this section, we will analyze two possible ways to extend the passivity preserving strategy for handling packets loss proposed in [6, 7] in the case in which master and slave sides are connected through the power scaled communication channel described in the previous section. Consider the communication strategy reported in (14.18) and replace packets lost in the communication with null packets, as it is proposed in [6, 7]. In case a packet is lost in the communication between master and slave (slave and master), we miss the contribution relative to the term $v_m(k-T)$ $(u_s(k-T))$ for building the packet $v_s(k)$ $(u_m(k))$. At this point two possible strategies can be adopted:

S1 Partially compute the incoming packet using only the contribution relative to the term $u_s(k-T)$ $(v_m(k-T))$.
S2 Set to zero the incoming packet discarding the contribution relative to the term $u_s(k-T)$ $(v_m(k-T))$.

In the following we will analyze $S1$ and $S2$ in order to verify which one is best suited for handling lost packets.

The following result can be proven (see Appendix A.3):

Proposition 3. *Consider a digital port-Hamiltonian based telemanipulation system where the interconnection between master and slave side is made through (14.18). When a packet is lost and strategy* $S1$ *is adopted, the communication channel injects extra energy into the system.*

Loosely speaking, when using strategy $S1$ for handling lost packets, the communication channel becomes an active subsystem which, consequently tends to destabilize the overall telemanipulation system. On the other hand, when adopting strategy $S2$, the following result can be proven (see Appendix A.4):

Proposition 4. *Consider a digital port-Hamiltonian based telemanipulation system where the interconnection between master and slave side is made through (14.18). If some packets are lost in the communication and strategy* $S2$ *is adopted, the communication channel introduces extra dissipation into the system.*

Summarizing, strategy $S1$ should not be adopted to handle lost packets since it transforms the communication channel into an active system and, consequently, it destabilizes the overall telemanipulation system. On the other hand, strategy $S2$ can be safely adopted since the loss of a packet causes power dissipation. Thus, the communication channel becomes a dissipative subsystem which does not alter the dissipativity of the overall telemanipulation scheme. When using strategy $S2$, packets loss causes a decrease of performances since part of the energy required to perform a certain task is not delivered but the overall system is still characterized by a stable behavior.

14.6 Simulations

The aim of this section is to provide some simulations in order to validate the obtained results.

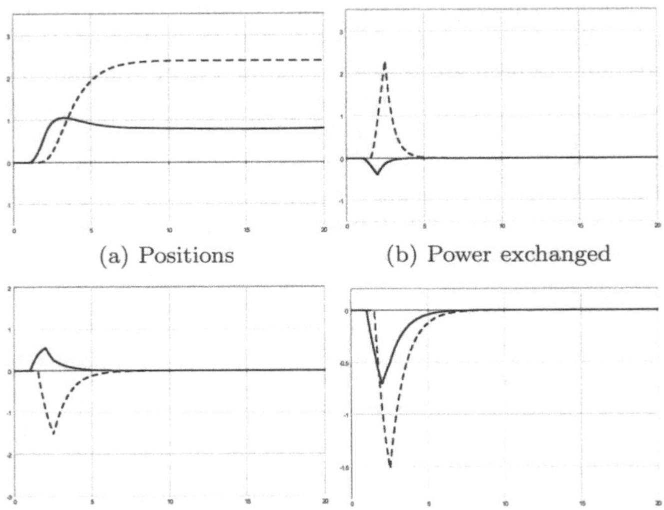

(a) Positions (b) Power exchanged

(c) Twists (d) Wrenches

Fig. 14.3. Power scaled telemanipulation in case of free motion: master (solid) and slave (dashed) behaviors

Consider a simple 1-DOF telemanipulation system. Master and slave robots are simple masses of 1 Kg and the digital controllers are obtained by discretizing through the technique proposed in [8, 7] a port-Hamiltonian system that is physically equivalent to the parallel of a spring and a damper. Master and slave sides are interconnected through a packet switching transmission line over which the discrete wave based communication strategy proposed in (14.18) is implemented. The communication delay is $T = 0.5$ *sec.* and the sample time is $T_s = 0.05$ *sec.*

In the first simulation the operator applies an impulsive force to the master; the scaling factors are $\alpha = 2$ and $\beta = 3$ and the communication strategy reported in (14.18) is used to implement the interconnection between master and slave. The simulation results are reported in Fig. 14.3

Both master and slave sides have an impedance causality (twist in/wrench out) and the position of the masses depends on the twist that is exchanged along the communication channel. In Fig. 3(a) we can see the positions of master and slave. Since the operator applies an impulsive force both master and slave reach a steady state position and we can see that, since the scaling factor for the exchanged twists is $\beta = 3$, we have that the position of the slave is three times bigger than that of the master as expected. In Fig. 3(b) the power flows at master and slave sides are reported. We can see that the power that is extracted from the master side is, as expected, amplified by a factor $\alpha\beta = 6$ and supplied, after a delay of 0.5 *sec.*, to the slave side. Thus we can see that we inject into the remote side an amount of power greater than what we extracted from the

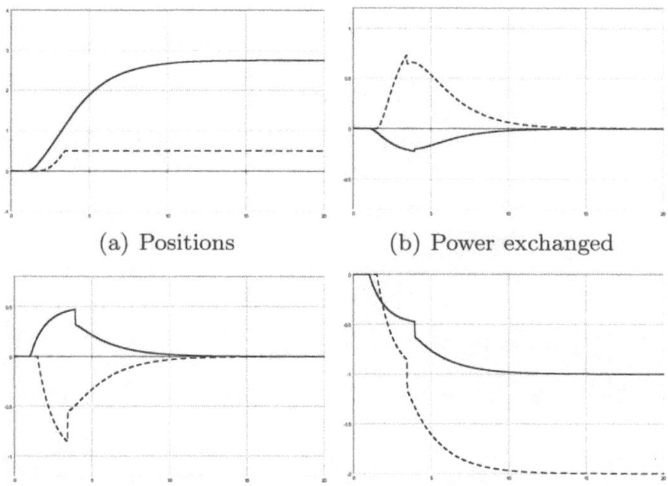

(a) Positions (b) Power exchanged

(c) Twists (d) Wrenches

Fig. 14.4. Power scaled telemanipulation in case of contact with a rigid wall: master (solid) and slave (dashed) behavior

master side and that therefore the communication channel acts as a non passive power amplifier. Nevertheless the behavior of the system is stable, as expected from Proposition 2. In Fig. 3(c) the twists exchanged between master and slave sides are reported. Since we are implementing a power scaled common effort interconnection we have that the twists have discording signs, however, because of the scaling, we have that the absolute value of t_s is three times bigger than t_m. Finally, in Fig. 3(d), the wrenches exchanged are reported. We can see that the wrench at the slave side is twice bigger than that at the master side and, therefore, the scaling is respected in the interconnection.

In the next simulation, a contact task is implemented. The operator applies a constant force to the master and the slave interacts with a rigid wall, implemented as the parallel of a stiff spring, $k = 1000\ N/m$, and a damper, $b = 100\ Nsec/m$, set at the position $x = 0.5\ m$. The communication delay is $T = 0.5\ sec.$ and the scaling factors are $\alpha = \beta = 2$. The results of the simulation are reported in Fig. 14.4. In Fig. 4(a) we can see the positions of master and slave. The slave stops when it meets the wall. The interaction force is scaled and fed back to counteract the force applied by the operator on the master robot which, therefore, stops as well. In Fig. 4(b) we can see that the power extracted from the master side is amplified by a factor $\alpha\beta = 4$ and supplied to the slave side and in Fig. 4(c) and Fig. 4(d) the twists and wrenches exchanged are reported. Notice that the wrench at the slave side is twice bigger than the wrench at the slave side because of the scaling imposed. Thus, the operator perceives a scaled version of the interaction of the slave with the remote environment. From this simulation we can see that the wave based power scaled common effort

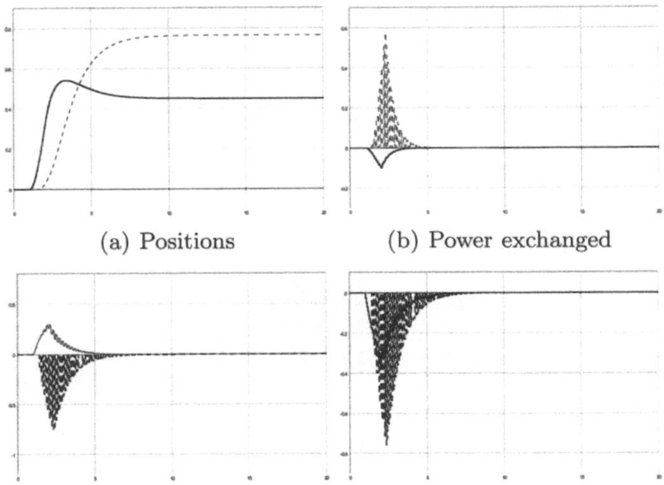

(a) Positions (b) Power exchanged

(c) Twists (d) Wrenches

Fig. 14.5. Power scaled digital telemanipulation in case of free motion: master (solid) and slave (dashed) behaviors

interconnection transmits to the master side a scaled version of the slave side and, nonetheless power scaling is NOT a passive operation, a stable behavior of the overall system is achieved.

In the next simulation we show the effect of packets loss on the scaled telemanipulation system. The operator applies an impulsive force to the master; the scaling factors are $\alpha = 2$ and $\beta = 3$, the loss rates are 30% and 50% in the communication between master and slave and viceversa respectively and the communication delay is $T = 0.3\ s$. The strategy $S2$, reported in Sec. 14.5 has been used to handle lost packets. The simulation results are reported in Fig. 14.5. In Fig. 5(a) we can see the positions of master and slave; both master and slave reach a steady state position since the force is applied for a limited interval of time. Since the scaling factors for the exchanged twists is $\beta = 3$ the position of the slave should be three times bigger than that of the master but, because of the high loss rate in the communication and because of the use of strategy $S2$ to handle lost packets, the content of lost packets is dissipated. Thus we have that power scaling is only partially achieved because of packet loss and, therefore, the scaling factor between the position is less than what expected. In Fig. 5(b) the power flows exchanged between master and slave sides are reported. The power extracted from the master side is scaled and then, after a delay $T = 0.3\ sec.$ supplied to the slave side. Nevertheless, in several sample periods, 0 power is supplied because lost packets are replaced by 0; the received packets, as expected, are scaled by a factor $\alpha\beta = 6$. In Fig. 5(c) the exchanged twists are reported. Since the adopted communication strategy implements a discrete power scaled common effort interconnection, the signs of the twists are discording. A lot of packets are lost in the communication and replaced by 0; consequently the twists are characterized by a very irregular behavior and, because of the dissipation introduced when using $S2$, we have that the scaling of the exchanged twists is only partially achieved, namely, the absolute value of t_s is less than three times bigger than t_m. In Fig. 5(d) the exchanged wrenches are reported and, for the same reasons as before, the desired scaling is only partially achieved. Summarizing, packets loss in the communication decreases the performances of the overall system and allows to achieve the desired wrench and twist scalings only partially; nevertheless, because of Proposition 4, the telemanipulation system keeps on being characterized by a stable behavior despite of packet loss.

14.7 Conclusions

In this chapter we have shown how to implement power scaling in digital port-Hamiltonian based bilateral telemanipulation. We have illustrated that it is possible to embed power scaling in the interconnection between master and slave sides while preserving a dissipative behavior of the overall telemanipulation scheme. Then, we have used scattering theory to make robust the power scaled interconnection with respect to any constant communication delay and we have obtained a novel wave based interconnection strategy that allows power scaling while preserving a dissipative behavior of the system. Furthermore, we have proposed a strategy to handle packets loss while preserving a stable behavior of the overall system.

In this way we have transformed the scattering based communication channel from a fixed entity, characterized by the fixed equations reported in (14.9), into a tunable entity characterized by (14.18) and by two parameters α and β that can be changed in order to provide a desired scaling. In this way, all the issues relative to power scaling are assigned to the interconnection. If we want to implement a power scaled bilateral telemanipulation scheme, we just need to take the master and slave robots with their (possibly embedded) control system and to interconnect them through a scattering based communication channel that has to be tuned in order to meet the different power scales at which the devices operate.

References

1. R.J. Anderson and M.W. Spong. Bilateral control of teleoperators with time delay. *IEEE Transactions on Automatic Control*, 34(5):494–501, 1989.
2. G. Niemeyer and J.-J. Slotine. Stable adaptive teleoperation. *IEEE Journal of Oceanic Engineering*, 16(1):152–162, 1991.
3. S. Stramigioli, A.J. van der Schaft, B. Maschke, and C. Melchiorri. Geometric scattering in robotic telemanipulation. *IEEE Transactions on Robotics and Automation*, 18(4), 2002.

4. A.J. van der Schaft. *L_2-Gain and Passivity Techniques in Nonlinear Control.* Communication and Control Engineering. Springer Verlag, 2000.
5. S. Stramigioli. *Modeling and IPC Control of Interactive Mechanical Systems: a coordinate free approach.* LNCIS. Springer, London, 2001.
6. C. Secchi, S. Stramigioli, and C. Fantuzzi. Digital passive geometric telemanipulation. In *Proceedings of IEEE International Conference on Robotics and Automation*, Taipei, Taiwan, September 2003.
7. S. Stramigioli, C. Secchi, A.J. van der Schaft, and C. Fantuzzi. Sampled data systems passivity and discrete port-hamiltonian systems. *IEEE Transactions on Robotics*, 21(4):574–587, 2005.
8. S. Stramigioli, C. Secchi, A.J. van der Schaft, and C. Fantuzzi. A novel theory for sample data system passivity. In *Proceedings to IEEE Conference on Intelligent Robotic Systems*, 2002.
9. P. Arcara and C. Melchiorri. Control schemes for teleoperation with time delay: a comparative study. *Robotics and Autonomous Systems*, 38(1), 2002.
10. J.E. Speich and M. Goldfarb. Implementation of loop-shaping compensators to increase the transparency bandwidth of a scaled telemanipulation system. In *Proceedings of IEEE International Conference on Robotics and Automation*, Washington, DC, USA, May 2002.
11. M. Boukhnifer, A. Ferreira, and J.-G. Fontaine. Scaled teleoperation controller design for micromanipulation over internet. In *Proceedings of IEEE International Conference on Robotics and Automation*, New Orleans, LA, USA, April 2004.
12. T. Itoh, K. Kosuge, and T. Fukuda. Human-machine cooperative telemanipulation with motion and force scaling using task-oriented virtual tool dynamics. *IEEE Transactions on Robotics and Automation*, 16(5), 2000.
13. D. Lee and P.I. Li. Passive bilateral feedforward control of linear dynamically similar teleoperated manipulators. *IEEE Transactions on Robotics and Automation*, 2003.
14. R.J. Anderson and M.W. Spong. Asymptotic stability for force reflecting teleoperators with time delays. In *Proceedings of IEEE International Conference on Robotics and Automation*, May 1989.
15. J.E. Colgate. Power and impedance scaling in bilateral manipulation. In *Proceedings of IEEE International Conference on Robotics and Automation*, Sacramento, California, April 1991.
16. R. Ortega, M. Perez, and J. Espinoza. Passivity-based PI control of switched power converters. *IEEE Transactions on Control Systems Technology*, 12(4), July 2004.
17. C. Batlle, A. Doria, and R. Ortega. Power flow control of a doubly-fed induction machine coupled to a flywheel. In *Proceedings of IEEE Conference on Control Applications*, Taipei, Taiwan, ROC, 2004.
18. D.C. Karnopp, D.L. Margolis, and R.C. Rosenberg. *System Dynamics: A Unified Approach.* John Wiley & Sons Inc., second edition, 1990.
19. R.M. Murray, Z. Li, and S.S. Sastry. *A mathematical introduction to robotic manipulation.* CRC press, 1994.
20. J.M. Selig. *Geometric Methods in Robotics.* Monographs in Computer Sciences. Springer Verlag, 1996. ISBN 0-387-94728-0.
21. C. Secchi, S. Stramigioli, and C. Fantuzzi. *Control of Interactive Roboti Interfaces: a port-Hamiltonian Approach.* Springer Tracts in Advanced Robotics. Springer, 2007.

A Proofs

A.1 Proof of Proposition 1

Since master and slave are continuous port-Hamiltonian systems, we have that the following power balances hold:

$$\begin{cases} w_H^T(t)t_H(t) + w_1(t)^T t_1(t) = \dot{H}_m(t) + P_{dm}(t) \\ w_E(t)^T t_E(t) + w_3^T(t)t_3(t) = \dot{H}_s(t) + P_{ds}(t) \end{cases} \tag{14.22}$$

where H_m, P_{dm} and H_s, P_{ds} represent the energy stored and the power dissipated by master and slave respectively. Since master and slave controllers are discretized port-Hamiltonian systems ([8, 7]), we have that

$$\begin{cases} w_2^T(k)t_2(k) + w_m(k)^T t_m(k) = dH_{cm}(k) + P_{dcm}(k) \\ w_4^T(k)t_4(k) + w_s^T(k)t_s(k) = dH_{cs}(k) + P_{dcs}(k) \end{cases} \tag{14.23}$$

where H_{cm}, P_{dcm} and H_{cs}, P_{dcs} represent the discrete energy stored and the discrete power dissipated by master and slave controllers respectively. Furthermore, thanks to the power preserving interconnection between the continuous and the discrete part ([8],[7]), we have a perfect match between continuous and discrete power flows, in the sense that, at each sampling instant,

$$\begin{cases} -w_1^T(t)t_1(t) = w_2^T(k)t_2(k) \\ w_4^T(k)t_4(k) = -w_3^T(t)t_3(t) \end{cases} \tag{14.24}$$

Using (14.12) we have that

$$\begin{aligned} &\alpha\beta w_H^T(t)t_H(t) + \alpha\beta w_m^T(k)t_m(k) + w_s^T(k)t_s(k) + w_E^T(k)t_E(k) = \\ &= \alpha\beta w_H(t)^T t_H(t) + w_E^T(t)t_E(t) \end{aligned} \tag{14.25}$$

Thus, combining (14.22), (14.23), (14.24) and (14.25), we have that:

$$\begin{aligned} &\alpha\beta w_H^T(t)t_H(t) + w_E^T(t)t_E(t)\, \underbrace{\alpha\beta\dot{H}_m(t) + \dot{H}_s(t) + \alpha\beta dH_{cm}(k) + dH_{cs}(k)}_{\text{Lower bounded}} + \\ &+ \underbrace{\alpha\beta P_{dm}(t) + P_{ds}(t) + \alpha\beta P_{dcm}(k) + P_{dcs}(k)}_{\geq 0} \end{aligned} \tag{14.26}$$

Thus we have that a *function* of the supplied power is either stored or dissipated, meaning that the system is dissipative.

A.2 Proof of Proposition 2

Using (14.18) and recalling that the standard Euclidean norm is used for wave variables we obtain by simple computations that:

$$\begin{aligned}&\tfrac{1}{2}\alpha\beta\|v_m(k)\|^2 - \tfrac{1}{2}\alpha\beta\|u_m(k)\|^2 + \tfrac{1}{2}\|u_s(k)\|^2 - \tfrac{1}{2}\|v_s(k)\|^2 = \\ &= \tfrac{1}{2}\alpha\beta\|v_m(k)\|^2 - \tfrac{1}{2}\alpha\beta\|v_m(k-T)\|^2 + \tfrac{1}{2}\|u_s(k)\|^2 - \tfrac{1}{2}\|u_s(k-T)\|^2\end{aligned} \tag{14.27}$$

By trivial computations it can be shown that, in general,

$$dI_{k-h}^{k} g = g(k) - g(k-h) \tag{14.28}$$

where $g(\cdot)$ is a generic sequence. Thus, we can write

$$\begin{aligned}&\tfrac{1}{2}\alpha\beta\|v_m(k)\|^2 - \tfrac{1}{2}\alpha\beta\|v_m(k-T)\|^2 + \tfrac{1}{2}\|u_s(k)\|^2 - \tfrac{1}{2}\|u_s(k-T)\|^2 = \\ &= dI_{k-T}^{k}(\tfrac{1}{2}\alpha\beta\|v_m\|^2 + \tfrac{1}{2}\|u_s\|^2) = dH_{ch}(k)\end{aligned} \tag{14.29}$$

where $H_{ch}(k) = I_{k-T}^{k}(\frac{1}{2}\alpha\beta\|v_m\|^2 + \frac{1}{2}\|u_s\|^2)$ is a lower bounded function. Thus, using (14.27) and (14.29), we have that the following power balance holds:

$$\tfrac{1}{2}\alpha\beta\|v_m(k)\|^2 - \tfrac{1}{2}\alpha\beta\|u_m(k)\|^2 + \tfrac{1}{2}\|u_s(k)\|^2 - \tfrac{1}{2}\|v_s(k)\|^2 = dH_{ch}(k) \tag{14.30}$$

From (14.30) we have that

$$\tfrac{1}{2}\|v_s(k)\|^2 - \tfrac{1}{2}\|u_s(k)\|^2 = -\alpha\beta(\tfrac{1}{2}\|u_m(k)\|^2 - \tfrac{1}{2}\|v_m(k)\|^2) - dH_{ch}(k) \tag{14.31}$$

Using the power decomposition reported in (14.7) we have that

$$w_s^T(k)t_s(k) = -\alpha\beta w_m^T(k)t_m(k) - dH_{ch}(k) \tag{14.32}$$

Therefore we have that

$$\begin{aligned}&\alpha\beta w_H^T(t)t_H(t) + \alpha\beta w_m^T(k)t_m(k) + w_s^T(k)t_s(k) + w_E^T(k)t_E(k) = \\ &= \alpha\beta w_H(t)^T t_H(t) + w_E^T(t)t_E(t) - dH_{ch}(k)\end{aligned} \tag{14.33}$$

Thus, combining (14.22), (14.23), (14.24) and (14.32) we have that

$$\begin{aligned}&\alpha\beta w_H^T(t)t_H(t) + w_E^T(t)t_E(t) = \\ &\underbrace{\alpha\beta\dot{H}_m(t) + \dot{H}_s(t) + \alpha\beta dH_{cm}(k) + dH_{cs}(k) + dH_{ch}(k)}_{\text{Lower bounded}} + \\ &+ \underbrace{\alpha\beta P_{dm}(t) + P_{ds}(t) + \alpha\beta P_{dcm}(k) + P_{dcs}(k)}_{\geq 0}\end{aligned} \tag{14.34}$$

Thus we have that a *function* of the supplied power is either stored or dissipated, meaning that the system is dissipative.

A.3 Proof of Proposition 3

Suppose that a packet is lost in the communication between master and slave. We have that, using strategy $S1$,

$$\begin{aligned} &\tfrac{1}{2}\alpha\beta\|v_m(k)\|^2 - \tfrac{1}{2}\alpha\beta\|u_m(k)\|^2 + \tfrac{1}{2}\|u_s(k)\|^2 - \tfrac{1}{2}\|v_s(k)\|^2 = \\ &= \tfrac{1}{2}\alpha\beta\|v_m(k)\|^2 - \tfrac{1}{2}\alpha\beta\|u_m(k)\|^2 + \tfrac{1}{2}\|u_s(k)\|^2 - \tfrac{1}{2}\|\tfrac{(\alpha-\beta)}{(\alpha+\beta)}u_s(k-T)\|^2 \end{aligned} \tag{14.35}$$

We can always write:

$$\begin{aligned} &\tfrac{1}{2}\alpha\beta\|v_m(k)\|^2 - \tfrac{1}{2}\alpha\beta\|u_m(k)\|^2 + \tfrac{1}{2}\|u_s(k)\|^2 - \tfrac{1}{2}\|v_s(k)\|^2 = \\ &= \tfrac{1}{2}\alpha\beta\|v_m(k)\|^2 - \tfrac{1}{2}\alpha\beta\|u_m(k)\|^2 + \tfrac{1}{2}\|u_s(k)\|^2 - \\ &-\tfrac{1}{2}\|\tfrac{(\alpha-\beta)}{(\alpha+\beta)}u_s(k-T) + \tfrac{2\alpha\beta}{(\alpha+\beta)}v_m(k-T) - \tfrac{2\alpha\beta}{(\alpha+\beta)}v_m(k-T)\|^2 = \\ &= \tfrac{1}{2}\alpha\beta\|v_m(k)\|^2 - \tfrac{1}{2}\alpha\beta\|u_m(k)\|^2 + \tfrac{1}{2}\|u_s(k)\|^2 - \\ &-\tfrac{1}{2}\|\tfrac{(\alpha-\beta)}{(\alpha+\beta)}u_s(k-T) + \tfrac{2\alpha\beta}{(\alpha+\beta)}v_m(k-T)\|^2 - \tfrac{1}{2}\|\tfrac{2\alpha\beta}{(\alpha+\beta)}v_m(k-T)\|^2 + \\ &+\tfrac{1}{2}\tfrac{4\alpha\beta}{(\alpha+\beta)}v_m(k-T)^T(\tfrac{\alpha-\beta}{\alpha+\beta}u_s(k-T) + \tfrac{2\alpha\beta}{\alpha+\beta}v_m(k-T)) = \\ &= dH_{ch}(k) + [\tfrac{1}{2}\tfrac{4\alpha^2\beta^2}{(\alpha+\beta)^2}\|v_m(k-T)\|^2 + \tfrac{1}{2}\tfrac{4\alpha\beta(\alpha-\beta)}{\alpha+\beta}v_m(k-T)^T u_s(k-T)] \end{aligned} \tag{14.36}$$

The term between brackets can be negative for some values of the wave variables and therefore the communication channel can inject some extra energy into the telemanipulation scheme. Thus, the resulting telemanipulation system becomes non dissipative and therefore characterized by a potentially unstable behavior.

A.4 Proof of Proposition 4

Proof. Suppose that a packet is lost in the communication between master and slave and that strategy $S2$ is adopted. We have that the (scaled) power flowing into the communication channel at time k is given by

$$\begin{aligned} &\tfrac{1}{2}\alpha\beta\|v_m(k)\|^2 - \tfrac{1}{2}\alpha\beta\|u_m(k)\|^2 + \tfrac{1}{2}\|u_s(k)\|^2 - \tfrac{1}{2}\|v_s(k)\|^2 \\ &= \tfrac{1}{2}\alpha\beta\|v_m(k)\|^2 - \tfrac{1}{2}\alpha\beta\|u_m(k)\|^2 + \tfrac{1}{2}\|u_s(k)\|^2 \end{aligned} \tag{14.37}$$

We can always write

$$\begin{aligned} &\tfrac{1}{2}\alpha\beta\|v_m(k)\|^2 - \tfrac{1}{2}\alpha\beta\|u_m(k)\|^2 + \tfrac{1}{2}\|u_s(k)\|^2 - \tfrac{1}{2}\|v_s(k)\|^2 \\ &= \tfrac{1}{2}\alpha\beta\|v_m(k)\|^2 - \tfrac{1}{2}\alpha\beta\|u_m(k)\|^2 + \tfrac{1}{2}\|u_s(k)\|^2 - \tfrac{1}{2}\|u_m(k)\|^2 + \tfrac{1}{2}\|u_m(k)\|^2 \end{aligned} \tag{14.38}$$

which, using (14.30), can be rewritten as

$$\frac{1}{2}\alpha\beta\|v_m(k)\|^2 - \frac{1}{2}\alpha\beta\|u_m(k)\|^2 + \frac{1}{2}\|u_s(k)\|^2 - \frac{1}{2}\|v_s(k)\|^2 =$$
$$= dH_{ch}(k) + \frac{1}{2}\|u_m(k)\| = dH_{ch}(k) + \frac{1}{2}\|\frac{\beta-\alpha}{\alpha+\beta}v_m(k-T) + \frac{2}{\alpha+\beta}u_s(k-T)\|^2 \tag{14.39}$$

Using an analogous procedure it is possible to model the energetic behavior of the scaled communication channel in case a packet is lost in the communication between slave and master. Let $\mathcal{L}_{ms} \subset \mathbb{Z}$ and $\mathcal{L}_{sm} \subset \mathbb{Z}$ be the sets of instants at which a packet is not received in the communication between master and slave and viceversa respectively and set:

$$l_{ms} = \begin{cases} 1 & k \in \mathcal{L}_{ms} \\ 0 & k \notin \mathcal{L}_{ms} \end{cases} \qquad l_{sm} = \begin{cases} 1 & k \in \mathcal{L}_{sm} \\ 0 & k \notin \mathcal{L}_{sm} \end{cases} \tag{14.40}$$

In general we can write the following power balance:

$$\frac{1}{2}\alpha\beta\|v_m(k)\|^2 - \frac{1}{2}\alpha\beta\|u_m(k)\|^2 + \frac{1}{2}\|u_s(k)\|^2 - \frac{1}{2}\|v_s(k)\|^2 =$$
$$= dH_{ch}(k) + l_{ms}\frac{1}{2}\|\frac{2\alpha\beta}{\alpha+\beta}v_m(k-T) + \frac{\alpha-\beta}{\alpha+\beta}u_s(k-T)\|^2 + \tag{14.41}$$
$$+l_{sm}\frac{1}{2}\|\frac{\beta-\alpha}{\alpha+\beta}v_m(k-T) + \frac{2}{\alpha+\beta}u_s(k-T)\|^2$$

Thus the behavior of the communication channel with respect to the scaled power is lossless in case no packet is lost and strictly passive in case of packet loss. The amount of power dissipated in case of packet loss is equivalent to the energy content of the packet not received and replaced with a null packet.

Proceeding as in Proposition 2 we can write:

$$\alpha\beta w_H^T(t)t_H(t) + w_E^T(t)t_E(t) =$$
$$= [\alpha\beta\dot{H}_m(t) + \dot{H}_s(t) + \alpha\beta dH_{cm}(k) + dH_{cs}(k) + dH_{ch}(k)] + \tag{14.42}$$
$$+[\alpha\beta P_{dm}(t) + P_{ds}(t) + \alpha\beta P_{dcm}(k) + P_{dcs}(k) + l_{ms}\frac{1}{2}\|v_s(k)\|^2 +$$
$$+l_{sm}\|u_m(k)\|^2]$$

where the value of $u_m(k)$ and of $v_s(k)$ is given by (14.18). The first term between brackets is lower bounded and represents the energy stored into the system while the second term into brackets is positive and represents the power dissipated by the overall system. Thus a function of the power injected into the system is either stored or dissipated, meaning that the overall telemanipulation scheme is dissipative. The effect of packets loss, using strategy $S2$, is nothing else than introducing further dissipation into the system.

15

Adaptive Synchronization of Bilateral Teleoperators with Time Delay*

Nikhil Chopra and Mark W. Spong

University of Illinois at Urbana-Champaign
Coordinated Science Laboratory
1308 West Main Street. Urbana, IL 61801, USA
{nchopra,mspong}@uiuc.edu

Summary. In this chapter we develop schemes for synchronization of bilateral teleoperators. It is well known that a feedback interconnection of two passive systems is passive. We discuss an extension of this property to the case when there are heterogeneous, constant communication delays in the interconnection. If the interconnected systems are output strictly passive, we show that their feedback interconnection is passive independent of the constant delays. We exploit this property to achieve delay-independent output synchronization. This result is then used in the problem of bilateral teleoperation to synchronize the master/slave velocities in free motion. We also develop an architecture that guarantees state synchronization of the master/slave robots in free motion independent of the constant delay. Experimental results are presented to verify the efficacy of the state-synchronizing architecture.

15.1 Introduction

In this chapter we address the problem of state synchronization in bilateral teleoperation. We refer the reader to [1] for a detailed survey of the various schemes developed for the problem of bilateral teleoperation. Henceforth, we restrict ourselves to the discussion of passivity-based methods in bilateral teleoperation.

The passivity-based-approach developed in [2] and [3] has been the cornerstone of teleoperator control for the last two decades. Subsequent schemes, building upon the above two approaches, have been proposed in [4, 5, 6, 7, 8, 9, 10, 11, 12] among others, to provide performance improvement via position feedback, impedance matching, robustness to time-varying delays and various other objectives.

However, most passivity-based architectures, notable exceptions being [6, 9, 11], dictate transfer of velocity information between the master and the slave. Consequently, mismatch of initial conditions can result in position drift between the master and the slave robots. Therefore, development of a stable high performance bilateral teleoperator necessitates transmission of position information between the master and the slave. This has been attempted in [6, 9], where the coupling gains were delay dependent,

* This research was partially supported by the Office of Naval Research under Grant N00014-02-1-0011, N00014-05-1-0186, and by the National Science Foundation under Grants ECS-0122412 and INT-0128656.

M. Ferre et al. (Eds.): Advances in Telerobotics, STAR 31, pp. 257–270, 2007.
springerlink.com © Springer-Verlag Berlin Heidelberg 2007

and in [11] where position information was encoded by transmitting the integral of the wave-variables [3].

The performance goals in bilateral teleoperation are:

G1: Demonstrate ultimate boundedness of master/slave trajectories, in both free and constrained motion, independent of the time delay.
G2: Synchronize the configuration variables of the master and the slave robot when the slave is allowed to move freely.
G3: Ensure that the force tracking error between the human operator force, and the environmental force experienced by slave on hard contact with the environment, is ultimately bounded.

In this chapter, we concentrate on the first and the second goal in free motion. We study feedback interconnections of passive systems with delays in Sec. 15.2 and develop an output synchronizing control law. The proposed control law is then used for constructing velocity and state-synchronizing algorithms for bilateral teleoperators in Sec. 15.3. Experimental results for the state-synchronizing architecture are presented in Sec. 15.4. Finally in Sec. 15.5, we summarize the results and discuss some future directions of work.

15.2 Passivity

The concept of passivity is one of the most physically appealing concepts of system theory [14] and, as it is based on input-output behavior of an system, is equally applicable to both linear and nonlinear systems. Most of the ideas presented in this section are from [15]. Consider a dynamical system represented by the state space model

$$\dot{x} = f(x,u) \tag{15.1}$$
$$y = h(x,u) \tag{15.2}$$

where f: $R^n \times R^p \rightarrow R^n$ is locally Lipschitz, h: $R^n \times R^p \rightarrow R^p$ is continuous, $f(0,0) = 0$, $h(0,0) = 0$ and the system has the same number of inputs and outputs.

Definition 1. *The dynamical system (15.1)-(15.2) is said to be passive if there exists a continuously differentiable non-negative definite scalar function* $V(x)$: $R^n \rightarrow R$ *(called the storage function) such that*

$$u^T y \geq \dot{V}, \qquad \forall (x,u) \in R^n \times R^p$$

Moreover, the system is said to be

- *strictly passive if* $u^T y \geq \dot{V} + S(x)$ *for some positive definite function* $S(x)$
- *lossless if* $u^T y = \dot{V}$
- *input strictly passive if* $u^T y \geq \dot{V} + u^T \psi(u)$*, where* $u^T \psi(u) > 0$ *for some function* ψ *and* $\forall u \neq 0$
- *output strictly passive if* $u^T y \geq \dot{V} + y^T \rho(y)$*, where* $y^T \rho(y) > 0$ *for some function* ρ *and* $\forall y \neq 0$

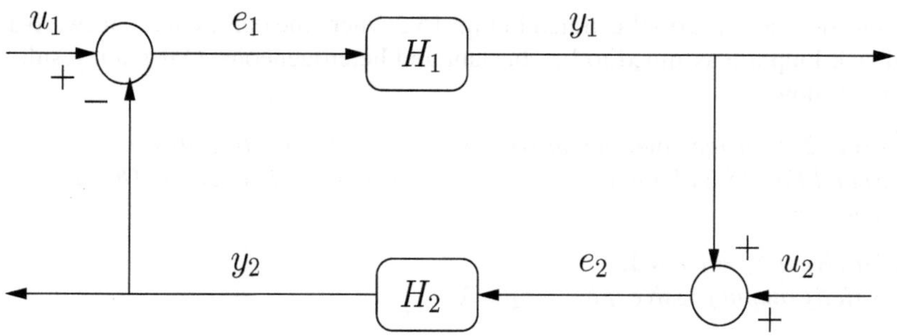

Fig. 15.1. A feedback interconnection

15.2.1 Feedback Interconnection of Passive Systems

At this point we recall a fundamental property of interconnection of passive systems. Assuming that the interconnection is well-posed (see [15]), consider the feedback connection of Fig. 15.1, where each of the feedback components is a time-invariant dynamical system represented by the state model

$$\begin{aligned} \dot{x}_i &= f_i(x_i, e_i) \\ y_i &= h_i(x_i, e_i) \end{aligned} \tag{15.3}$$

The closed-loop (composed of the components H_1 and H_2) then takes the form

$$\begin{aligned} \dot{x} &= f(x, u) \\ y &= h(x, u) \end{aligned} \tag{15.4}$$

where $x = \begin{pmatrix} x_1 \\ x_2 \end{pmatrix}$, $u = \begin{pmatrix} u_1 \\ u_2 \end{pmatrix}$, $y = \begin{pmatrix} y_1 \\ y_2 \end{pmatrix}$. A fundamental result on the feedback interconnection of passive system is the following

Theorem 1. *The feedback connection of two passive systems is passive.*

We refer the reader to [15] for a proof of this result. A similar property follows when the two systems are output strictly passive with

$$e_i^T y_i \geq \dot{V}_i + \delta_i y_i^T y_i \;\; \delta_i > 0 \;\; i = 1, 2 \tag{15.5}$$

In this case it is possible to show that

$$u^T y \geq \dot{V} + \delta y^T y$$

where $V(x) = V_1(x_1) + V_2(x_2)$ and $\delta = \min\{\delta_1, \delta_2\}$.

The effect of delays on linear interconnections of passive systems has been studied in [16] and in the context of large-scale systems in [17]. We next study the extension of the aforementioned interconnection properties to the case when the outputs are delayed. The subsequent result is closely related to the work in [16] where the notion of quasi-dominance was used.

Consider the scenario illustrated in Fig. 15.2 where the delays in the forward and the feedback loop are assumed to be constant and heterogeneous. Our main result in this section follows

Theorem 2. *Consider two output strictly passive systems described by (15.1), (15.2), (15.5) and Fig. 15.2 along with the constant communication delays. Then the feedback interconnection is*

1. *Passive if* $\delta_1 = \delta_2 = 1$.
2. *Strictly output passive if* $\delta_1 = \delta_2 > 1$

Proof. Consider the case when $\delta_1 = \delta_2 = 1$. Then,

$$e_i^T y_i \geq \dot{V}_i + y_i^T y_i \quad i = 1, 2 \tag{15.6}$$

Using the feedback connection of Fig. 15.2, we have

$$\begin{aligned}
&\dot{V}_1 + \dot{V}_2 = e_1^T y_1 + e_2^T y_2 - y_1^T y_1 - y_2^T y_2 \\
&= (u_1 - y_2(t - T_2))^T y_1 + (u_2 + y_1(t - T_1))^T y_2 - y_1^T y_1 - y_2^T y_2 \\
&= u_1^T y_1 + u_2^T y_2 - y_2(t - T_2)^T y_1 + y_1(t - T_1)^T y_2 - y_1^T y_1 - y_2^T y_2 \\
&\leq u_1^T y_1 + u_2^T y_2 + \frac{1}{2}(y_2(t - T_2)^T y_2(t - T_2) + y_1^T y_1) \\
&+\frac{1}{2}(y_1(t - T_1)^T y_1(t - T_1) + y_2^T y_2) - y_1^T y_1 - y_2^T y_2 \\
&\leq u_1^T y_1 + u_2^T y_2 - \frac{1}{2}(y_2^T y_2 - y_2(t - T_2)^T y_2(t - T_2)) \\
&-\frac{1}{2}(y_1^T y_1 - y_1(t - T_1)^T y_1(t - T_1)) \\
&\leq u_1^T y_1 + u_2^T y_2 - \frac{1}{2}\frac{d}{dt}\Big(\int_{t-T_1}^{t} y_1^T y_1 dk + \int_{t-T_2}^{t} y_2^T y_2 dk\Big) \\
&\Rightarrow \dot{V}_1 + \dot{V}_2 + \dot{V}_{channel} \leq u_1^T y_1 + u_2^T y_2 = u^T y
\end{aligned}$$

where

$$V_{channel} = \frac{1}{2}\frac{d}{dt}\Big(\int_{t-T_1}^{t} y_1^T y_1 dk + \int_{t-T_2}^{t} y_2^T y_2 dk\Big)$$

Therefore, the feedback interconnection is passive with $V(x) = V_1(x_1) + V_2(x_2) + V_{channel}$ as the storage function.

Inorder to prove the second claim, it follows from the above calculations that when $\delta_1 = \delta_2 > 1$,

$$\begin{aligned}
\dot{V}_1 + \dot{V}_2 + \dot{V}_{channel} &\leq u_1^T y_1 + u_2^T y_2 - (\delta_1 - 1)y_1^T y_1 - (\delta_2 - 1)y_2^T y_2 \\
&\leq u^T y - \delta_c y^T y
\end{aligned}$$

where $\delta_c = \min\{(\delta_1 - 1), (\delta_2 - 1)\}$. Hence the feedback interconnection is output strictly passive when $\delta_1 = \delta_2 > 1$.

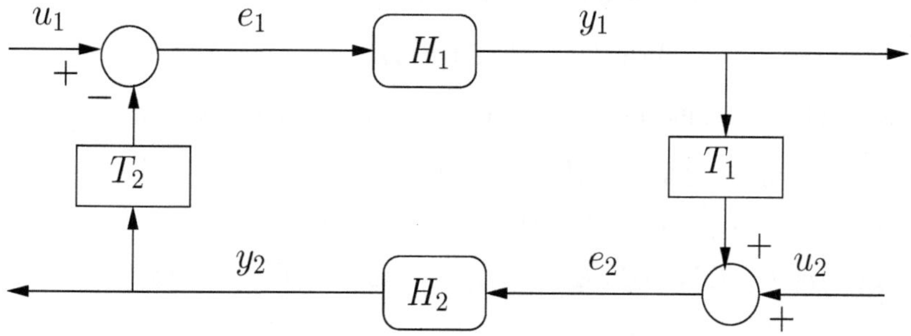

Fig. 15.2. A feedback interconnection with delays

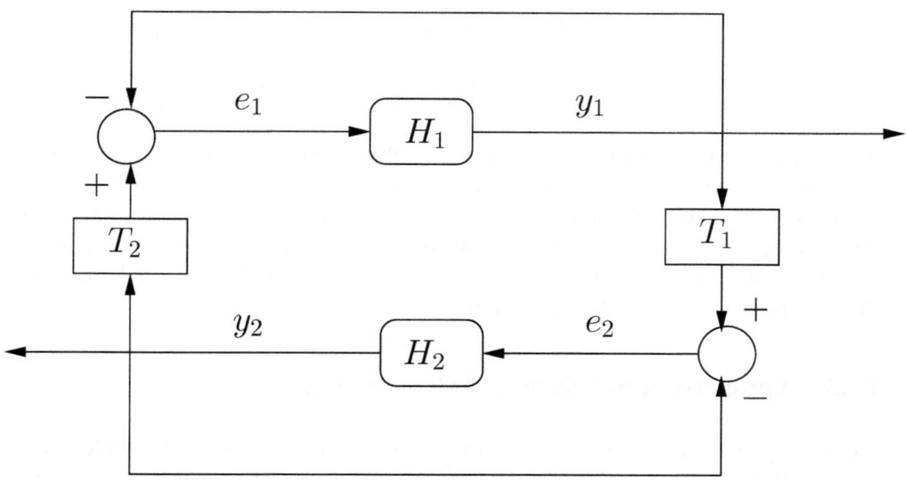

Fig. 15.3. An output synchronizing feedback interconnection

It is well known [14] that output strict passivity can be induced in a passive system with the choice of the control input $u_i = -y_i$. Using this fact and Theorem 2, we propose an output synchronizing control law (see related work in [18, 19]) for the coupled passive systems. The control schematic is illustrated in Fig. 15.3.

Theorem 3. *Consider two passive systems described by (15.1), (15.2) and Fig. 15.3 along with the constant communication delays. Then,*

1. *The signals* $(y_2(t-T_2)-y_1), (y_1(t-T_1)-y_2) \in \mathcal{L}_2[0,\infty)$.
2. *If* $\dot{y}_1, \dot{y}_2 \in \mathcal{L}_\infty[0,\infty)$, *the agents output synchronize in the sense that*

$$\lim_{t\to\infty} |y_2(t-T_2)-y_1| = \lim_{t\to\infty} |y_1(t-T_1)-y_2| = 0$$

Proof. Passivity of the individual systems implies that

$$\dot{V}_1 + \dot{V}_2 \leq e_1^T y_1 + e_2^T y_2$$

Observing Fig. 15.3, the above differential inequality can be rewritten as

$$\begin{aligned}
&\dot{V}_1 + \dot{V}_2 \leq (y_2(t-T_2) - y_1)^T y_1 + (y_1(t-T_1) - y_2)^T y_2 \\
&\leq y_2(t-T_2)^T y_1 + y_1(t-T_1)^T y_2 - y_1^T y_1 - y_2^T y_2 \\
&\leq \frac{1}{2}\Big(-(y_2(t-T_2)-y_1)^T(y_2(t-T_2)-y_1)-(y_1(t-T_1)-y_2)^T(y_1(t-T_1)-y_2) \\
&+y_2(t-T_2)^T y_2(t-T_2) + y_1(t-T_1)^T y_1(t-T_1) - y_1^T y_1 - y_2^T y_2\Big) \\
&\leq \frac{1}{2}\Big(-(y_2(t-T_2)-y_1)^T(y_2(t-T_2)-y_1)-(y_1(t-T_1)-y_2)^T(y_1(t-T_1)-y_2)\Big) \\
&-\dot{V}_{channel} \\
&\Rightarrow \dot{V}_1 + \dot{V}_2 + \dot{V}_{channel} \leq \frac{1}{2}\Big(- (y_2(t-T_2) - y_1)^T(y_2(t-T_2) - y_1) \\
&\qquad -(y_1(t-T_1) - y_2)^T(y_1(t-T_1) - y_2)\Big)
\end{aligned}$$

Integrating the above equation, it is easy to see that the signals $(y_2(t-T_2)-y_1), (y_1(t-T_1) - y_2) \in \mathcal{L}_2[0,\infty)$, and hence the first claim holds.

To prove the second claim, it is important to note that a $\mathcal{L}_2[0,\infty)$ signal with a bounded derivative converges asymptotically to the origin (pp. 116, [13]). If $\dot{y}_1, \dot{y}_2 \in \mathcal{L}_\infty[0,\infty)$, the derivatives of the signals $(y_2(t-T_2)-y_1), (y_1(t-T_1)-y_2)$ are bounded. The second claim follows from the above discussion.

15.3 Application to Bilateral Teleoperation

We next apply the above ideas in the context of bilateral teleoperation. Neglecting friction or other disturbances, the Euler-Lagrange equations of motion for n-link master and slave robots are given as [20]

$$\begin{aligned}
M_m(q_m)\ddot{q}_m + C_m(q_m, \dot{q}_m)\dot{q}_m + g_m(q_m) &= F_h + \tau_m \\
M_s(q_s)\ddot{q}_s + C_s(q_s, \dot{q}_s)\dot{q}_s + g_s(q_s) &= \tau_s - F_e
\end{aligned} \tag{15.7}$$

where q_m, q_s are the $n\times 1$ vectors of joint displacements, $\dot{q}_m$, $\dot{q}_s$ are the $n\times 1$ vectors of joint velocities, τ_m, τ_s are the $n\times 1$ vectors of applied torques, $M(q)$ is the $n\times n$ symmetric positive definite manipulator inertia matrix, $C(q,\dot{q})$ is the $n\times n$ matrix of Centripetal and Coriolis torques and $g(q) = \frac{\partial P}{\partial q}$ is the gradient of the gravitational potential energy $P(q)$. Also, F_h is the human operator force and F_e is the environmental force acting on the slave robot when it contacts the environment. We list here fundamental properties of the master and slave robots that we use in the subsequent analysis

- **Property 1:** The inertia matrix $M(q)$ is symmetric positive definite and there exists a positive constant m such that $mI \leq M(q)$.

- **Property 2:** The Lagrangian dynamics are linearly parameterizable which gives us that
$$M(q)\ddot{q} + C(q,\dot{q})\dot{q} + g(q) = Y_1(q,\dot{q},\ddot{q})\theta$$
where θ is a constant p-dimensional vector of inertia parameters (such as link masses, moments of inertia etc.) and Y_1 is an $n \times p$ matrix of known functions of the generalized coordinates (q_m, q_s) and their higher derivatives.
- **Property 3:** Under an appropriate definition of the matrix C, the matrix $\dot{M}$ - $2C$ is skew symmetric

We study synchronization of bilateral teleoperators in the free motion scenario, and therefore develop synchronization schemes when $F_h, F_e = 0$. These forces can be easily included in the subsequent analysis provided they satisfy the passivity property with respect to the appropriate outputs. At the end of the subsequent synchronization schemes, we will point the reader to appropriate references where the human operator and environment forces have been taken into consideration and briefly discuss the schemes.

15.3.1 Master-Slave Velocity Synchronization

We first develop an algorithm that guarantees velocity synchronization for the bilateral teleoperator. The master/slave robots are said to velocity synchronize if

$$\lim_{t\to\infty} |\dot{q}_m(t-T_1) - \dot{q}_s| = 0 \;\; ; \;\; \lim_{t\to\infty} |\dot{q}_s(t-T_2) - \dot{q}_m| = 0 \tag{15.8}$$

Let the preliminary master/slave torques be given as

$$\tau_m = \bar{\tau}_m + g_m \;\; ; \;\; \tau_s = \bar{\tau}_s + g_s \tag{15.9}$$

where $\bar{\tau}_m, \bar{\tau}_s$ are the output-synchronizing torques that will be designed subsequently. The master/slave dynamics are now given as

$$\begin{aligned} M_m(q_m)\ddot{q}_m + C_m(q_m,\dot{q}_m)\dot{q}_m &= \bar{\tau}_m \\ M_s(q_s)\ddot{q}_s + C_s(q_s,\dot{q}_s)\dot{q}_s &= \bar{\tau}_s \end{aligned} \tag{15.10}$$

The skew-symmetry property (*Property 3*) implies passivity of (15.10) from input $\bar{\tau}_i$ to output $\dot{q}_i$ [20]. This can be demonstrated by choosing

$$V_i(q_i,\dot{q}_i) = \frac{1}{2}\dot{q}_i^T M_i(q_i)\dot{q}_i \quad i = m, s \tag{15.11}$$

as the positive semi-definite storage function for the system. Differentiating V_i along trajectories of (15.10) we have

$$\begin{aligned} \dot{V}_i &= \dot{q}_i^T(-C_i\dot{q}_i + \bar{\tau}_i) + \frac{1}{2}\dot{q}_i^T\dot{M}_i\dot{q}_i \\ &= \dot{q}_i^T\bar{\tau}_i \qquad \text{(Using Property 3)} \end{aligned}$$

and hence the dynamics are passive with $(\bar{\tau}_i, \dot{q}_i)$ as the input-output pair.

In context of Fig. 15.3, the dynamical system (15.10) can be written as

$$H_1 \begin{cases} M_m(q_m)\ddot{q}_m + C_m(q_m,\dot{q}_m)\dot{q}_m = \bar{\tau}_m \\ y_m = \dot{q}_m \end{cases} \tag{15.12}$$

$$H_2 \begin{cases} M_s(q_s)\ddot{q}_s + C_s(q_s,\dot{q}_s)\dot{q}_s = \bar{\tau}_s \\ y_s = \dot{q}_s \end{cases} \tag{15.13}$$

Following Theorem 3, the velocity-synchronizing torques are given as

$$\bar{\tau}_m = \dot{q}_s(t-T_2) - \dot{q}_m \ ; \ \bar{\tau}_s = \dot{q}_m(t-T_1) - \dot{q}_s \tag{15.14}$$

Velocity synchronization for the teleoperator system can be demonstrated (see Theorem 3) by choosing

$$V = \frac{1}{2}\Big(\dot{q}_m^T M_m(q_m)\dot{q}_m + \dot{q}_s^T M_s(q_s)\dot{q}_s + \int_{t-T_1}^{t} \dot{q}_m^T\dot{q}_m dk + \int_{t-T_2}^{t} \dot{q}_s^T\dot{q}_s dk\Big)$$

as the positive semi-definite storage function for the teleoperator system. Using Theorem 3,

$$\dot{V} = -\frac{1}{2}\Big((\dot{q}_s(t-T_2)-\dot{q}_m)^T(\dot{q}_s(t-T_2)-\dot{q}_m)+(\dot{q}_m(t-T_1)-\dot{q}_s)^T(\dot{q}_m(t-T_1)-\dot{q}_s)\Big)$$

Therefore, the signals $(\dot{q}_s(t-T_2)-\dot{q}_m), (\dot{q}_m(t-T_1)-\dot{q}_s) \in \mathcal{L}_2[0,\infty)$. Additionally, as $\dot{V}$ is negative semi-definite, $\dot{q}_m, \dot{q}_s \in \mathcal{L}_\infty[0,\infty)$. Noting that $||C_i(q_i,\dot{q}_i)|| \le k_c||\dot{q}_i||$, where $k_c > 0$ and $||\cdot||$ denotes the induced norm for a matrix and the Euclidean norm for a vector, from the system dynamics (15.12), (15.13), $\ddot{q}_m, \ddot{q}_s \in \mathcal{L}_\infty[0,\infty)$. Invoking the second claim of Theorem 3,

$$\lim_{t\to\infty} |\dot{q}_s(t-T_2) - \dot{q}_m| = \lim_{t\to\infty} |\dot{q}_m(t-T_1) - \dot{q}_s| = 0$$

Remark 1. In addition to velocity synchronization, it is possible to guarantee that the master/slave velocities converge to the origin. This can be achieved by choosing the coupling controls as

$$\begin{aligned} \bar{\tau}_m &= (\dot{q}_s(t-T_2) - \dot{q}_m) - \delta_d\dot{q}_m \\ \bar{\tau}_s &= (\dot{q}_m(t-T_1) - \dot{q}_s) - \delta_d\dot{q}_s \qquad \delta_d > 0 \end{aligned} \tag{15.15}$$

Remark 2. The human operator and the environment forces can be included in the above analysis provided the input-output pairs $(-F_h, \dot{q}_m), (F_e, \dot{q}_s)$ are passive [6, 9].

In context of goal G3, using the coupling control (15.15), and on being manipulated by a passive human operator, good force reflection will not be generated on contact with a passive environment. This is due to the asymptotic convergence of the torques $\bar{\tau}_m, \bar{\tau}_s$ to the origin. To overcome this difficulty, and also achieve position synchronization, transmission of position information was advocated in [6, 9]. The proposed schemes resulted in static force reflection on contact with the environment. However, to ensure stability, the position gains were required to be delay-dependent. We next propose a scheme that guarantees delay-independent state synchronization of the master/slave robots in free-motion.

15.3.2 Master-Slave State Synchronization

In free motion, the bilateral teleoperator is said to state synchronize if

$$\begin{aligned} \lim_{t\to\infty} |q_m(t-T_1) - q_s| &= \lim_{t\to\infty} |\dot{q}_m(t-T_1) - \dot{q}_s| = 0 \\ \lim_{t\to\infty} |q_s(t-T_2) - q_m| &= \lim_{t\to\infty} |\dot{q}_s(t-T_2) - \dot{q}_m| = 0 \end{aligned} \tag{15.16}$$

It is well known that passivity can be induced in a Lagrangian system with a different choice of output and a preliminary feedback control. This is know as *Feedback Passivation* [14]. We use this idea to achieve passivity with respect to an output from which both the position and velocity information are available. To this end, choose a preliminary control input for the master and slave robots as

$$\begin{aligned} \tau_m &= \bar{\tau}_m - \hat{M}_m(q_m)\lambda\dot{q}_m - \hat{C}_m(q_m,\dot{q}_m)\lambda q_m + \hat{g}_m(q_m) \\ \tau_s &= \bar{\tau}_s - \hat{M}_s(q_s)\lambda\dot{q}_s - \hat{C}_s(q_s,\dot{q}_s)\lambda q_s + \hat{g}_s(q_s) \end{aligned} \tag{15.17}$$

where $\hat{M}_i, \hat{C}_i, \hat{g}_i \quad i = m, s$ are the estimates of the respective matrices available at that instant, λ is a constant positive definite matrix, and $\bar{\tau}_m, \bar{\tau}_s$ are the synchronizing torques that will be described below. As the dynamics are linearly parameterizable ($Property\,2$), the motor torques can also be written as

$$\tau_m = \bar{\tau}_m - Y_m(q_m,\dot{q}_m)\hat{\theta}_m \ ; \ \tau_s = \bar{\tau}_s - Y_s(q_s,\dot{q}_s)\hat{\theta}_s$$

where Y_m, Y_s are known functions of the generalized coordinates, and $\hat{\theta}_m$, $\hat{\theta}_s$ are the time-varying estimates of the manipulators' actual constant p dimensional inertial parameters given by θ_m, θ_s respectively. The master and slave dynamics (15.7) (with F_h, F_e=0) reduce to

$$\begin{aligned} \dot{q}_m &= -\lambda q_m + r_m \\ M_m\dot{r}_m + C_m r_m &= Y_m\tilde{\theta}_m + \bar{\tau}_m \\ \dot{q}_s &= -\lambda q_s + r_s \\ M_s\dot{r}_s + C_s r_s &= Y_s\tilde{\theta}_s + \bar{\tau}_s \end{aligned} \tag{15.18}$$

where r_m, r_s are the new outputs of the master and slave robots respectively, and $\tilde{\theta}_m = \theta_m - \hat{\theta}_m$, $\tilde{\theta}_s = \theta_s - \hat{\theta}_s$ are the estimation errors. Let the time-varying estimates of the uncertain parameters evolve as

$$\dot{\hat{\theta}}_m = \Gamma_m Y_m^T r_m \ ; \ \dot{\hat{\theta}}_s = \Gamma_s Y_s^T r_s \tag{15.19}$$

where Γ_m and Γ_s are constant positive definite matrices. Therefore, in context of Fig. 15.3, the bilateral teleoperator system can be viewed as an interconnection of two systems given by

$$H_1 \left\{ \begin{aligned} \dot{q}_m &= -\lambda q_m + r_m \\ M_m\dot{r}_m + C_m r_m &= Y_m\tilde{\theta}_m + \bar{\tau}_m \\ \dot{\hat{\theta}}_m &= \Gamma_m Y_m^T r_m \\ y_m &= r_m \end{aligned} \right. \tag{15.20}$$

$$H_2 \begin{cases} \dot{q}_s = -\lambda q_s + r_s \\ M_s \dot{r}_s + C_s r_s = Y_s \tilde{\theta}_s + \bar{\tau}_s \\ \dot{\tilde{\theta}}_s = \Gamma_s Y_s^T r_s \\ y_s = r_s \end{cases} \tag{15.21}$$

Consider a positive semi-definite storage function for the master/slave robot as

$$V_i(q_i, r_i, \tilde{\theta}_i) = \frac{1}{2}\left(r_i^T M_i(q_i) r_i + \tilde{\theta}_i^T \Gamma_i^{-1} \tilde{\theta}_i\right) \quad i = m, s \tag{15.22}$$

The derivative of this storage function, along trajectories of H_i, is given as

$$\begin{aligned} \dot{V}_i &= r_i^T(-C_i r_i + Y_i \tilde{\theta}_i + \bar{\tau}_i) + \frac{1}{2} r_i^T \dot{M}_i r_i - \tilde{\theta}_i^T Y_i^T r_i \\ &= r_i^T \bar{\tau}_i \end{aligned}$$

Thus, H_1, H_2 are passive with $(\bar{\tau}_m, r_m), (\bar{\tau}_s, r_s)$ as the input-output pairs respectively.

To state synchronize the master/slave robots, the synchronizing torques for the master and slave robots are given as

$$\bar{\tau}_s = (r_m(t - T_1) - r_s) \;\; ; \;\; \bar{\tau}_m = (r_s(t - T_2) - r_m) \tag{15.23}$$

It is to be noted that

$$\begin{aligned} r_m(t - T_1) - r_s &= (\dot{q}_m(t - T_1) + \lambda q_m(t - T_1)) - (\dot{q}_s + \lambda q_s) \\ &= \dot{e}_m + \lambda e_m \end{aligned} \tag{15.24}$$

where $e_m = q_m(t - T_1) - q_s$. Similarly $r_s(t - T_2) - r_m = \dot{e}_s + \lambda e_s$ where $e_s = q_s(t - T_2) - q_m$. (15.24) represents an exponentially stable linear system with input $r_m(t - T_1) - r_s$. It follows that if $(r_m(t - T_1) - r_s), (r_s(t - T_2) - r_m)$ are $\mathcal{L}_2[0, \infty)$ signals that converge asymptotically to zero, then

$$\lim_{t \to \infty} |e_m| = \lim_{t \to \infty} |e_s| = \lim_{t \to \infty} |\dot{e}_m| = \lim_{t \to \infty} |\dot{e}_s| = 0$$

from which state synchronization follows immediately.

Using (15.22), the positive semi-definite storage function for the bilateral teleoperator described by (15.20), (15.21), (15.23) and Fig. 15.3 is given as

$$V = V_m + V_s + \frac{1}{2}\left(\int_{t-T_1}^{t} r_m^T r_m dk + \int_{t-T_2}^{t} r_s^T r_s dk\right)$$

From Theorem 3, the derivative of V along trajectories of the system is given as

$$\dot{V} = -\frac{1}{2}\Big((r_s(t-T_2)-r_m)^T(r_s(t-T_2)-r_m)+(r_m(t-T_1)-r_s)^T(r_m(t-T_1)-r_s)\Big)$$

and hence, the signals $(r_m(t - T_1) - r_s), (r_s(t - T_2) - r_m) \in \mathcal{L}_2[0, \infty)$. As $\dot{V}$ is negative semi-definite, $r_m, r_s, \tilde{\theta}_s, \tilde{\theta}_m \in \mathcal{L}_\infty[0, \infty)$. Consequently, from (15.20), (15.21), $\dot{q}_m, q_m, \dot{q}_s, q_s \in \mathcal{L}_\infty[0, \infty)$ and therefore, the derivatives $\dot{r}_m, \dot{r}_s \in \mathcal{L}_\infty[0, \infty)$. The second claim of Theorem 3 then gives us that

$$\lim_{t \to \infty} |r_s(t - T_2) - r_m| = \lim_{t \to \infty} |r_m(t - T_1) - r_s| = 0$$

State synchronization follows from the earlier discussion.

Remark 3. If the human-operator force is given as $F_h = -K_h\dot{q}_m$, then it is possible to show that in addition to state-synchronization, the master/slave velocities go the origin [21].

Remark 4. If the human operator and the environment models are given as

$$F_h = \alpha_o - \alpha_m r_m \quad ; \quad F_e = \alpha_s r_s \tag{15.25}$$

where $\alpha_o, \alpha_m, \alpha_s$ are bounded positive constants, then the trajectories are ultimately bounded independent of the constant delay [21].

15.4 Experiments

In this section, we test the proposed state synchronization scheme under time-varying delays and packet losses. The experiments were performed on two direct-drive, planar, two-degree-of freedom nonlinear robots exchanging information across a stochastic Internet model. Force sensors, located on the end-effectors, measure the force exerted by the operator/environment. The controllers and the Internet model were implemented using Wincon 3.3, which is a Windows application used for running real-time Simulink models.

In the experiments, time-varying delays from the master to the slave and vice versa, were fluctuating between (0.448, 0.544)s. The mean delay was 0.48s with a standard deviation of 0.022s. The packet loss rate in the experiments ranged from 45-55 percent. In the free motion experiment, as seen in Fig. 15.4, the tracking performance is good in the face of time-varying delays and packet losses. The steady state errors were 0.0853 rad and 0.1125 rad for the first and second link respectively.

In the next experiment, the motion of the slave (the second link) is constrained by an aluminium wall, approximately during the (8,17)s of motion. The trajectories of the

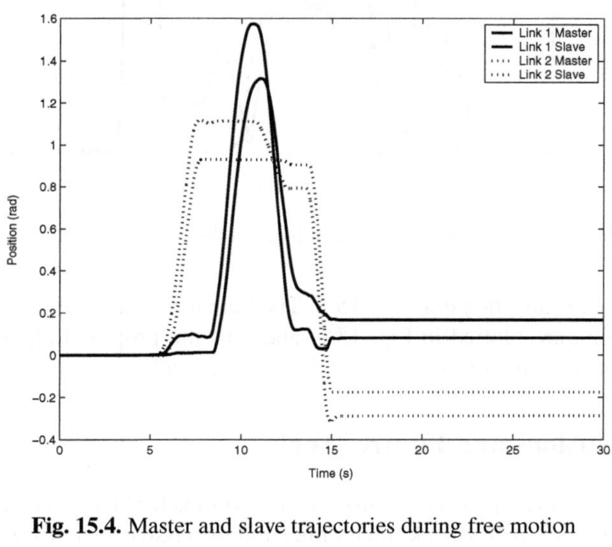

Fig. 15.4. Master and slave trajectories during free motion

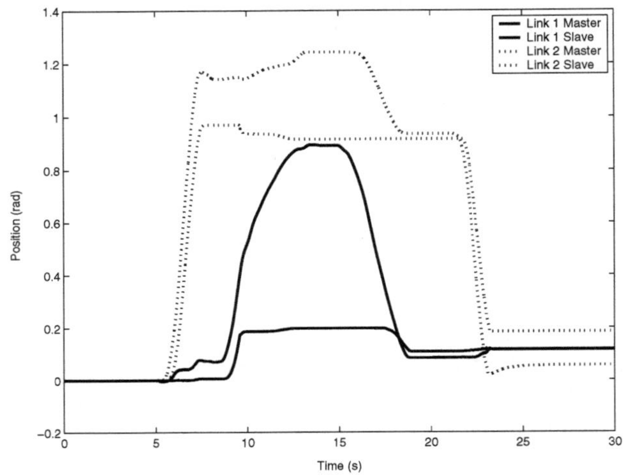

Fig. 15.5. Master and slave trajectories in the constrained motion experiment

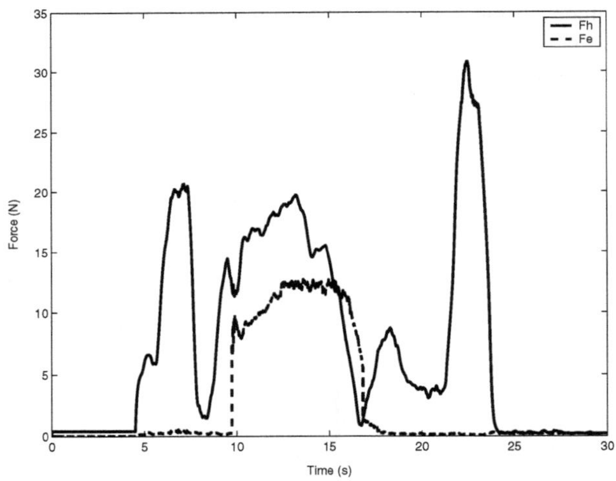

Fig. 15.6. Human operator and environment force in the constrained motion experiment

master/slave robots are shown in Fig. 15.5. The human operator/environment forces at the end effectors, are plotted in Fig. 15.6, and thus the proposed algorithm provides reasonable force tracking on contact with the environment.

15.5 Conclusions and Future Work

In this chapter we developed synchronization algorithms for bilateral teleoperators. The problem of passivity in feedback interconnection of passive systems with delays was studied. If the individual systems are output strictly passive, it was shown that their feedback interconnection is passive independent of the constant delays. This property was then exploited to output synchronize the individual systems independent of the constant delays.

The synchronization result was first applied to the problem of bilateral teleoperation to synchronize the master/slave velocities in free motion. To improve tracking performance, a second architecture was also proposed that guaranteed state synchronization of the master/slave robots in free motion independent of the constant delay. Experimental results were also presented to test the state-synchronizing architecture.

There is considerable work still to be done. The realistic case of time-varying delays needs to be addressed. The approach in [4] is likely to maintain passivity of the state-synchronizing teleoperator, provided the rate of change of the delay is bounded. However, state synchronization with time-varying delays is an open problem. The problem of static force reflection in the state-synchronizing architecture, with a non-passive human operator is also interesting. Finally, teleoperating teams of robots, rather than a single slave robot, will open up new challenges in the field.

Acknowledgements. The authors would like to thank Erick. J. Rodríguez-Seda for providing the experimental data.

References

1. P.F. Hokayem and M.W. Spong. Bilateral teleoperation: an historical survey. *Automatica*, 42(12):2035–2057, 2006.
2. R.J. Anderson and M.W. Spong. Bilateral control of teleoperators with time delay. *IEEE Transactions on Automatic Control*, 34(5):494–501, 1989.
3. G. Niemeyer and J.J.E. Slotine. Stable adaptive teleoperation. *IEEE Journal of Oceanic Engineering*, 16(1):152–162, 1991.
4. N. Chopra, M.W. Spong, S. Hirche, and M. Buss. Bilateral teleoperation over the internet: the time varying delay problem. In *American Control Conference*, pages 155–160, 2003.
5. N. Chopra, M.W. Spong, and R. Lozano. Adaptive coordination control of bilateral teleoperators with time delay. In *IEEE Conf. on Decision and Control*, pages 4540–4547, 2004.
6. N. Chopra, M.W. Spong, R. Ortega, and N.E. Barabanov. On tracking performance in bilateral teleoperation. *IEEE Transactions on Robotics*, 22(4):861–866, 2006.
7. S. Munir. *Internet-based tele-operation.* PhD thesis, Department of Mechanical Engineering, Georgia Institute of Technology, 2001.
8. K. Kosuge, H. Murayama, and K. Takeo. Bilateral feedback control of telemanipulators via computer network. In *IEEE/RSJ International Conference on Intelligent Robots and Systems*, pages 1380–1385, 1996.
9. D.J. Lee and M.W. Spong. Passive bilateral control of teleoperators under constant time delay. *IEEE Transactions on Robotics*, 22(2):269–281, 2006.
10. R. Lozano, N. Chopra, and M.W. Spong. Passivation of force reflecting bilateral teleoperation with time varying delay. In *Mechatronics'02*, Entschede, Netherlands, 2002.
11. G. Niemeyer and J.J.E. Slotine. Towards force-reflecting teleoperation over the internet. In *IEEE Conference on Robotics and Automation*, pages 1909–1915, 1998.
12. S. Stramigioli, A.V.D. Schaft, B. Maschke, and C. Melchiorri. Geometric scattering in robotic telemanipulation. *IEEE Transactions on Robotics and Automation*, 18(4):588–596, 2002.

13. R. Lozano, B. Brogliato, O. Egeland, and B. Maschke. *Dissipative Systems Analysis and Control: Theory and Applications*. Springer-Verlag, London, 2000.
14. R. Sepulchre, M. Janković, and P. Kokotović. *Constructive Nonlinear Control*. Springer-Verlag, London, 1997.
15. H.K. Khalil. Nonlinear systems. In *Upper Saddle River*. Prentice Hall, New Jersey, 2002.
16. P.J. Moylan. A connective stability result for interconnected passive systems. *IEEE Transactions on Automatic Control*, 25:812–813, 1980.
17. R.M. Lewis and B.D.O. Anderson. Necessary and sufficient conditions for delay-independent stability of linear autonomous systems. *IEEE Transactions on Automatic Control*, 25:735–739, 1980.
18. A.Y. Pogromsky. Passivity based design of synchronizing systems. *International Journal of Bifurcation and Chaos*, 8:295–319, 1998.
19. N. Chopra and M.W. Spong. Passivity-based control of multi-agent systems. In Sadao Kawamura and Mikhail Svinin, editors, *Advances in Robot Control: From Everyday Physics to Human-Like Movements*, pages 107–134. Springer Verlag, 2006.
20. M.W. Spong, S. Hutchinson, and M. Vidyasagar. *Robot Modeling and Control*. John Wiley & Sons, Inc., New York, 2006.
21. N. Chopra, M.W. Spong, and R. Lozano. Synchronization of bilateral teleoperators with time delay. Submitted, 2006.

16

Bilateral Control of Teleoperation Systems Through State Convergence

José M. Azorín[1], Rafael Aracil[2], Nicolás M. García[1] and Carlos Pérez[1]

[1] Universidad Miguel Hernández de Elche
Departamento de Ingenieria de Sistemas Industriales
Avda. de la Universidad s/n, 03202 Elche (Alicante), Spain
{jm.azorin,nicolas.garcia,carlos.perez}@umh.es

[2] Universidad Politécnica de Madrid
Dtpo. Automática, Ingeniería Electrónica e Informática Industrial
C/. José Gutierrez Abascal, 2. 28006 Madrid, Spain
aracil@etsii.upm.es

Summary. This chapter describes a new bilateral control method of teleoperation systems. This method is based on the state space formulation and it can be applied to any teleoperation system where the master and the slave would be represented by nth-order linear differential equations. The control system allows that the slave manipulator follows the master. The tracking is achieved by state convergence between the master and the slave. The method is able also to establish the desired dynamics of this convergence and the dynamics of the slave manipulator. Furthermore a simple design procedure is provided to obtain the control system gains. The advantage of this design procedure is that it is only necessary to solve a set of equations to calculate the control system gains. The control by state convergence can be also applied to teleoperation systems with communication time delay. In this case, the Taylor expansion is used to approximate the time delay. The state convergence allows that the slave manipulator follows the master in spite of the time delay in the communication channel. Experimental results with a teleoperation system of one DoF are presented to verify the control method.

16.1 Introduction

The state space technique represents the modern approach to control system theory and its applications [1]. This approach is very convenient for representation of high-order dimensional and complex systems. However, not many references can be found in the literature that model and control teleoperation systems using the state space framework. Some works have used the state space approach to control teleoperation systems with time delay [2] [3] [4].

In order to introduce the state space framework, a teleoperation system of 1 DoF is going to be modeled on the state space. The dynamic model of an element with one DoF is:

$$J\ddot{\theta}(t) + b\dot{\theta}(t) + mgl\sin\theta(t) = u(t) \tag{16.1}$$

M. Ferre et al. (Eds.): Advances in Telerobotics, STAR 31, pp. 271–288, 2007.
springerlink.com © Springer-Verlag Berlin Heidelberg 2007

where $J = \frac{1}{3}ml^2$ is the inertia of the element, m is the mass of the element, g is the gravity acceleration, l is the lenght of the element, $\theta(t)$ is the rotate angle, b is the viscous friction coefficient and $u(t)$ is the control torque applied. The simplified linear model is:

$$J\ddot{\theta}(t) + b\dot{\theta}(t) = u(t) \tag{16.2}$$

This model can be represented on the state space considering as state variables the position and the velocity:

$$x_1(t) = \theta(t) \tag{16.3}$$
$$x_2(t) = \dot{\theta}(t) \tag{16.4}$$

From (16.2) and considering the previous state variables, the next state space representation can be obtained:

$$\dot{x}_1(t) = x_2(t) = \dot{\theta}(t) \tag{16.5}$$
$$\dot{x}_2(t) = \ddot{x}_1(t) = \ddot{\theta}(t) = -\frac{b}{J}x_2(t) + \frac{1}{J}u(t) \tag{16.6}$$
$$y(t) = x_1(t) \tag{16.7}$$

These equations can be represented in the following matrix way:

$$\begin{bmatrix} \dot{x}_1(t) \\ \dot{x}_2(t) \end{bmatrix} = \begin{bmatrix} 0 & 1 \\ 0 & -\frac{b}{J} \end{bmatrix} \begin{bmatrix} x_1(t) \\ x_2(t) \end{bmatrix} + \begin{bmatrix} 0 \\ \frac{1}{J} \end{bmatrix} u(t)$$
$$y(t) = \begin{bmatrix} 1\,0 \end{bmatrix} \begin{bmatrix} x_1(t) \\ x_2(t) \end{bmatrix} \tag{16.8}$$

$$\begin{aligned} \dot{x}(t) &= Ax(t) + Bu(t) \\ y(t) &= Cx(t) \end{aligned} \tag{16.9}$$

Therefore, a teleoperation system of 1 DoF will be represented in the state space modeling the master and the slave through (16.9) in each case.

On the other hand, the main problems of the majority of the control architectures for teleoperation systems are that they do not provide a simple design procedure to calculate the control system gains, and they can not establish the desired dynamic behavior. In addition, most of them only consider master and slave dynamic models represented by second order linear differential equations [5] [6].

This chapter presents a new design and control method of bilateral teleoperation systems. This method considers master and slave manipulators modeled by nth-order linear differential equations. The design method is based on the state space formulation and it allows that the slave follows the master through state convergence. The method is able also to establish the desired dynamics of this convergence and the dynamics of the slave manipulator. In addition, only a set of equations must be solved to obtain the control system gains. This design and control method can be also applied to teleoperation systems with communication time delay.

The control system presented is only valid during the contact phase of the slave with the environment. Separate controllers should be designed for free motion phase, with switching between these controllers occurring according to the identified phase. In addition, a control system must be designed for each DoF of the master and slave manipulators. On the other hand, in spite of that master and slave manipulators of one DoF are habitually modeled as mechanical systems by 2nd-order linear differential equations, in this chapter it is considered that the master and slave are modeled by nth-order linear differential equations. In this way, the control method can be applied to other kind of manipulators (e.g. pneumatic or hydraulic manipulators), or to manipulators modeled by transfer functions of nth-order obtained by means of a identification process [7].

The chapter is organized as follows. Sec. 16.2 describes the modeling of the teleoperation system. The control method of teleoperation systems through state convergence is explained in Sec. 16.3. In Sec. 16.4 the state convergence methodology is applied to teleoperation systems with time delay. Sec. 16.5 shows the experimental results obtained when a teleoperation system of one DoF has been controlled using the state convergence method. Finally, laste section presents the conclusions of this chapter.

16.2 Modeling of the Teleoperation System

The modeling of the teleoperation system has been made on the state space considering all the possible interactions that could appear in the operator-master-slave-environment set, Fig. 16.1. In this figure F_m represents the force that the operator applies to the master, and the master and the slave system are represented in the state space like:

$$\begin{aligned} \dot{x}_m(t) &= A_m x_m(t) + B_m u_m(t) \\ y_m(t) &= C_m x_m(t) \end{aligned} \tag{16.10}$$

$$\begin{aligned} \dot{x}_s(t) &= A_s x_s(t) + B_s u_s(t) \\ y_s(t) &= C_s x_s(t) \end{aligned} \tag{16.11}$$

where u_m and u_s are the master and slave control signals, x_m and x_s are the master and slave state vectors, and y_m and y_s are the master and slave positions.

The matrices that appear in the model are:

- G_2: influence in the slave of the force that the operator applies to the master.
- K_m: feedback matrix of the master state.
- K_s: feedback matrix of the slave state. It allows to consider the interaction force of the slave with the environment.
- R_m: interaction slave - master. It allows modeling the force reflection to the master.
- R_s: interaction master - slave.

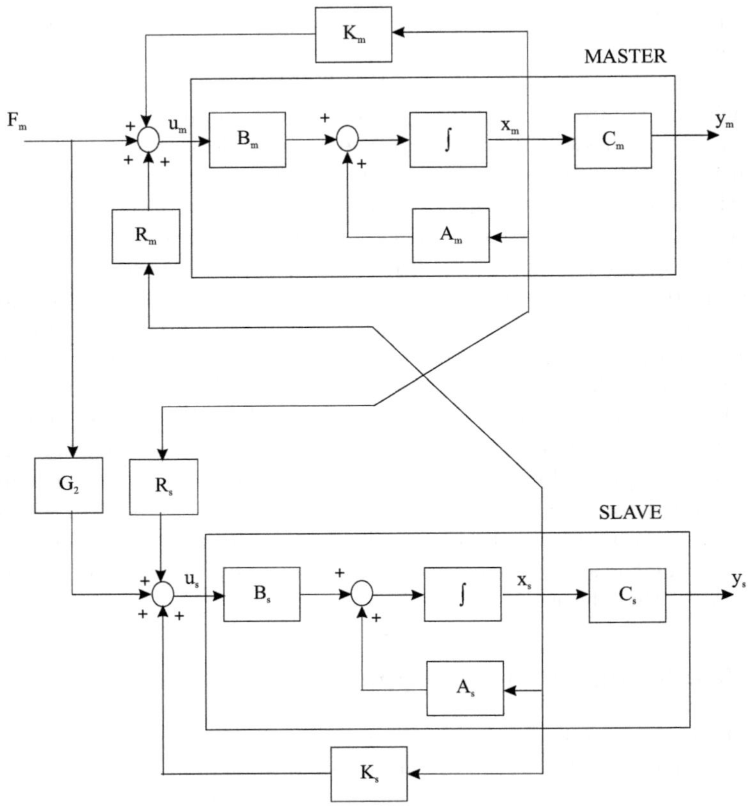

Fig. 16.1. Modeling of the teleoperation system

From the model shown in the Fig. 16.1, it could be observed that the master control signal, $u_m(t)$, and the slave control signal, $u_s(t)$, are respectively:

$$u_m(t) = K_m x_m(t) + R_m x_s(t) + F_m(t) \tag{16.12}$$

$$u_s(t) = K_s x_s(t) + R_s x_m(t) + G_2 F_m(t) \tag{16.13}$$

Replacing the master and slave control signal, (16.12) and (16.13), in the master and slave state equation, (16.10) and (16.11), the next state equations are obtained:

$$\dot{x}_m(t) = (A_m + B_m K_m) x_m(t) + B_m R_m x_s(t) + B_m F_m(t) \tag{16.14}$$

$$\dot{x}_s(t) = (A_s + B_s K_s) x_s(t) + B_s R_s x_m(t) + B_s G_2 F_m(t) \tag{16.15}$$

The state equations (16.14) and (16.15) can be represented in a matrix way as:

$$\begin{bmatrix} \dot{x}_s(t) \\ \dot{x}_m(t) \end{bmatrix} = \begin{bmatrix} A_{11} & A_{12} \\ A_{21} & A_{22} \end{bmatrix} \begin{bmatrix} x_s(t) \\ x_m(t) \end{bmatrix} + \begin{bmatrix} B_1 \\ B_2 \end{bmatrix} F_m(t) \tag{16.16}$$

where

$$A_{11} = A_s + B_s K_s \tag{16.17}$$
$$A_{12} = B_s R_s \tag{16.18}$$
$$A_{21} = B_m R_m \tag{16.19}$$
$$A_{22} = A_m + B_m K_m \tag{16.20}$$
$$B_1 = B_s G_2 \tag{16.21}$$
$$B_2 = B_m \tag{16.22}$$

16.2.1 Master and Slave Representation

The master and the slave are represented by nth-order linear differential equations. To explain the design method, two different kinds of differential equations are considered: differential equations that do not contain zeros, and differential equations that contain zeros. The reason of this distinction is that, in the former case, the method will be always applicable. However, in the latter case, it will be applicable only if the master and slave equations satisfy some properties. In both cases, the structure of the matrices that appear in the model is the next:

- $K_m = \begin{bmatrix} k_{m1}, k_{m2}, \cdots, k_{mn} \end{bmatrix}$
- $K_s = \begin{bmatrix} k_{s1}, k_{s2}, \cdots, k_{sn} \end{bmatrix}$
- $R_m = \begin{bmatrix} r_{m1}, r_{m2}, \cdots, r_{mn} \end{bmatrix}$
- $R_s = \begin{bmatrix} r_{s1}, r_{s2}, \cdots, r_{sn} \end{bmatrix}$
- $G_2 = g_2$

Differential Equations That Do Not Contain Zeros

The representation of the master through a differential equation that do not contain zeros is given by:

$$\frac{d^n y_m(t)}{dt^n} + a_{mn-1}\frac{d^{n-1} y_m(t)}{dt^{n-1}} + \cdots + a_{m1}\frac{dy_m(t)}{dt} + a_{m0} y_m(t) = b_{m0} u_m(t) \tag{16.23}$$

and the representation of the slave is analogous.

The state space representation of the master and the slave is obtained considering as state variables the position and its $n-1$ first derivatives [8]. Therefore the state space representation of the master is the following:

$$\begin{bmatrix} \dot{x}_{m1}(t) \\ \dot{x}_{m2}(t) \\ \vdots \\ \dot{x}_{mn}(t) \end{bmatrix} = \begin{bmatrix} 0 & 1 & \cdots & 0 \\ \vdots & \vdots & \ddots & \vdots \\ 0 & 0 & \cdots & 1 \\ -a_{m0} & -a_{m1} & \cdots & -a_{mn-1} \end{bmatrix} \begin{bmatrix} x_{m1}(t) \\ x_{m2}(t) \\ \vdots \\ x_{mn}(t) \end{bmatrix} + \begin{bmatrix} 0 \\ 0 \\ \vdots \\ b_{m0} \end{bmatrix} u_m(t) \tag{16.24}$$

$$y_m(t) = \begin{bmatrix} 1 \; 0 \cdots 0 \end{bmatrix} \begin{bmatrix} x_{m1}(t) \\ x_{m2}(t) \\ \vdots \\ x_{mn}(t) \end{bmatrix} \tag{16.25}$$

and the state space representation of the slave is analogous.

The state variables of the master and the slave can be obtained from the position of the master and the slave applying derivatives. Another option to get the state variables would be designing state observers.

Differential Equations that Contain Zeros

The representation of the master through a differential equation that contain zeros is given by:

$$\frac{d^n y_m(t)}{dt^n} + a_{mn-1}\frac{d^{n-1}y_m(t)}{dt^{n-1}} + \cdots + a_{m0}y_m(t) = b_{mn-1}\frac{d^{n-1}u_m(t)}{dt^{n-1}} + \cdots + b_{m1}\frac{du_m(t)}{dt} + b_{m0}u_m(t) \tag{16.26}$$

and the representation of the slave is similar.

In this case, the master and the slave are represented in the state space using the controller canonical form [1]. Therefore the state space representation of the master is the next:

$$\begin{bmatrix} \dot{x}_{m1}(t) \\ \dot{x}_{m2}(t) \\ \vdots \\ \dot{x}_{mn}(t) \end{bmatrix} = \begin{bmatrix} 0 & 1 & \cdots & 0 \\ \vdots & \vdots & \ddots & \vdots \\ 0 & 0 & \cdots & 1 \\ -a_{m0} & -a_{m1} & \cdots & -a_{mn-1} \end{bmatrix} \begin{bmatrix} x_{m1}(t) \\ x_{m2}(t) \\ \vdots \\ x_{mn}(t) \end{bmatrix} + \begin{bmatrix} 0 \\ 0 \\ \vdots \\ 1 \end{bmatrix} u_m(t) \tag{16.27}$$

$$y_m(t) = \begin{bmatrix} b_{m0} \; b_{m1} \; \cdots \; b_{mn-1} \end{bmatrix} \begin{bmatrix} x_{m1}(t) \\ x_{m2}(t) \\ \vdots \\ x_{mn}(t) \end{bmatrix} \tag{16.28}$$

and the state space representation of the slave is similar.

The state variables of the master and the slave must be obtained designing state observers.

16.2.2 Modeling of the Environment

Through the matrix K_s it is possible to consider in the teleoperation system shown in Fig. 16.1 the interaction force of the slave with the environment, and by means of matrix R_m it is possible to feedback this contact force to the master.

Differential Equations That Do Not Contain Zeros

If the environment is modeled by means of a stiffness (k_e) [9], and it is considered that the slave position coincides with the deformation of the environment, the reaction force that acts against the slave is given by:

$$f_s(t) = k_e y_s(t) = k_e C_s x_s(t) = k_e x_{s1}(t) \tag{16.29}$$

In this case, to incorporate in the teleoperation system model the interaction of the slave with the environment, the structure of the matrix K_s must be:

$$K_s = \left[k'_{s1} - k_e, \, k_{s2}, \, \cdots, \, k_{sn} \right] \tag{16.30}$$

In this way, the slave control signal (16.11) will be:

$$\begin{aligned} u_s(t) = K_s x_s(t) + R_s x_m(t) + G_2 F_m(t) = \\ \left[k'_{s1} - k_e, \, k_{s2}, \, \cdots, \, k_{sn} \right] x_s(t) + R_s x_m(t) + G_2 F_m(t) = \\ \left[k'_{s1}, \, k_{s2}, \, \cdots, \, k_{sn} \right] \begin{bmatrix} x_{s1}(t) \\ x_{s2}(t) \\ \vdots \\ x_{sn}(t) \end{bmatrix} - k_e x_{s1}(t) + R_s x_m(t) + G_2 F_m(t) = \\ K'_s x_s(t) - f_s(t) + R_s x_m(t) + G_2 F_m(t) \end{aligned} \tag{16.31}$$

where the interaction force of the slave, $f_s(t)$, will oppose to the slave control signal, modeling that the slave is interacting with the environment.

And, to consider force feedback from the slave to the master, the structure of the matrix R_m must be:

$$R_m = \left[r_{m1}, \, r_{m2}, \, \cdots, \, r_{mn} \right] = \left[k_f k_e, \, 0, \, \cdots, \, 0 \right] \tag{16.32}$$

where k_f is the force feedback gain. In this way the feedback signal to the master is the interaction force of the slave with the environment:

$$R_m x_s(t) = \left[k_f k_e, \, 0, \, \cdots, \, 0 \right] \begin{bmatrix} x_{s1}(t) \\ x_{s2}(t) \\ \vdots \\ x_{sn}(t) \end{bmatrix} = k_f k_e x_{s1}(t) = k_f f_s(t) \tag{16.33}$$

Differential Equations That Contain Zeros

If, as the previous case, the environment is modeled by means of a stiffness (k_e), the reaction force that acts against the slave, in this case, is given by:

$$f_s(t) = k_e y_s(t) = k_e C_s x_s(t) = \left[k_e b_{s0}, \, k_e b_{s1}, \, \cdots, \, k_e b_{sn-1} \right] x_s(t) \tag{16.34}$$

Therefore, to incorporate in the teleoperation system model the interaction of the slave with the environment, the structure of the matrix K_s must be:

$$K_s = \left[k'_{s1} - k_e b_{s0}, \, k'_{s2} - k_e b_{s1}, \, \cdots, \, k'_{sn} - k_e b_{sn-1} \right] \tag{16.35}$$

In this way, the slave control signal (16.11) will be:

$$\begin{aligned} u_s(t) = K_s x_s(t) + R_s x_m(t) + G_2 F_m(t) = \\ \left[k'_{s1} - k_e b_{s0}, \, \cdots, \, k'_{sn} - k_e b_{sn-1} \right] x_s(t) + R_s x_m(t) + G_2 F_m(t) = \\ \left[k'_{s1}, \, \cdots, \, k'_{sn} \right] x_s(t) - \left[k_e b_{s0}, \, \cdots, \, k_e b_{sn-1} \right] x_s(t) + R_s x_m(t) + G_2 F_m(t) = \\ K'_s x_s(t) - f_s(t) + R_s x_m(t) + G_2 F_m(t) \end{aligned} \tag{16.36}$$

where the interaction force of the slave, $f_s(t)$, will oppose to the slave control signal, modeling that the slave is interacting with the environment.

And, to consider force feedback from the slave to the master, the structure of the matrix R_m must be:

$$R_m = \begin{bmatrix} r_{m1}, r_{m2}, \cdots, r_{mn} \end{bmatrix} = \begin{bmatrix} k_f k_e b_{s0}, k_f k_e b_{s1}, \cdots, k_f k_e b_{sn-1} \end{bmatrix} \quad (16.37)$$

where k_f is the force feedback gain. In this way the feedback signal to the master is the interaction force of the slave with the environment:

$$R_m x_s(t) = \begin{bmatrix} k_f k_e b_{s0}, \cdots, k_f k_e b_{sn-1} \end{bmatrix} x_s(t) = k_f k_e C_s x_s(t) = k_f f_s(t) \quad (16.38)$$

16.3 Control Method Through State Convergence

From the teleoperation system model shown in Fig. 16.1, it can be observed that there are $3n+1$ control gains that are necessary to obtain:

- $K_m = \begin{bmatrix} k_{m1}\ k_{m2}\ \cdots\ k_{mn} \end{bmatrix}$
- $K_s = \begin{bmatrix} k_{s1}\ k_{s2}\ \cdots\ k_{sn} \end{bmatrix}$
- $R_s = \begin{bmatrix} r_{s1}\ r_{s2}\ \cdots\ r_{sn} \end{bmatrix}$
- $G_2 = g_2$

To calculate these control gains it is necessary to solve $3n+1$ design equations. With these control gains the slave manipulator will follow the master, and the dynamics of the error and the slave will be established. In the following subsections the $3n+1$ design equations are shown considering the two different representations of the master and the slave.

16.3.1 Differential Equations That Do Not Contain Zeros

Considering that the characteristic polynomial desired of the slave and the error are, respectively, $p(s) = s^n + p_{n-1}s^{n-1} + \cdots + p_1 s + p_0$ and $q(s) = s^n + q_{n-1}s^{n-1} + \cdots + q_1 s + q_0$, the $3n+1$ control gains are obtained solving the following equations (see Appendix for details):

$$g_2 = \frac{b_{m0}}{b_{s0}} \quad (16.39)$$

$$-a_{s0} + b_{s0}k_{s1} - b_{m0}r_{m1} + b_{s0}r_{s1} + a_{m0} - b_{m0}k_{m1} = 0 \quad (16.40)$$

$$\cdots$$

$$-a_{sn-1} + b_{s0}k_{sn} - b_{m0}r_{mn} + b_{s0}r_{sn} + a_{mn-1} - b_{m0}k_{mn} = 0 \quad (16.41)$$

$$-a_{s0} + b_{s0}k_{s1} + b_{s0}r_{s1} = -p_0 \quad (16.42)$$

$$\cdots$$

$$-a_{sn-1} + b_{s0}k_{sn} + b_{s0}r_{sn} = -p_{n-1} \quad (16.43)$$

$$-a_{m0} + b_{m0}k_{m1} + b_{s0}r_{s1} = -q_0 \quad (16.44)$$

$$\cdots$$

$$-a_{mn-1} + b_{m0}k_{mn} + b_{s0}r_{sn} = -q_{n-1} \quad (16.45)$$

16.3.2 Differential Equations That Contain Zeros

In this case, considering the same representation of the characteristic polynomial desired of the slave and the error as previously, the $3n+1$ control gains are obtained solving the following equations (see Appendix for details):

$$g_2 = \frac{b_{mn-1}}{b_{sn-1}} \tag{16.46}$$

$$\begin{aligned} &b_{m0}b_{sn-1}a_{s0} - b_{m0}b_{sn-1}k_{s1} + b_{mn-1}r_{m1}b_{m0} \\ &-b_{s0}b_{sn-1}r_{s1} - b_{s0}b_{mn-1}a_{m0} + b_{s0}b_{mn-1}k_{m1} = 0 \end{aligned} \tag{16.47}$$

$$\cdots$$

$$\begin{aligned} &b_{mn-1}b_{sn-1}a_{sn-1} - b_{mn-1}b_{sn-1}k_{sn} + b_{mn-1}r_{mn}b_{mn-1} \\ &-b_{sn-1}b_{sn-1}r_{sn} - b_{sn-1}b_{mn-1}a_{mn-1} + b_{sn-1}b_{mn-1}k_{mn} = 0 \end{aligned} \tag{16.48}$$

$$a_{s0}b_{m0} - k_{s1}b_{m0} - r_{s1}b_{s0} = p_0 b_{m0} \tag{16.49}$$

$$\cdots$$

$$a_{sn-1}b_{mn-1} - k_{sn}b_{mn-1} - r_{sn}b_{sn-1} = p_{n-1}b_{mn-1} \tag{16.50}$$

$$b_{sn-1}r_{s1} + b_{mn-1}a_{m0} - b_{mn-1}k_{m1} = q_0 b_{mn-1} \tag{16.51}$$

$$\cdots$$

$$b_{sn-1}r_{sn} + b_{mn-1}a_{mn} - b_{mn-1}k_{mn} = q_{n-1}b_{mn-1} \tag{16.52}$$

In addition, to assure that there is not any error between the master and slave output, the next conditions must be verified:

$$b_{s0} = \frac{b_{m0}}{b_{m1}}b_{s1}, \cdots, b_{sn-2} = \frac{b_{mn-2}}{b_{mn-1}}b_{sn-1} \tag{16.53}$$

If these conditions are not satisfied, there will be a constant error between the master and slave output.

16.3.3 Comments

Previously it has been considered that the master and the slave are represented by linear differential equations of the same order. However, the state convergence methodology can be also applied to teleoperation systems where the master and the slave are modeled by differential equations of different order [7]. In this case the order of the smaller differential equation must be increased adding the necessary pole-zero pairs to design the control system. Then, state observers that estimate the number of state variables fixed by the higher differential equation must be designed in order to apply the control.

As it has been appointed in the introduction, a control system must designed for each DoF of the master and slave. In addition, to apply the control system it is necessary to obtain the operator force and the interaction force of the slave with the environment. This can be achieved using force/torque sensors attached to the last link of the master and the slave. However, to obtain the influence of the operator force and the interaction force in the rest of the links, the Jacobian matrix must be used.

16.4 Teleoperation System with Time Delay

The state convergence methodology can be also applied to control teleoperation systems with communication time delay [10]. Using this method, the slave will follow the master in spite of the time delay, and the desired dynamics of the convergence and the slave will be established.

The teleoperation system with time delay has been modeled on the state space as it was explained in Sec. 16.2. But, a constant time delay of T seconds has been considered between the local and the remote environment. The state equation of the teleoperation system with time delay is the next (see Appendix for details):

$$\begin{bmatrix} \dot{x}_s(t) \\ \dot{x}_m(t) \end{bmatrix} = \begin{bmatrix} A_{11} & A_{12} \\ A_{21} & A_{22} \end{bmatrix} \begin{bmatrix} x_s(t) \\ x_m(t) \end{bmatrix} + \begin{bmatrix} B_1 \\ B_2 \end{bmatrix} F_m(t) \tag{16.54}$$

where

$$A_{11} = S(A_s + B_s K_s - T B_s R_s B_m R_m) \tag{16.55}$$
$$A_{12} = S(B_s R_s - T B_s R_s (A_m + B_m K_m)) \tag{16.56}$$
$$A_{21} = M(B_m R_m - T B_m R_m (A_s + B_s K_s)) \tag{16.57}$$
$$A_{22} = M(A_m + B_m K_m - T B_m R_m B_s R_s) \tag{16.58}$$
$$B_1 = S(B_s G_2 - T B_s R_s B_m) \tag{16.59}$$
$$B_2 = M(B_m - T B_m R_m B_s G_2) \tag{16.60}$$

and

$$M = (I - T^2 B_m R_m B_s R_s)^{-1} \tag{16.61}$$
$$S = (I - T^2 B_s R_s B_m R_m)^{-1} \tag{16.62}$$

From the state equation (16.54), using the state convergence methodology, it is possible to get $3n+1$ design equations that must be solved to calculate the $3n+1$ control gains. In this case, to obtain the design equations it is also necessary to distinguish between master and slave modeled by differential equations that do not contain zeros, and master and slave modeled by differential equations that contain zeros. In the first case, the control method is always applicable. However, in the second case, only it is applicable if the master and slave zeros verify some conditions.

It is necessary to make two comments about the assumptions considered to derive the design method when there is communication time delay. First, the

Taylor expansion of first order has been used to approximate the time delay. The proposed control method is oriented to teleoperation systems where a "small" time delay exists. In these cases, the Taylor expansion of first order can be used to approximate the time delay. In teleoperation systems where there are time delays bigger than 1 s is more adequate to use "teleprogramming" techniques [11] to control the system. On the other hand, the operator force (F_m) has been considered as constant to obtain the design method. To consider the operator force as constant is equivalent to assume that the slave will receive the operator force ($F_m(t)$) instead of the time delayed operator force ($F_m(t-T)$), see Fig. 16.1. So there will be an error between the force that the slave really receives ($F_m(t-T)$) and the force that it has been considered in the design phase ($F_m(t)$). As the time delay was bigger and the operator force changed more abruptly, the force error will be bigger. This force error will produce a position error between the master and the slave that will increase as the force error increases. However, the proposed control system is oriented to cases where there is an "small" time delay in the communication channel. Therefore the force error will not be important and the position error will be small.

16.5 Experimental Results

The control method by state convergence has been tested on a experimental teleoperation system of one DoF, Fig. 16.2, considering a constant time delay [10].

The next state equations of the master and the slave have been experimentally identified:

$$\dot{x}_m(t) = \begin{bmatrix} 0 & 1 \\ 0 & -7.1429 \end{bmatrix} x_m(t) + \begin{bmatrix} 0 \\ 0.2656 \end{bmatrix} u_m(t) \tag{16.63}$$

Fig. 16.2. Experimental teleoperation system of one DOF

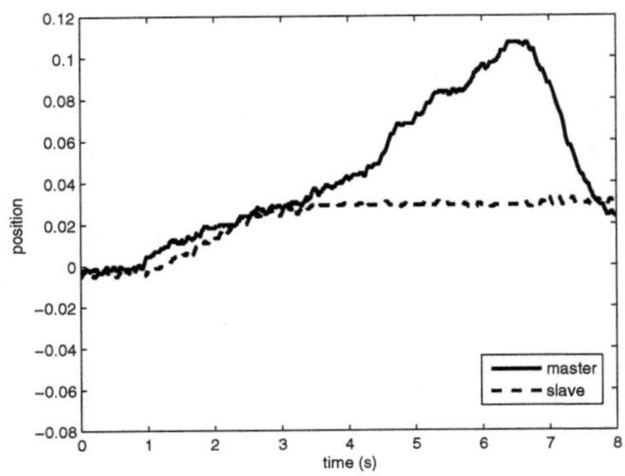

Fig. 16.3. Mater and slave position

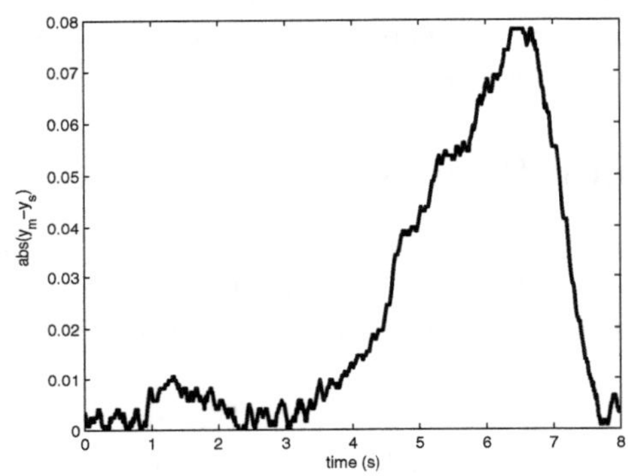

Fig. 16.4. Position error between the master and slave

$$y_m(t) = \begin{bmatrix} 1 & 0 \end{bmatrix} x_m(t) \tag{16.64}$$

$$\dot{x}_s(t) = \begin{bmatrix} 0 & 1 \\ 0 & -6.25 \end{bmatrix} x_s(t) + \begin{bmatrix} 0 \\ 0.2729 \end{bmatrix} u_s(t) \tag{16.65}$$

$$y_s(t) = \begin{bmatrix} 1 & 0 \end{bmatrix} x_s(t) \tag{16.66}$$

where y_m and y_s are the master and slave positions.

These state equations have been obtained from second order differential equations that do not contain zeros, so it is possible to achieve that the slave follows the master, and to establish the dynamics of the slave-master error and the slave. To solve the design equations it is necessary to know the design parameters. The next design parameters have been considered for the experiment:

- A environment stiffness (k_e) of 20 Nm/rad.
- A time delay (T) of 0.1 s.

- A force feedback gain (k_f) of 0.1.
- The error and slave desired dynamics are given by the next characteristic polynomial: $q(s) = p(s) = s^2 + 10s + 25$.

To apply the control method it is necessary to obtain the contact force of the slave with the environment ($f_s(t)$), and the operator force ($F_m(t)$). The contact force $f_s(t)$ has been directly obtained from the slave motor current intensity. The force that the operator exerts over the master has been calculated as follows:

$$F_m(t) = k_{op} y_m(t) \tag{16.67}$$

where k_{op} has been obtained experimentally (k_{op}=100 Nm/rad).

The control gains of the teleoperation system obtained solving the design equations are the following:

- $K_m = \begin{bmatrix} -96.1403 & -10.5588 \end{bmatrix}$
- $K_s' = \begin{bmatrix} -69.6486 & -13.9337 \end{bmatrix}$
- $R_s = \begin{bmatrix} -1.9459 & 0 \end{bmatrix}$
- $G_2 = 0.9729$

The controlled teleoperation system without considering the time delay approximation is delay-independent asymptotically stable. The delay-independent stability has been verified applying the criteria proposed in [12].

In Fig. 16.3 can be observed as the slave follows the master in spite of the time delay until the slave can not continue its interaction with the environment because of the interaction force. Fig. 16.4 shows the error between the master and the slave position.

Fig. 16.5 shows the operator force exerted over the master ($F_m(t)$), the master control signal ($u_m(t)$), and the slave control signal ($u_s(t)$). It can be observed

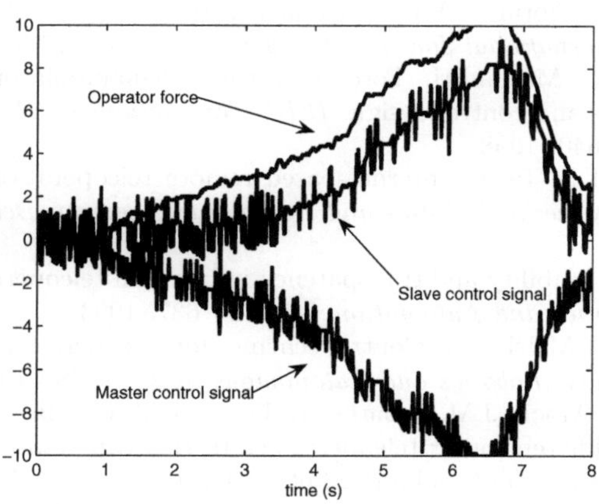

Fig. 16.5. Operator force exerted over the master, master control signal, and slave control signal

as the master control signal annuls the force operator to achieve that the slave stops its interaction with the environment. It can be also appreciated that, when the slave can not continue its interaction with the environment, the slave control signal has the biggest values.

16.6 Conclusions

In this chapter a design and control method of teleoperation systems where the master and slave are modeled by nth-order linear differential equations has been presented. The method is based on the state space formulation and it allows that the slave follows the master through state convergence. The method is also able to establish the dynamics of the convergence and the slave. To obtain the control gains only a design equations system must be solved.

To describe the method, it has been necessary to distinguish between two kinds of linear differential equations: differential equations that do not contain zeros, and differential equations that contain zeros. The reason is that, in the former case, the method is applicable to any teleoperation system. However, in the former case, the method is applicable only if the master and slave zeros verify some conditions. If these conditions are not satisfied, there will be an error between the master and slave output.

The state convergence methodology can be applied to control teleoperation systems with or without time delay in the communication channel. Some experiments have been also performed to test the control method in teleoperation systems with varying time delay [13].

References

1. Z. Gajic and M. Lelic. *Modern control systems engineering.* Prentice Hall, 1996.
2. R. Oboe and P. Fiorini. A design and control environment for internet-based telerobotics. *International Journal of Robotics Research*, 17(4):433–449, 1998.
3. A. Eusebi and C. Melchiorri. Force reflecting telemanipulators with time-delay: stability analysis and control design. *IEEE Transactions on Robotics and Automation*, 14(4):635–640, 1998.
4. K. Brady and T.J. Tarn. Internet-based remote teleoperation. In *IEEE International Conference on Robotics and Automation, Leuven, Belgium*, IEEE, pages 65–70, 1998.
5. D.A. Lawrence. Stability and transparency in bilateral teleoperation. *IEEE Transactions on Robotics and Automation*, 9(5):624–637, 1993.
6. P. Arcara and C. Melchiorri. Control schemes for teleoperation with time delay: a comparative study. *Robotics and Autonomous Systems*, 38:49–64, 2002.
7. J.M. Azorin, R. Aracil, J.M. Sabater, M. Ferre, N.M. Garcia, and C. Perez. Bilateral control of different order teleoperators. In H.I. Christensen, editor, *European Robotics Symposium 2006*, volume 22 of *STAR*, pages 119–127, 2006.
8. K. Ogata. *Modern Control Engineering.* Prentice Hall, 1998.
9. F. Janabi-Sharifi. Collision: modelling, simulation and identification of robotic manipulators interacting with environments. *Journal of Intelligent and Robotic Systems*, 13 (1):1–44, 1995.

10. J. M. Azorin, O. Reinoso, R. Aracil, and M. Ferre. Generalized control method by state convergence of teleoperation systems with time delay. *Automatica*, 40 (9):1575–1582, 2004.
11. J. Funda. *Teleprogramming: toward delay-invariant remote manipulator.* PhD thesis, University of Pennsylvania, Philadelphia, USA, 1991.
12. J.H. Su, I.K. Fong, and C.L. Tseng. Stability analysis of linear systems with time delay. *IEEE Transactions on Automatic Control*, 39:1341–1344, 1994.
13. J. M. Azorin, O. Reinoso, J. M. Sabater, and R. Aracil. Control through state convergence of teleoperation systems with varying time delay. In *Proceedings of the 1st International Conference on Informatics in Control, Automation and Robotics (ICINCO 2004)*, volume 2, pages 379–382, 2004.

Appendix

Procedure to Obtain the Design Equations for Differential Equations that do not Contain Zeros

First, the next linear transformation is applied to the system (16.16):

$$\begin{bmatrix} x_s(t) \\ x_s(t) - x_m(t) \end{bmatrix} = \begin{bmatrix} I & 0 \\ I & -I \end{bmatrix} \begin{bmatrix} x_s(t) \\ x_m(t) \end{bmatrix} \tag{16.68}$$

Applying this state transformation, the next state equation is obtained:

$$\dot{\tilde{x}}(t) = \tilde{A}\tilde{x}(t) + \tilde{B}F_m(t) \tag{16.69}$$

where

$$\tilde{x}(t) = \begin{bmatrix} x_s(t) \\ x_s(t) - x_m(t) \end{bmatrix} \tag{16.70}$$

$$\tilde{A} = \begin{bmatrix} \tilde{A}_{11} & \tilde{A}_{12} \\ \tilde{A}_{21} & \tilde{A}_{22} \end{bmatrix} = \begin{bmatrix} A_{11} + A_{12} & -A_{12} \\ (A_{11} - A_{21}) + (A_{12} - A_{22}) & -(A_{12} - A_{22}) \end{bmatrix} \tag{16.71}$$

$$\tilde{B} = \begin{bmatrix} \tilde{B}_1 \\ \tilde{B}_2 \end{bmatrix} = \begin{bmatrix} B_1 \\ B_1 - B_2 \end{bmatrix} \tag{16.72}$$

Let $x_e(t)$ be the error between the slave and the master, $x_e(t) = x_s(t) - x_m(t)$. The error state equation between the slave and the master will be:

$$\dot{x}_e(t) = \tilde{A}_{21}x_s(t) + \tilde{A}_{22}x_e(t) + \tilde{B}_2F_m(t) \tag{16.73}$$

If $\tilde{A}_{21}$ and $\tilde{B}_2$ were null in (16.73), the error would evolve as an autonomous system. In this case, if the error poles were placed in the left part of the plane s, the error between the master and the slave would be eliminated, and the slave would follow the master. To achieve that the error evolves like an autonomous system, the following equations must be verified:

$$\tilde{B}_2 = B_1 - B_2 = 0 \tag{16.74}$$

$$\tilde{A}_{21} = A_{11} - A_{21} + A_{12} - A_{22} = 0 \tag{16.75}$$

By mathematical transformation of (16.74), (16.39) is obtained. And by transforming (16.75), (16.40) – (16.41) are obtained. Therefore if these $n+1$ equations are verified, it will be achieved that the error evolves as an autonomous system.

When the error evolves as an autonomous system, the characteristic polynomial of the system is:

$$\det\left(sI - (A_{11} + A_{12})\right)\det\left(sI - (A_{22} - A_{12})\right) = 0 \tag{16.76}$$

The first determinant defines the slave dynamics, while the second defines the dynamics of the slave-master error. Doing operations in the first determinant of (16.76), (16.42) – (16.43) are obtained to establish the desired dynamics of the slave. And, doing operations in the second determinant of (16.76), (16.44) – (16.45) are obtained to fix the desired dynamics of the error.

Procedure to Obtain the Design Equations for Differential Equations that Contain Zeros

In this case, the next linear transformation has been applied to the system (16.16):

$$\begin{bmatrix} x_s(t) \\ E_s x_s(t) - E_m x_m(t) \end{bmatrix} = \begin{bmatrix} I & 0 \\ E_s & -E_m \end{bmatrix} \begin{bmatrix} x_s(t) \\ x_m(t) \end{bmatrix} \tag{16.77}$$

where

$$E_s = \begin{bmatrix} b_{s0} & 0 & \cdots & 0 \\ 0 & b_{s1} & \cdots & 0 \\ \vdots & \vdots & \ddots & \vdots \\ 0 & 0 & \cdots & b_{sn-1} \end{bmatrix} \tag{16.78}$$

and E_m is similar to E_s.

Let $x_e(t)$ be the error between the slave and the master, $x_e(t) = E_s x_s(t) - E_m x_m(t)$. It can be verified that if the error converges to zero, the slave output will follow the master output.

Applying the state transformation (16.77), the next state equation is obtained:

$$\dot{\tilde{x}}(t) = \tilde{A}\tilde{x}(t) + \tilde{B}F_m(t) \tag{16.79}$$

where

$$\tilde{x}(t) = \begin{bmatrix} x_s(t) \\ E_s x_s(t) - E_m x_m(t) \end{bmatrix} \tag{16.80}$$

$$\tilde{A} = \begin{bmatrix} \tilde{A}_{11} & \tilde{A}_{12} \\ \tilde{A}_{21} & \tilde{A}_{22} \end{bmatrix} \tag{16.81}$$

$$\tilde{B} = \begin{bmatrix} \tilde{B}_1 \\ \tilde{B}_2 \end{bmatrix} = \begin{bmatrix} B_1 \\ E_s B_1 - E_m B_2 \end{bmatrix} \tag{16.82}$$

and

$$\tilde{A}_{11} = A_{11} + A_{12}E_{sm} \tag{16.83}$$
$$\tilde{A}_{12} = -A_{12}E_m^{-1} \tag{16.84}$$
$$\tilde{A}_{21} = (E_sA_{11} - E_mA_{21}) + (E_sA_{12} - E_mA_{22})E_{sm} \tag{16.85}$$
$$\tilde{A}_{22} = -(E_sA_{12} - E_mA_{22})E_m^{-1} \tag{16.86}$$

$$E_{sm} = \begin{bmatrix} b_{s0}/b_{m0} & 0 & \cdots & 0 \\ 0 & b_{s1}/b_{m1} & \cdots & 0 \\ \vdots & \vdots & \ddots & \vdots \\ 0 & 0 & \cdots & b_{sn-1}/b_{mn-1} \end{bmatrix} \tag{16.87}$$

The error state equation between the slave and the master will be:

$$\dot{x}_e(t) = \tilde{A}_{21}x_s(t) + \tilde{A}_{22}x_e(t) + \tilde{B}_2F_m(t) \tag{16.88}$$

To achieve that the error evolves like an autonomous system, the following equations must be verified:

$$\tilde{B}_2 = E_sB_1 - E_mB_2 = 0 \tag{16.89}$$

$$\tilde{A}_{21} = (E_sA_{11} - E_mA_{21}) + (E_sA_{12} - E_mA_{22})E_{sm} = 0 \tag{16.90}$$

From (16.89), (16.46) is obtained. And, operating in (16.90), (16.47) – (16.48) are obtained. In this case, to achieve that (16.90) would be null, it is necessary also that the conditions in (16.53) are verified.

When the error evolves as an autonomous system, the characteristic polynomial of the system is:

$$\det\left(sI - (A_{11} + A_{12}E_{sm})\right)\det\left(sI - (E_mA_{22} - E_sA_{12})E_m^{-1}\right) = 0 \tag{16.91}$$

The first determinant defines the slave dynamics, while the second defines the dynamics of the slave-master error. Doing operations in the first determinant of (16.91), (16.49) – (16.50) are obtained to establish the desired dynamics of the slave. And, doing operations in the second determinant of (16.91), (16.51) – (16.52) are obtained to fix the desired dynamics of the error.

Procedure to Obtain the State Equation of the Teleoperation System with Time Delay

From the model shown in Fig. 16.1, if a constant time delay T is considered, the master control signal, $u_m(t)$, and the slave control signal, $u_s(t)$, are respectively:

$$u_m(t) = K_mx_m(t) + R_mx_s(t-T) + F_m(t) \tag{16.92}$$

$$u_s(t) = K_sx_s(t) + R_sx_m(t-T) + G_2F_m(t-T) \tag{16.93}$$

Replacing the master and slave control signal, (16.92) and (16.93), in the master and slave state equation, (16.10) and (16.11), the next state equations are obtained:

$$\dot{x}_m(t) = (A_m + B_m K_m)x_m(t) + B_m R_m x_s(t-T) + B_m F_m(t) \quad (16.94)$$
$$\dot{x}_s(t) = (A_s + B_s K_s)x_s(t) + B_s R_s x_m(t-T) + B_s G_2 F_m(t-T) \quad (16.95)$$

In order to obtain the state equation of the teleoperation system, the time delayed signals in the state (16.94) and (16.95) have been approximated using the Taylor expansion of first order:

$$\begin{aligned} x_s(t-T) &= x_s(t) - T\dot{x}_s(t) \\ x_m(t-T) &= x_m(t) - T\dot{x}_m(t) \\ F_m(t-T) &= F_m(t) - T\dot{F}_m(t) \end{aligned} \quad (16.96)$$

Using the time delay approximation (16.96) in the state (16.94) and (16.95), and considering that the operator force (F_m) is constant, these equations are transformed in:

$$\begin{aligned} \dot{x}_m(t) = &\ (A_m + B_m K_m)x_m(t) + B_m R_m x_s(t) \\ &-T B_m R_m \dot{x}_s(t) + B_m F_m(t) \end{aligned} \quad (16.97)$$
$$\begin{aligned} \dot{x}_s(t) = &\ (A_s + B_s K_s)x_s(t) + B_s R_s x_m(t) \\ &-T B_s R_s \dot{x}_m(t) + B_s G_2 F_m(t) \end{aligned} \quad (16.98)$$

The state equation (16.54) is obtained representing the state equations (16.97) and (16.98) in a matrix way.

17

Re-configurable Control Scheme for Guiding Telerobotics

Adrián Mora and Antonio Barrientos

Universidad Politécnica de Madrid
Dpto. Automática, Ing. Electrónica e Inf. Industrial
C/. José Gutierrez Abascal, 2. 28006 Madrid
amora@etsii.upm.es, antonio.barrientos@upm.es

Summary. In telerobotics two control modes are usually implemented for guiding: position control and rate control. Numerous works have been carried out comparing them. This chapter introduces a new re-configurable system for guiding robots. It is based on the fact that guiding performance depends directly on the task requirement. The system presented is able to change its control scheme during task execution in order to accommodate itself to the task requirement at all times. An architecture for the re-configurable system is proposed. It has been experimentally implemented and tested. Its performance is compared to conventional force-position and force-rate bilateral control schemes. Findings show that the re-configurable system obtains the best results in all analysed variables.

17.1 Introduction

Past works [2], [7] have found that for each telemanipulation task requirement there is a specific system configuration that obtains better results. In [7] it was found that for several tasks in endoscopic surgery (e.g. catheter insertion or membrane puncturing), there is one type of control of the telemanipulation system that performs better than any other. The authors proposed that in surgery the system must change its control parameters according to the task performed at any given moment. In [2] it was found that the interaction between task type and the others factors analysed (kinematics coupling and master position) had a significant effect on performance. Another work [17] compared different master configurations with two different tasks finding that a specific master configuration produces better performance than others. These works support the idea that each task requirement has a system configuration that obtains better results.

The control mode for guiding significantly affects task performance. Two control modes are usually implemented: position control and rate control. Several works [5, 8] have been carried out in order to address which of them produces better results. In [8], the authors found that when master and slave workspaces are similar, position control is 1.5 times faster than rate control for a simulated pick and place task. However, rate control obtains better results when the slave workspace is larger than the master workspace. In [5], several types of bilateral

M. Ferre et al. (Eds.): Advances in Telerobotics, STAR 31, pp. 289–301, 2007.
springerlink.com © Springer-Verlag Berlin Heidelberg 2007

control schemes were studied. The authors compared position control vs. rate control and found that the former obtains better results. Another work related to position and rate control is [14]. The authors compared these two modes of guiding and studied the transparency of the bilateral control scheme under position and rate control. They found that the four-channel control architecture is transparent under ideal conditions. The results showed satisfactory position-force and velocity-force tracking in the position and rate modes of operation, respectively.

Our previous works [10] were designed to find out which system configuration obtains better results for each type of task under different conditions (time delay, operator's training, etc.). In relation to guiding control, it was found that each guiding mode (position and rate) obtains better results for different types of task, i.e. position control is better for tasks in which short and precise movements such as insertion are important. However, rate control is better for tasks in which long and precise movements are necessary in an extremely rigid environment. We found that complex telemanipulation tasks such as endoscopic surgery or satellite panel change can be divided into several basic subtasks, each with specific requirements and different suitable control configurations. This motivates the approach of a re-configurable teleoperation control system adapting the guiding control strategy to the subtask requirements. Differing from other chapters in Part II of the book, this chapter evaluates rate control for guiding, finding that for some task it obtains satisfactory results.

Some types of similar systems have been proposed in the literature [13], [16]. The system proposed in [16] has different control modes: from manual (bilateral) to autonomous mode with some intermediate modes. The operator decides when it is necessary to change from one control mode to another. These works are based principally on the concept of supervisory control proposed by Sheridan [15]. In contrast, in the system proposed in this chapter, the control mode is always manual control (bilateral); the system changes its control scheme configuration in order to attain better task performance.

The re-configurable system is based on an event-based control architecture in which a discrete automaton is used to control the changes in the system configuration. This type of hybrid control has been used in several works [1, 4, 6, 9, 12]. In [1], a switching control mode is used to perform teledrilling task. The experiments were carried out with constant time delay. The hybrid automaton is used to switch among four control modes: free motion, velocity-restricted motion, force control and move back. The results were satisfactory and the system was stable under time delay. In [9] a complex teleoperation system for cooperation between robots is implemented. The system used internet as communication channel. An event-based controller was implemented in order to synchronize the two robots in the presence of time delay. Several works have studied the stability of these types of control schemes. In [6] a tool to analyze hybrid systems was developed. One of the main advantages of these event-based controllers is that they can eliminate or reduce the problems caused by time delay [12]. In this chapter, no time delay is introduced in the system control loop. The event-based controller is used to determine the change in the system configuration.

17.2 Re-configurable System Design

The system proposed is one that is able to change its configuration during task execution. Configuration means the value of factors that affect system behaviour such as: control parameters (bilateral control, control gains, etc.) or operator information parameters (force feedback, type of images, etc.). The re-configurable system is able to change the value of these factors during task execution. Fig. 17.1 presents an overview of the system proposed.

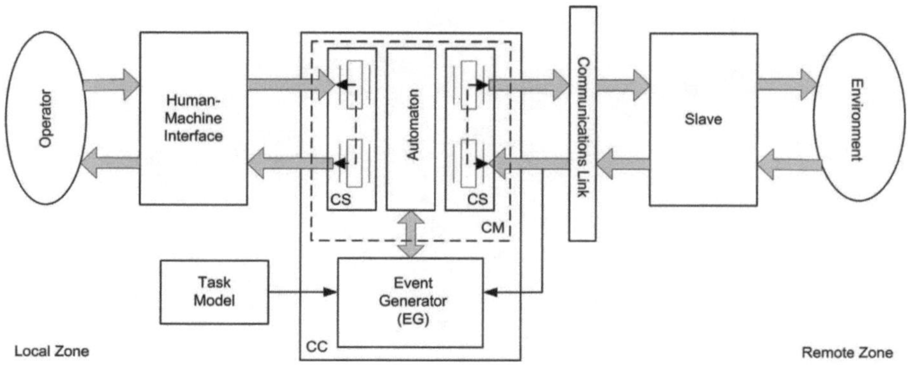

Fig. 17.1. Re-configurable system architecture. It is based on a Configuration Control (CC) module introduced in the local zone. It is in charge to control the change in configuration.

In Fig. 17.1 the two main parts of a telemanipulation system can be seen: remote and local zones. The local zone includes a Configuration Control (CC). The CC is composed of two parts: Configuration Manager (CM) and Event Generator (EG). The CM is in charge of producing the change in system configuration. The CM has to guarantee that the system does not become unstable during the change in configuration. The EG is in charge of generating the appropriate events to the CM in order to produce the correct change in configuration. The EG must decide *what* configuration needs to be adopted and *when* the change is to be made. The CM and the EG are described below.

17.2.1 Configuration Manager - CM

The CM is composed of two principal parts: the communications supervisor (CS) and the automaton. The communications supervisor (CS) is in charge of disconnecting the local zone and the remote zone; during such disconnection this part of the CM maintains the stability of the system by sending the appropriate commands to the slave and master devices. The main objective of these commands is to maintain the slave in the same position during the disconnection.

The other part of the CM is a state-based automaton. Each state corresponds to a possible system configuration. There are five possible states for the automaton, as shown in Fig. 17.2. Table 17.1 contains a description of the states.

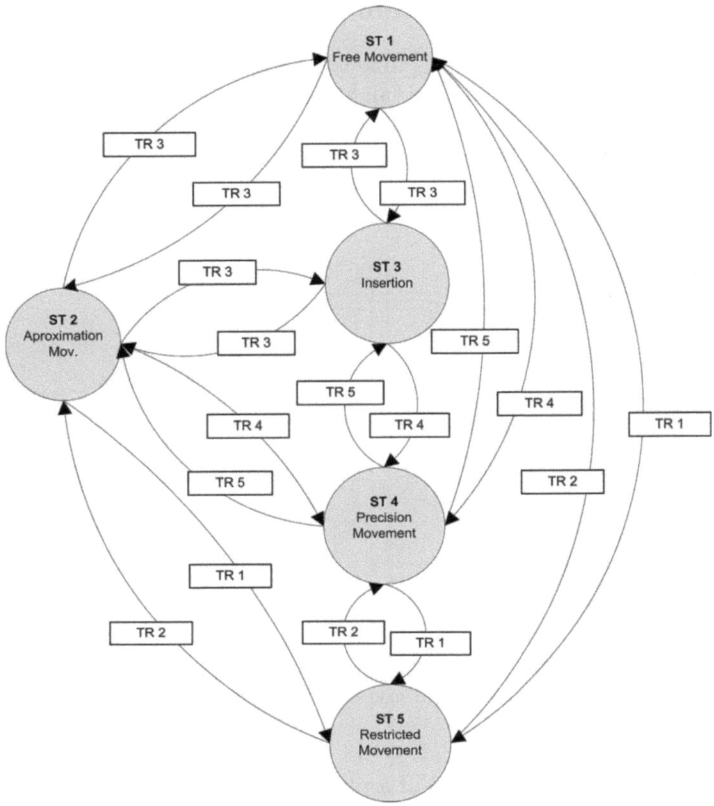

Fig. 17.2. Automaton states introduced in the Configuration Manager (CM). Each sate of the automaton corresponds with a possible system configuration.

As the table shows, each state (system configuration) is used for a corresponding subtask. These subtasks were selected from previous studies [10], in which it was found that for each of these subtasks there is a system configuration that produces better results. The configuration for each subtask was obtained based on exhaustive experimentation in which more than 15 system factors were studied.

This chapter introduces a system with changes related to the system control (guiding control). This is because these changes are the most critical from the point of view of system stability. However, the re-configurable system must take into account other factors such as: images, frame rate, bandwidth, etc.

The main challenge in this system is the change from one configuration to another. This is because during the process of changing configuration many problems must be addressed, such as stability and performance. For this reason, specialized algorithms for transitions between configurations were developed: change from position control to rate control (TR 1) and vice versa (TR 2).

These transition algorithms have to change not only the movement control but also the bilateral control, changing from force-position ((17.1) and (17.2)) to force-rate ((17.3), (17.4), and (17.5)) and vice versa.

Table 17.1. States Adopted by the Automaton

State	Configuration	Subtask
ST1	Position Control - High Position Gain ($k_p > 1.2$)	Free Movement
ST2	Position Control - Medium Position Gain ($0.8 < k_p < 1.2$)	Approach Movement
ST3	Position Control - Medium Position Gain ($0.5 < k_p < 0.8$)	Insertion Movement
ST4	Position Control - Low Position Gain ($k_p < 0.5$)	Precision Movement
ST5	Rate Control	Restricted Movement

$$f_m = k_f f_e \tag{17.1}$$

$$X_s(s) = k_p X_m(s) \tag{17.2}$$

$$f_m = (X_m(s) - X_{ref}(s))(k_r + b_h s) + f_{rz} \tag{17.3}$$

$$f_{rz} = k_f f_e \tag{17.4}$$

$$\dot{X}_s(s) = k_v X_m(s) \tag{17.5}$$

Where f_m is the force applied in the master, k_f force gain, f_e environment reaction force, X_s slave position, k_p position gain, X_m master position, k_v rate gain and b_h human arm stiffness. In the force-position scheme the force feedback to the operator is proportional to the force exerted by the environment, (17.1), while the slave position commands are proportional to the master position, (17.2). In contrast, in the force-rate scheme, the master force is composed of two parts: the first corresponds to a self-centred force that allows the master to behave as a Joystick,$((X_m(s) - X_{ref}(s))(k_r + b_h s))$, and the second is the force due to the environment reaction force, $(k_f.f_e)$, (17.3) and (17.4). For rate control the master's position X_m multiplied by a rate gain k_v was used as a slave's rate reference $\dot{X}_s$, (17.5). Due to this transformation, it is necessary to implement the self-centred force as described above.

The process that allows the transition is similar in both cases: (1) the remote zone and local zone are decoupled by means of the communications supervisor. (2) The transition algorithm (TA) is applied. (3) Remote zone and local zone are connected again. The transition algorithm (TA) must guarantee the stability of the system during the connection.

Different types of TAs were developed according to the requirements of the transitions (Fig. 17.2). In some cases it is necessary to make a change in the master position during the transition, e.g. to change from position control to rate control the master must be moved to its centre position and then it is necessary to apply the self-centred force as described above. This is made possible by a virtual attraction force implemented in the master as a spring with increasing stiffness over time:

$$F(t) = -k(t)d \; , \tag{17.6}$$

where $k(t)$ is the spring stiffness and d is the distance between the current master position and the desired master position. If the change is from position to rate control the desired master position is its centre. But if it is from rate to position control, the desired master position depends on the current slave position; therefore it must be calculated based on the current slave position (X_s) and new position gain (k_p) that will be applied.

Three more TAs were developed, all related to the change in position gain when position control is used. The first one was for change when there is not a great difference between the position gains (TR 3). The last two transitions were used when there is a great difference between the position gains: change from high gain to low gain (TR 4) and from low gain to high gain (TR 5).

The TA used for TR 3 was based on a virtual coordinate system translation in order to obtain a transparent change in position gain. The other two TAs were similar to the ones used for TR 1 and TR 2, because it was necessary to make a change in the master position.

17.2.2 Event Generator - EG

The event generator is in charge of decision-making. This part of the system has to decide *what* configuration it is necessary to adopt and *when* it is necessary to change the configuration.

To know *what* the appropriate system configuration is, the EG has to rely on a task model. Based on the task model, a sequence of subtasks is defined. This sequence defines which is the right system configuration for each subtask of the complete task. With the sequence generated, the EG knows which configuration the system has to adopt. During task execution, the EG executes the sequence of states.

The second part of the EG, determining *when* to change the configuration, can have different solutions. It is necessary to take into account the negative effect that an unexpected change in system configuration might have on the operator. That is why all changes in system configuration must be decided or validated by the operator. A first approach would be for the operator to decide when it is necessary to make a change in system configuration. The second alternative would be to develop advanced algorithms for the system decision-making. Using these algorithms the system can propose the change to the operator, but the operator must acknowledge the change in order to validate it.

Currently, the operator makes the decision as to when to change the configuration. The operator decides when it is necessary to make the change and by means of a button in his interface, he instructs the EG to make the change.

If the EG knows that it is necessary to change the system configuration it generates the appropriate event to inform the CM that it must change the configuration. In the event generated, the EG commands the configuration needed.

On this basis, the CM activates the clutches and informs the automaton to change to the appropriate state (the automaton uses the appropriate TA according to the current and final state). When the configuration has changed the CM informs the EG that the configuration has been successfully changed.

17.3 Performance of the Re-configurable Control Scheme

System: The re-configurable system was implemented using the platform for experimentation with telerobotics systems developed by the authors [11]. With this platform several types of telerobotics systems can be implemented easily. The platform is a distributed system in which each element relies on a server connected to Ethernet LAN. Additionally, there is a client in charge of the interconnection between the devices (servers). In the client the Configuration Control (CC) was implemented.

In this experiment, a 6 DOF PUMA 560 robot was used as a slave. An F/T sensor was installed on the robots wrist. Additionally, an on-board colour camera was installed on the wrist, pointing at the robot's hand. A second colour camera was also used, this one with an overview of the remote zone (Fig. 17.3). As a master a PHANToM device was used. This device has 6 DOF and force feedback in 3 position axes, with a servo loop of 1000 Hz providing good force sensation. The operator interface has a monitor, the control interface and the master device (Fig. 17.3). The images from the on-board and overview cameras were shown on the monitor. The slave robot and the task layout comprised the remote zone (Fig. 17.3).

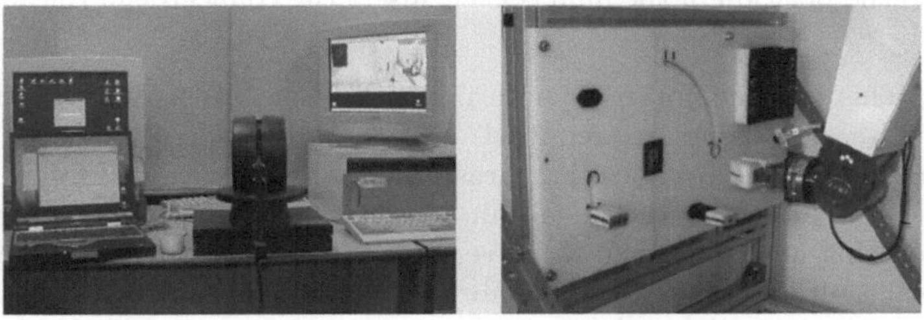

Fig. 17.3. Left: Local Zone, comprised of the master device and the display with the images from cameras. Rigth: Remote Zone, comprised of robot slave and the task board.

Operator: The operator was trained in the use of the re-configurable system and in the system's configurations and transitions, to become familiar with them.

Task: A complex prototype task was implemented. It had three different phases, each with different characteristics (Fig. 17.4).

Phase 1: This phase includes precise and restricted movements in a rigid environment. The operator had to move the slave to the reference point. Then he

Fig. 17.4. Left: Task Layout, the three phases of the task can be observed: groove movement, insertion and membrane-puncturing. Center/right: Linear/circular groove detail.

had to move it to the initial position of the tool (Fig. 17.4), grasp it and move it along the linear groove to the hole at the end in order to release the tool from the groove. Then the operator had to move the slave to the initial hole of the circular groove (Fig. 17.4), insert the tool, and move the tool along the circular groove to the end. Next he had to come back along the circular groove and release the tool from it through the hole. Finally, the operator had to place the tool in its initial position. To do so, he had to come back to the linear groove, insert the tool, move it along the groove to the end, and then release the tool, leaving it in its initial position.

Phase 2: This is a connector insertion task. The connector used is a standard IEC power connector (Fig. 17.4). The operator had to move the slave to the initial connector position, grasp and release it from the socket. Then he had to move it to the position of the final socket, insert the connector in the socket and leave it there.

Phase 3: This is a membrane-puncturing task with an elastic environment in which precise movements are required. The operator had to move the slave to the initial, puncturing tool position, grasp the tool and release it from its socket. Then he had to move to the membrane position and puncture the membrane. The puncturing movement had to be very precise and not touch a piece of foam rubber placed 10mm behind the membrane. When the puncturing was finished the operator had to release the puncturing tool in its socket. To finish the task the operator had to move the slave to the reference point. The membrane was made with a 3mm rubber sheet simulating muscular tissue. The puncturing tool has a diameter of 1.5mm and its length is 60mm. The socket is a 2mm diameter hole, meaning that it is a very restricted insertion.

The complete task was done sequentially: Phase 1, Phase 2 and Phase 3. The task workspace is approximately three times the master workspace.

Configuration Control Implementation: As mentioned above the Configuration Control (CC) was implemented in the central client. The automaton was completely implemented as shown in Fig. 17.2. The Transition Algorithms (TAs) were also implemented. The operator decides when it is necessary to make the change in configuration, but it is the Configuration Manager (CM) that changes the system configuration, maintaining system stability at all times.

Analysed Variables: The dependent variables measured were: completion time, SOSF (Sum Of Squared Forces) and insertion forces. Completion time gives an idea of the overall system performance; it gives a measure of the system manoeuvrability. The SOSF is a measure of the energy used by the system and a parameter of the forces exerted over the environment. Finally, insertion forces give a measure of the maximum forces exerted during insertion. A peak of force is required in order to insert the connector correctly. The data registered included: completion time, slave position and slave forces. Completion time was measured starting at the moment the operator moved the slave away from the reference point until it came back to the reference point. All position and force data were registered at a frequency of 100Hz.

Additional Systems: In order to compare the re-configurable system with conventional systems, two additional non-configurable (one configuration throughout task execution) systems were implemented: force-position and force-rate bilateral control. The first system had a force-position bilateral control scheme and the second one had a force-rate bilateral control scheme.

17.4 Results

All the data were processed and analysed using a means comparison for each variable. For each analysed variable several statistical tests were done in order to verify the results. First, the data were validated in order to determine that there was no significant difference in variance: Cochran's test and Bartlett's test were performed ([3]). These tests showed that there was no significant difference among variances for all variables. ($p\text{-value} > 0.05$), (Table 17.2). Then ANOVA tests were performed in order to see if there was any significant difference among the means compared. The ANOVA test showed that there were statistical differences among means in all variables, ($p\text{-value} < 0.05$), (Table 17.2). Finally, multiple range tests were done in order to identify which means were statistically different from the others.

Completion Time: Multiple range tests showed that all means are different. As shown in Fig. 17.5, the re-configurable system gave the best performance. The poorest result was obtained with force-rate control scheme. As expected, the force-rate scheme was the slowest, because with rate control the long movements

Table 17.2. Statistical Analysis Summary

Variable	Cochran Test (p-value)	Bartlett Test (p-value)	ANOVA Test (p-value)
Completion Time	0.5540	0.6727	0.0001
SOSF	0.6414	0.4528	0.0001
Insertion Forces	0.9543	0.8072	0.0001

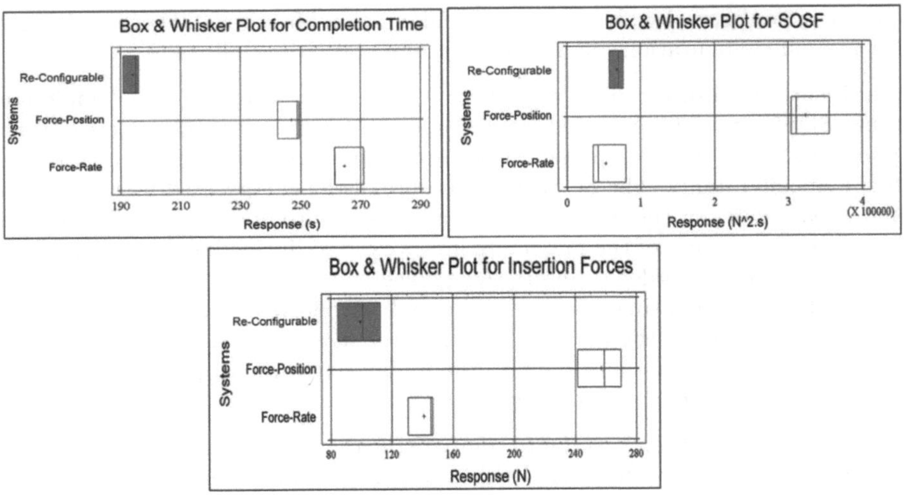

Fig. 17.5. Box Plots. Up-left: Completion Time. Up-Right: SOSF. Down: Insertion Forces. Clearly, regarding all analyzed variables, the Re-configurable system performs best.

between different points of the task are slow. Further, with this type of control, correction movements for insertion are difficult to carry out because two movements are necessary: one for moving and another for stopping.

SOSF: Multiple range tests showed that there is no difference between the force-rate system and the re-configurable system. Fig. 17.5 shows that the best performance was obtained with the force-rate system but there is no significant difference with regard to the re-configurable system. In contrast with the completion time variable, the poorest result was obtained with force-position control. This is because with position control phase 1 of the task is very difficult; this kind of restricted movement is hard to carry out with this type of control movement. On the other hand, with rate control this phase was efficiently accomplished due to the behaviour of this bilateral control scheme in rigid environments.

Insertion Forces: Multiple range tests show that all means are different. Fig. 17.5 shows that the best performance was obtained with the re-configurable system. The poorest result was obtained with force-position control. As with SOSF, rate control produces better results because in the insertion (phase 3) the final movement is very restricted. Although the approach is better with position control, the insertion itself (forces) is better with rate control.

17.5 Discussion

As can be observed, the re-configurable system gives the best performance in all analysed variables, indicating that it is a viable and efficient system. The force-rate system obtains the best results (the same as those of the re-configurable system) for the SOSF variable, as was expected. This means that with rate control the forces exerted in the environment are less than those exerted with other types of control. With force-position control the forces exerted were high. This is because with this kind of control scheme, control of the system in phase 1 is very difficult; the system can easily become unstable. In contrast, force-position control obtains good results for completion time.

As the results suggest, each guiding control scheme obtains better results in a specific variable: position control in completion time and rate control in forces. However, the main advantage of the re-configurable system is that it obtains good results in all variables. This means that for each task requirement (generally associated with a measured variable) the system is able to obtain good results.

The operator expressed a favourable opinion about the re-configurable system. While it is necessary to become used to the transitions, not much training is needed. As an additional test, another operator (experienced in telemanipulation; no experience with re-configurable control) was asked to try the re-configurable system. His opinion was also favourable. He was able to do the complete task without major difficulties. The fact that the operators themselves make the decision about when to make the change in configuration by pressing a button seems to be a good strategy.

There are a number of factors that affect the system. Some of the most important ones are: the type of force feedback, image resolution, frame rate, etc. The re-configurable system implemented in this work uses only movement control and position gain as variable factors because they affect the slave movement. Slave movement is considered the most critical aspect in a re-configurable system and was therefore chosen as the validation criterion. A generalized re-configurable system could also implement changes in other factors, e.g. low resolution images for free movement but high resolution during approach/manipulation, thus reducing required communication bandwidth.

In the current implementation of the re-configurable system, the operator decides about when the system has to change its configuration, but the system changes it autonomously. Some doubts about efficiency arise if the system makes this decision and changes the configuration itself. The system changing the configuration by itself could cause problems for the operator because the change might be unexpected and the operator might lose control. Another problem arises if the system changes the configuration at the wrong moment (in the middle of a subtask) leading to overall task failure. One solution could be shared decision, i.e. the re-configurable system proposes when it is necessary to make the change but the operator has to confirm the decision e.g. by a button or voice command.

17.6 Conclusion

This chapter presents a new re-configurable control scheme for telemanipulation systems. This system is able to change its control scheme in order to accommodate the different tasks requirements. A hybrid control scheme was implemented based on an discrete automaton in order to implement the configuration control. The main advantage of this system is that it obtains the best result for all analysed variables, as opposed to conventional control schemes which obtain better results for a specific variable: position control for completion time and rate control for forces.

The critical part of the system i.e. the change in configuration has obtained satisfactory results. The transition algorithms developed have obtained good results. While it is necessary to become used to the transitions, not much training is needed. When the operator has become used to the transitions, he feels comfortable with the system. Thus, the obtained results suggest that the main transitions (those involving change in master position) does not increase the completion time as it was expected.

References

1. H. Baier, M. Buss, and G. Schmidt. Control Mode Switching for Teledrilling Based on a Hybrid System Model. In *Proc. of Advanced Intelligent Mechatronics, AIM 97*, 1997.
2. O. Ben-Porat, M. Shosham, and J. Meyer. Control Design and Task Performance in Endoscopic Teleoperation. *Presence: Teleoperators and Virtual Environments*, 9:256–267, 2000.
3. Box G.E.P, W.G. Hunter, and J.S. Hunter. Statistics for Experimenters, An introduction to Design, Data Analysis, and Model Building. *John Willey and Sons, Inc.*, 1996.
4. M. Buss and G. Schmidt. Hybrid System Behavior Specification for Multiple Robotic Mechanics. In *Proc. of 1996 IEEE/RSJ International Conference on Intelligent Robots and Systems, IROS*, 1996.
5. H. Das, M. Zak , W.S. Kim, A.K. Bejczy, and P.S. Schenker. Operator Performance with Alternative Manual Control Modes in Teleoperation. *Presence: Teleoperators and Virtual Environments*, 1:201–217, 1992.
6. G. Hirzinger and P.H. Ho. Hytec: The cornell hybrid technology tool. Lecture Notes in Computer Science 999: Hycrid Systems II. Springer 265–293, 1995.
7. A. Kazi. Operator Performance in Surgical Telemanipulation. *Presence: Teleoperators and Virtual Environments*, 10:495-510, 2001.
8. W.S. Kim, F. Tendick, R.S. Ellis, and L.W. Stark. A Comparison of Position Control and Rate Control for Telemanipuations with Consideration of Manipulator System Dynamics. *IEEE Journal of Robotics and Automation*, 3:426–436, 1987.
9. W-t. Lo, Y. Liu, I.H. Elhajj, N. Xi, Y. Wang, and T. Fukuda. Cooperative Teleoperation of a Multirobot System with Force Reflection via Internet. *IEEE/ASME Transactions on Mechatronics*, 9:661–670, 2004.
10. A. Mora and A. Barrientos. Functional Factors in Telemanipulation System: Effects of System Bandwidth in Task Performance. In *Proc. of the Mechatronics and Robotics*, 2004
11. A. Mora and A. Barrientos. Platform for experimentation with Tele-Robotics systems. In *Proc. of 2002 IEEE/RSJ International Conference on Intelligent Robots and Systems, IROS 2002*, 2:1632–1637, 2002.

12. Z. Nichol, Y. Liu, P. Suchyta, M. Prokos, A. Goradia, and N. Xi. Super-Media Enhanced Internet-Based Real-Time Teleoperation. In *Proc. of Hands-On International Mechantronics and Automation Conference*, 2005.
13. T. Sato and S. Harai. MEISTER: A Model Enhanced Intelligent and Skiful Teleoperation Robot Systems. In *Robotics Research - The Fourth International Symposium - The MIT Press*, 1998
14. S.E. Saculdean, M. Zhu, W-H. Zhu, and K. Hastrudi-Zaad. Transparent Bilateral Teleoperation under Position and Rate Control. *The International Journal of Robotics Research*, 19:1185–1202, 2000.
15. T. B. Sheridan. Telerobotics, Automation, Control and Human Supervisory Control. *MIT Press. Cambridge, Massachusetts*, 1992.
16. Y. Yokokohji, A. Ogawa, H. Hasanume, and T. Yoshikawa. Operation Modes for Cooperating with Autonomous Functions in Intelligent Teleoperation Systems. In *Proc. of IEEE International Conference on Robotics and Automation*, 3:510–515, 1993.
17. W.K. Yoon, S. Tachihara, Y. Tsumaki, and M. Uchiyama. Evaluation of the Different Master Device Approaches for a Model-Based Space Teleoperation System. In *Proc of 10th International Conference on Advanced Robotics*, 2001.

18

Teleprograming: Capturing the Intention of the Human Operator

Miguel Hernando[1] and Ernesto Gambao[2]

[1] Universidad Politécnica de Madrid
Dpto. Eletrónica, Automática e Informtica Industrial
C/. Ronda de Valencia, 3. 28012 Madrid, Spain
miguel.hernando@upm.es

[2] Universidad Politécnica de Madrid
Dpto. Automática, Ingeniera Electrónica e Informtica Industrial
C/. José Gutierrez Abascal, 2. 28006 Madrid, Spain
gambao@etsii.upm.es

Summary. The presence of a significant time delay in the communications between the local and remote zone of a teleoperated system causes two undesirables effects: (1) dynamic instability and (2) unmanageability. Instability usually appears in bilateral control schemes as a consequence of force feedback and makes the system useless with 0.1 or more seconds of time delay. Robot teleprogramming was proposed as an intermediate solution between supervised control systems and direct teleoperation when a significant delay appears in the communications between the local and the remote zones. In this chapter a teleprogramming architecture with an analyzer for the operators intention is proposed and tested. A task analyzer observes the movements made by the operator in a virtual environment. Through the force and geometric information obtained (attending only to geometric features), a set of symbolic commands are generated. These commands are then transmitted to an interpreter which pipes them through the communication system to the remote zone. The remote system receives and translates the information to absolute references in the remote model that is continuously actualized by the perception system.

18.1 Introduction

Most of the teleoperation architectures that have been recently developed have focused in objectives related to the controllability of the system. Their main objective is to conjugate the transparency with the instability due to the bilateral master/slave coupling. Many of them have been centred in particular solutions to the time delay problem, a common drawback of systems where a bilateral Teleoperation system is needed. That is the case of spatial systems, like planetary exploration or tasks of maintenance and control of orbital systems from bases placed on earth [1],[2] or underwater systems, in which due to the type of communication technology the transmission bandwidth is very low. Lately there is a growing interest in obtaining an efficient bilateral control for systems with variable time-delay like Teleoperation through Internet [3].

M. Ferre et al. (Eds.): Advances in Telerobotics, STAR 31, pp. 303–320, 2007.
springerlink.com © Springer-Verlag Berlin Heidelberg 2007

The time-delay generates an effect of instability and inmanageability in the system. From classic control theory is easy to derive that a delay in a control loop is an important cause of instability. A pure delay decreases the phase of the system in a factor equal to the product between the frequency and the time-delay value. Also, with the presence of a pure delay, as the static gain increases the system deviates rapidly from a stable condition.

But along with a real and intrinsic instability of the control system, from the beginnings the presence of a inmanageability effect has been already observed. Ferrel conducted in 1962 the first experiments with an unilateral system under time delay in the visual feedback [4] . The 'move-and-wait' strategy was first conceived and employed as a solution to overcome that inmanageability. Evidently this effect is much more important in the bilateral control systems.

Since then, many proposals to overcome time-delay have appeared in the literature. Proposals can be divided into two different types of approaches: those based on the more traditional manual teleoperation approach [5] and those based on the supervisory control concept conceived by Ferrel and Sheridan [4] and further developed by Sheridan [6].

Usually both techniques can be considered complementary, and in fact it is usual to find both in mostly of the developed systems. One of the main reasons for those hybrid approaches is the technical difficulty of developing a complete supervised control system.

There are many sources where to find summarised information on techniques for time-delayed teleoperation. [7] is a very recommended and well-documented study by Sheridan, although it focuses very much on predictors displays and supervisory control instead on manual teleoperation. Also, although it contains the foundations, many new methods and techniques have been proposed since its publication.

Following the generalized formalism presented by Keddar [9], a control architecture in which there are more control strategies than state variables, is closer to a supervisory control system. While on the contrary, if the state variables are more important than strategies, the system is closer to the traditional bilateral teleoperation system as it is shown in Fig. 18.1. The present work is framed in a teleprogrammed system and, as is reflected in [Keddar], it is an architecture of type:

$$U_m = F(strategy) \ \ and \ \ U_s = G(P_m, Strategy) \tag{18.1}$$

Thus, the operator acts on a simulated slave system, and therefore the feedback information is artificially generated by a simulator. This is reflected in the first expression of (18.1), in witch the state of the master device (U_m) is a function (F) of an artifice. The second expression means that the remote slave manipulator (U_s) will be moved by a complex controller that will try to adapt the references sent from the local system (P_m)with the information about the real environment obtained by the sensors placed on the remote system.

Although the teleprogramming term was proposed by Paul and Funda in 1990, the work of Hirzinger, Bejczy, Kotoku and Machida, are considered as precursors of this way for overcoming the communications delay problem [8][10][11][1].

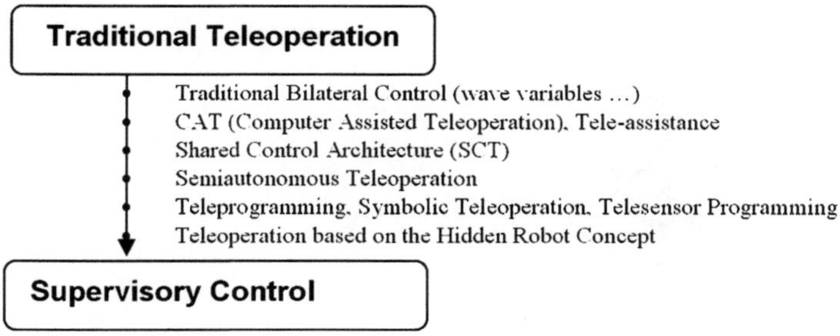

Fig. 18.1. Teleoperation systems ordered by their intelligence

Multiple tools and strategies have been defined in the 90s dealing with different aspects of the teleprogramming task: For the analysis of the operator intention Neural Networks, Petri nets, etc. have been proposed [12][13]. For simplifying the analysis and former execution of the task, synthetic fixtures have been used [14]. Different control strategies for different parts of the task were defined to control the remote manipulator [15]. In [16] an on-line path planner has been used in a supervised control system. In [15] the use of references related to objects was proposed for simplifying the system. Many visual aids have been developed to help the operator [14][17] and recently photorealistic 3D models have been used to improve the quality of the predictive display [18]. Several works focus on the effect of errors during teleprogramming [19], or about the necessity of including or not previous information about the task in the system [20]. In this chapter, a general teleprogramming architecture implementation is briefly described, and two main aspects are described in more detail: the local analysis and remote use of the operator intention without previous information about the task, and a fast virtual contact force generator for the predictive Simulator. The outline of the rest of the chapter is as follows. First, to have a better understanding of the problem, section 2 is devoted to give a brief overview of the main features of the implemented teleprogramming architecture. Section 3 contains a brief description of the predictive simulator and the contact force generation algorithm that will serve to better understand the explanation of the task analyzer and the task manager later presented in section 4. Finally, the analysis of simulated and experimental results is performed in section 5.

18.2 Teleprogramming Architecture

The control architecture with its components and the links among themselves is shown in Fig. 18.2. The architecture modularity allows a parallel development of its components. A good definition of the functionality, inputs and outputs of each element makes feasible the use of the system while not all the components are full-developed.

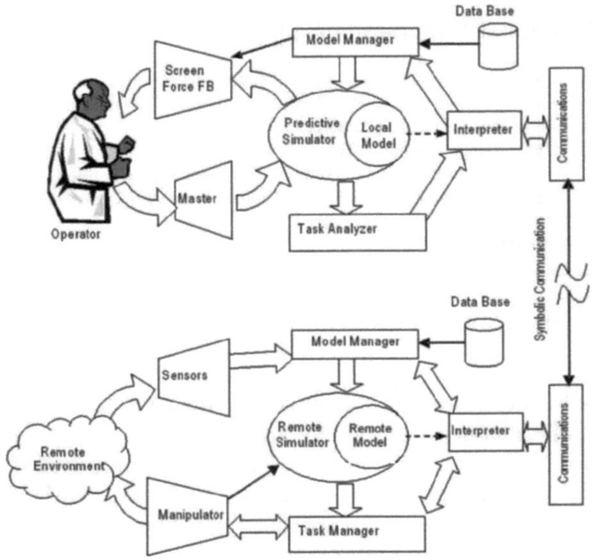

Fig. 18.2. General Teleprogramming Architecture

In the normal performance of the proposed architecture the operator works in a predictive simulator through the input devices. A local model includes all the information about the environment, objects, and the manipulator. The Task Analyzer observes the operations that are being done by analyzing the evolutions of the objects represented in the model. As a consecuence, the operator actions are translated into a set of commands for the remote robot that are transmitted to the Interpreter. This component translates the absolute references of the commands to symbolic references and then pipes them through the communication system to the remote zone. (e.g.: a command like "*Move the TCP 5 cm over the plane 5 of the element 3*" will be translated to "*Move the TCP 5 cm over the upper and biggest plane of the object labelled as 'my box'*").

The remote system receives the commands and translates the symbolic references to absolute references of the remote model. This remote model is continuously updated by the perception system that is included in the Model Manager. The commands are then transmitted to the Task Manager, which is responsible of their execution in the remote system. During the execution the remote system reacts at environment variations due to the data obtained from the robot sensors and the continuously updated remote model.

If the local and remote models are equal, the Model Managers will be completely transparent. Moreover, if the remote model and the real world are identical, the Task Analyzer and the Task Manager are also transparent to the operator. In this situation, the architecture moves on to a common teleoperation system. Obviously this last condition implies a perfect perception system and the absence of time delay in the communication channel.

To cope with possible execution errors on the remote system, a feedback channel is necessary. This feedback allows closing the control loop between both zones. The main task of the feedback control loop, in the architecture, is to avoid future execution errors by modifying the local model with the information transmitted by the remote system during the normal operation. The simplest strategy is to generate an execution error when the remote system is not able to solve the consequences of the model discrepancies. In this case, the normal way of working of the system is interrupted, the local model is updated, and the operator have to repeat the robot movements from the exception point. A more difficult strategy is to modify the local model during the normal operation. Due to the time difference between the remote and the local system, these modifications have to be done taking into account the current actions of the human operator in the local system.

The three control loops of the developed architecture are shown in Fig. 18.3. It is important to underline the architecture ability of working even if the communication between the local and the remote system is interrupted.

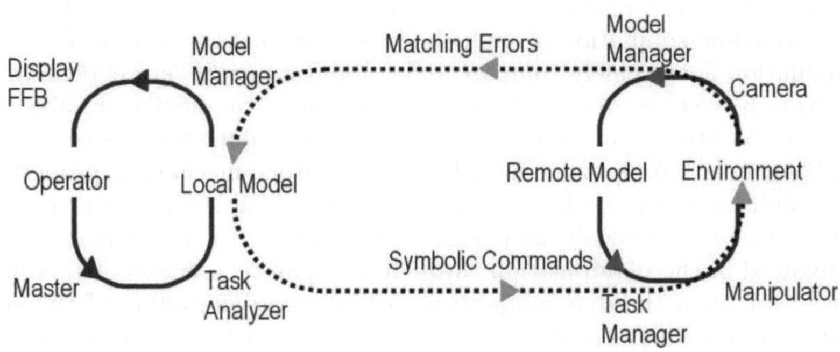

Fig. 18.3. Control loops of the teleprogramming architecture

18.3 Predictive Simulator

The Predictive Simulator is the mathematical and graphical apparatus that allows the interaction between the local model and the human operator. An specific aspect of the local model of a predictive simulator is the presence of different time references of the represented objects. Let suppose that time T is the time delay of the communication channel and that t_0 is the current time. The objects of the environment that are not directly affected by the actions of the operator only could be updated by the remote sensorial measures. This means that they are updated with an interpretation made in the remote system in the moment $t_0 - T$. On the contrary, the elements that are directly affected by the actions of the operator are immediately updated. Given that the local model is a representation of the remote environment, these immediately updated elements -e.g.: the virtual robot- are representations of the future that will take place in the remote system in $t_0 + T$. Therefore, in the same model coexist representations of what has happen in the remote system (variations of the environment), and predictions of what will happen in the future (the movement of the manipulator and its effects). If the remote environment remains almost without variations during the execution of the task, a valid strategy will be not to update the information about the remote environment in the local model.

The main inputs to the predictive Simulator are the motion references for the robot that is being teleprogrammed. These references could be position commands or velocity commands either in the configuration space or in the cartesian space.

The outputs in a predictive simulator have to be the same than the outputs to the operator in a common teleoperated system. Therefore, the visual feedback is especially important and many studies had been carried out on this topic since the initial steps on teleprogramming [17]. The other typical output from the simulator is the force that is being done during the task execution. In a teleprogramming system the knowledge about the contact forces is especially important for the interpretation of the operator's intention [16]. Nevertheless, the real time dynamic simulation of multi-body systems entails several drawbacks. Such simulation is extremely complex and computationally very expensive [21]. Currently, only systems of fixed topology and of low complexity are able to be simulated in real-time. Moreover, a lot of additional information is required for this kind of simulation: dynamic model of the actuators; mechanical properties of the robot links and objects; friction models for all possible one-to-one contacts.

For this reason, almost all the predictive simulators use an estimation of the forces involved in the manipulation through a kinematic analysis. This solution is less realistic than the dynamic approach but a correct tuning of the elastic constants (always lower) results in a set of force and torque references good enough to understand what is desired to make. In a Kinematic approach of the simulation of contact forces, the inputs are the values desired for the robot joints. In this case there are two different simulation modes:

- *Force-position approach:* The operator reacts to the virtual contact forces and as a consequence stops the motion of the virtual robot. The virtual robot penetrates the collided object and the contact forces are computed as proportional to the overlapped volumes between the object and the robot. Although this approach is more realistic and easier to implement, this scheme is more unstable than the position-position approach [22]. (Fig. 18.4.a.)
- *Position-position approach:* The volume overlapping is not allowed. The force is proportional to the error between the commanded position and the position achieved by the virtual robot. A specific implementation of this approach is explained in 18.3.1. The resultant scheme is more stable, and faster to compute than the previous one.(Fig. 18.4.b.)

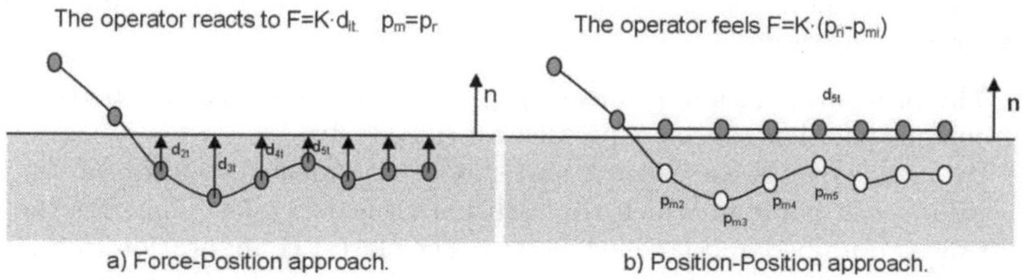

Fig. 18.4. Modes for the simulation of contact forces based on the kinematic approach.The filled circles represent the virtual robot position while the empty circles represent the commanded robot position.

An efficient collision detection algorithm and a very fast distance computation subsystem is needed for any of these approaches. Complex environments and/or exhaustive object descriptions are not feasible with the force-position scheme due to the complexity of the overlapping computation. Furthermore, the recommended updating frequency is 1 KHz [23]. Therefore the computation of the force have to be done and updated in the master device in 1 ms. This is the main reason for the selection of the second scheme in the developed predictive simulator. In order to generate the virtual forces based on the position error between the master device and the simulated robot, it is necessary to decouple the master reference and the virtual slave position. Therefore the virtual robot is able to consider during the motion the restrictions imposed by the presence of obstacles in the environment. The way in which the obstacles affect the movement of the virtual slave is the base of the correct operation of the algorithm, and it is exposed in detail next.

18.3.1 Algorithm for the Restriction of the Virtual Slave Motion

Let consider a single control cycle. The desired increase in the robot position is sampled in n steps to make sure that possible intermediate states of collision are detected. By initial conditions of the algorithm the robot configuration is free of collision. According to this, the objective robot configuration is represented by:

$$\mathbf{q}_{obj} = \mathbf{q}_{ini} + n \cdot \Delta\mathbf{q} \tag{18.2}$$

where $\Delta\mathbf{q}$ is a vector of joint increments that added n times to the current robot configuration $\mathbf{q}_{ini}$, moves the robot to the objective position $\mathbf{q}_{obj}$.

These increments are added to the starting point in a consecutive way until reaching the goal position. If no collision is detected, there will be no error between the commanded position and the achieved position of the virtual robot. Nevertheless, in case of detecting collision in some of the steps in which the trajectory is divided, the following algorithm is executed in order to obtain the virtual robot position:

(i) The previous position is named $\mathbf{q}_{ant}$ and is free of collision. Therefore, the current position can be expressed as: $\mathbf{q}_{ant} + \Delta\mathbf{q}$.
(ii) By means of a binary search and supposed a linear evolution of the joint values, the point in which the collision situation takes place for the first time is obtained and labelled as $\mathbf{q}_{col}$. The closest configuration to $\mathbf{q}_{col}$ that is free of collision is obtained in the same process and it is labelled as $\mathbf{q}_{free}$.
(iii) In $\mathbf{q}_{col}$ the point $\mathbf{p}$ of the surface of the virtual robot where occurs the contact is obtained. The same point but belonging to the object that is colliding with the robot determines the vector $\mathbf{n}$ perpendicular to the object surface.
(iv) The vector $\mathbf{D}$ is computed. It represents the displacement of the point $\mathbf{p}$ that take place if the robot moves from $\mathbf{q}_{ant}$ to $\mathbf{q}_{ant} + \Delta\mathbf{q}$.:

$$\mathbf{D} = \mathbf{p}(\mathbf{q}_{ant} + \Delta\mathbf{q}) - \mathbf{p}(\mathbf{q}_{ant}) \tag{18.3}$$

As a function of $\mathbf{D}$ a new vector $\mathbf{u}$ is defined as:

$$\mathbf{u} = \mathbf{D} - 1.1(\mathbf{D} \cdot \mathbf{n}) \cdot \mathbf{n} \tag{18.4}$$

Due to the 1.1 factor, this vector is almost parallel to the object surface but slightly oriented to the free space.
(v) The Jacobian Matrix ($\mathtt{J}$) for the point $\mathbf{p}$ is computed, but only considering the robot joints that affect the movement of $\mathbf{p}$. Therefore, if i is the number of considered joints, and given that $\mathbf{p}$ is a three-dimensional point, then $\mathtt{J} \in \mathtt{R}^{3\times i}$.

The system have to approximately compute the vector of joint increments $\boldsymbol{\delta}$ that moves the point $\mathbf{p}$ of the virtual robot, to the point $\mathbf{p} + \mathbf{u}$. Therefore, the value of $\boldsymbol{\delta}$ have to satisfice the equation:

$$\mathbf{u} = \mathtt{J} \cdot \boldsymbol{\delta} \tag{18.5}$$

(18.5) is easy to solve if $\mathtt{J}$ is a square and non-singular matrix and therefore invertible. Nevertheless, in this case, usually it is not invertible, because the number of degrees of freedom affected by the collision use to be more than three (e.g.: a robot of six degrees of freedom touching an object with the clamp), and in some cases lower than three (e.g.: if the first link or the second is colliding).

This problem is solved by means of the pseudoinverse matrix having originated multiple works to simplify its obtaining or to deal with the special cases that this computation generates. In addition to the complexity of its computation, it is necessary to identify singular situations and to deal with them. For that reason a simpler and robust numerical method has been used in the present algorithm. It is important to underline that the solution $\boldsymbol{\delta}$ is a part of a bigger algorithm and therefore the numerical method is faster but not as accurate as the pseudoinverse approximation.

In order to obtain the vector of increases, the exposed method will try to diminish its module. For that, three robot joints are selected in a consecutive way, forming with the corresponding columns of the jacobian matrix J a new invertible square matrix $\tilde{\mathtt{J}} \in \mathtt{R}^{3\times3}$, that allows the computing of $\boldsymbol{\delta}$. These joints are selected in the following way:

a) The first selected joint (index k) is the joint that causes a displacement of the point $\mathbf{p}$ closest to $\mathbf{u}$:

$$\left\|\mathbf{u}^t \cdot \mathbf{J_k}\right\| = \max_{r=0\ldots i}\left(\left\|\mathbf{u}^t \cdot \mathbf{J_r}\right\|\right) \tag{18.6}$$

where $\mathbf{J_r}$ is the r-column vector of the matrix J.

b) The second column of $\tilde{\mathtt{J}}$, labelled with the sub index l, is selected among the remained columns of J and obtains the movement of $\mathbf{p}$ more perpendicular to the one determined by the column k previously selected. Therefore this column fulfils:

$$\|\mathbf{J_l} \times \mathbf{J_k}\| = \max_{\substack{r=0\ldots i \\ r\neq k}}\left(\|\mathbf{J_r} \times \mathbf{J_k}\|\right) \tag{18.7}$$

c) Finally the column m that is more perpendicular to the two previous ones is selected:

$$\left\|(\mathbf{J_l} \times \mathbf{J_k})^t \cdot \mathbf{J_m}\right\| = \max_{\substack{r=0\ldots i \\ r\neq k,l}}\left(\left\|(\mathbf{J_l} \times \mathbf{J_k})^t \cdot \mathbf{J_r}\right\|\right) \tag{18.8}$$

There is a lower threshold for each selection. Therefore it is possible that no index k, m and l can be selected. The algorithm count the number of successful selections made. If only two columns have been selected, a third column is computed as the vectorial product of the first and the second column.

In the normal case, the matrix $\tilde{\mathtt{J}}$ is constructed with the union of the three selected columns :

$$\tilde{\mathtt{J}} = (\mathbf{J_k}|\mathbf{J_l}|\mathbf{J_m}) \tag{18.9}$$

If only two columns have been selected, the matrix $\tilde{\mathtt{J}}$ would be:

$$\tilde{\mathtt{J}} = (\mathbf{J_k}|\mathbf{J_l}|\mathbf{J_k} \times \mathbf{J_l}) \tag{18.10}$$

(vi) Obtained the matrix $\tilde{\mathtt{J}}$ the vector $\boldsymbol{\delta}$ is computed through the inverse of this matrix. Being δ_i the increment of the value of the robot joint i, its value is computed as follows:

$$\delta_i = \begin{cases} 0 & i \neq k, l, m \\ (\tilde{\mathtt{J}}^{-1} \cdot \mathbf{u})\mathbf{i} & i = k \\ (\tilde{\mathtt{J}}^{-1} \cdot \mathbf{u})\mathbf{j} & i = l \\ (\tilde{\mathtt{J}}^{-1} \cdot \mathbf{u})\mathbf{k} & i = m \end{cases} \tag{18.11}$$

If only the index k could be selected, the following expresion will be used instead:

$$\delta_i = \sqrt{\frac{||\mathbf{u}||}{||\mathbf{J_k}||}} \quad i = k \tag{18.12}$$

(vii) Using again a binary search, the robot is moved towards the configuration given by:

$$\mathbf{q} = \mathbf{q_{ant}} + \boldsymbol{\delta} \tag{18.13}$$

If no collision is detected, the obtained configuration is asigned to $\mathbf{q_{ant}}$.

(viii) The value of the increment for each step is recomputed:

$$\Delta\mathbf{q} = \frac{\mathbf{q_{obj}} - \mathbf{q_{ant}}}{n} \tag{18.14}$$

This procedure is repeated for the n initially defined steps. Fig. 18.5 shows the effect of this algorithm in the movement of a point in a two dimensional example. The destination point (black) is located inside an object.

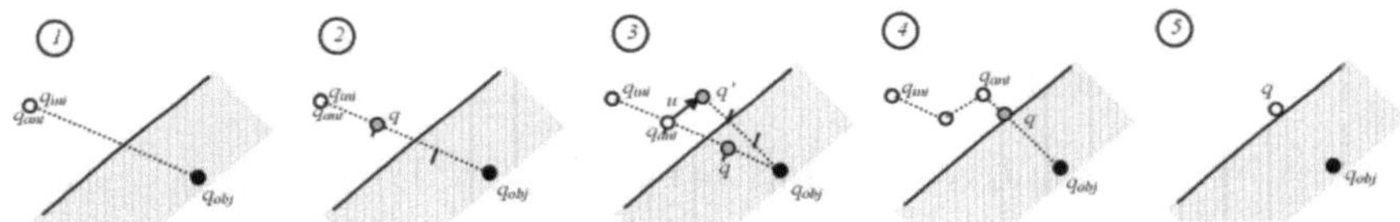

Fig. 18.5. Algorithm of restriction of the virtual slave motion

Finally the force reflected to the operator through the master device is easily computed by:

$$f_{tcp} = K \cdot (\mathbf{p}(\mathbf{q}) - \mathbf{p}(\mathbf{q_{obj}})) \tag{18.15}$$

Fig. 18.6 shows the algorithm behaviour for the surface and edge following. The inner and simpler line represents the movement made by the master, while the blue one is the trayectory described by the TCP of the virtual robot. The vectors represent the computed contact forces during the execution of the movement. This experiment was made with a PHANToM master device teleoperating a virtual Puma 560 robot.

18.4 Task Analyzer

The most critical part of a teleprogramming architecture is the direct control of the task. The Task Analyzer and the Task Manager modules are responsible of this part. Both elements are virtually communicated and can be seen as an unique structure. The proposed control architecture is based on the Atlantis Architecture [24], that makes use of the Subsumption Architecture in the lower actuation level.

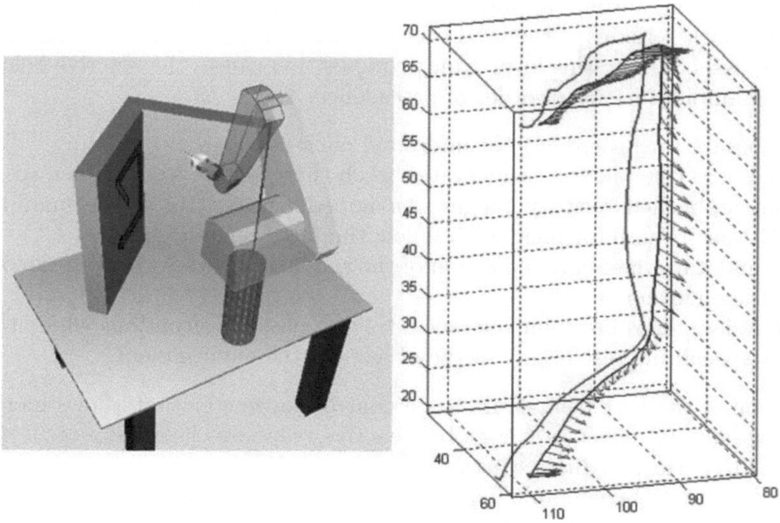

Fig. 18.6. Experiment for testing the behaviour of the Algorithm

The main features of the proposed control architecture are:

- Four actuation layers.
- The information managed by a layer is based on the commands and error values of the adjoining lower layer.
- Each level is able to detect and recognize possible failures that will be communicated to the adjoining upper layer.
- All the requests are sent to the adjoining upper level, acting always in this hierarchical way.
- The four layers act asynchronously and independently, allowing different actuation cycles: level I forms the reactive layer, level II forms the planning layer that controls the operations sequence, level III forms the task layer that uses spatial reasoning to achieve the objectives, and finally the level IV forms the learning layer.

The most advanced and tested control layer is the reactive layer. Indirectly this layer gives the basic information for the identification of the operator's intention and for the description of the robot movement. Therefore a more detailed description of this part of the architecture is described next.

18.4.1 The Reactive Layer

The main goal of this level is to provide a fast reaction capacity to the unexpected events of the remote environment . The control loop must be fast. In the experiments the outputs are updated at 40 Hz. This frequency is limited by the specific characteristics of the used robot. The main objectives of the reactive behaviour associated to this level, which composition causes the reactive behaviour of the manipulator, are in order of precedence:

- To assure that there are not undesired contacts.
- To assure that the forces generated when there is a desired contact are inside the established security ranges. An impedance control of the manipulator TCP has been implemented to follow the force references.
- To maintain the relative position of the robot tool with respect to the objects of the environment.
- To follow the robot joint references solving possible situations where there is not special interest in any of the objects of the environment.

Each one of the four behaviours has a control parameter that allows increasing or decreasing its effect in the global reactive response. For instance, it will be possible to perform a direct control of the manipulator by means of a cancellation of the first three behaviours. This way of working is shown in Fig. 18.7. One of the more interesting elements of the reactive layer is the Relative Following behaviour. This algorithm is the main responsible of the adaptative ability to changes of the remote environment. Together with the other basic behaviours, a non explicit interpretation of the operator intention is obtained. As it was previously explained, any basic behaviour consists of an analyzer and a follower. For each iteration of the main local cycle, the nearest objects to the robot TCP are analyzed. Since the distance calculation is computationally very expensive, a proximity algorithm, based on evolving boxes, has been used. When an object is selected, the relative position between the object and the TCP is analyzed and

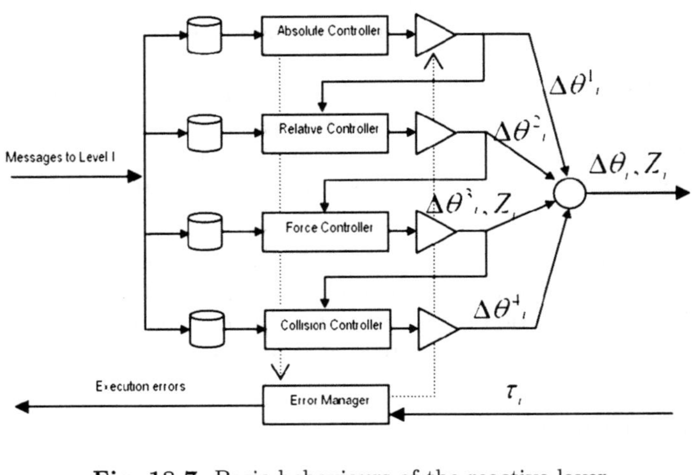

Fig. 18.7. Basic behaviours of the reactive layer

transmitted to the remote zone. This relation is obtained considering only the geometric features of the object, not using any previously defined information. Such a way of working allows the remote system to obtain the relative position of the TCP from the data obtained from the perception system.

18.4.2 The Relative Analyzer

Selected one object, the relative analysis is a quantification of the relation between the TCP position and the planes, edges and vertexes of the object. Only the planes of the object in witch the TCP is located in front of the outer face are analyzed. For these planes an interest index I_p is computed:

$$I_p = \frac{dist_p^{tcp}}{\sqrt{A_p}} \tag{18.16}$$

This relation has been experimentally obtained, after trying different formulas combining the distance to the plane $dist_p^{tcp}$ and the plane area A_p. A lower value of I_p represents a higher interest. Based on the second order geometric moments and the geometric centre of mass a new coordinate system is defined for the plane. The geometric moments are used to determine the asymmetry feature of the plane. Specifically, if a plane has the 2nd order geometric moment of one axis 3/2 times higher than the moment of the other axis, the feature of asymmetry is defined. If this property is defined, the relative position between the TCP and the plane is defined through the coordinates of the TCP respect the self-defined reference system. Otherwise, the distance to the gravity centre and the distance to the plane are used to quantify the relative position. Independently, the most interesting edge is obtained through a closest distance criterion. The edge is defined through the vectors $\mathbf{v_1}$ and $\mathbf{v_2}$ of both extremes. The relative position is

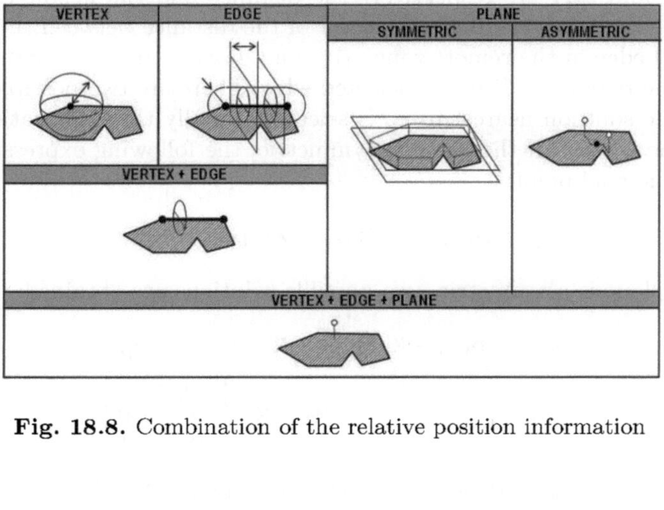

Fig. 18.8. Combination of the relative position information

quantified through the edge axis projection a of the vector that links the TCP with the middle point of the edge:

$$a = \left| \left(\mathbf{p_{tcp}} - \frac{\mathbf{v_1} + \mathbf{v_2}}{2} \right) \cdot \frac{\mathbf{v_2} - \mathbf{v_1}}{|\mathbf{v_2} - \mathbf{v_1}|} \right| \tag{18.17}$$

Finally for the vertex, the interest index and the relative position are obtained directly from the distance to the TCP. Fig. 18.8, shows the relative position obtained through the combination of the information extracted from one plane - symmetric or asymmetric -, one edge and one vertex.

18.4.3 The Relative Follower

Relative Follower in the remote zone proposes a goal point for the TCP based on the relative information sent by the relative analyzer of the local zone. The point proposed due to the vertex information $\mathbf{p_v}$ is obtained first:

$$\mathbf{p_v} = \mathbf{p_{tcp}} + (|\mathbf{p_{tcp}} - \mathbf{v}| - \mathbf{d_{ref}}) \cdot \frac{\mathbf{p_{tcp}} - \mathbf{v}}{|\mathbf{p_{tcp}} - \mathbf{v}|} \tag{18.18}$$

where $\mathbf{v}$ is the vertex identified by the interpreter of the remote zone, and $\mathbf{d_{ref}}$ is the distance received from the relative analyzer of the local zone. In the same way, a goal point is obtained from the information relative to an edge:

$$\Delta\mathbf{p_d} = \mathbf{v_d} \cdot (d - d_{ref}) \quad E_d = |d - d_{ref}| \tag{18.19}$$

$$\Delta\mathbf{p_a} = (a - a_{ref}) \cdot \frac{\mathbf{v_2} - \mathbf{v_1}}{|\mathbf{v_2} - \mathbf{v_1}|}) \quad E_a = |a - a_{ref}| \tag{18.20}$$

$$\mathbf{p_a} = \frac{\Delta\mathbf{p_a} \cdot E_a + \Delta\mathbf{p_d} \cdot E_d}{E_a + E_d} \tag{18.21}$$

where d_{ref} and a_{ref} are respectively the distance and the height transmitted by the analyzer, $\mathbf{v_d}$ is the director vector of the distance between the TCP and the selected edge in the remote zone $v_1\bar{v}_2$, and d and a are the distance and the height of the remote TCP to the selected edge. There are two possible solutions but only the solution nearest to p_v is selected. Finally the information relative to the plane is used. If the plane is symmetric, the following expression is used to obtain the goal point:

$$\mathbf{p_p} = \mathbf{p_{tcp}} + \mathbf{z} \cdot (\mathbf{z_{ref}} - \mathbf{z} \cdot (\mathbf{p_{tcp}} - \mathbf{O_p})) \tag{18.22}$$

But if the plane is asymmetric, two possible solutions are obtained:

$$\mathbf{p_{p,1}} = \mathbf{p_{tcp}} + \mathtt{M}^t \left(\mathbf{v_{ref}} - \mathtt{M}^t \cdot (\mathbf{p_{tcp}} - \mathbf{O_p}) \right) \tag{18.23}$$

$$\mathtt{M}_2 = \mathtt{M} \cdot \begin{bmatrix} -1 & 0 & 0 \\ 0 & -1 & 0 \\ 0 & 0 & 1 \end{bmatrix} \tag{18.24}$$

$$\mathbf{p_{p,2}} = \mathbf{p_{tcp}} + \mathtt{M_2}^t \left(\mathbf{v_{ref}} - \mathtt{M_2}^t \cdot (\mathbf{p_{tcp}} - \mathbf{O_p}) \right) \tag{18.25}$$

were M is the orientation matrix of the self defined plane reference system. The solution with the lowest index C_i is then selected:

$$C_i = \|\mathbf{p_{p,i}} - \mathbf{p_v}\| + \|\mathbf{p_{p,i}} - \mathbf{p_v}\| \tag{18.26}$$

The goal position for the remote TCP is obtained through a weighted combination of $\mathbf{p_p}$, $\mathbf{p_a}$ and $\mathbf{p_v}$:

$$\mathbf{p_c} = \frac{E_{cp} \cdot \mathbf{p_p} + E_{ca} \cdot \mathbf{p_a} + E_{cv} \cdot \mathbf{p_v}}{E_{cp} + E_{ca} + E_{cv}} \tag{18.27}$$

where E_{cp} is defined for the plane as:

$$E_{cp} = |\mathbf{p_p} - \mathbf{p_{tcp}}| \cdot I_p \tag{18.28}$$

where I_p is a confidence index of the identification of the plane done by the remote zone interpreter. E_{ca} and E_{cv} are obtained in the same way for the edge and the vertex respectively.

18.5 Results

The first two levels of the exposed teleprogramming architecture have been tested using virtual and real experiments. Four types of virtual experiments have been used to adjust the behaviour performance. In a first type of experiments, a 9 DOF virtual manipulator have to follow the perimeter of a box when there is a discrepancy of the box position in the local and remote model. In a second type of experiments, the box is not only moved but also rotated. Two of these experiments are shown in Fig. 18.9. In a third type of experiments, the teleprogramming system has been able to maintain the relative position of the manipulator with respect to a box even if the box is moved in the remote environment while performing the task. The collision avoiding behavior has been successfully tested in a fourth experiment using the box as an obstacle when the manipulator tries to follow the table perimeter. If the box is moved during the

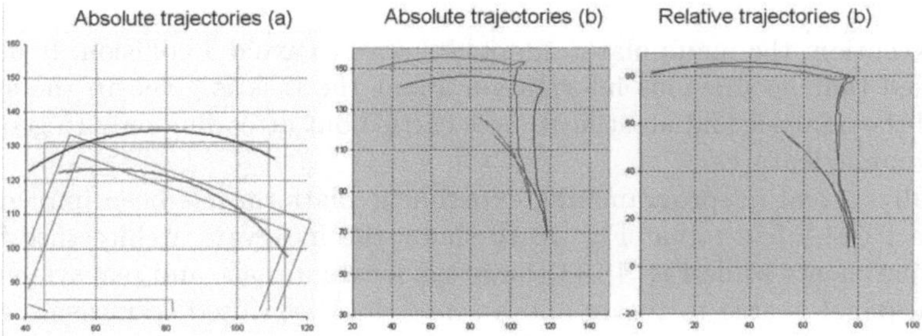

Fig. 18.9. Absolute and relative trajectories of two teleprogramming experiments (a) and (b)

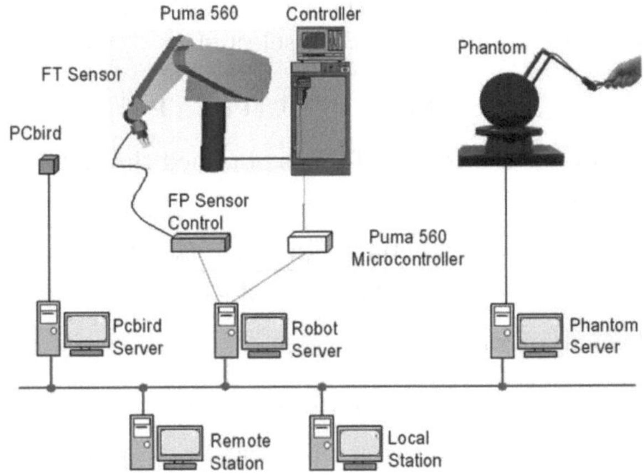

Fig. 18.10. Hardware architecture for the real experiments

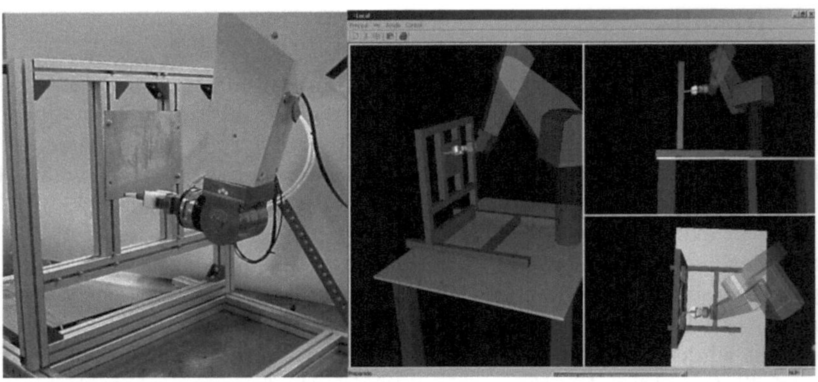

Fig. 18.11. Remote Zone (left) and Predictive Display (right) of the real teleprogramming experiments

task execution, the manipulator adapts its pose to avoid a collision. It must be remarked that no previous information about the task is given to the system, and all the analysis and identification is carried out attending only to geometric and dynamic features.

Finally, a real teleprogramming experiment platform has been implemented using a PUMA 560 robot. Fig. 18.10 shows the hardware architecture for the real experiments while Fig. 18.11 shows the local interface and real system. The vertical panel is able to rotate and a sensor has been used to measure the rotated angle. This sensor is affected with the proximity of metal, and therefore its measure has not a good accuracy value (+/- 2cm). Therefore, the force control behavior becomes fundamental for the correct execution of the task. In the experiments with the real system, the operator must maintain contact with a surface or must have punctual contacts in previously planned positions. Several letters and symbols have been painted on the small rectangular panel positioned in the center of the moving structure. These simple tasks have been successfully tested with delays from 10 to 200 seconds and with different panel angles in the remote system.

18.6 Conclusion

In this chapter several issues related to teleprogramming control architecture for time-delayed teleoperated systems have been described. Some of the taken solutions for solving problems that arise during the implementation of this technique have been presented. More specifically, the algorithm implemented for the real time generation of virtual contact forces in the predictive simulator is showed in detail. The presented control architecture is based on the Cognitive Architecture Atlantis. Thus, a modular development of the whole system is allowed. Finally an indirect analysis of the intention of the movements performed by the human operator is explained. This analysis is carried out without any information about the task that is going to be performed. The lower levels of the exposed teleprogramming architecture have been tested using virtual and real experiments as is briefly described at the end of this chapter.

References

1. G. Hirzinger, B. Brunner, J. Dietrich and J. Heindl. Sensor-Based Space Robotics-ROTEX and Its Telerobotic Features. *IEEE Trans. on Robotics and Automation*, 9:649–663, 1993.
2. W.K. Yoon, T. Goshozono, H. Kawabe, M. Kinami, Y. Tsumaki, M. Uchiyama, M. Oda, and T. Doi. Model-Based Space Robot Teleoperation of ETS-VII Manipulator. *Transactions on Robotics and Automation*, Vol. 20:602–612, 2004.
3. I. Belousov I, S. Chebukov, and V. Sazonov. Web-based Teleoperation of the Robot Interacting with Fast Moving Objects. In *Proc. of the 2005 IEEE International Conference on Robotics and Automation*, 2005.
4. W.R. Ferrel and T.B. Sheridan. Supervisory control of remote manipulation. *IEEE Spectrum*, 1967.
5. G. Niemeyer and J. Slotine. Telemanipulation with time delays. *International Journal of Robotics Research* Vol. 23:873–890 , 2004.
6. T.B. Sheridan. Telerobotics and Human Supervisory Control. *The MIT Press*, 1992.
7. T.B. Sheridan. Space Teleoperation Through Time Delay: Review and Prognosis. *IEEE Transactions on Robotics and Automation* Vol. 9:592–606, 1993.
8. A.K. Bejczy, S. Venema, and W.S. Kim. Role of Computer Graphics in Space Telerobotics: Preview and Predictive Displays. In *Proc. SPIE Cooperative intelligent Robotics in Space* Vol. 1387:365–377, 1990.
9. A. Kheddar. Teleoperation Based on the Hidden Robot Concept. *IEEE Transactions on Systems, Man, and Cybernetics* Vol. 31:1–13, 2001.

10. T. Kotoku, K. Tanie, and A. Fujikawa. Force reflecting bilateral master-salve teleoperation system in virtual environment. In *Proc. Int. Symposium on Artificial Intelligence, Robotics and Automation in Space*: pages 295–298, 1990.
11. K. Machida, Y. Toda, and T. Iwata. Graphic-simulator-augmented teleoperation system for space applications. *Journal of Spacecraft and Rockets* Vol. 21:64–69, 1990.
12. Y.J. Cho, T. Kotoku, and T. Tanie. Discrete-Event-Based Planning and Control of Telerobotic Part-Mating Process with Communication Delay and Geometric Uncertinty. In *Proc. IEEE/RSJ/GI Int. Conference on Intelligent Robotics*, 1995.
13. A.K. Bejczy, P. Fiorini, W.S. Kim, and P. Schenker. Toward Integrated Operator Interface for Advanced Teleoperation under Time Delay. In *Proc. IEEE/RSJ/GI Int. Conference on Intelligent Robotics*, 1994.
14. C. Sayers. Remote Control Robotics. Springer-Verla, NewYork, 1998.
15. C.H. Spenny and D.L. Schneider. Object Resolved Teleoperation. In *Proc. IEEE International Conference on Robotics and Automation*, 1997.
16. T. Backmon, and L.W. Stark. Model-based Supervisory Control in Telerobotics. *Presence* Vol 5(2):205–233, 1996.
17. J.E. Lloyd, J.S. Beis, K.D. Pai, and D.G. Lowe. Model-based Telerobotics with vision. In *Proc. IEEE International Conference on Robotics and Automation*, 1997.
18. T. Burkert , J. Leupold, and G. Passig. A Photorealistic Predictive Display. *Presence* Vol. 13:22–43, 2004
19. M. Skubic and R.A. Volz. Learning Force-Based Assembly Skills from Human Demostration for Execution in Unstructured Environments. In *Proc. IEEE International Conference on Robotics and Automation*, 1998.
20. A. Kheddar, T. Tanie, and P. Coiffet. Detection of discrepancies and sensory-based recovery for virtual reality based telemanipulation systems. In *Proc. IEEE International Conference on Robotics and Automation*, 1998.
21. D. Baraff. Interactive simulation of solid rigid bodies. *IEEE Computer Graphics and Applications* 15:63-75, 1995.
22. S. Hasegawa and M. Sato. Real-time Rigid Body Simulation for Haptic Interactions Based on Contact Volume of Polygonal Objects. *EUROGRAPHICS* 23:529–538, 2004.
23. A. Gregory, M.C. Lin, S. Gottshalk, and R. Taylor. A framework for fast and accurate collision detection for haptic interaction. In *Proc. IEEE Virtual Reality Conference*, pages 38–45, 1999.
24. E. Gat. Integrating planning and reacting in a heterogeneous architecture for Mobile robots. *SIGART Bulletin* Vol. 2:17–74, 1991.
25. W.K. Yoon, S. Tachihara, Y. Tsumaki, and M. Uchiyama. Evaluation of the Different Master Device Approaches for a Model-Based Space Teleoperation System. In *Proc. International Conference on Advanced Robotics*, 2001.

Part III

Applications

19

DLR's Advanced Telerobotic Concepts and Experiments for On-Orbit Servicing

Detlef Reintsema, Klaus Landzettel, and Gerd Hirzinger

German Aerospace Center (DLR)
Institute of Robotics and Mechatronics
Oberpfaffenhofen, D-82234 Wessling, Germany
{Detlef.Reintsema,Klaus.Landzettel,Gerd.Hirzinger}@dlr.de

Summary. Space robotics will become a key technology for the exploration of outer space and the operation and maintenance of space stations, satellites and other platforms, saving costs and relieving man from dangerous tasks. But we do not have to wait until robots are really autonomous or intelligent, since by modern teleoperation and telepresence we are able to remotely control robot systems from the ground in the sense of "prolonging man's arm into space". Humans, with their several hundred thousand years of evolution, will not adapt themselves to the hostile space environment, whilst robots, which have only been developed for just over 40 years, can be much more easily adapted to such an environment. As presented within this work, few pioneering telerobotic experiments like *ROTEX*, the first remotely controlled space robot system, *ETS-VII*, the first free-floating space robot experiment, or *ROKVISS*, Germany's recent advanced space robot experiment on the International Space Station, have been proposed and conducted on the way towards a space robot assistant system for the usage as an artificial astronaut to perform On-Orbit Servicing (OOS) tasks.

19.1 Introduction

Although there exists a large application field for telerobotics in space, the ongoing worldwide telerobotic activities concentrate on three specific missions and application fields: *on-orbit assembly and servicing*, (science) *payload tending* and *planetary surface robotics*. Within these fields of application the advanced robotics tasks to be handled range from preparation and operational support of scientific experiments (both internal and external) on a space station, to inspection, maintenance, and repair of orbital infrastructure and systems, towards recovery and cleaning up space debrise, as well as exploration and coverage of planets and orbs in general.

In general, the design and deployment of space robotic components is caused by the specific manipulation and mobility aspects of the mission's requirements. For instance, the increasing number of launched satellites per year calls for solutions:

- to overcome malfunctions or system failures and keep space infrastructure operational, as well as extend the operational lifetime of ageing space systems and satellites like stations or telecommunication systems.

M. Ferre et al. (Eds.): Advances in Telerobotics, STAR 31, pp. 323–345, 2007.
springerlink.com © Springer-Verlag Berlin Heidelberg 2007

- to keep free operational space for telecommunication systems in geosynchronized orbit, as well as to avoid the endangering of space systems in LEO and of the public dwelling in the habited parts of Earth. Examples for such dangerous stranded space systems in the past are Skylab and MIR. In the future, the uncontrolled and accidental de-orbiting of other huge satellites is expected, where parts of these will hit the surface of the Earth.

Robotics scientists are developing teleoperated human-like robots for orbital servicing (*On-Orbit-Servicing*, OSS), to potentially eliminate the need for dangerous and expensive astronaut servicing. As the following sections will show, nowadays robotics technology is mature enough to act as a useful tool in supporting the astronaut during usual work. In addition, advanced robotics technology can relieve the human from servicing tasks to perform tasks best suited to human decision-making and flexibility that is unlikely in the near-term to be matched by autonomous or teleoperated robots. DLR's achievements on advanced telerobotics concepts in On-Orbit-Servicing within the last 20 years give an impression about the near future space robotics capabilities in case of On-Orbit-Servicing (see Sec. 19.4).

19.2 The Space Robotics On-Orbit Servicing Profile

When comparing human skills to those of present-day robots, of course human beings in general, are by far superior, but when comparing the skill of an astronaut in a clumsy space-suit with that of the best available robot technology, then the differences are becoming smaller. For intra-vehicular laboratory activities (IVA) e.g. a robot basically would have to be compared to the full human skills and mobility. To be honest, many of the manual operations to be done in a space-laboratory environment are fairly simple standard operations, like handling parts, opening and closing doors, pulling drawers, pushing buttons etc., which have to be done just by stepping through extensive, written procedures. Real intuition and manual skills are particularly requested in non-nominal situation and repair situations. Although it is not clear today when a multi-fingered robot hand might be as skilled as the human hand and when (if ever) a robot might show real intelligence and autonomy, it nevertheless is obvious that even with today's technology and the available telerobotic concepts based on close cooperation between man (e.g. the ground operator) and machine, there are many tasks in space, where robots can replace or at least augment human activities with reduced cost from a mid-term perspective.

Launch access to orbit, orbital and target inspection maneuvering are mandatory before a robot can serve for proximity operations like in-orbit assembly, maintenance of equipment, and replenishment of consumables towards system upgrade and repair (see Fig. 19.1) [4]. In general, robot servicing can be easily achieved through the exchange of modular components (Orbit Replaceable Units, ORU). To minimize the complexity of servicing tasks, ORUs are widely used to provide and serve for an easy module-based replacement concept. Major maintenance tasks are the re-supply of consumables (fluids, material and

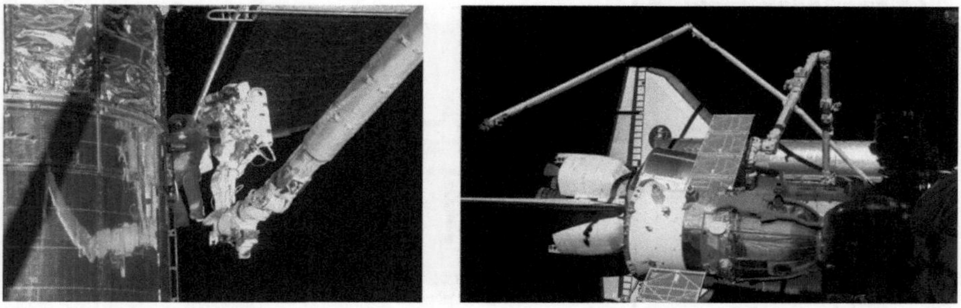

Fig. 19.1. Typical manipulation profiles to serve for by astronauts are associated with the assembly of space structures or satellite servicing missions as maintenance of the Hubble Space Telescope (left). In general, most of the servicing technology has been pushed from the International Space Station (ISS) programme. A number of cargo-handling manipulators are proposed to operate on the ISS as the European Robotic Arm (ERA), the Japanese Remote Manipulator System (JEMRMS), and even the US Mobile Servicing System (MSS) including the Space Station Remote Manipulator System (SSRMS, Canadarm2) (right). [Photos by courtesy of NASA].

data storage) to the target, and cleaning/resurfacing/decontamination charging. Diagnoses and correction of faults or failures are typically tailored by ORUs examinations and replacements.

The first general purpose robotic manipulator[1] designed for specific use in the harsh environment of space was the Shuttle Remote Manipulator System (SRMS, Canadarm). SRMS has successfully flown on over 70 Shuttle flights and has retrieved more than a dozen satellites [3]. The SRMS was the first robotic manipulator to demonstrate the principle of robotic servicing. The Canadarm2 and the original Canadarm (see Fig. 19.1, right side) could work together, handing payloads to one another.

19.3 Pioneering Orbital Telerobotics Experiments

Future On-Orbit-Servicing systems seem to be designed as unmanned spacecrafts which will be remotely operated by a human operator on the ground or an astronaut on board of an orbiter spacecraft or space station. In spite of the fact that all these proposed new systems are still in the design of preparational phase, a lot of know-how concerning the operation and the design of space robot servicer systems were gained during previous space robotics missions. Within the last 20 years three pioneering space robot experiments have been performed with the aim at demonstrating key robotics technology for unmanned On-Orbit-Servicing activities: **ROTEX** (1993) - the first remotely controlled space robot system,

[1] "Manipulator" and "robot arm" have similar concepts. "Manipulator" means a machine similar to a human arm that can do various operation, whereas "robot arm" has a nuance of an arm of a "robot" with certain degree of intelligence.

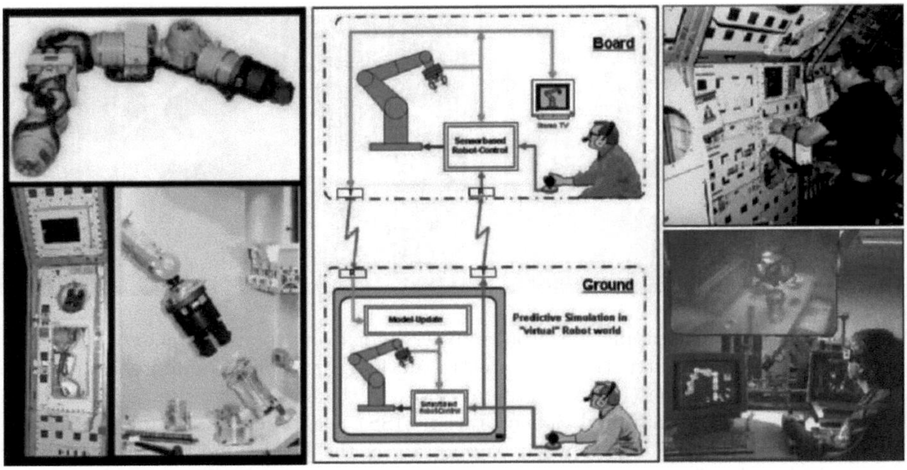

Fig. 19.2. The first remotely controlled robot in space, a multi-sensory robot on board of a shuttle worked in autonomous modes, teleoperated by astronauts, as well as in different telerobotic control modes by a human operator from ground

ETS-VII (1998) - the first free-flying space robot system, and **ROKVISS** (2005) - the first high-fidelity telepresence controlled high performance light-weight robot on the ISS.

19.3.1 ROTEX - The First Remotely Controlled Space Robot System

With the Spacelab-Mission D2, flown at the end of April 1993, the first remotely controlled robot arm was carried into the earth orbit inside the cargo of the Shuttle COLUMBIA. The advanced Space Robot Technology Experiment **ROTEX** was the beginning of Germany's participation in space automation and robotics. For the first time in the history of space flight a small, multi-sensory robot proposed and realized by DLR's Institute of Robotics and Mechatronics (DLR-RM) was conducted to demonstrate the feasibility of servicing prototype capabilities by a remotely controlled robot inside the Spacelab-Module. The basic goals of **ROTEX** had been (1) the verification of joint control (including friction motion) under zero gravity, (2) the evaluation of DLR's sensor-based 6 DoF hand-controller (one-hand operation) under zero gravity, and (3) the performance demonstration of a complex, multi-sensory robot system with powerful man-machine-interfaces, in a variety of operational modes, including on-line teleoperation and off-line programming from the ground (see Fig. 19.3) [11].

ROTEX was operated within an enclosed workcell, which was integrated into a rack of the Spacelab-D2 Module (left side of Fig. 19.2). Key operational modes for the remote robot control were performed from both, an on-board workstation, integrated into another rack, and from DLR's ground control center in

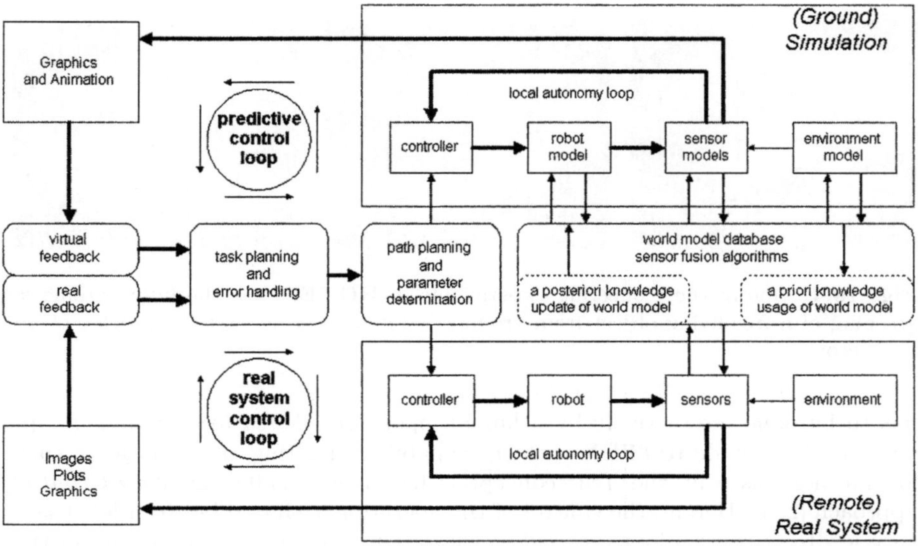

Fig. 19.3. The telerobotic concepts of **ROTEX**: *shared local autonomy*, *shared control* and optional switching between *task level*, *sensor-based tele-programming* and *on-line teleoperation* using predictive graphics (see also Fig. 19.12)

Oberpfaffenhofen (right side of Fig. 19.2). The defined workcell was prepared to demonstrate the most different applications not only by restricting the performed prototype tasks to internal servicing operations, but also aiming at assembly and external servicing (e.g. grasping a floating satellite) as assembling a mechanical truss structure from identical cube-link parts, connecting/disconnecting an electrical plug, to simulate an ORU exchange using a bayonet closure mechanism, and grasping a floating object.

The **ROTEX** manipulator was a small robot arm built up by six joints to be able to reach in all directions and grasp objects within the enclosed workcell. The multi-sensory gripper of **ROTEX** was equipped with a number of sensors, including two 6-axis force-torque wrist sensors, located at the back of the gripper to prevent the robot from overloading, a gripping assembly, containing an array of nine lasers for distance-measuring, tactile sensors and grasping force control. A tiny pair of stereo television cameras within the gripping assembly gave a direct view of the manipulated object. In addition, a fixed pair of video cameras provided stereo images of the whole workcell.

The basic operational modes of **ROTEX** based on a unified *shared local autonomy* control concept, that distributes intelligence between the operator and the teleoperator in the sense of a task-directed approach, termed as *tele-sensor-programming* (TSP) approach (compare Fig. 19.3) [13, 8]. Presuming that sufficient information about the actual environment is available from sensors, partial tasks can be executed independently on the machine level. Local sensory feedback control loops are executed by the robot system, while global task planning

Fig. 19.4. One of the pioneering experiments in **ROTEX** was the fully automatic grasping of a free-flying object from ground despite a varying round-trip-delay of up to 7 seconds

has to be done interactively by a human operator (*shared local autonomy* approach). To provide **ROTEX** with the capability of acting and reacting in such an autonomous way, the TSP concept is based on a sufficient *shared control* approach [2, 1]. Herein, the control of the operation is shared between local sensory feedback control loops, closed at the robot's site (i.e. on-board and in the predictive ground simulation), and gross commands, which are generated by the robot control system of the operator. Basic or gross commands are refined autonomously, providing the robot with a modest kind of sensory intelligence [33], and may be originated from a human operator, handling the control/sensor ball, or alternatively from an intelligent path planning module [12].

Even if the essential telerobotic control concept of **ROTEX** had used the integrated sensors, on-board sensory feedback was restricted to force-torque and range finder signals, only due to on-board processor limitations. In general, feedback to the human operator during on-line teleoperation was provided via the visual system (stereo video images). In addition, to deal with time delays of up to 5-7 seconds, a powerful delay-compensating predictive 3D-stereo-graphic simulation of the workcell and the robot, including the robot's sensory behavior, had provided the human operator on ground.

The most considered experiment was the autonomous catching of a free-floating object (see Fig. 19.4). This experiment may be treated as a precursor mission to the long-term goal of capturing a non cooperative, tumbling satellite in free space (see also TECSAS 19.4.1). Due to the lack of space proven image processing hardware, the control loop of the capturing experiment was closed via an image processing system on the ground. The current total signal round trip time was determined during experiment execution and regarded by the prediction of the free floating parts position. This principle may be directly applied to nowadays satellite repair missions in order to keep the required hardware effort for the space craft as low as possible [12, 15].

19.3.2 ETS-VII Satellite - The First Free-Floating Space Robot

In November 1997, the 7th satellite of the ETS satellite series, proposed and realized by the Japanese space agency JAXA (formerly NASDA), was launched.

Fig. 19.5. The **ETS-VII** satellite was designed to verify key technologies on autonomous rendezvous, docking (left) and robotics technologies in space (right). To experiment the rendezvous-and-docking maneuver *HIKOBOSHI* had opened the docking mechanism holding *ORIHIME* and started to separate from each other at the low speed of 2cm/sec. After separation *HIKOBOSHI* and *ORIHIME* flew in formation for several minutes at a constant distance. On command *HIKOBOSHI* started approaching *ORIHIME* along the flight direction (1cm/sec) and captured *ORIHIME* with the docking mechanism [45]. [Photos by courtesy of JAXA].

The **ETS-VII** satellite was designed to verify key technologies on autonomous rendezvous, docking and robotics technologies in space [41, 49, 50]. To demonstrate an autonomous rendezvous and docking maneuver, the **ETS-VII** system (see Fig. 19.5, left) was composed of two unmanned spacecrafts, a servicer satellite (*Hikoboshi*) and a smaller client subsatellite (*Orihime*). The first free-floating space robot arm was mounted on the Earth pointing surface of the servicer satellite to perform several teleoperation experiments.

The robotics experiment system of **ETS-VII** (see Fig. 19.5, right) was composed of a 6 DoF robot arm and various payloads, designed for prototypical teleoperated experiments like manipulating a simulated ORU [43, 42, 47]. To handle payloads like small equipments, or to grasp a floating object like the target satellite, specific tools as the taskboard or the target satellite handling tool were used. Up to six CCD cameras were installed on board to monitor the **ETS-VII** system, especially the telerobotics experiments. A pair of hand cameras were used as an on-board vision sensor for visual servicing control of the robot arm.

The **ETS-VII** robot arm provided three kinds of control modes: a joint angle/velocity control mode, a compliance control mode (including force control, active limp and impedance control) and an arm tip position/attitude control mode. To evaluate the performance of the robot's compliance capabilities, a peg-in-hole task had been experimented. Errors, which occurred in position and attitude of the peg, were corrected by a compliance controller, a local sensory feedback control loop approach similiar to the one of **ROTEX**.

A time delay of approximately seven seconds had requested telerobotic concepts like *shared control*, *predictive display* and *visual guidance* as **ROTEX**

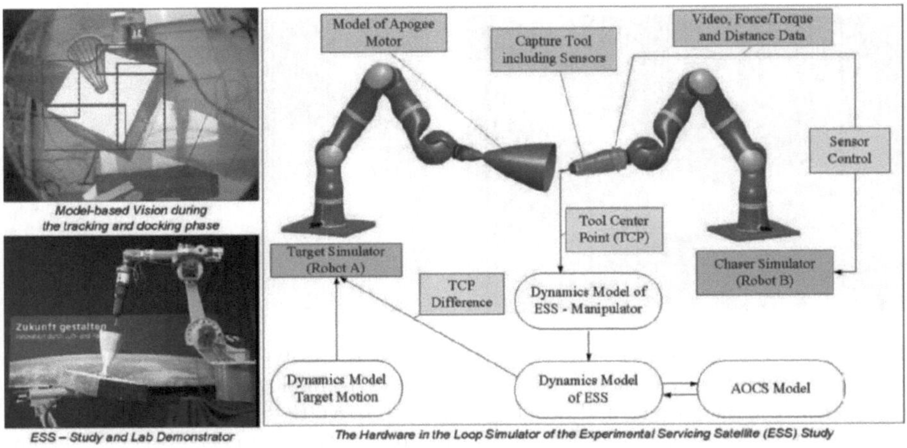

Fig. 19.6. In the course of the ESS study, DLR-RM had investigated the dynamical behavior of a free-flying servicing satellite (carrying a robot arm) to find out the influence of the robot arm motion to the pose (position & orientation) of the servicer satellite [26, 27, 37]. A hardware-in-the-loop (HIL) simulator, which mainly consist of a two-robot system and the relevant dynamic models of the involved subsystems, was developed allowing the simulation of the mission critical final approach and capturing phases, taking the robot's motion into consideration.

had already introduced and proved, to enable and to support (tele-)operation controlled from the ground. Thus, it is obvious to use similar concepts within **ETS-VII** and to distinguish between an *automatic tele-programming mode* and a *telemanipulation mode* to command and operate the robot arm from the ground. In general, during telemanipulation experiments, the robot arm was manually controlled by a set of two hand controllers, to separate the translation and rotation commanding.

The advanced telerobotics experiments of **ETS-VII** demonstrated the first bilateral ground-space teleoperation, limited due to a large time delay [44, 50, 51, 52]. Several experiments, as a slope tracing task and a peg-in-hole task were teleoperated by a model-based bilateral control approach, using a 6DoF haptic interface as master. Task performance was compared between the bilateral mode and the unilateral mode with force telemetry data, visually displayed on the screen. As experienced with the **ETS-VII** system, kinesthetic force feedback to the operator is helpful to improve the performance of the task even under such a long time delay.

19.3.3 GETEX - The German ETS-VII Technology Experiment

The invitation to participate in the Japanese **ETS-VII** project led up to the verification of the dynamical considerations and lab experiments of the nationally funded *Experimental Servicing Satellite* (ESS) study under real mission conditions (see Fig. 19.6). In April 1999, DLR-RM received the permission from JAXA to remotely program and control the **ETS-VII** robot from the ground control station in Tsukuba/Japan [24]. The so-called **GETEX** (*GErman Technology EXperiment*) project, performed in cooperation with the Institute of Robotics Research (IRF) at the University of Dortmund [16], was very successful as was the whole **ETS-VII** mission. The **ETS-VII** robot was operated

- to verify the dynamic models for the interaction between a robot arm and its free-flying carrier [19]. The results gained from this crucial experiment represent an important step towards a free-flying service satellite [21, 22].
- to verify the sensor-based task-level programming approach TSP, including on-board autonomy via selected image features and force-torque information;
- to demonstrate a world model update approach, using real video images from space, and relating them with virtual images from the 3D CAD model.

To gain a reasonable amount of experimental data for the dynamics verification, the major part of the **GETEX** experiment time was allocated to dynamic motion experiments, which consisted of a series of motions, carried out by the manipulator while the attitude control system of the **ETS-VII** carrier was *switched off*.

In general, a robot, who is mounted on a spacecraft, generates linear and angular momentum onto the spacecraft while moving around. Without any attitude control the satellite position remains uncontrolled. Thus, the attitude control system will permanently produce forces and torques, to compensate the arm motion disturbances. The spacecraft may then be considered as inertial in the co-ordinates of an orbit-fixed system, and the problem of robot motion planning can be solved, using the same methods as for terrestrial, fix base manipulators. Due to the linear momentum conservation, which states that the center of mass of the system comprising the robot and the satellite is constant, the motion of a manipulator mounted on the satellite will lead to a compensating motion of the satellite. The amount of satellite translation produced, depends on the masses of the bodies, constituting the system.

For space robotic systems which are neither position nor attitude controlled, the angular momentum conservation law leads further to a rotation of the spacecraft by an amount, which results from the mass and inertia properties of the manipulator links and the spacecraft. It is generally assumed that no external forces act on such free-floating robots [38, 39]. The free-floating mode of operation is of interest for space robots not only for the reason that attitude control fuel may be saved, it will also be of importance during repair missions, when the servicing satellite is very close or in contact to the target satellite: any action of the attitude control system of either of the two satellites during this phase could lead to a collision and thus to a potential damage on the two spacecrafts.

As long as the tasks, performed by the robot, are described in robot-fixed coordinates, the fact that the satellite position remains uncontrolled has no influence. If, however, the task is described in relation to an orbit-fixed co-ordinate system, as for example, capturing of a defect satellite, the satellites motion has to be taken into account (see Fig. 19.7).

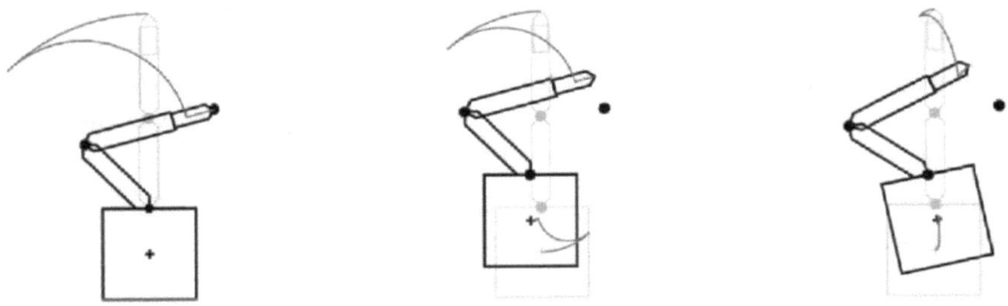

Fig. 19.7. The influence of the satellite attitude control mode on the path described by the robot end-effector - the same joint motion is carried out by a robot with a fixed base (left), an attitude controlled robot (middle) and a free-floating robot (right)

19.3.4 ROKVISS - High Performance Light-Weight Robotics in Space

Space represents an extremely harsh operational environment for robotics technology, that limits the on-orbit life and performance capability of space robotics technologies, in particular, the electronics components due to radiation surroundings. The break for (low-cost) intelligent space robotics technology was the absence of innovative, high-performance and survivable electronics space components. Even if radiation-hardened (rad-hard) circuit versions are available, their tolerance levels are not always compatible with the space requirements. So far, the technology for space robotics applications has to deal with bulky and most expensive rad-hard components, which limits the performance and capability of space robots, especially of telerobotics applications in space. The ongoing telerobotics experiment **ROKVISS** - *Robotics Component Verification on the International Space Station (ISS)* - was prepared to demonstrate the feasibility of innovative, high-performance, and survivable electronics components for a new kind of intelligent robotics joint technology, as required for On-Orbit-Servicing applications.

ROKVISS was proposed to verify and demonstrate the performance and capabilities of DLR's modular light-weight, torque-controlled robotic joints under real space conditions [17]. Composed by a lot of common electronics components, which are identical to those used in DLR's seven joint light weight robot [34, 31], a latch-up protection power supply circuit was designed and integrated into the robotics joint electronics to prevent burn out of the robotics joint hit, and hence to protect the integrated sensors and common industrial electronics components. Thus, **ROKVISS** investigates the functions and capabilities of robotic joints, and will identify their dynamic and friction behavior over a long time duration, with the aim to get rid of bulky and most expensive rad-hard components for space application in favor of highly integrated circuits [17].

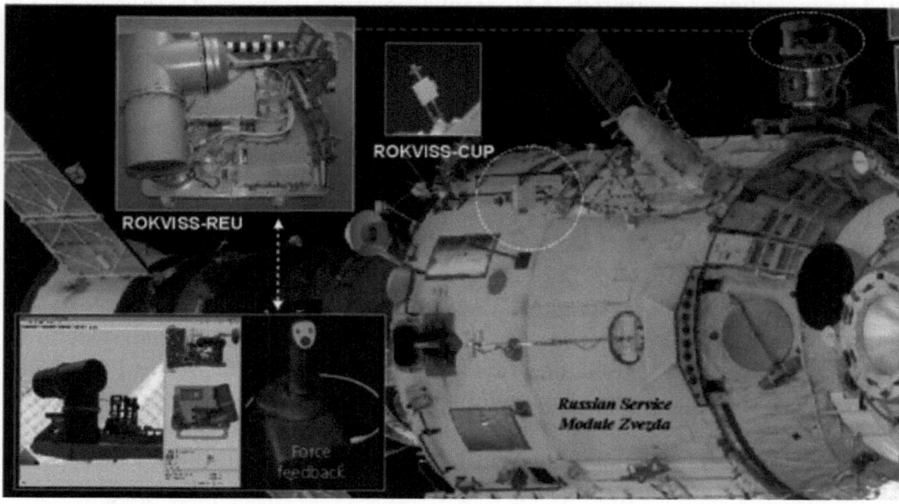

Fig. 19.8. At the end of a 5 hours space walk the astronauts succeeded in mounting the external flight unit (REU) and S-Band antenna (CUP) of ROKVISS, and connecting the necessary cables to the on-board controller. [ISS-Photo by courtesy of NASA].

ROKVISS is the second space robot mission after **ROTEX**, proposed by DLR's Institute of Robotics and Mechatronics[2]. During a space walk in January 2005, the external flight unit (REU) and a dedicated S-Band antenna were mounted outside at Zvezda, the Russian Service Module of the ISS (Fig 19.8). Inside the service module, CUP was connected to the on-board controller (OBC) of the REU. The REU of ROKVISS (see right-hand side of Fig. 19.10) is composed by a small robot arm with two torque-controlled joints, a power supply, and a video supply to control a stereo camera, an earth observation camera, and an illumination system. A mechanical contour device is placed within the working environment of the robot arm to verify the robot's functions and performance. Since February 2005, ROKVISS has been operated by DLR-RM in close collaboration with ZUP, the ISS ground control station in Moscow. After one year of successful operation the **ROKVISS** hardware is still working, and the mission will be prolongated for another year.

For the first time in space robotics a direct radio link contact between on-ground and on-board controllers is used for ground-controlled teleoperation instead of a delayed inter-orbit communication link, as realized within **ROTEX** or **ETS-VII**, who used a tracking and data relay satellite [11, 41]. Thus, the OBC of ROKVISS has direct access to CUP, a dedicated S-Band communication

[2] ROKVISS was realized by DLR's Institute of Robotics and Mechatronics (DLR-RM) in cooperation with the German space companies EADS-ST, Kaiser-Threde, and vHS (von Hrner & Sulger) and close collaboration of the Russian Federal Space Agency ROSKOSMOS and RKK Energia.

system including a separate boom antenna pointing to the earth (Fig 19.8). On the earth the **ROKVISS** ground controller (OGC) has also a direct access to a transceiver system (Cortex Data EGSE) of DLR's tracking station in Weilheim [17]. To get access to the OBC, at first there must be a direct radio link established. The use of a direct radio link limits the access time to the on-board system down to time windows of 8 minutes length, when the ISS passes through the tracking space of the ground antenna.

Deviant of past space robotics experiments, the **ROKVISS** communication protocol is served by OBC/OGC instead of using external support equipment. Both are responsible for (de-)modulation of **ROKVISS**-data into frames, which are transferred over the S-Band channel. In spite of ESA requirements the transfer frame protocol is compliant to the CCSDS telemetry/telecommand standards, which have been tailored, as described in [17], to get rid of protocol overhead, and to meet the specific real-time requirements of the telepresence mode (a proposed sample rate for transfer of 500 Hz and a jitter of at most 1 ms). In general, the dedicated S-band communication link provides an overall data rate of 256 kbit/s for telecommand (uplink-channel) and 4 Mbit/s for telemetry (downlink-channel) data, including 3,5 Mbit/s for videodata. Due to a lean, dedicated point-to-point protocol implementation, the measured round-trip times are less than 20ms depending on the position of the ISS.

During one year of operation the robot joints have been extensively tested and identified (dynamics, joint parameters) by repetitively performing predefined robot tasks in an automatic mode, or even by direct operator interaction. For the design of the joint level controller (position, torque, and impedance control capabilities) efficient and reliable autonomous off-line identification methods were developed, which allow the stiffness, damping and friction parameter identification at joint level. Starting with the model and corresponding identification measurements, a modified time-efficient, on-line identification procedure based upon Recursive Least Squares algorithms is used for identification. To investigate how the space environment affects the behavior of two interacting bodies, the **ROKVISS** experiment contour is used to measure the energy dissipation occurring during intermittent impact events, as well as the frictional forces, acting between two bodies while they are moving w.r.t. each other in a lasting contact situation.

The telerobotic concept of **ROKVISS** is based on DLR-RM's *shared control / shared local autonomy* techniques (see Sec. 19.3.1) as introduced with **ROTEX** and improved as MARCO supervisory control concept within **GETEX** [10]: all predefined tasks of **ROKVISS** can be executed by sending a path or a force trajectory to the on-board system. Feedback to the man-machine-interface (MMI) of the MARCO telerobotic ground control station is provided via the on-board camera system and the system's housekeeping data. Due to limited direct link experiment time, the following automatic mode experiments have been performed (independent of direct radio link): Predefined trajectories (a) without force contact, (b) with force contact (i.e. contour tracking or movement against spring load) and (c) with a change from non-contact to contact

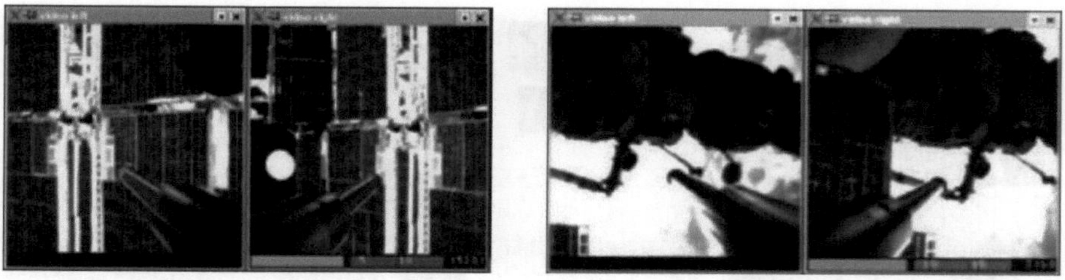

Fig. 19.9. Stereo video images taken by the integrated cameras of ROKVISS showing the robot arm pointing at the solar panel (left) and the Soyuz (right), docked to the ISS

condition (contact dynamics experiment). In contrast to the automatic mode experiments, the teleoperation experiments are conducted via direct operator interaction (see ROTEX 19.3.1) instead of time-line-based mission activation.

Another important issue of **ROKVISS** is to verify a high-fidelity force-reflecting telemanipulation concept, based on the dedicated high-speed direct radio link, to show the feasibility of immersive telepresence methods for future satellite servicing tasks. Differently to the teleoperation mode the human operator is "immersively" included into the control loop consuming stereo video images (like Fig. 19.9) in conjunction with the measured robot joint / torque values, which are fed back to the MMI on the ground (see Fig. 19.10). The **ROKVISS** robot is controlled by the human operator using DLR-RM's force-feedback joystick to generate force and position commands, to drive the robot joints into the desired state. To cope with longer and varying time-delays, sophisticated bilateral control schemes are in use. Among common bilateral control schemes new approaches as time domain passivity and wave-variables, which compensate the measured or simulated time-varying delay due to the orbit of the ISS, were developed to investigate the stability, efficiency and feasibility of these bilateral control approaches under real mission conditions [7, 28, 29, 30].

19.3.5 DLR's Telerobotics Technologies for On-Orbit Servicing

The experience gained with **ROTEX** laid an excellent basis for projects and studies on space robotics. **ROTEX** and **ETS-VII** were pure experimental systems, to show and improve the principles of ground control under time delay constraints. The **ROKVISS** technology experiment investigates innovative advances in space robotics and demonstrates current capabilities in state-of-the-art A&R technology, demonstrating the principle readiness of space robots for On-Orbit-Servicing applications. As the previous space robotics experiments showed, the performance needed by telemanipulation requires a hierarchically and modularly structured shared automation concept tunable to the special operational case, which, in addition, allows human interference on different levels of supervisory and decision control [18, 20, 23].

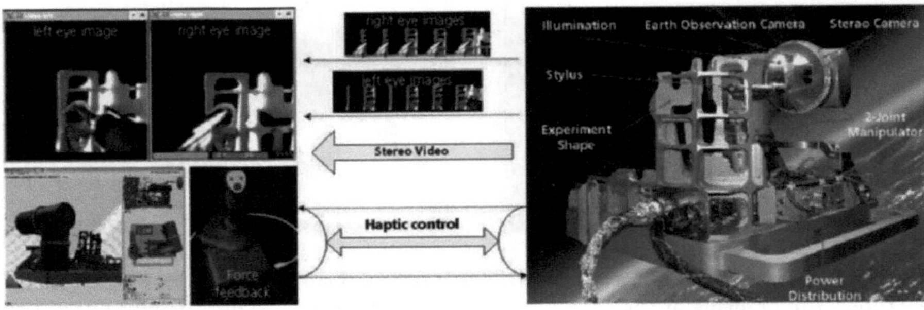

Fig. 19.10. High-fidelity telepresence control: during direct radio link contact the ROKVISS manipulator can be commanded by an operator on ground. A stereo camera is mounted on the 2nd joint: the stereo video images, together with the current robot joint and torque values, are fed back in real-time to an operator at the ground station where the operator controls the manipulator via a force-feedback-control device.

DLR's Light-Weight Robot Technology

Service robotics requires robots, which are able to manipulate objects in an unknown, changing environment autonomously, as far as possible, or by a human in interactive supervisory control. Basic requirements have to be met by the arms and hands, which differ substantially from industrial manipulators: space and service robotics require arms of light weight to meet the needs of human friendly interaction as well as for mobility and safety reasons. A load to weight ratio of 1:1 similar to that of the human arm is desired. Interaction with unknown environments needs compliant arms and fingers, facilitated by the information of many different sensors and advanced control strategies. To integrate different multi-sensory components as arms, hands, and maybe mobile platforms into mobile robotics applications, sophisticated mechatronic concepts and an adequate, flexible architecture of the control system are needed. To provide arm and hand systems that meet these requirements is one goal of a long term project at DLR-RM. Especially, the last years, our focus on space robotics was caused by strong considerations, how to push robotic technologies towards space servicing applications. We developed a new generation of light weight robots with an unbeatable weight to load ratio as well as impressive control features, which make the system easy-to-use and safe for space as well as terrestrial servicing applications [17, 31, 34].

The design philosophy of DLR's light weight robot is to achieve a manipulator similar to the kinematic redundancy of the human arm, with a load to weight ratio of better than 1:2, a total system-weight of less than 20 kg for arms with a reach space of up to 1,5 m, no bulky wiring on the robot (and no electronics cabinet as it comes with every industrial robot), and a high dynamic performance. As all modern robot control approaches are based on commanding, joint torques, joint torque control, allowing programmable impedance, stiffness and damping, has been a must for our light-weight robotics concept since the beginning. Also,

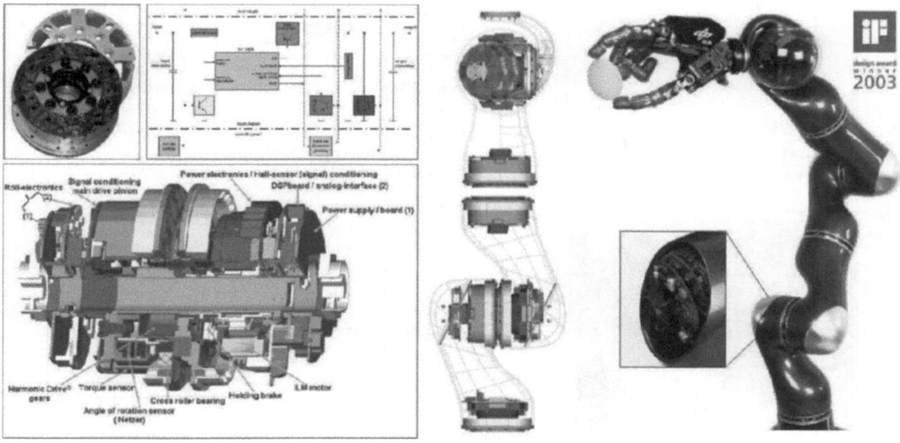

Fig. 19.11. The DLR's operational light-weight robot technology (right) is based on a modular drive concept (bottom left) with integrated torque sensors (top left); The latch-up protection power supply circuit (top middle) were designed and integrated into the robotics joint electronics for space applicability

the use of precise motor position sensing and link angular sensing has been considered as absolutely essential.

The current robot arm concept (see Fig. 19.11) aimed at a completely modular assembly system with only a few basic components concerning joint mechanics, electronics and links. It is based on a fully modular joint-link-assembly system, with only a few basic components, namely three one-dof robot joint-link types and a two-dof wrist joint. This modularity concept was supported by SIMPACK, a powerful kinematics-dynamics analysis and design software. In general the modularity concept gives a number of advantages, e.g. rotation symmetric components, few single parts, short force transmission from bearing to off-drive connection, identical design for pitch and roll joints, big hollow shaft in all joints with up to 30 mm diameter, enabling the placement of cables and plug links inside the arm.

In addition to the mechanical modularity, the robot electronics was also designed in a modular manner. Each joint unit contains the power electronics for the motor and a motor current controller board, analog and digital signal processing hardware for the sensor signals and a DSP board for decentralized joint control. A DC-DC converter board provides the different voltages required by the joint electronics. A high-speed optical serial bus (SERCOS) connects the joints with the central computer. The only additional external connections are the power supply wires for the electronics and the motor.

Telerobotic Control and Telemanipulation Concept

Satellite repair in LEO or GEO should be performed under direct control of a human operator, due to the fact, that such missions can not be prepared in detail

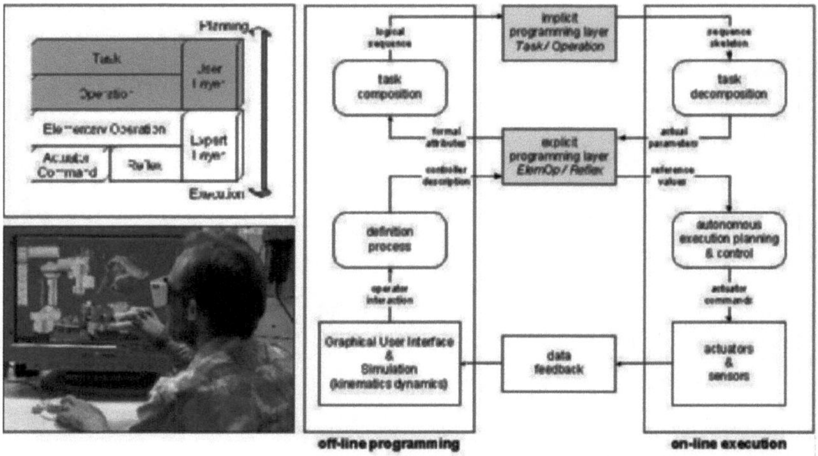

Fig. 19.12. The programming and execution layers of MARCO.

off-line on-ground. On the other hand, planetary exploration missions must be performed by a task-directed programming and control system, which requires intelligent sensor processing methods, sophisticated planning algorithms, and robust FDIR (failure detection, isolation and recovery) techniques.

DLR has developed MARCO (*Modular Automation and Robotics Controller architecture*) [9], which integrates both worlds, direct telemanipulation as well as off-line programming, combined with features for autonomous execution. The goal for the development of this high-level programming system was to design a unified concept for a flexible, highly interactive, on-line programmable teleoperation station, also usable as an off-line programming tool, which includes all the sensor-based control features as introduced and already tested in ROTEX [14], but in addition, providing the opportunity to program a robot on an implicit, task-directed level.

For that reason MARCO is based on a *2in2-layer-concept* (see Fig. 19.12, top right), which represents the hierarchical control structure from the planning to the executive layer. On the implicit (Payload Expert) level the instruction set is reduced to what has to be done. No specific robot actions will be considered at this task-oriented level. On the other hand, the robot system has to know how the task can be successfully executed, which is described in the explicit (Robotic Expert) control layers. In particuälr, each layer offers special *services / operations* to the layer immediately above. The overall process chain from planning and programming down to execution is shwon in Fig. 19.12.

19.4 DLR's Envisaged On-Orbit Servicing Applications

The previous orbital telerobotic experiments have evaluated and demonstrated key robotics technologies for prototypical On-Orbit Servicing tasks. For real

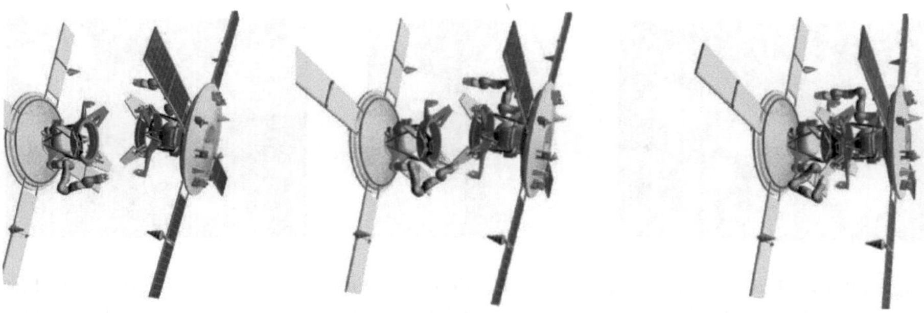

Fig. 19.13. The capturing approach of the floating, non-cooperative target system: (1) Close approach phase, (2) capture of the target by chaser with the aid of manipulator, and (3) pulling along guide elements of Docking Units and fixation with the help of locks of Docking Units

On-Orbit-Servicing as a business case the feasibility of these technologies have to be demonstrated first by realistic, most common servicing application scenarios as extending the ageing of a functional satellite or repairing a malfunctioned one.

19.4.1 TECSAS - Technology Demonstration for On-Orbit Servicing

A German-Russian space robotics project is going on: the **TECSAS** (*TEChnology SAtellite for demonstration and verification of Space systems*) project aims at the in-orbit qualification of the key robotics elements (both hardware and software) for advanced space maintenance and servicing system, especially w.r.t. docking and robot-based capturing procedures. It is planned, in close cooperation with the Russian Babakin Space Center, to perform such sensitive operations as rendezvous and close approach maneuvers, which will be necessary for further servicing activities. The Russian Multi-Purposes Orbital Boost Platform, as the base module for the mission, provides the insertion into initial parking orbit, as well as supports rendezvous and docking maneuvers. For docking and capturing operations we prefer to use our own robotics means (manipulator, controller, etc.), including MARCO to provide teleoperation and supervisory control from ground.

The entire mission will be performed utilizing the following steps: far rendezvous, close approach, inspection fly around, formation flight, capture, stabilization and calibration of the compound as well as compound flight maneuver, manipulation of the target, active ground control via telepresence, passive ground control during autonomous operations (monitoring), and controlled de-orbiting of the compound. For the capturing of the floating target satellite, the control modes will be applied as developed in the ESS study and verified during the **ETS-VII** mission. Even a high-fidelity telepresence mode as demonstrated within **ROKVISS** could support the critical phases of the capturing process. After capturing, the manipulator can be used as an active damping system for compound

Fig. 19.14. The capturing approach of a floating, cooperative target system: (1) Far approach phase, (2) close approach phase with a relative motion of zero between the target and the servicer, and (3) compound satellite system after docking

stabilization issues: since the platform, including the robot system, and the captured target system build one compound system, the dynamic behavior of the complete system can be influenced by moving the robot. Also for de-orbiting maneuvers the manipulator can be used as a support system: it can be considered as a passive link, building the mechanical interface between the chaser and the target system. The geometry of the compound system can be influenced by the robots attitude. Additionally, the robot can be considered as an active link, controlling the thrusters vector pointing to the common center of mass.

19.4.2 CX-OLEV - The First Step Towards On-Orbit-Servicing as Business Case

The priority of the **TECSAS** experiment concerns to the technological aspects of space robot servicing. A technology satellite for demonstration and verification of elementary robot-based on-orbit servicing tasks is already in preparation, to verify and to demonstrate the servicing of (non-)cooperative targets. At the same time a less complex robotics servicing task, concerning cooperative targets only, is on the way to become the first real business case towards space robotics OOS: attaching a "tugboat" to a satellite, whose propellant is declining; the lifetime of valuable telecommunication satellites could be prolonged for several years. Telecommunications satellites typically cost at least $250 million - and they are designed for an average useful on-orbit life of 10-15 years. Once their on-board propellant load is depleted, the satellites are boosted into a disposal orbit and decommissioned, even though their revenue-generating communications relay payloads continue to function.

Our industrial partner Orbital Recovery Corp. has initiated its *ConeXpress-Orbital Live Extension System* (**CX-OLEV**), which will significantly prolong the operating lifetimes of these valuable telecommunications satellites. The **CX-OLEV** system will operate as an orbital tugboat, supplying the propulsion, navigation and guidance to keep a telecommunications satellite in its proper orbital slot for many additional years. Another application of **CX-OLEV** could be the rescue of a spacecraft placed in a wrong orbit by their launch vehicles, or which became stranded in an incorrect orbital location during positioning maneuvers. The system is designed to easily mate with all telecommunications satellites now in space or on the drawing boards.

It will rendezvous with the telecommunications satellite and link up using our proprietary docking device that connects to the telecommunication satellite's apogee kick motor, as we have proposed within the ESS technology study (see Sec. 19.3.3). Apogee kick motors are used by nearly every telecommunications satellite for orbital boost and station-keeping, and they provide a strong, easily accessible interface point for the **CX-OLEV** linkup that is always within the satellite's center of gravity. DLR-RM's contributions to the **CX-OLEV** system are closely related to the experiences gained during previous telerobotic experiments:

- to design and develop a Capture Tool, including locking mechanism, sensors and control software, similiar to the ESS capture tool (see Sec. 19.3.3);
- to provide a telemanipulation software to guide the Capture Tool, which is mounted to the **CX-OLEV** system, into the apogee-motor of the target satellite;
- to deploy and perform the Ground Control System and capture strategies.

19.5 Conclusion

As previous pioneering telerobotic experiments showed, the achievements and deployments in Automation and Robotics (A&R) during the last 20 years have reached a profession, which can be used in near future to support, to unburden, even to replace the human by safe critical or monotonous work as a matter of routine. It seems that in future space robotics will become a key technology for the exploration of outer space and the operation and maintenance of space stations, satellites and other platforms, saving costs and relieving man from dangerous tasks. Within future spaceflights and on-orbit servicing missions, intelligent robotics technology will be used for assisting in and carrying out different tasks inside spacecrafts or space laboratories (internal usage) as well as in free space (external usage). But we do not have to wait until robots are really autonomous or intelligent, since by modern teleoperation and telepresence we are able to remotely control robot systems from the ground in the sense of "prolonging man's arm into space".

Ongoing development of space robotics technology has to focus on intelligent, sensor-controlled, light-weight robots using modular, flexible grippers, articulated hands and tool systems for high versatility. Even improved man-machine interfaces for teleoperation and supervisory control concepts have to be deployed for efficient cooperation and coordination of multi-arm and multi-robot system. But in particular, concerning the manipulative skills, present hardware and software concepts, requested to cover the range from high-fidelity telepresent manipulation up to (partly) autonomous operation is still available for the step towards the first business case in On-Orbit-Servicing.

References

1. S. Hayati and S.T. Venkataraman. Design and Implementation of a Robot Control System with Traded and Shared Control Capability. In *Proc. of the IEEE Int. Conf. on Robotics and Automation (ICRA)*, 3:1310–1315, 1989.
2. L. Conway, R. Volz, and M. Walker. Tele-Autonomous Systems: Methods and Architectures for Intermingling Autonomous and Telerobotic Technology. In *Proc. of the IEEE Int. Conf. on Robotics and Automation (ICRA)*, 2:1121–1130, 1987.
3. D. King. On-orbit servicing business; an industry vision. In *Proc. of the 1st Bilateral DLR-CSA Workshop on On-Orbit Servicing of Space Infrastructure Elements via Automation & Robotics Technologies (OOS 2002) "Defining a Way Forward"*, Cologne, Germany, 2002.
4. A. Ellery. Handling technology: technology of robotic in-orbit servicing. In *Proc. of the 1st Bilateral DLR-CSA Workshop on On-Orbit Servicing of Space Infrastructure Elements via Automation & Robotics Technologies (OOS 2002) "Defining a Way Forward"*, Cologne, Germany, 2002.
5. J.-C. Piedboeuf. On-Orbit Servicing and Beyond: A Canadian Perspective. In *Proc. of the 1st Bilateral DLR-CSA Workshop on On-Orbit Servicing of Space Infrastructure Elements via Automation & Robotics Technologies (OOS 2002) "Defining a Way Forward"*, Cologne, Germany, 2002.
6. J. Andary and P. Spidaliere. Development test flight of the Flight Telerobotic Servicer: design description and lessons learned. In *Proc. of the IEEE Int. Conf. on Robotics and Automation (ICRA)*, 2:1121–1130, 1993.
7. J. Artigas, C. Preusche, and G. Hirzinger. Wave Variables based Bilateral Control with a Time Delay Model for Space Robot Applications. In *Robotik 2004, VDI-Bericht, München, Germany*, 1841:101–108, 2004.
8. B. Brunner, K. Arbter, and G. Hirzinger. Task Directed Programming of Sensor Based Robots. In *Proc. of the IEEE/RSJ Int. Conf. on Intelligent Robots and Systems (IROS)*, 2:1080–1087, 1994.
9. B. Brunner, K. Landzettel, B.-M. Steinmetz, and G. Hirzinger. Tele-Sensor-Programming - A task-directed programming approach for sensor-based space robots. In *ICAR - The 7th International Conference on Advanced Robotics*, Sant Feliu de Guixols, Catalonia (Spain), 1995.
10. B. Brunner, K. Landzettel, G. Schreiber, B.-M. Steinmetz, and G. Hirzinger. A universal Task-Level Ground Control and Programming System for Space Robot Applications - The MARCO Concept and its Application to the ETS-VII Project. In *Proc. of the 5th International Symposium on Artificial Intelligence, Robotics and Automation in Space (iSAIRAS), ESTEC*, Noordwijk, The Netherlands, pages 507–514, 1999.
11. G. Hirzinger, K. Landzettel, J. Heindl, and J. Dietrich. ROTEX - The First Robot in Space. In *The 5th European Symposium on Space Environmental Control Systems and 24th International Conference on Environmental Systems (ICES)*, Friedrichshafen, Germany, 1994.
12. Ch. Fragerer and G. Hirzinger. Predicitve Telerobotic Concept for Grasping a Floating Object. In *International Federation of Automatic Control, Spacecraft Automation and On-Board Autonomous Mission Control*, Darmstadt, Germany, 1992.
13. G. Hirzinger, B. Brunner, J. Dietrich and J. Heindl. Sensor-Based Space Robotics - ROTEX and Its Telerobotic Features. In *Proc. of the IEEE Int. Conf. on Robotics and Automation (ICRA)*, 9:649–663, 1993.

14. G. Hirzinger. Multisensory Shared Autonomy and Tele-Sensor-Programming - Key Issuses in Space Robotics. In *IAS - 3 International Conference in Intelligent Autonomous Systems*, Pittsburgh, PA, USA, 1993.
15. G. Hirzinger, K. Landzettel, and Ch. Fagerer. Telerobotics with large time delays - the ROTEX experience. *In Proc. of the IEEE/RSJ Int. Conf. on Intelligent Robots and Systems (IROS)*, München, Germany, 1994.
16. E. Freund and J. Rossmann. Space Robot Commanding and Supervision by means of Projective Virtual Reality: The ERA Experiences. In *Proc. of the 7th Conference on Telemanipulator and Telepresence Technologies*, pages 312–322, 2000.
17. G. Hirzinger, K. Landzettel, D. Reintsema, C. Preusche, A. Albu-Schäffer, B. Rebele, and M. Turk. ROKVISS - Robotics Component Verification on ISS. In *Proc. of the 8th International Symposium on Artificial Intelligence, Robotics and Automation in Space (iSAIRAS)*, München, Germany, 2005.
18. G. Hirzinger, B. Brunner, R. Lampariello, K. Landzettel, J. Schott, and B.-M. Steinmetz. Advances in Orbital Robotics. In *Proc. of the IEEE Int. Conf. on Robotics and Automation (ICRA)*, pages 898–907, 2000.
19. G. Hirzinger, B. Brunner, R. Lampariello, K. Landzettel, G. Schreiber et al. A Unified Ground Control and Programming Methodology for Space Robotics Applications - Demonstrations on ETS-VII. In *Proc. of the International Symposium on Robotics (ISR 2000)*, pages 422–427, 2000.
20. G. Hirzinger, K. Landzettel, B. Brunner, M. Fischer, C. Preusche, D. Reintsema, A. Albu-Schäffer, G. Schreiber, and B.M. Steinmetz. DLR's Robotics Technologies for On-Orbit Servicing. In *Advanced Robotics, Special Issue on Service Robots in Space (1)*, 18:139–174, 2004.
21. R. Lampariello and G. Hirzinger. Freeflying Robots -Inertial Parameters Identification and Control Strategies. In *Proc. of ASTRA 2000 - The 6th ESA Workshop on Advanced Space Technologies for Robotics and Automation*, Noordwijk, The Netherlands, 2000.
22. G. Hirzinger, K. Landzettel, B. Brunner, I. Schaefer, M. Fischer et al. DLR's Robotics Lab - Recent developements in Space Robotics. In *Proc. of the 5th International Symposium on Artificial Intelligence, Robotics and Automation (iSAIRAS)*, Noordwijk, The Netherlands, 1999.
23. K. Landzettel, B. Brunner, R. Lampariello, C. Preusche, D. Reintsema, and G. Hirzinger. System Prerequisites and Operational Modes for On-Orbit-Servicing. In *The International Symposium on Space Technology and Science (ISTS)*, Miyazaki, Japan, 2004.
24. K. Landzettel, B. Brunner, K. Deutrich, G. Hirzinger, G. Schreiber, and B.-M. Steinmetz. DLR's Experiments on the ETS VII Space Robot Mission. In *Proc. of the 9th International Conference on Advanced Robotics (ICAR)*, Tokyo, Japan
25. R. Lampariello, S. Agrawal, and G. Hirzinger. Optimal Motion Planning of Free-Flying Robots. In *Proc. of the IEEE Int. Conf. on Robotics and Automation (ICRA)*, 2003.
26. B. Rebele, R. Krenn, and B. Schäfer. Grasping Strategies and Dynamic Aspects in Satellite Capturing by Robotic Manipulator. In *Proc. of ASTRA 2002 - The 7th ESA Workshop on Advanced Space Technologies for Robotics and Automation*, Noordwijk, The Netherlands, 2002.
27. K. Arbter, J. Langwald, G. Hirzinger, G.Q. Wei, and P. Wunsch. Proven Techniques for Robust Visual Servo Control. In *Proc. of the IEEE Int. Conf. on Robotics and Automation (ICRA)*, 1998.
28. B. Hannaford and J.H. Ryu. Time Domain Passivity Control of Haptic Interfaces. *IEEE - Transactions on Robotics and Automation*, 18:1–10, 2002.

29. J.H. Ryu, B. Hannaford, C. Preusche, and G. Hirzinger. Time Domain Passivity Control with Reference Energy Behavior. In *Proc. of the IEEE/RSJ Int. Conf. on Intelligent Robots and Systems (IROS)*, 3:2932–2937, 2003.
30. C. Preusche, G. Hirzinger, J.H. Ryu, and B. Hannaford. Time Domain Passivity Control for 6 Degrees of Freedom Haptic Displays. In *Proc. of the IEEE/RSJ Int. Conf. on Intelligent Robots and Systems (IROS)*, 3:2944–2949, 2003.
31. G. Hirzinger, N. Sporer, A. Albu-Schäffer, M. Hähnle, R. Krenn, A. Pascucci, and M. Schedl. DLR's torque-controlled light weight robot III - are we reaching the technological limits now?. In *Proc. of the IEEE Int. Conf. on Robotics and Automation (ICRA)*, pages 1710–1716, 2002.
32. G. Hirzinger, N. Sporer, M. Schedl, J. Butterfaß, and M. Grebenstein. Robotics and Mechatronics in Aerospace. In *The 7th International Workshop on Advanced Motion Control (AMC)*, pages 19–27, 2002.
33. G. Hirzinger, B. Brunner, J. Dietrich, and J. Heindl. ROTEX - The First Remotely Controlled Robot in Space. In *Proc. of the IEEE Int. Conf. on Robotics and Automation (ICRA)*, 3:2604–2611, 1994.
34. G. Hirzinger, J. Butterfaß, M. Grebenstein, I. Schaefer, N. Sporer, M. Fischer, H. Liu, A. Albu-Schäffer, M. Schedl, and P. Neumann. Space Robotics - Driver for a new mechatronic Generation of light-weight arms and multifingered hands. In *Proc. of the IEEE/ASME International Conference on Advanced Intelligent Mechatronics (AIM)*, pages 1160–1168, 2001.
35. G. Hirzinger, K. Landzettel, B. Brunner, M. Fischer, C. Preusche, D. Reintsema, A. Albu-Schäffer, G. Schreiber, and B.-M. Steinmetz. DLR's robotics technologies for on-orbit servicing. *Advanced Robotics, RSJ - The International Journal of the Robotics Society of Japan*, 18(2):139–174. VSP, Netherlands, ISSN 0169–1864, 2004.
36. B. Schäfer and M. Lösch. Simulation of Elastic Space Manipulator Dynamics During Satellite Capturing. In *Proc. of the 4th International Conference of the European Association for Structural Dynamics (EURODYN '99)*, Prague, Czech Republic, 1999.
37. B. Schäfer, M. Lösch, and K. Landzettel. Simulation of Manipulator Deployment and Satellite Capturing Dynamics. In *Proc. of the 5th ESA Workshop on Advanced Space Technologies for Robotics and Automation (ASTRA 98)*, Noordwijk, The Netherlands, 1998.
38. R.W Longman, R.E. Lindberg and M.F. Zedd. Satellite-mounted Robot Manipulators - New kinematic and Reaction Moment Compensation. *International Journal of Robotics Research* (3), 1987.
39. S. Dubowsky and E. Papadopoulos. The Kinematics, Dynamics and Control of Free-Flying and Free-Floating Space Robotic Systems. *IEEE Transactions on Robotics and Automation* (5), 1993.
40. H.G. McCain and J.F. Andary. The Flight Telerobotic Servicer Project and systems overview. In *Electronics and Aerospace Conference*, 18(2):139–174, 1988.
41. M. Oda and D. Toshitsugu. Teleoperation System of ETS-VII Robot Experiment Satellite. In *Proc. of the IEEE/RSJ Int. Conf. on Intelligent Robots and Systems (IROS)*, 3:1644–50, 1997.
42. M. Oda, T. Doi and K. Wakata. Tele-manipulation of a satellite mounted robot by an on-ground astronaut. In *Proc. of the IEEE Int. Conf. on Robotics and Automation (ICRA)*, 2:1891–96, 2001.
43. M. Oda. Space Robot Experiments on NASDAs ETS-VII Satellite. In *Proc. of the IEEE Int. Conf. on Robotics and Automation (ICRA)*, 2:1390–1395, 1999.

44. W.-K. Yoon, T. Goshozono, H. Kawabe, M. Kinami, Y. Tsumaki, M. Uchiyama, M. Oda, and T. Doi. Model-based teleoperation of a Space robot on ETS-VII using a haptic interface. In *Proc. of the IEEE Int. Conf. on Robotics and Automation (ICRA)*, 1:407–12, 2001.
45. I. Kawano, M. Mokuno, et. al. Result of Autonomous Rendezvous Docking Experiment of Engineering Test Satellite VII. *Journal of Spacecraft and Rockets*, 38(1):105–111, 2001.
46. I. Kawano, M. Mokuno, T. Miyano, and T. Suzuki. Analysis and Evaluation of GPS Relative Navigation Using Carrier Phase for RVD Experiment Satellite of ETS-VII. In *ION GPS 2000 - The 13th International Technical Meeting of the Satellite Division of the Institute of Navigation*, pages 1655–60, 2000.
47. G. Visentin and F. Didot. Testing Space Robotics on the Japanese ETS-VII Satellite. In *ESA Bulletin*, 99:61–65, 1999.
48. I. Kawano, et. al. First Results of Autonomous Rendezvous Docking Experiments on NASDA's ETS-VII Satellite. In *Proceedings of the 49th International Astronautical Congress*, Melbourne, Australia, 1998.
49. Y. Fukushima, N. Inaba, and M. Oda. Capture and berthing experiment of a massive object using ETS-VII's space robot - World's first on-orbit satellite capture experiment by space robot system. In *Proceedings of AIAA/AAS Astrodynamics Specialist Conference*, Denver, CO, 2000.
50. K. Yoshida, K. Hashizume, and S. Abiko. Zero Reaction Maneuver: Flight Validation with ETS-VII Space Robot and Extension to Kinematically Redundant Arm. In *Proc. of the IEEE Int. Conf. on Robotics and Automation (ICRA)*, 1:441–46, 2001.
51. T. Imaida, Y. Yokokohji, T. Doi, M. Oda, and T. Yoshikawa. Ground-Space Bilateral Teleoperation Experiment Using ETS-VII Robot Arm with Direct Kinesthetic Coupling. In *Proc. of the IEEE Int. Conf. on Robotics and Automation (ICRA)*, 1:1031–1038, 2001.
52. Y. Yokokohji, T. Imaida, Y. Iida, T. Doi, M. Oda, and T. Yoshikawa. Bilateral Teleoperation: Towards Fine Manipulation with Large Time Delay. *(ISER)*, pages 11–20, 2000.

20

Underwater Telerobotics for Collaborative Research

Pere Ridao, Marc Carreras, Emili Hernandez, and Narcis Palomeras

University of Girona,
Computer Vision and Robotics Research Group
Institute of Informatics and Applications
Edifici PIV, Campus Montilivi, 17071 Girona, Spain
{pere,marcc,emilihb,npalomer}@eia.udg.es

Summary. Underwater robotics constitutes one of the most representative application fields of telerobotics. Over the last few decades, Remotely Operated Vehicles (ROVs) have played a very important role in undersea exploration/intervention, reducing the need of manned submersibles. In the future, Autonomous Underwater Vehicles (AUVs) equipped with acoustic modems and modern telerobotics technologies will do the job. This chapter explores underwater telerobotics reporting on the most representative scenarios involving ROVs and AUVs. A scenario involving the remote training of skilled researchers is analyzed in detail. In this context the tools needed for remote experimentation with the underwater robots developed by the University of Girona (Spain) are explained. These tools include a hardware in the loop (HIL) simulator called NEPTUNE and a control architecture called O^2CA^2.

20.1 Introduction

Over the last 50 years, underwater telerobotics have evolved into a very powerful tool for revealing the secrets hidden under the sea. It has played a very important role in the discovery of our historic heritage through surveying archaeological sites unaffected by the passage of history. Extraordinary forms of life which have evolved in the very hostile environment of profoundly deep water can now be studied, providing a very important source of information for understanding how life on Earth started. Other applications in marine science include the surveying and monitoring of underwater volcanoes, improving our knowledge of how volcanic islands are formed. Moreover, they help reduce the risk of navigation hazards for ships. In rescue missions, underwater robots help in the search for black boxes when an airplane crashes or in the recovery of the oil spilled from a sunken tanker. All these applications require some kind of human presence for observation and/or intervention under the sea. Since humans are not adapted to the underwater environment, we have engineered systems for observing and operating under the sea. Nowadays, a professional diver can reach depths up to 100 m using mixed gas and 300 m using saturation diving. Nevertheless, these distances are very far from the average ocean depth (3700 m) and even more from the deepest ocean depth (11000 m). Moreover,

M. Ferre et al. (Eds.): Advances in Telerobotics, STAR 31, pp. 347–359, 2007.
springerlink.com © Springer-Verlag Berlin Heidelberg 2007

even above 300 m, professional divers are exposed to a very hostile and dangerous environment. For these reasons, underwater robots have played a crucial role from the beginning of the development of telerobotics. Probably, the first ROV was POODLE Fig. 20.1 left, a torpedo like vehicle developed in 1953 by an underwater photographer called Dimitri Rebickoff. Nevertheless, the CURV (Cable Controlled Underwater Recovery Vehicle) is considered to be the first operative ROV (see Fig. 20.1 right). This robot, developed in 1960 by the US Navy, became very famous when it was used in 1966 to recover the hydrogen bomb accidentally dropped from a B52 airplane off Palomares on the Almeria coast of Spain. During the 80s and 90s, deep-ocean ROVs were developed by some of the most prestigious oceanographic institutions. DOLPHIN3K [1] and KAIKO [2] were developed by the Japan Agency for Marine-Earth Science and Technology (JAMSTEC), JASON/ARGO [3] by Woods Hole Oceanographic Institution (WHOI) and VICTOR 6000 [4] by the Institut Francais de Recherche pour l'Exploitation de la Mer (IFREMER), among others. In 1985, a joint expedition of WHOI and IFREMER discovered the Titanic [18] sitting on the ocean floor at a depth of 3798 m. A year later, Dr. Ballard's team (WHOI) used the JASON Jr ROV operated from the ALVIN manned submersible to explore the Titanic wreck. In 1994, the ROV KAIKO (JAMSTEC) [2] navigated the deepest rift in the ocean, the Mariana Trench (10.909 meters) and in 1999 [5] another JAMSTEC ROV, the DOLPHIN3K, was used for the recovery of the main engine of the H-I1 Rocket dropped 380 nautical miles off Izu-Ogasawara Islands in 3000 m of water. Although in all the above mentioned applications an umbilical cord allowed the use of high speed communication links, the cable constrains the ROV to a certain area around the mother ship. To remove this constraint, modern approaches to underwater telerobotics are based on the use of AUVs linked to the surface remote station through low speed acoustic modems introducing a non-neglectable time delay in the control loop similar to the one present in space based telerobotics. On the other hand, with the popularization of the Internet during last decade, new applications have arisen such as the network based sharing of robotic testbeds in the context of international projects through the use of virtual worlds.

In this chapter the architecture of the most common underwater telerobotics applications is studied first. Then, we focus on a detailed example of underwater telerobotics applied to joint research in the field of control architectures for autonomous underwater robots. In this context, a real time hardware in the loop graphical simulator for AUVs, called NEPTUNE, is presented as well as O^2CA^2, a distributed object oriented control architecture used for programming URIS AUV. The chapter ends with a simple application example of how to carry out a remote experiment through the Internet.

20.2 Underwater Telerobotics Applications

Over the last few years, underwater telerobotics have been applied in different application scenarios, deep ocean exploration/intervention being the most

Fig. 20.1. (left) POODLE: the first ROV (1953), (right) CURV: The first intervention ROV (1960)

classical. A deep ocean ROV (see Fig. 20.2) is a heavy class hovering vehicle deployed from a mother ship by a hydraulic winch. The ship must be equipped with a dynamic positioning system in order maintain its position during the ROV operation. The robot is tied to the ship via an umbilical cable which provides power, data and video channels. For shallow water operations (up to 400 m) it is possible to tie the cable directly to the robot control room onboard the ship. Nevertheless, for deep water operation, it is necessary to uncouple the robot and ship movements. This is achieved through the use of a hard ballast weight towed behind the ship through a primary umbilical cable. Ocasionally the system being towed is equipped with cameras and lights to provide images of the ROV's operation to the pilot [3]. The robot is then tied to the towed weight with a neutrally buoyant umbilical cord. For the operation of the robot, a navigation system is needed in order to know the robot's position. There are various alternatives based on the use of acoustic transponders, the most common being long base lines (LBL) or ultra short base lines (USBL). From the telerobotics point of view, simple strategies involving close-loop control by means of a joystick are used for ROV teleoperation with visual feedback from remote imagery.

The second scenario appears when the ROV pilot station is not located on the ship but in another part of the world. This is the case in the experiment carried out in the CNR-IAN [6] where the ROMEO ROV was teleoperated from Rome through the Internet while exploring the under ice benthic fauna of the Antarctica. In this case, a 64 Kbps satellite channel connected the robot control system in Antarctica to the CNR-IAN web server in Genoa and from there to the rest of the world through the Internet. A simple Human Computer Interface was setup with two main roles: piloting and exploring. As is common, only one user was allowed to enter the piloting mode while several were allowed to enter the observing mode.

The third scenario arises when the umbilical cable is removed and the wired communication link is replaced with an acoustic modem. Now the challenges include a low communication speed (aprox.10 Kbps), a long round trip for

Fig. 20.2. Deep ocean ROV VICTOR 6000 (IFREMER)

messages (aprox. 10 s at 7K depth), and a very unreliable channel (frequent breaks). To deal with these constraints the researchers have proposed to avoid common high speed global control loops and replace them with local control loops combined with a global low speed control loop. Hence an Autonomous Underwater Vehicle (AUV) takes the place of the ROV, but now an augmented reality system is needed to complement the real imagery which becomes very constrained due to the very low bandwidth. In this context, researchers of the WHOI have proved how the synthetic fixtures [7] concept can be used to recover an object from the seafloor under the above mentioned conditions. Synthetic fixtures are visual and force clues related to the task being carried out which help the operator to achieve the task in a very uncertain context.

The last scenario is related to promoting world wide collaboration and joint-research using the Internet. In this context, the researchers of the Instituto Superior Tecnico (IST) in Lisbon interconnected their mission control system called CORAL to the ROMEO control architecture developed by the researchers at the CNR-IAN [8], making it possible to control from Lisbon the mission of a ROV working in Genoa. Before carrying out the real experiments, they used a Hardware In the Loop (HIL) graphical simulator to test their approach [9]. Other teams, like the Autonomous System Laboratory in Hawaii and the researchers of the University of Tokio, have interconnected their respective graphical simulators DEVECS [10] and MVS [11] and have been able to carry out a multiple vehicle mission in a virtual environment. They used two real robots: (1) ODIN located in Hawaii and (2) the Twin-Burger placed in a water tank at the university of Tokio. In this case, both vehicles were able to interact virtually with each other and the researchers on both sides were able to observe the multi-vehicle mission in a virtual representation. This is also the case with the NEPTUNE HIL simulator developed at the University of Girona which provides facilities to carry out remote experiments across the Internet and gather the results to promote joint projects between laboratories and the sharing of underwater robotic testebs. The rest of this chapter is devoted to the description of this system.

20.3 Telerobotics as a Tool for Collaborative Research in Underwater Robotics

Mobile robotics have become very wellknown over the last few years and nowadays there are many different manufacturers providing robot platforms, sensor suites and software for a variety of applications. Hence, it is simple and not very expensive to have access to a robot for research purposes. In the underwater robotics counterpart the panorama is very different. While there is a very open market of ROVs, only a few companies are manufacturing and selling AUVs, normally aimed at big oceanographic institutions and offshore industries. Prices are very high since the target market is very small. For this reason, there are only a few AUV prototypes available for research and experimentation world wide. Some of them have been extensively tested in scientific missions while others are basically research platforms. Therefore, researchers in this field often have to move to other institutions to carry out their experimental research work. On the other hand, several researchers often work with different parts of an AUV having a need to schedule the robot's time. Moreover, these vehicles are often heavy, requiring several people to operate them making experimentation very expensive. Hence, the reduction of the number of hours the real vehicle is needed for experimentation is desirable. For these reasons, it is necessary to have a development methodology which provides facilities for: (1) exhaustive experimentation, (2) concurrent engineering, (3) remote experimentation and (4) a reduced turn around time. Graphical simulators are the key tool in achieving this methodology and the principal research centers in underwater robotics have developed their own software for Hardware In the Loop (HIL) simulation. At the University of Girona, we are setting up a new methodology for carrying out collaborative research with our partners world wide. The method is described in next section.

20.4 Networked Collaborative Research at the Unversity of Girona

At the university of Girona, Telerobotics are used for remote training and collaborative research. It is based on the use of two main tools: (1) an HIL Simulator called NEPTUNE and (2) the Object Oriented Control Architecture for Autonomy, O^2CA^2. Using these tools, a researcher should follow the following methodology to carry out a remote experiment. First, a researcher placed at the remote site develops his control algorithm within the O^2CA^2 framework on a local computer. This algorithm is tested with the help of NEPTUNE. The online simulation of the controller allows the researcher to be sure that his code is logically and temporally correct. Then the controller is uploaded to the control computer of the robot in Girona. Now, a remote experiment is carried out using the robot operating in Girona with the help of NEPTUNE (located at the remote site) which reproduces a virtual view of the experiment using the telemetry transmitted through the Internet. The logging mechanism of the O^2CA^2 is

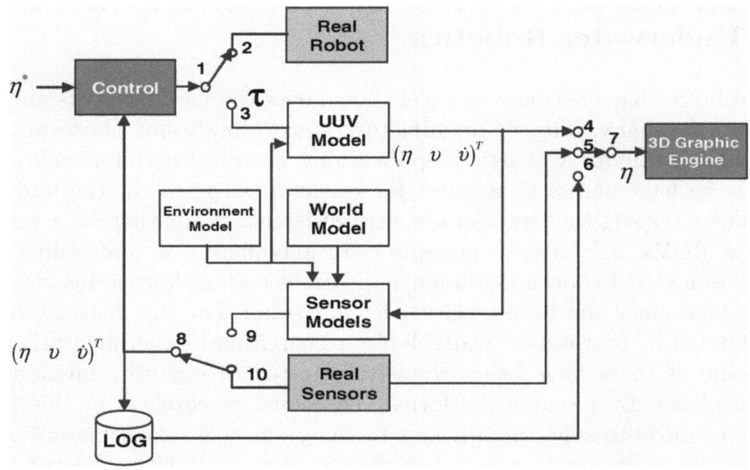

Fig. 20.3. Block diagram of a robot simulator. η': Desired trajectory; η: Position and attitude; ν: Velocity; $\dot{\nu}$: Acceleration; τ: Force and Torque

used to generate a file of results that can be downloaded and evaluated. Due to the difference between the model of the robot and the real robot, the tested algorithm needs to be tuned before reaching its optimal performance. In the following sections, NEPTUNE and O^2CA^2 are described in more detail.

20.4.1 NEPTUNE HIL Simulator

Fig. 20.3 shows the main components of a generic underwater robot simulator. The system's input is the desired trajectory (η') to be followed or tracked. This is also the control block input whose output is the generalized force and torque (τ) that must be exerted by the thrusters and/or the control planes. The control output can be routed towards the real robot actuators or towards the Unmanned Underwater Vehicle (UUV) model. In the first case, when the UUV moves, the real sensors will sense this movement. In the second case, the UUV Model will predict the evolution of the robot's position (η), velocity (ν) and acceleration ($\dot{\nu}$) which will then be used by the sensor models to predict the sensor readings. A world model and an environmental model are also needed to simulate sensors like sonar or currentimeters. The output coming from the sensors (model/real) is used to close the feedback as well as for logging purposes. Finally, the robot's position read from the sensors or the UUV model is used to display the robot within the virtual world using the 3D graphic engine.

NEPTUNE follows this architecture and provides capabilities for on-line and hardware in the loop simulation, being able to simulate multiple UUVs. For the robot simulation, the well known hydrodynamical model proposed in [12] is used. The geometrical models of the world and the robots can be imported from

VRML files. NEPTUNE is able to simulate the most representative sensor suite (sonar, vision, navigation, attitude ...) commonly available in AUVs and it is built as a distributed object oriented application which runs on the GNU/Linux operating system. Refer to [13] for a detailed description of NEPTUNE.

20.4.2 The O^2CA^2 Control Architecture

O^2CA^2 is a hybrid control architecture which merges a behavioral layer with a mission planning layer. The behavioral layer is composed of a set of parallel behaviors coordinated through a hierarchical network of hybrid coordination modules [14]. As usual, each behavior has its own goal which it tries to accomplish. Each one senses the environment through the sensing sub-system and generates an appropriate response. All behavior responses are homogeneous having two components: (1) a desired velocity vector and (2) an activation level expressing how important it is for the behavior to take entire control of the robot. Each coordination module has two inputs: (1) a dominant input (high priority) and (2) a non-dominant input (lower priority), as well as one output. Whenever the behavior connected to the dominant input asks for full control (activation level equal to one) the non-dominant behavior is suppressed. Nevertheless, if the activation level is less than one, the responses from both behaviors cooperate through a weight average function based on both activation levels to generate the output of the coordination module. This output is again a desired velocity together with a corresponding activation level that is routed towards another coordination module or towards the velocity controller. By appropriately connecting the desired behaviors through a hierarchy of coordination modules, it is possible to fix the priorities which ensure the expected functionality while maintaining the robot's safety. The interested author can find more details in [15].

20.5 Case Study

In order to illustrate how remote training is achieved, in this section we will consider the case of a researcher in Italy who is interested in developing a new controller for the URIS robot. Hereafter we describe, step by step, the procedure that should be followed by this researcher in order to test his controller in Italy on our robot located in Girona (Spain). It is assumed that the researcher has local access to: (1) a desktop computer compatible with the onboard control computer of the robot and (2) an installed version of NEPTUNE and O^2CA^2. In the following paragraphs, the URIS robot as well as the water tank and the navigation system used for the experiments are presented. The processes for running local HIL simulations of the controller as well as the execution of remote testing experiments are then detailed.

20.5.1 URIS AUV

The URIS robot (see Fig. 20.4 left) was developed at the University of Girona with the aim of building a small-sized AUV. The hull is composed of a stainless steel sphere with a diameter of 350mm, designed to withstand pressures of 4 atmospheres (30 meters depth). On the outside of the sphere there are two video cameras (forward and down looking) and four thrusters (two in X direction and two in Z direction). Due to the stability of the vehicle in pitch and roll, the robot has four degrees of freedom (DOF): surge, heave, sway and yaw. Except for the sway, the other DOFs can be directly controlled. The robot has two onboard PC-104 computers. One runs the low level and high level controllers and the other runs the computer vision algorithms. Both computers are connected through an Ethernet network and run the O^2CA^2 control architecture on a GNU/Linux-RTAI operating system. An umbilical cable is used for communication, power and video signal transmissions. The URIS hydrodynamic model is known and has been reported on in [16] and therefore is used for its HIL simulation [13] within NEPTUNE.

Fig. 20.4. (left) The URIS robot. (right) URIS in the water tank.

20.5.2 Experimental Setup

A small water tank located in the lab is used for experimenting with URIS. The tank has a 4.5 m. diameter and a 1 m. depth (see Fig. 20.4 right). Despite its small size, the dimensions are enough to run experiments in yaw, pitch and surge. Only heave experiments cannot be run due to the lack of depth. A specially designed coded pattern is located on the bottom of the water tank. Using a down looking camera mounted on the robot, URIS is able to process the coded pattern images and estimate its position, attitude and velocity in real time [17]. Two more cameras are used in the setup. One provides a top view of the robot. The other is an underwater camera located in the water tank.

20.5.3 Behaviours Used During the Experiment

Fig. 20.5 shows the architecture used for the experiments. The most dominant (priority) behaviour is AvoidWall. This behaviour uses the absolute position of the robot to compute the distance to the wall of the water tank. If the distance

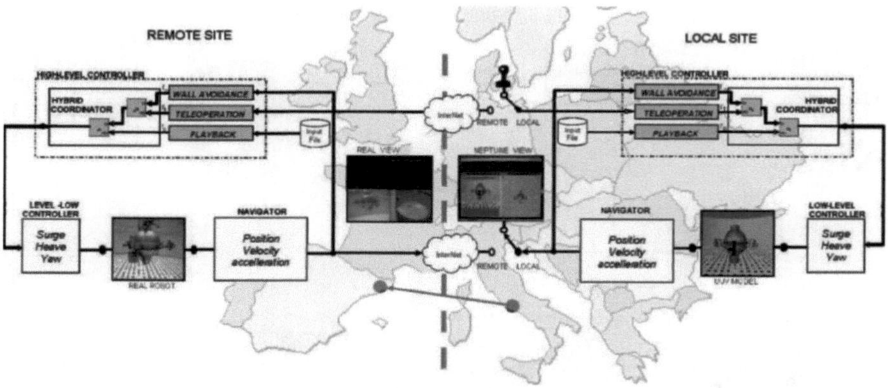

Fig. 20.5. Block diagram of a remote experiment through Internet. First an online simulation is done at the local site. Then, a real experiment is done in the remote site while NEPTUNE reproduces a real time virtual view.

is small the AvoidWall asks for control of the robot through its corresponding activation level and generates a desired velocity vector directed towards the centre of the tank. The behaviour with medium priority, Teleoperation, is used to drive the robot following joystick commands. This is done by converting the joystick readings into velocity stepoints. Whenever the joystick is moved, the activation level of this behaviour is set to one (normally set at zero). Finally, a Playback behaviour (non dominant) is used to read velocity setpoints from an input file and send them to the low level controller in order to test the experimental controller. A backup low level controller is also used to take control of the robot in case the robot gets into an unsafe region around the walls or the bottom of the tank. This is just for safety reasons in case a badly conditioned controller is tested. With this hierarchy of behaviours, the robot is expected to track the velocity setpoints read from the input file until the operator gets control of the experiment with the joystick. In case of a collision risk during the experiment or during the teleoperation, the AvoidWall behaviour will drive the robot toward a safe region. Finally, a navigator module responsible for computing the robot's position, velocity and acceleration is also used in the experiments.

20.5.4 Online Simulation

In order to run an HIL simulation with the URIS robot, several steps have to be followed. First (see Fig. 20.6), a CAD software has to be used to model the world and the robot, and to export these models into the VRML files "Water_tank.WRL" (step 1) and "URIS.WRL" (step 2). Then, using the coefficients identified in [16] the "URIS.DYN" file, which defines the robot hydrodynamics, is set up (step 3). At the next step (4), the "URIS.UUV" file is created tying together the robot shape ("URIS.WRL") and its dynamics ("URIS.DYN").

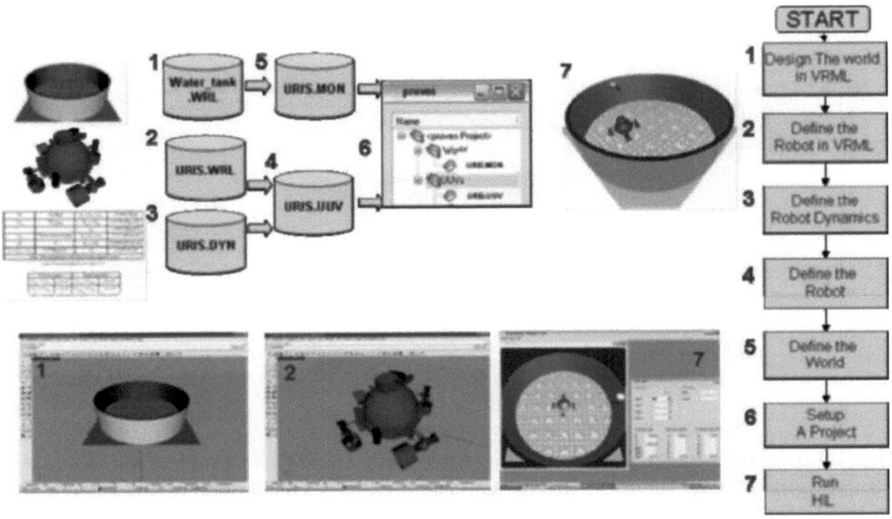

Fig. 20.6. Procedure for online simulation

"URIS.UUV" also defines the radius of the smaller sphere which contains the robot (used for collision detection), its initial position as well as a list of sensors to be simulated for the robot (number of sonar range sensors, their accuracy, their position and attitude ...). Step (5) consists of creating the "URIS.MON" file which defines the components of the virtual world in which the robot will move. It links the VRML file ("Water_tank.WRL") which contains the 3D topography of the world together with a list of objects (also VRML files) with their scale factors, radius of influence (for obstacle detection) and their position-attitude vectors. At this moment, all the files needed have been defined and a simulation project (step 7) can be setup within NEPTUNE. This projects ties the virtual world ("URIS.MON") and all the AUVs that will be involved in the simulation together, in our case one ("URIS.UUV"). Finally, NEPTUNE starts the simulation and the O^2CA^2 control architecture can be tied to the simulator in order to run an online simulation of the experiment. This simulation corresponds to the operation of the local modules represented in the right part of Fig. 20.5.

20.5.5 Remote Experiment

Once the online simulation has been executed satisfactorily, a remote experiment involving the real robot has to be executed. In order to do this, first the code of the controller is uploaded to the control computer of the robot (see Fig. 20.7). NEPTUNE is then run at the local site connected to the O^2CA^2 running onboard the robot. Since the control architecture is distributed, the object corresponding to the joystick runs in a local machine. The experiment is then started and the remote navigator object is read periodically to update the virtual view shown

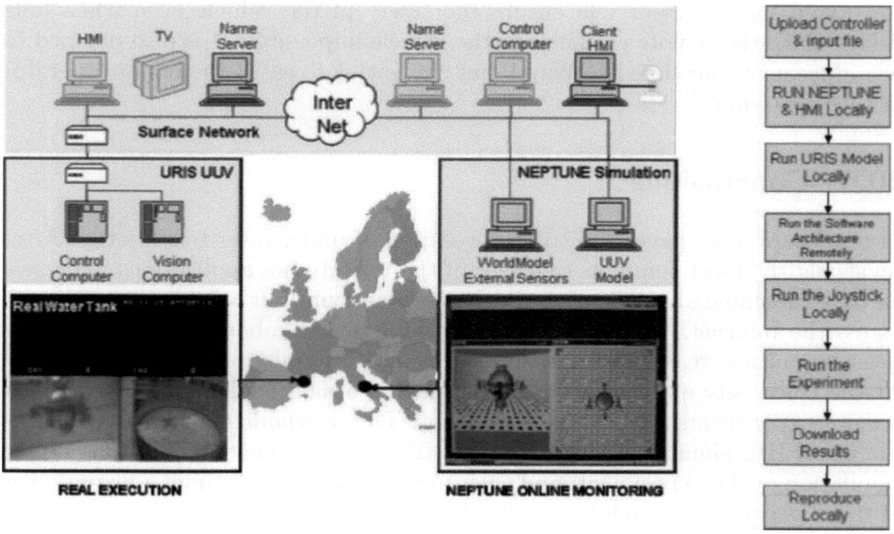

Fig. 20.7. Procedure for remote experimentation

in NEPTUNE. It is worth noting that the robot is working autonomously and only navigation data (position) is transferred through the network. The logging system of URIS is enabled, recording the whole experiment. At the end, the result file gathered during the experiment is downloaded and evaluated by the researcher. Since there are differences between the simulated and the real experiment, some sort of tuning phase involving several remote experiments is usually needed.

20.5.6 Rationale

It is worth noting that, in the telerobotics approach presented above, the control loop is closed locally at the remote site where the robot is operating. The AUV is totally autonomous and once started, performs its operation without human intervention. The only information traversing the network is telemetry (position and attitude), used to represent a virtual view of the remote experiment, and the operator commands via the joystick. In our experience, working with high speed communication links commonly available at universities and research centers, and taking into account the reduced data traffic generated by our approach, this is not a big problem. Although common communication problems (long delays, packet-loss) will degenerate the remote experience, the fact that the experiment is autonomously executed ensures that its results will not degenerate and will therefore be useful. In fact, once completed, the experiment can be reproduced virtually at the local site using the log files. Nevertheless, long delays during the joystick tele-operation will make the robot guidance difficult. In these situations, the AvoidWall behavior will ensure the safety of the vehicle even when long delays make the remote guidance of the vehicle impossible. It is also planned to introduce new behaviors like Way-Point navigation to assist the remote operator in such situations.

20.6 Conclusions

In this chapter we have studied the evolution of underwater telerobotics. After reviewing the most common scenarios of the telerobotics applications undersea, we have presented an application which allows carrying out collaborative research across the Internet. This joint work with underwater robots allows sharing expensive and not readily available experimental testbeds for research purposes. At the University of Girona, we are developing a methodology for carrying out remote experiments with our URIS robot. This methodology is based on the use of an HIL simulator called NEPTUNE, first to execute a real-time online simulation of the experiment, and then to represent, in real-time, a virtual view of the experiment through the feedback of the navigation data (position) across the Internet. The Object Oriented Control Architecture for Autonomy together with NEPTUNE allow a researcher to carry out a safe experiment with a real robot from a remote site.

Acknowledgements

This research was sponsored by the Spanish commission MCYT (DPI2005-09001-C03-01).

References

1. M. Nomoto and M. Hattori. A deep rov dolphin 3k: Design and peformance analysis. *Oceanic Engineering, IEEE Journal* of 11 (3) 373–391, 1986.
2. M. Kyo, E. Hiyazaki, S. Tsukioka, H. Ochi, Y. Amitani, T. Tsuchiya, T. Aoki, and S. Takagawa. The sea trial of "kaiko", the full ocean depth research rov. In *OCEANS '95. MTS/IEEE. 'Challenges of Our Changing Global Environment'. Conference Proceedings.*, pages 1991–1996, 1995.
3. D. Yoerger, J. Newman, and J. Slotine. Supervisory control system for the jason rov. *Oceanic Engineering, IEEE Journal* of 11 (3) 393–400, 1986.
4. J. Cadiou, S. Coudray, P. Leon, and M. Perrier. Control architecture of a new deep scientific rov: Victor 6000. In *OCEANS '98 MTS/IEEE Conference Proceedings*, pages 492–497, 1998.
5. H. Momma, M. Watanabe, K. Mitsuzawa, K. Danno, M. Ida, M. Arita, and I. Ujino. Search and recovery of the h-ii rocket flight no. 8 engine. In *Underwater Technology, 2000. UT 00. Proceedings of the 2000 International Symposium on*, pages 19–23, 2000.
6. G. Bruzzone, R. Bono, G. Bruzzone, M. Caccia, M. Cini, and P. Coletta. Internet-based satellite teleoperation of the romeo rov in antarctica. In *Proceedings of the 10th Mediterranean Conference*, 2002.

7. C. Sayers, R. Paul, L. Whitcomb, and D. Yoerger. Teleprogramming for subsea teleoperation using acoustic communication. *Oceanic Engineering, IEEE Journal of*, 23 (1) 60–71, 1998.
8. G. Bruzzone, R. Bono, M. Caccia, G. Veruggio, C. Ferreira, C. Silvestre, P. Oliveira, and A. Pascoal. Internet mission control of the romeo unmanned underwater vehicle using the coral mission controller, In *OCEANS '99 MTS/IEEE. Riding the Crest into the 21st Century*, pages 1081–1087, 1999.
9. G. Bruzzone, R. Bono, M. Caccia, and G. Veruggio. A simulation environment for unmanned underwater vehicles development, In *Oceans 2001 MTS/IEEE*, Honolulu, Havaii, USA, 2001.
10. S. C. et al. Distributed virtual environment collaborative simulator for underwater robots, In *IEEE/RSJ Int. Conf. on Robots and Systems*, Takamatsu, Japan, pages 861–866, 2000.
11. Y. Kuroda, K. Aramaki, and T. Ura. Auv testing using real/virtual synthetic world, In *Proc. of IEEE Symp. on Autonomous Underwater Vehicle Technology '96*, pages 365–372, 1996.
12. T. Fossen. Marine Control Systems: Guidance, Navigation and Control of Ships, Rigs and Underwater Vehicles. *Marine Cybernetics AS*, Trondheim, December 2002.
13. P. Ridao, E. Batlle, D. Ribas, and M. Carreras. Neptune: A hil simulator for multiple uuvs. In *Oceans'04 MTS/IEEE*, Kobe, Japan, 2004.
14. M. Carreras, P. Ridao, J. Batlle, and T. Nicosevici. Efficient learning of reactive robot behaviors with a neural-q_learning approach, In *IEEE/RSJ International Conference on Intelligent Robots and Systems*, Lausanne, Switzerland, 2002.
15. M. Carreras, P. Ridao, R. Garcia, and J. Batlle. Behavioural Control of unmanned underwater vehicles. *Advances in Unmanned Marine Vehicles*, 1st Edition, Vol. Chapter 4, pages 67–83, 2006.
16. P. Ridao, A. Tiano, A. El-Fakdi, M. Carreras, and A. Zirilli. On the identification of non-linear models of unmanned underwater vehicles, *Control Engineering Practice Vol 12/12*, pages 1483–1499, 2004.
17. M. Carreras, P. Ridao, R. Garcia, and T. Nicosevici. Vision-based localization of an underwater robot in a structured environment, In *IEEE International Conference on Robotics and Automation ICRA'03*, Taipei, Taiwan, 2003.
18. J.L. Michel and R.D. Ballard. The RMS Titanic 1985 discovery expedition, In *Oceans'94 MTS/IEEE* , Brest, France, 1994.

21

Robot Assisted Force Feedback Surgery

Tobias Ortmaier[1,3], Barbara Deml[2], Bernhard Kübler[1], Georg Passig[1], Detlef Reintsema[1], and Ulrich Seibold[1]

[1] German Aerospace Center (DLR)
Institute of Robotics and Mechatronics
Oberpfaffenhofen, D-82234 Wessling, Germany
georg.passig@dlr.de

[2] University of the Armed Forces Munich (UniBW)
Human Factors Institute
D-85577 Neubiberg, Germany
barbara.deml@unibw.de

[3] KUKA Roboter GmbH
86165 Augsburg, Germany
tobiasortmaier@kuka-roboter.de

Summary. Minimally invasive surgery characterizes a sophisticated operation technique in which long, slender instruments are inserted into the patient through small incisions. Though providing crucial benefits compared to open surgery (i.e. reduced tissue traumatization) it is also faced with a number of disadvantages. One of the major problems is that the surgeon cannot access the operating field directly and, therefore, can neither palpate tissue nor sense forces sufficiently. Furthermore, the dexterity of the surgeon is reduced as the instruments have to be pivoted around an invariant point.

To overcome some of the drawbacks, telepresence constitutes a promising approach. The surgical instruments can be equipped with miniaturized force/torque sensors and contact forces can be displayed to the surgeon using a suitable man-machine interface. Furthermore, instruments can be built with additional degrees of freedom at the distal end, providing full dexterity inside the patient's body. Thanks to telepresence the surgeon regains direct access to the operating field, similar to open surgery.

In this chapter a prototypical force reflecting minimally invasive robotic surgery system based on two surgical robots is presented. The robots are equipped with a sensorized scalpel and a stereo laparoscope for visual feedback. The operator console consists of a PHANToM force feedback device and a stereoscopic display. Experimental results of a tissue dissection task revealed significant differences between manual and robot assisted surgery. At the end of the chapter some conclusions based on the experimental evaluation are drawn, showing that both, manual and robotic minimally invasive surgery have specific advantages.

21.1 Introduction and Motivation

In conventional open surgery the surgeon has full access to the operation area and thus can use all senses for the demanding task of surgery. In contrast to this, in minimally invasive surgery the access is restricted as the surgeon works with long instruments through small incisions.

M. Ferre et al. (Eds.): Advances in Telerobotics, STAR 31, pp. 361–379, 2007.
springerlink.com © Springer-Verlag Berlin Heidelberg 2007

In this section the peculiarities of manual minimally invasive surgery (MIS) are described and advantages as well as disadvantages are discussed. Subsequently, a short introduction in minimally invasive robotic surgery (MIRS) is given which illustrates the research needs.

21.1.1 Minimally Invasive Surgery

Minimally invasive surgery is an operation technique which was established in the 1980s. In contrast to conventional, open surgery there is no direct access to the operating field and the surgeon employs long, slender instruments. These are inserted into the patient through narrow incisions which are typically slightly bigger than the instrument diameter (see Fig. 21.1).

The main advantages of MIS, compared to open surgery, are reduced pain and trauma, shorter hospitalisation, shorter rehabilitation time and cosmetic advantages. However, MIS is faced with at least three major disadvantages [1]: (a) As the surgeon does not have direct access to the operating field the tissue cannot be palpated any more. (b) Because of the relatively high friction in the trocar[1] and due to the torques which are necessary to rotate the instrument around the entry point, the contact forces between instrument and tissue can hardly be sensed. This is especially true when the trocar is placed in the intercostal space (between the ribs). (c) As the instruments have to be pivoted around an invariant fulcrum point (see Fig. 21.1), intuitive direct hand-eye coordination is lost. Furthermore, due to kinematic restrictions only four degrees of freedom (DoF) remain inside the body of the patient. Therefore, the surgeon cannot reach any point in the work space at arbitrary orientation. This is a main drawback of MIS, which makes complex tasks like knot tying very time consuming and requires intensive training [2, 3]. As a consequence MIS did not prevail as desired by patients as well as by surgeons and while most standard cholecystectomies (gall bladder removal) are performed minimally invasively in the industrialised world, MIS is hardly used in any other procedure to this extent.

21.1.2 Minimally Invasive Robotic Surgery

Robotic and mechatronic systems become a key technology for coping with the drawbacks of manual MIS. Together with telemanipulation techniques they enable a surgeon to regain full access to the operation field. Minimally invasive robotic surgery provides at least five potential advantages: (a) Small force/torque sensors placed near the instrument tip can measure manipulation forces/torques directly and thus, provide kinesthetic feedback when displayed to the surgeon [4]. (b) Actuated instruments with two additional DoF give back full dexterity inside the human body. (c) The undesired reverse hand motion can be avoided by appropriate control algorithms [5]. (d) More accurate movements are possible as the surgeon's hand motion can be scaled down before being transmitted to

[1] The trocar is a surgical instrument, which makes it possible to create incisions in a visceral cavity (i.e. thorax, abdominal cavity) and keep it open with the aid of a tube.

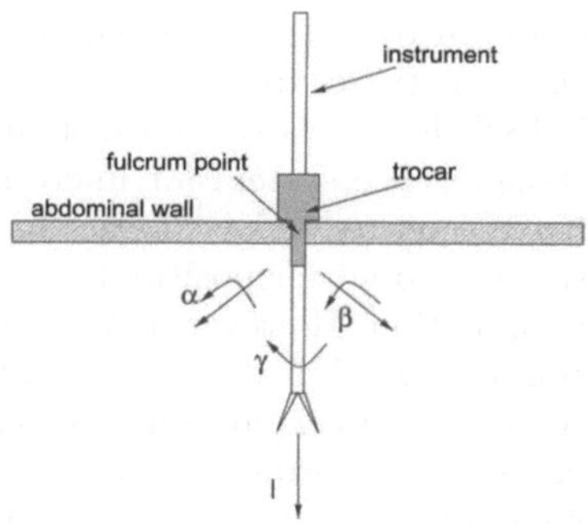

Fig. 21.1. Schematic exposition of the situation in MIS: The instrument is moved around an invariant fulcrum point. In consequence the surgeon can command only four degrees of freedom (α, β, γ, l) inside the patient's body.

the robot. Additionally, the surgeon's tremor can be reduced by low-pass filters. (e) Furthermore, autonomous functions such as motion compensation can be realized by MIRS telepresence systems [6, 7, 8]. Thus, surgeons could perform new operation techniques like endoscopic minimally invasive bypass surgery on the beating heart [9, 10].

Robotic systems for minimally invasive surgery are particularly used in urology, visceral, and heart surgery. Despite of first encouraging successes (e. g. a completely endoscopic radical prostectomy as described in [11]) the cost-benefit-ratio of this operation technique is still subject to discussion [12]. In case of heart surgery, robot assisted interventions are only employed for a small number of highly selected patients [13, 14, 15]. In order to become applicable to a wider range of patients, more sophisticated visualization and navigation techniques [16] as well as improved manipulator mechanics [17, 18] are necessary. Besides, it is likely that the implementation of force feedback may also yield an important additional benefit in terms of further enhancement and broader application.

21.2 Related Work

Since the early 1990s more than 35 surgical robotic systems have been developed [19]. In the field of minimally invasive robotic surgery especially three commercial systems are to be mentioned: the Zeus system (Computer Motion Inc. [20]), the daVinci system (Intuitive Surgical Inc. [21]), and the Laprotek system (endoVia Medical Inc. [22]). At the end of 2006 almost 470 installations of the daVinci system are recorded [23]. The Zeus and the Laprotek system were also in clinical use, but both are no longer commercially available. In addition to these systems, the robotic tele-surgical workstation for laparoscopy (University of Berkeley, University of San Francisco; California) has to be pointed out [24]. None of these systems provides kinesthetic feedback and thus prototypical force feedback systems are currently only available at research laboratories. The following paragraphs provide an overview of research activities in the area of telesurgery systems with kinesthetic feedback.

In Korea a group at KAIST (Korea Advanced Institute of Science and Technology) has developed a telepresence system for micro surgical tasks [25]. It is designed for six DoF force/torque reflection at the master console. The slave consists of an industrial six DoF robot for positioning a modified six DoF Stewart platform for micro manipulation. However, it has to be mentioned that the system does not provide full manipulability (i.e. 6 DoF) for laparoscopic surgery due to kinematic restrictions at the fulcrum point. Nevertheless, it is one of the few systems which realizes full force/torque feedback at all.

A further approach for measuring grasping forces is addressed in the work of Hu [26]. Here, conventional laparoscopic tools are equipped with strain gauge sensors and the sensed forces are displayed by a PHANToM (SensAble Technologies Inc.), a rather widespread kinesthetic device also used in this work. As this tool is not yet fixed to a robot and as the grip is not actuated, the current setup requires two users: one to actuate the surgical instrument and the second one to feel the grasping forces at the PHANToM. At least two further issues have to be mentioned: First, no contact forces can be measured at present. Second, as the strain gauge sensors are placed at the proximal end the grasping forces are superposed by friction.

A force reflecting master-slave system for minimally invasive surgery is described in [27]. In this bilateral system, two modified PHANToMs are used: one serves as force-reflecting master, the other one is equipped with a custom-built instrument and constitutes the slave robot. Master and slave are coupled via a virtual-reality peripheral network. To control the position of the instrument tip an artificial neural network is applied which supports an online adaption to different load conditions at the instrument tip. Unfortunately, the strain gauge sensors are placed at the proximal end, too, so that the measured contact forces are again distorted by the friction between instrument and trocar.

A force controlled laparoscopic surgical robot without distal force sensing is presented in [28]. A standard force sensor is integrated into a trocar and thus remains outside the patient. This makes it easier to guarantee the required standards of sterilization and requires less effort to miniaturize components. Due to the specific installation of the sensor the measurement is not deteriorated by friction between trocar and instrument. To calculate the contact forces only gravity compensation is necessary. Currently, the robot runs in co-manipulation mode which means that the surgeon and the robot manipulate the same instrument. Initial experimental in vivo and in vitro results are encouraging.

A system for the evaluation of force feedback in MIRS is presented in [29, 30]. The usage of commercially available instruments of the daVinci System together with industrial standard robots provides a simple set-up for experiments in MIRS. Only two forces perpendicular to the instruments are measured with strain gauges applied to the outside of the instrument shaft near the wrist. Due to this sensor position, actuation forces for the instrument wrist can not be separated from contact forces. However, friction in the trocar point does not influence the sensor readings. Force display and position command is realized by two PHANToMs. The system is not designed for clinical use since the problem of sterilisability is not adressed and the use of industrial robots in a clinical environment is difficult.

The design and realization of a pair of kinesthetic forceps for virtual reality (VR) microsurgery training is described in [31]. The proposed forceps have two actuated DoF and thus, it is possible to rotate around the instrument axis or to open and to close the gripper. To provide full force feedback (i.e. in 6 DoF) the forceps are fixed to a kinesthetic 6 DoF DELTA device [32]. The displayed forces are rendered in real-time by a VR system and so is the visual feedback.

21.3 Experimental Setup

In the following sections the telesurgery scenario developed at the Institute of Robotics and Mechatronics of the German Aerospace Center (DLR) is presented (for a more detailed description see [7]). The teleoperator supports manipulations in 4 DoF inside the patient and provides visual as well as force/torque sensor data. The remote sensor data are displayed at the operator console.

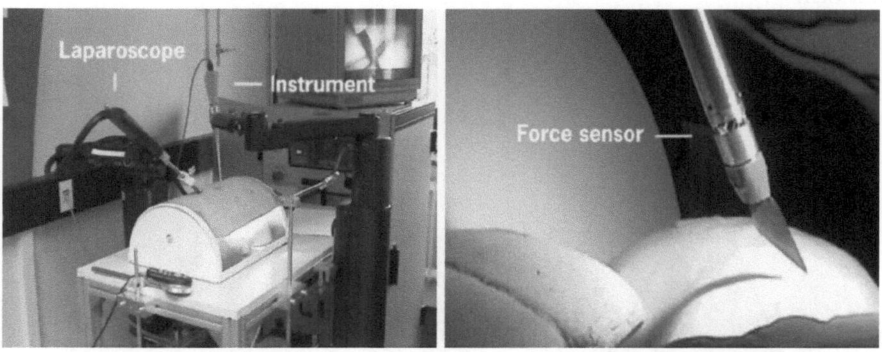

(a) Teleoperator. (b) Sensorized scalpel.

Fig. 21.2. DLR telesurgery scenario: teleoperator and sensorized scalpel

21.3.1 Teleoperator

The teleoperator consists of two surgical robots, an Aesop 3000 DS and an Aesop 1000 DS (both from Computer Motion Inc.). While the Aesop 3000 DS is equipped with an operating instrument, the Aesop 1000 DS provides stereo view from the surgical site using a 3D-laparoscope, a rigid endoscope (see Fig. 21.2, left).

Each of the two robots has four active and two passive joints. The passive joints are equipped with encoders and do neither contain a motor nor a brake. They are necessary to guarantee that no significant forces are exerted at the (a priori unknown) invariant fulcrum point. The camera-robot runs in an automatic mode: As the instrument is equipped with a color mark near the distal end, the surgical instrument can be detected by color segmentation within the stereo camera images of the laparoscope. The camera-robot is then able to follow the instrument automatically. Thus, the surgeon can focus on the operation and is not distracted by unnecessary tasks. For further details on automatic camera guidance see [33, 34].

Surgical Instruments

The surgical instrument (see Fig. 21.2, right) which was developed at DLR is equipped with a miniaturized force/torque sensor (10 mm in diameter) [4] and can easily be attached to the robot. The sensor itself is fixed at the distal end of the instrument in order to guarantee a collocated measurement of the contact forces. A force-torque transducer based on a Stewart Platform is well suited for this application. Advantages include high stiffness, adaptable properties, annular shape, and scalability. The geometry and the properties of a Stewart Platform transducer are described completely by the set of variables R, L, α, β and γ shown in Fig. 3(a). The parameters L and R denote the link length and base radius respectively. Further geometrical parameters derived thereof are the platform radius r, the radius of the link intersection a and the joint separation at the platform i'.

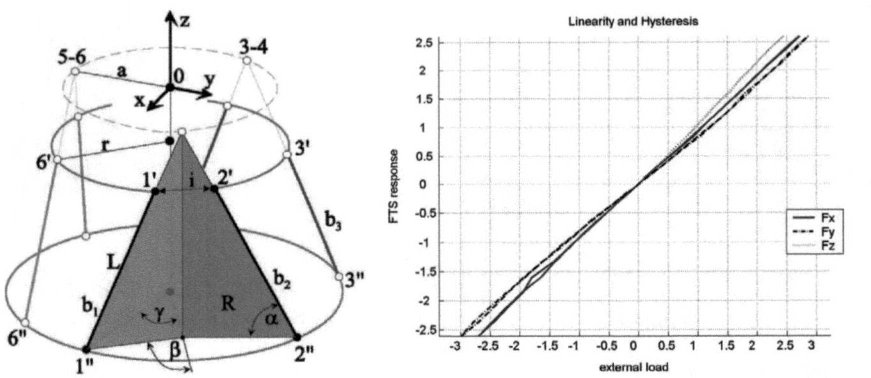

(a) Illustration of geometrical parameters.
(b) Transducer response to externally applied forces after calibration.

Fig. 21.3. Parameters of a force sensor based on a Stewart Platform and its response to externally applied forces

The characteristic matrix $\mathbf{A} \in \mathbb{R}^{6\times6}$ describing the transformation of measured link forces $\boldsymbol{F}_{int} = [F_1, F_2, F_3, F_4, F_5, F_6]^T$ to externally applied loads $\boldsymbol{F}_{ext} = [F_x, F_y, F_z, M_x, M_y, M_z]^T$

$$\boldsymbol{F}_{ext} = \boldsymbol{A} \cdot \boldsymbol{F}_{int}, \tag{21.1}$$

is calculated as follows [35]:

$$\mathbf{A} = -\frac{1}{2} \cdot \begin{bmatrix} -2n & 2n & \sqrt{3}m+n & \sqrt{3}m-n & -\sqrt{3}m+n & -\sqrt{3}m-n \\ -2m & -2m & m-\sqrt{3}n & m+\sqrt{3}n & m+\sqrt{3}n & m-\sqrt{3}n \\ -2q & -2q & -2q & -2q & -2q & -2q \\ 2aq & 2aq & -aq & -aq & -aq & -aq \\ 0 & 0 & aq\sqrt{3} & aq\sqrt{3} & -aq\sqrt{3} & -aq\sqrt{3} \\ -2an & 2an & -2an & 2an & -2an & 2an \end{bmatrix}, \quad \begin{aligned} m &= \cos(\alpha)\,\cos(\beta), \\ n &= \cos(\alpha)\,\sin(\beta), \\ q &= \sin(\alpha). \end{aligned}$$

To find a sensor geometry that is well conditioned and optimized for the force range expected in a surgical application, the following optimization method is used. The radius of the base R and the link length L are determined by the space available in the instrument. For all geometrically valid combinations (non-intersecting links) of R, L, α, β and γ, $\boldsymbol{A}^{-1}$ is calculated. Various sets of maximally expected external loads $[F_x, F_y, F_z, M_x, M_y, M_z]^T$ are selected, containing loads in the 6 principal directions. Every member of the load set is pre-multiplied by $\boldsymbol{A}^{-1}$, yielding the corresponding set of internal leg forces $[F_1, F_2, F_3, F_4, F_5, F_6]^T$. The variance s^2 of the internal leg force set is a measure of the isotropy of the sensor structure with respect to the external load set. This however is not an isotropy in the classical definition, since the external loads in the principal directions need not to be equal. For the load set $F_{x,y,z} = 20\,\mathrm{N}$, $M_{x,y} = 200\,\mathrm{Nmm}$, $M_z = 100\,\mathrm{Nmm}$ the following parameters were selected as optimal sensor geometry: $R = 4.2\,\mathrm{mm}, L = 3.9\,\mathrm{mm}, \alpha = 57°, \beta = 90°, \gamma = 36°$, yielding a variance of $s^2 = 236\,\mathrm{N}^2$. Using appropriate design of flexural hinges and leg cross-section, properties of the transducer structure are in good agreement with the prediction of the ideal analytical model [4].

The force/torque sensor was calibrated by applying known external loads up to 4 N and 100 Nmm. The measurement range is 20 N and 200 Nmm, respectively. Exemplary calibration results are shown in Fig. 3(b).

To reduce the influence of noise the electronic measuring equipment is placed inside the instrument shaft. The digital resolution is approximately 9 bits, the sample rate is 800 Hz. Further details on the sensor design are presented in [4], see also Fig. 21.4 for details. On the same basis of a miniaturized force/torque sensor, a new instrument with two additional DoF and an actuated end-effector at the distal end was built (see Fig. 21.5) [36, 37]. Grasping forces at the pair of forceps can be measured independently. The instrument is currently being tested with respect to position and force measurement accuracy, thermal stability, etc.

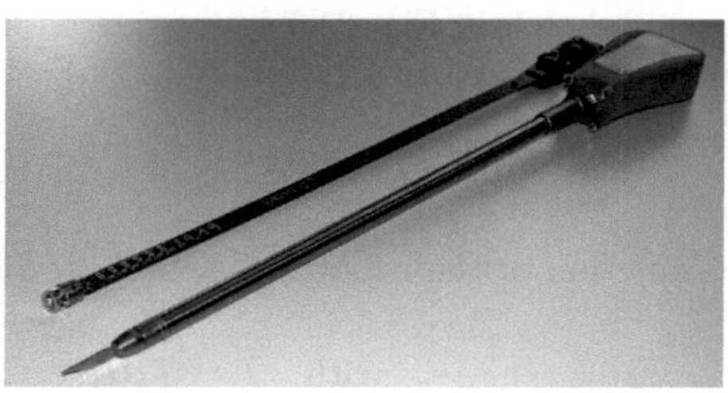

Fig. 21.4. Sensorized scalpel with force/torque sensor at tip and electronics

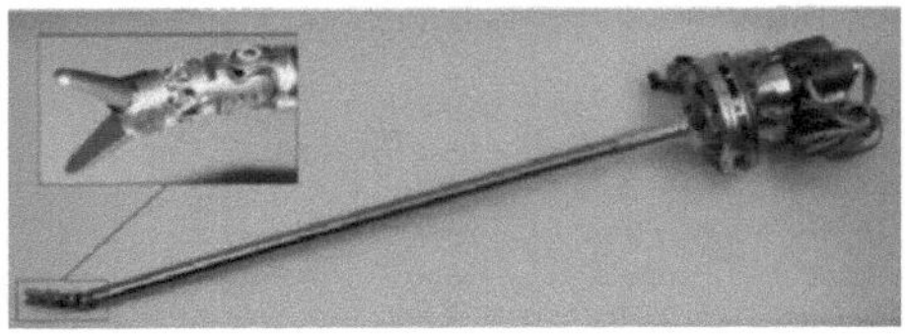

Fig. 21.5. Sensorized pair of forceps, with force/torque sensor and two additional DoF at distal end as well as drive unit at proximal end

Robot Control

While the camera-robot runs autonomously, the instrument-robot is teleoperated. This robot is commanded via serial connection (RS 232, 38.4 kbaud). The control software is implemented in C/C++ on a SUN Ultra 60 UPA/PCI. The position control architecture is designed to meet the specific kinematic requirements of an invariant point and considers the passive joints as well (see Fig. 21.6). The dexterity of the entire system is increased as correct hand-eye coordination is realized. The information flow can be described as follows: The trocar position estimation provides the position $\boldsymbol{t}$ of the entry point which is used together with the joint positions $\boldsymbol{\theta}$ to compute the inverse Jacobian matrix $\boldsymbol{P}^{-1}$. On this basis the simplified dynamics can be calculated, which is a prerequisite for a self-adjusting controller. By a self-adjusting controller (see below) the different joint dynamics can be taken into account resulting in Cartesian dynamic behavior independent of the working position. The controller is described by the transfer functions F_{CX}, F_{CY}, and F_{CZ} which are tuned for the x-, y-, and z-direction separately. The subscript D (see Fig. 21.6) denotes the desired values, whereas the subscripts p and a indicate the passive and the active joints, respectively. The transfer function $F_D = e^{-\frac{T_d}{2}s}$ represents the delay transfer function caused by the serial connection ($T_d = 22$ ms). The delay time T_d dominates the delay of the communication network between master and slave (TCP/IP via Ethernet) which is less than 1 ms.

A self-adjusting controller was chosen, as the Cartesian dynamics of the robot depends on the current configuration. This is due to the fact, that the robot joints have different dynamics which are mapped via the (position dependent) Jacobian into the Cartesian space [38]. The control loop for the simplified one degree of freedom case as shown in Fig. 21.7 is considered in the following.

The controller transfer function for each Cartesian DoF of the robot is written as

$$F_C = K\frac{1+T_1 s}{1+T_2 s} . \tag{21.2}$$

The velocity transfer function of the robot for each Cartesian DoF is:

$$F_P = \frac{1}{1+Ts} \quad \text{with} \quad T = T(\boldsymbol{\theta}) . \tag{21.3}$$

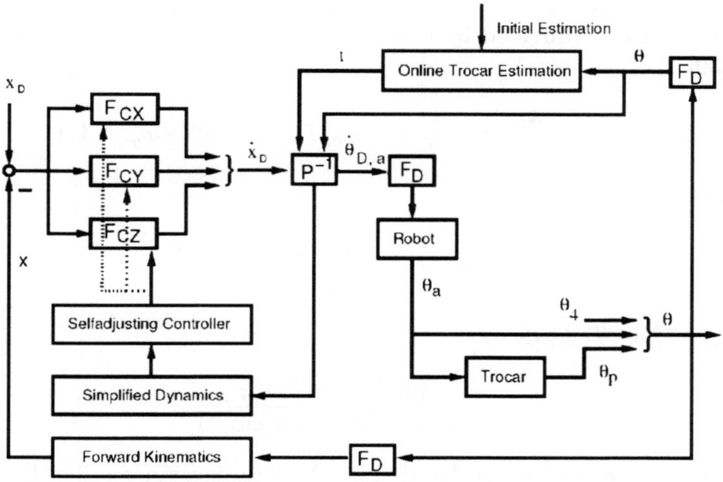

Fig. 21.6. Teleoperator position control loop

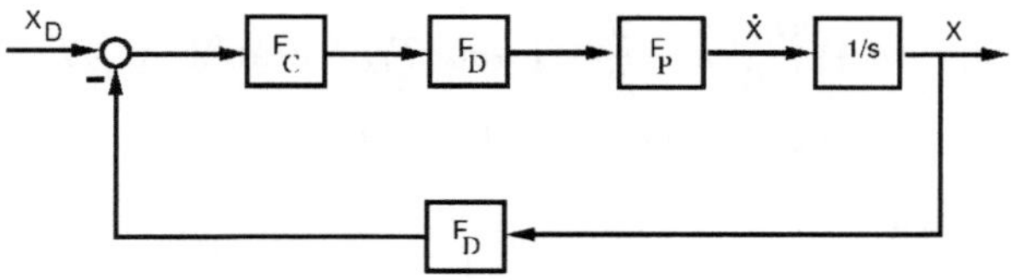

Fig. 21.7. Closed control loop

Note that T is not only position dependent but also differs for the Cartesian DoF of the robot. In the following, the equations to tune the parameters for F_C for one DoF are derived. For other DoF the same equations hold.

The gain of the open loop equation is as follows:

$$A = K\frac{\sqrt{1+(wT_1)^2}}{\sqrt{1+(wT_2)^2}}\frac{1}{\sqrt{1+(wT)^2}}\frac{1}{w} = KA^* \; . \tag{21.4}$$

The corresponding phase is:

$$\phi = \text{atan2}(wT_1, 1) + \text{atan2}(-wT_2, 1) + \text{atan2}(-wT, 1) - \frac{\pi}{2} - wT_D \; . \tag{21.5}$$

Considering the structure of the controller F_C three parameters have to be determined: K, T_1, and T_2. First $T_2 = kT_s$ is chosen, with T_s being the sample time of the digital implementation, to move the negative part of the phase of $\frac{1}{1+T_2 s}$ as far as possible towards high w. One way to compute T_1 is to choose $T_1 = T$ to compensate the pole of the plant; this works well for $T_D \ll T$ only. An adaption law able to handle the general case is: For a small w_{g1} a desired phase margin

ϕ_{R1} (e.g. 80°) is chosen that provides good damping. The necessary phase shift $\Delta\phi_1$ that has to be provided by the controller F_C at w_{g1} can be computed as follows [39]:

$$\Delta\phi_1 = -\pi + \frac{\phi_{R1}}{180^\circ}\pi + \frac{\pi}{2} - (-w_{g1}T_D + \operatorname{atan2}(-w_{g1}T, 1)) \ . \tag{21.6}$$

The phase of F_C at w_{g1} is:

$$\angle F_C = \angle \frac{1 + jw_{g1}T_1}{1 + jw_{g1}T_2} \ . \tag{21.7}$$

Solving (21.7) for T_1 leads to:

$$T_1 = \frac{\tan(\Delta\phi_1) + T_2 w_{g1}}{w_{g1}(1 - T_2 w_{g1}\tan(\Delta\phi_1))} \ . \tag{21.8}$$

The last parameter to be calculated is K. A phase-margin ϕ_{R2} for F_{openloop} at the gain crossover-frequency w_{g2} is chosen and (21.9) is solved for $w = w_{g2}$:

$$\begin{aligned}\Delta\phi_2 &= 0\\ &= -\pi + \frac{\phi_{R2}}{180^\circ}\pi + \frac{\pi}{2} - (-w_{g2}T_D + \operatorname{atan2}(-w_{g2}T, 1) + \\ &\quad + \operatorname{atan2}(w_{g2}(T_1 - T_2), 1 + T_1T_2w_{g2}^2)) \ .\end{aligned} \tag{21.9}$$

Finally, with (21.4):

$$K = \frac{1}{A^*(w_{g2})} \ , \tag{21.10}$$

because w_{g2} is the gain crossover-frequency. As T differs by a large range, but changes slowly, the system can be considered as quasi-linear and the control law is stable.

The automatic control law is configured such that the phase margin of the closed position loop remains between 80° and 60°. This adjustment guarantees a well damped behavior over the entire workspace and reduces the risk of overshooting. The cross-over frequency of the position control loop is about 0.5 Hz, which causes an undesired phase shift between the desired position and the actual position of the robot. See [5, 38] for more details.

21.3.2 Operator Console

The operator console (see Fig. 21.8, left) consists of a stereo display (25 Hz active stereo with shutter glasses[2]) and a PHANToM (SensAble Technologies Inc.). This kinesthetic device provides 6 DoF for position and orientation sensing and uses 3 translational DoF for force feedback. Additionally, the forces can be displayed in the stereo video stream at the TCP of the surgical instrument by

[2] In order to avoid flickering a monitor allowing 50 Hz stereo images was used for the experiments presented in Sec. 21.4.

means of a 3D arrow (see Fig. 21.8, right). This enables the surgeon to receive feedback of the manipulation forces, even if no kinesthetic feedback device is available. Furthermore, to provide the surgeon with information on the robot configuration, a robot model can be displayed, thus, no additional bandwidth demanding video transmission is necessary.

The position x_P obtained by the PHANToM is scaled by the factor k_S before being sent to the teleoperator. This ensures a correct hand-eye coordination and provides position scaling in order to manipulate the instrument tip more precisely. The roll axis of the instrument is commanded by the last rotational DoF of the PHANToM. The force F_T and the position x_T measured at the end effector (instrument tip) of the teleoperator are transferred to the operator console. Thereby a force F_P is calculated:

$$F_P = k_F F_T + k_C(x_T - k_S x_P) \tag{21.11}$$

and displayed at the PHANToM. For $k_F = 1$ and $k_C = 0$ the force F_P corresponds to the measured force F_T. By changing the values for k_F the displayed forces are sensed as scaled. The component $k_C(x_T - k_S x_P)$ causes a position coupling between operator and teleoperator, whereby x_P represents the desired position of the instrument tip. This position coupling constitutes a safety feature as it prevents the user to command too fast motions which cannot be executed by the teleoperator. The PHANToM control loop runs with 1 kHz. Further details of the teleoperation concept are presented in [40].

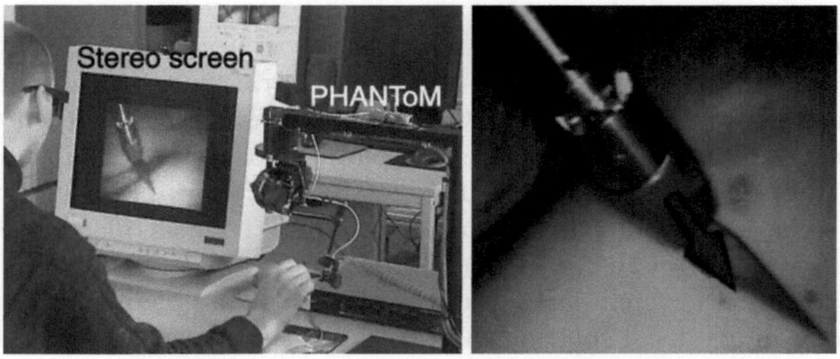

(a) Operator. (b) Visual force feedback.

Fig. 21.8. DLR telesurgery scenario: operator console

Additionally to the presented operator station based on a PHANToM two different concepts for the man-machine interface where realized. The new prototypical device to feedback grasping forces shown in the left part of Fig. 21.9 offers a very natural way of force display.

In contrast a user interface where the surgeon moves standard surgical instruments without force feedback was evaluated, too. Optical markers are attached to these instruments and are tracked by stereo cameras (see Fig. 21.9b). Therefore, the current pose of the instruments can be reconstructed. The captured

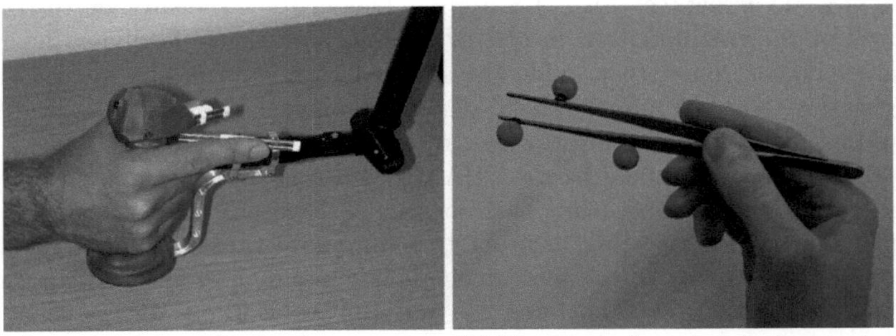

(a) Device to feedback grasping forces. (b) Tracking of surgical instruments.

Fig. 21.9. Prototypes for minimally invasive robotic surgery under development

instrument motion is then transmitted to the robot or to a virtual reality simulation. Kinesthetic feedback is not possible with such an approach, but forces can be displayed in the (stereo) video stream by augmented reality techniques (see Fig. 21.8 in Sec. 21.3). Evaluating such a system and comparing the results with the setup described above allows for a comparison of different force feedback modalities and man-machine interfaces.

21.3.3 Communication

The communication between teleoperator and operator console is established by a TCP/IP protocol via Ethernet. For network transparency a Common Object Request Broker Architecture (CORBA) middle-ware layer is used and thus communication is independent of a certain platform or a specific implementation [41]. The implemented architecture provides streams for positions and forces as well as channels for event based commands (e.g. to connect/disconnect master and slave or to open/close a gripper) [42].

To reduce overall system latency in the experiments, video transmission was realized as a simple local solution. A pair of framegrabbers capture the analog camera signal which is displayed by an OpenGL based visualization using shutter glasses. Advanced solutions for the transmission of compressed videostreams are readily available, and therefore not subject to this research. The images are not rectified for visualization since the laparoscope has only minor radial distortion.

21.4 Force Feedback: An Experimental Evaluation

In order to gain a deeper insight into the importance of force feedback for MIRS systems an experimental evaluation was carried out. The technical set-up used here differs slightly from that described in Sec. 21.3: the automatic camera guidance was turned off as it was not subject to evaluation. Additionally, an active stereo screen based on shutter glasses with a refresh rate of 100 Hz was integrated into the robotic surgery system. The stereo image itself was updated with 50 Hz, thus flickering was avoided. To realize force feedback, the PHANToM based solution was used as an operator console.

As surgeons spend about 25-35% of their operation time on dissecting tissue [43] a representative surgical task was realized: An artery that was covered by tissue and that could be recognized as elevation only should be dissected as fast and as un-injured as possible. Although this task is most often carried out by a dissection hook, here, a scalpel was chosen for assessing manipulation errors (i. e. injuries of blood vessels) more exactly. In order to create comparable conditions no organic material was used, but tissue was replaced with modeling material and arteries were substituted by cellular rubber (see Fig. 21.10). Although the experimental material does not correspond to the visco-elastic characteristics of real tissue the force profiles of the artificial models may be assumed to be similar to natural tissue [44].

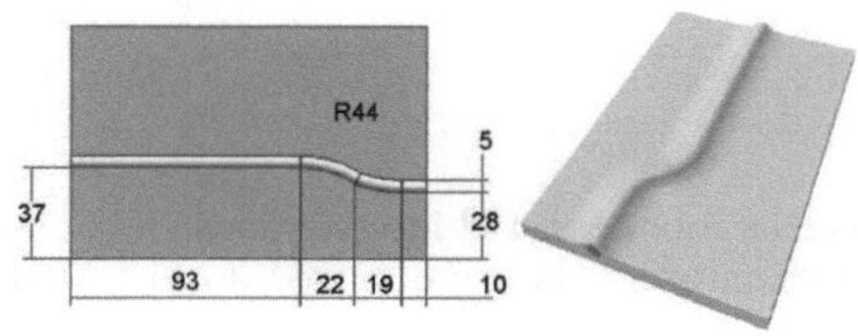

Fig. 21.10. Experimental scenario: An artificial artery was to be dissected under standardized conditions (dimensions in mm)

Surgeons mainly performing minimally invasive procedures were recruited as participants. After a sufficient practice session all of the 25 participants were asked to carry out a manual intervention as well as robot assisted interventions with and without force feedback. Learning or fatigue effects were counterbalanced by a Latin Square Design [45]. As dependent variables speed as well as accuracy were recorded. Thereby, speed was operationalized by the amount of surface which was dissected within four minutes. To guarantee an objective assessment all items were photographed and the dissected surface was measured in pixel. Accuracy was measured by the extent of injury which was made up of the length, the depth, and the amount of tissue and artery transection. Therefore, all items were judged by three independent raters. According to a scale proposed by [46] the inter-rater reliability (Cohen's Kappa: 0.82) can be described as almost perfect. In order to explore differences between manual and robot assisted interventions all trials were filmed and an observational video analysis was carried out by three independent raters. As according to [46] the inter-rater agreement (Cohen's Kappa: 0.72) can be described as substantial the video analysis is assumed to be reliable.

In order to explore whether MIS, MIRS without force feedback, and MIRS with force feedback differ, an analysis of variances (ANOVAs) was carried out [47]. For the speed-variable amount of dissected surface as well as for one of the accuracy-variables, depth of artery transection, significant differences could be detected on a 0.05-level. A Bonferroni Post-Hoc Test revealed that the participants dissected significantly more surface (55.1 %) when the intervention was accomplished manually instead of robot assistedly without force feedback. Within MIRS without force feedback the participants were a little bit faster (9.4 %) compared to MIRS with force feedback though both robot conditions do not differ significantly. To sum up, the surgeons managed the task the fastest manually while robot assisted surgery caused a significant deceleration (see Fig. 21.11 left).

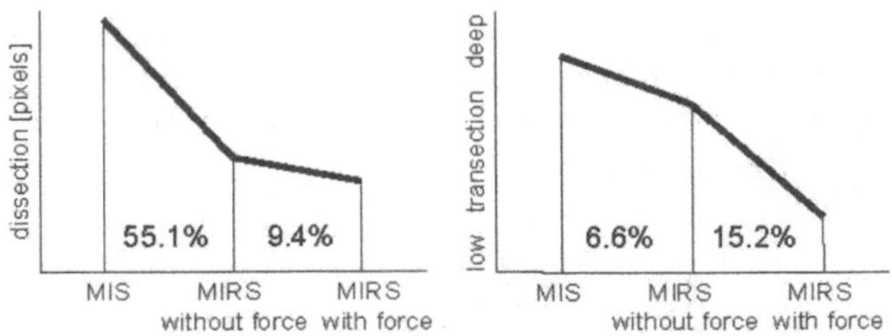

Fig. 21.11. Manual versus robot assisted interventions: While MIS was accomplished much faster (left), significantly less injury occurred during 'MIRS with force feedack' (right)

The results are reverse when the accuracy-variable is regarded (see Fig. 21.11 right): In average the deepest artery transection occurred when the task was executed manually. While during MIRS without force feedback the task was performed a little bit more carefully (6.6 %) compared to a manual intervention, artery transection could be reduced only significantly (15.2 %) when force feedback was available. To sum up, the surgeons managed the task the most precise when force feedback was available.

At a first glance the results may be associated with a speed-accuracy trade-off. It may be argued that during robotic interventions a more accurate result was to be observed only due to the fact that the surgeons worked slower and not because of the availability of force feedback. This assumption cannot be held for two reasons: First, the extent of injuries was related to the dissected surface and thus the ANOVA-calculations are not based on absolute, but on relative, standardized values. Second, further interesting insights are gained by the video analysis: When manual and robot assisted interventions are compared, a t-Test reveales that significantly more cutting operations appeared within MIRS (see Fig. 21.12). As it is obvious that the risk of injury increases especially during cutting, the fact that within MIRS significantly less injury occurred can only be explained by the availability of force feedback.

The video analysis provides also insight concerning the training demands of robot assisted surgery. Irrespective of whether the task was performed manually or robot assistedly the same techniques, 'scraping', 'lifting', and 'cutting', were applied (see Fig. 21.12). Though both conditions differ quantitatively there are no qualitative different operations. A remarkable difference is only apparent when dynamics of movement is considered: During MIS the surgeons tended to operate more often length- as well as crosswise to the artery. Besides, the direction was changed more frequently and the participants tended to move more often back and forth compared to MIRS. In consequence, robot assisted surgery does not afford new operating techniques but requires a more continuous working style which has to be trained by experienced surgeons.

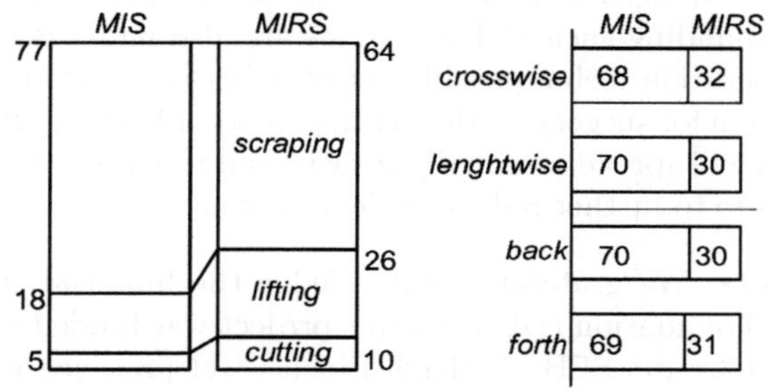

Fig. 21.12. Operating techniques (left) and dynamic of movements (right) within MIS and MIRS (in %). The following definitions apply: crosswise and lengthwise refer to the artery direction, back and forth are related to person's view direction. Irrespective of whether the task was performed manually or robot assisted the same techniques were applied (scraping, lifting, cutting).

21.5 Conclusions

Manual and robot assisted minimally invasive surgery techniques turn out to have certain advantages as well as disadvantages: Whenever it is essential to reduce operation time a manual technique seems to be most suitable, whenever any unnecessary traumatization is to be avoided a robot assisted technique with force feedback will be more appropriate.

Though all in all the participants tended to work more carefully within the robot surgery setting, a significant reduction of injuries was to be observed only when force feedback was available. While this fact is comprehensible it is less obvious why at the same time operating time doubles; especially when it is considered that the robot surgery stands out by a more intuitive hand-eye coordination. This observation is probably due to the low bandwidth in the position control loop of the robot (see Sec. 21.3). This fact causes a phase shift between the desired position and the current position of the instrument. Consequently, the users tended to work slower to guarantee an accurate positioning of the instrument (see Fig. 21.11). Nevertheless, despite of this possible drawback the present force reflecting setup reduces unintentional injuries successfully.

In general, evaluation results depend on the experimental system under consideration. Therefore, in order to derive a more general conclusion, further enhancement of the components used for the MIRS system is desirable. To increase the bandwidth of the telesurgery system (i. e. to reduce the time delay between the surgeon's commands and the motion of the robots) an increased dynamics of the surgical robot and a shorter communication delay between surgical robot and operator console are needed. Therefore, a new kinematically redundant surgical robot with fast dynamics and high sample rate (3 kHz) was developed [48]. To provide full dexterity inside the patient the latest DLR instruments (see Fig. 21.5) are designed to integrate a miniaturized force/torque sensor as well as to provide two additional DoF and an actuated pair of forceps at the distal end. Suitable user interfaces for commanding such MIRS systems are also under development (see Fig. 21.9). New operation techniques like automatic camera guidance [33, 34] or motion compensation for surgery on the beating heart [49, 50, 51] will be available in the future. This will spread minimally invasive surgery procedures drastically, and, thus, contribute to further reduce patient trauma.

Acknowledgments. We gratefully acknowledge the financial support by the German Research Foundation (DFG) as this project was funded within the Collaborative Research Center SFB 453 'High-Fidelity Telepresence and Teleaction'.

References

1. M.R. Treat. *Computer-Integrated Surgery*, chapter A Surgeon's Perspective on the Difficulties of Laparoscopic Surgery, pages 559–560. MIT Press, 1995.
2. M.A. Helmy. A comparative study between laparoscopic versus open appendicectomy in men. *J Egypt Soc Parasitol*, Aug. 2001.
3. S. Sauerland, R. Lefering, and EA. Neugebauer. Laparoscopic versus open surgery for suspected appendicitis. *Cochrane Database Syst Rev*, 2002.
4. U. Seibold and G. Hirzinger. A 6-axis force/torque sensor design for haptic feedback in minimally invasive robotic surgery. In *Proceedings of the 2nd VDE World Microtechnologies Congress*, Munich, Germany, October 2003.
5. T. Ortmaier and G. Hirzinger. Cartesian control issues for minimally invasive robot surgery. In *Proceedings of the IEEE/RSJ International Conference on Intelligent Robots and Systems IROS 2000*, Takamatsu, Japan, October 2000.
6. T. Ortmaier, H. Weiss, and G. Hirzinger. Minimally invasive robotic surgery: Foundations and perspectives. In *IEEE International Conference on Robotics and Automation: Workshop on Recent Advances in Medical Robotics,*, Taipei, Taiwan, September 2003.
7. T. Ortmaier. *Motion Compensation in Minimally Invasive Robotic Surgery.* VDI Verlag, 2003. PhD Thesis.
8. Y. Nakamura. Virtual stillness and small size robot system that occupies less space in or. In *International Conference on Robotics and Automation (ICRA) 2003, Workshop Recent Advances in Medical Robotics*, Taipei, Taiwan, 2003.
9. V. Falk, A. Diegeler, T. Walther, B. Vogel N. Löscher, C. Ulmann, T. Rauch, and F. W. Mohr. Endoscopic coronary artery bypass grafting on the beating heart using a computer enhanced telemanipulation system. *Heart Surg Forum*, 2:199–205, 1999.

10. D. H. Boehm, H. Reichenspurner, C. Detter, M. Arnold, H. Gulbins, B. Meiser, and B. Reichart. Clinical use of a computer-enhanced surgical robotic system for endoscopic coronary artery bypass grafting on the beating heart. *Thorac Cardiovasc Surg*, 48(4), 2000.
11. A. Tewari, A. Srivasatava, M. Menon. A prospective comparison of radical retropubic and robot-assisted prostatectomy: experience in one institution. *BJU Int*, 92(3):205–10, 2003.
12. V. Dotzel, D. Wetzel, D. Wilhelm, A. Schneider, G. Wessels, and H. Feussner. Robotic and navigation systems: surgical practicability and benefit for the patient? *Zentralbl Chir*, 128(3):227–31, March 2003.
13. W.D. Boyd, K.D. Stahl. Janus syndrome: a perspective on a new era of computer-enhanced robotic cardiac surgery. *J Thorac Cardiovasc Surg*, 126(3):625–30, September 2003.
14. V. Falk, S. Jacobs, J.F. Gummert, T. Walther, and F.W. Mohr. Computer-enhanced endoscopic coronary artery bypass grafting: the davinci experience. *Semin Thorac Cardiovasc Surg*, 15(2):104–11, April 2003.
15. R.J. Novick, S.A. Fox, B.B. Kiaii, L.W. Stitt, R. Rayman, K. Kodera, A.H. Menkis, and W.D. Boyd. Analysis of the learning curve in telerobotic, beating heart coronary artery bypass grafting: a 90 patient experience. *Ann Thorac Surg*, 76(3), September 2003.
16. P. Bergmann, S. Huber, H. Segl, H. Maechler, U. Reiter, G. Reiter, R. Rienmueller, P. Oberwalder, and B. Rigler. Cardiac MR in robotic heart surgery for preoperative identification of the target vessel and precise port placement – a theoretical model. *Thorac Cardiovasc Surg*, 51(4):204–10, August 2003.
17. V. Falk, S. Jacobs, J.F. Gummert, and T. Walther. Robotic coronary artery bypass grafting (CABG) – the Leipzig experience. *Surg Clin North Am*, 83(6):1381–6, December 2003.
18. S. Jacobs, D. Holzhey, B.B. Kiaii, J.F. Onnasch, T. Walther, F.W. Mohr, and V. Falk. Limitations for manual and telemanipulator-assisted motion tracking – implications for endoscopic beating-heart surgery. *Ann Thorac Surg*, 76(6):2029–35, December 2003.
19. R.H. Taylor and D. Stoianovici. Medical robotics in compter-integrated surgery. *IEEE Transactions on Robotics and Automation*, 19(5):765–781, October 2003.
20. J.M. Sackier and Y. Wang. *Computer-Integrated Surgery*, chapter Robotically Assisted Laparoscopic Surgery: From Concept to Development, pages 577–580. MIT Press, 1995.
21. G. Guthart and J. Salisbury. The intuitive telesurgery system: Overview and application. In *Proceedings of the 2000 IEEE International Conference on Robotics and Automation*, San Francisco, U.S.A., April 2000.
22. H. Düpree. Laprotek – Master Slave Systeme in der Viszeralchirurgie. In *2. Jahrestagung der Deutschen Gesellschaft für Computer- und Roboterassistierte Chirurgie (CURAC)*, Nürnberg, Germany, November 2003.
23. Intuitive Surgical Inc. Webpage. http://www.intuitivesurgical.com/, 2005.
24. M.C. Cavusoglu, W. Williams, F. Tendick, and S.S. Sastry. Robotics for telesurgery: Second generation Berkeley/UCSF laparoscopic telesurgical workstation and looking towards the future applications. In *Proceedings of the 39th Allerton Conference on Communication, Control and Computing*, Monticello, Italy, October 2001.
25. D.S. Kwon, K.Y. Woo, S.K. Song, W.S. Kim, H.S. Cho. Microsurgical telerobot system. In *Proc. of the IEEE/RSJ Int. Conf. on Intelligent Robots and Control Systems*, 1998.

26. T. Hu, A. Castellanos, G. Tholey, and J. Desai. Real-time haptic feedback in laparoscopic tools for use in gastro-intestinal surgery. In *Medical Image Computing and Computer-Assisted Intervention - MICCAI 2002: 5th International Conference, Proceedings*, pages 66–74, Tokyo, Japan, September 2002.
27. M. Tavakoli, R.V. Patel, and M. Moallem. A force reflective master-slave system for minimally invasive surgery. In *Proc. of the IEEE/RSJ International Conference on Intelligent Robots and Systems IROS 2003*, Las Vegas, USA, October 2003.
28. N. Zemiti, T. Ortmaier, M. Vitrani, and G. Morel. A force controlled laparoscopic surgical robot without distal force sensing. In *Proc. of the ISER 2004; 9th International Symposium on Experimental Robotics*, Singapore, June 2004.
29. H. Mayer, I. Nagy, A. Knoll, E. Schirmbeck, R. Bauernschmitt. Robotic system to evaluate force feedback in minimally invasive computer aided surgery. In *Proc. of the 2004 ASME Design Engineering Technical Conferences*, Salt Lake City, Utah, USA, 2004.
30. E.U. Schirmbeck, H. Mayer, I. Nagy, A. Knoll, R. Lange, and R. Bauernschmitt. Evaluation of force feedback in minimally invasive robotic surgery. In *Fachtagung "Biomedizinische Technik"*, Technische Universität Ilmenau, September 2004.
31. E. Burdet, R. Gassert, F. Mani, F. Wang, C. Teo, and H. Bleuler. Design of a haptic forceps for microsurgery training. In *Proceedings of the 4th International Conference EuroHaptics 2004*, Munich, Germany, June 2004.
32. ForceDimension. Webpage. http://www.forcedimension.com, June 2004.
33. G.-Q. Wei, K. Arbter, and G. Hirzinger. Real-time visual servoing for laparoscopic surgery. *IEEE Engineering in Medicine and Biology*, 16(1), January/February 1997.
34. K. Omote, H. Feussner, A. Ungeheuer, K. Arbter, G.-Q. Wei, J. R. Siewert, and G. Hirzinger. Self-guided robotic camera control for laparoscopic surgery compared with human camera control. *The American Journal of Surgery*, 117:321–324, April 1999.
35. M. Sorli and S. Pastorelli. Six-axis reticulated structure force/torque sensor with adaptable performances. *Mechatronics*, 5(6):585–601, 1995.
36. B. Kübler, U. Seibold, and G. Hirzinger. Development of actuated and sensor integrated forceps for minimally invasive robotic surgery. *International Journal of Medical Robotics and Computer Assisted Surgery*, 1(3):96–107, 2005.
37. U. Seibold, B. Kuebler, H. Weiss, T. Ortmaier, and G. Hirzinger. Sensorized and actuated instruments for minimally invasive robotic surgery. In *Proceedings of the 4th International Conference EuroHaptics 2004*, Munich, Germany, June 2004.
38. T. Ortmaier and G. Hirzinger. Cartesian control of robots with working-position dependent dynamics. In *Proceedings of the 6th International IFAC Symposium on Robot Control – Syroco 2000*, Vienna, Austria, September 2000.
39. C. Natale, R. Koeppe, and G. Hirzinger. An automatic procedure for force controller design. In *IEEE/ASME International Conference on Advanced Intelligent Mechatronics*, Atlanta, Georgia, USA, sep 1999.
40. C. Preusche, T. Ortmaier, and G. Hirzinger. Teleoperation Concepts in Minimally Invasive Surgery. In *Proceedings of 1. IFAC Conference on Telematics Application in Automation and Robotics*, Weingarten, July 2001. VDI/VDE - GMA.
41. Object Management Group (OMG). *The Common Object Request Broker: Architecture and Specification*, 1998. OMG Document 98-02-33.
42. D. Reintsema, J. Vogel, G. Hirzinger, and T. Unterschütz. CORBA – Ein standardisierter Software-Bus für verteilte Anwendungen in der Robotik. In *Industrielle Automation und Internet/Intranet-Technologie*, VDI-Berichte 1515, pages 181–192. VDI Verlag, 1999.

43. C. E. H. Scott-Conner. *The SAGES manual: fundamentals of laparoscopy and GI endoscopy.* Springer, New York, 1999.
44. C. R. Wagner, N. Stylopoulos, and R. Howe. The role of force feedback in surgery: Analysis of blunt dissection. In *10th Symposium on Haptic Interfaces for Virtual Environment and Teleoperator Systems*, pages 73–79, 2002.
45. W. J. Diamond. *Practical Experiment Designs: for Engineers and Scientists.* John Wiley, New York, 2001.
46. J. Landis and G. Koch. The measurement of observer agreement for categorical data. *Biometrics*, 33:159–174, 1977.
47. L. K. Edwards. *Applied analysis of variance in behavioral science.* Dekker, 1993.
48. T. Ortmaier, H. Weiss, U. Hagn, M. Grebenstein, M. Nickl, A. Albu-Schäffer, C. Ott, S. Jörg, R. Konietschke, L. Le-Tien, and G. Hirzinger. A Hands-On-Robot for Accurate Placement of Pedicle Screws. In *To be presented at: IEEE International Conference on Robotics and Automation (ICRA)*, Orlando, Florida, USA, May 2006.
49. T. Ortmaier, M. Groeger, D. H. Boehm, V. Falk, G. Hirzinger. Motion estimation in beating heart surgery. *IEEE Transactions on Biomedical Engineering*, 52(10):1729–1740, 2005.
50. Y. Nakamura, K. Kishi, and H. Kawakami. Heartbeat synchronization for robotic cardiac surgery. In *IEEE International Conference on Robotics and Automation (ICRA)*, pages 2014–2019, Seoul, Korea, May 2001.
51. R. Ginhoux, J.A. Gangloff, M.F. de Mathelin, L. Soler, M.M. Arenas Sanchez, and J. Marescaux. Beating heart tracking in robotic surgery using 500 Hz visual servoing, model predictive control and an adaptive observer. In *IEEE International Conference on Robotics and Automation (ICRA)*, pages 274–279, New Orleans, USA, April 2004.

Telerobotic Control by Virtual Fixtures for Surgical Applications

Ming Li, Ankur Kapoor, and Russell H. Taylor

Johns Hopkins University
Department of Computer Science
3400 N. Charles St.
Baltimore, MD 21218, USA
{liming,kapoor,rht}@cs.jhu.edu

Summary. We present a new method to generate spatial motion constraints for surgical robots that provide sophisticated ways to assist the surgeon. Surgical robotic assistant systems are human-machine collaborative systems (HMCS) that work interactively with surgeons by augmenting their ability to manipulate surgical instruments in carrying out a variety of surgical tasks. The goal of "virtual fixtures" (VF) is to provide anisotropic motion behavior to the surgeon's motion command and to filter out tremor to enhance precision and stability. Our method uses a weighted, linearized, multi-objective optimization framework to formalize a library of virtual fixtures for task primitives. We set the objective function based on user input that can be obtained through a force sensor, joystick or a master robot. We set the linearized subject function based on five basic geometric constraints. The strength of this approach is that it is extensible to include additional constraints such as collision avoidance, anatomy-based constraints and joint limits, by using an instantaneous kinematic relationship between the task variables and robot joints. We illustrate our approach using three surgical tasks: percutaneous needle insertion, femur cutting for prosthetic implant and suturing. For the percutaneous procedures we provide a remote center of motion (RCM) point that provides an isocentric motion that is fundamental to these types of procedures. For femur cutting procedures we provide assistance by maintaining proper tool orientation and position. For the suturing task we address the problem of stitching in endoscopic surgery using a circular needle. We show that with help of VF, suturing can be performed at awkward angles without multiple trials, thus avoiding damage to tissue.

22.1 Introduction

Robotic surgical assistance is an emerging technology of human-computer cooperation to accomplish delicate and difficult surgical tasks. Examples of surgical assistant systems can be seen in laparoscopic surgery [1, 2, 3, 4, 5], microsurgery [6, 7, 8], orthopedic surgery [9] and sinus surgery [10]. The Intuitive daVinci robotic surgical system [3] and the Computer Motion Zeus robotic surgical system [4] are two commercialized surgical telemanipulators which are capable of performing remote telerobotic laparoscopic surgery. Although most of the teleoperated systems are admittance-controlled microsurgical robots ([6, 11, 12, 13, 14]), based on

M. Ferre et al. (Eds.): Advances in Telerobotics, STAR 31, pp. 381–401, 2007.
springerlink.com © Springer-Verlag Berlin Heidelberg 2007

force–reflecting master–slave configurations, some of the teleoperated robotic augmentation systems employ a passive input device for operator control [7], while others are joystick controlled [10].

Most surgical procedures in which robots are called for assistance are characterised by restricted access to the workspace as well as constrained manipulation of the surgical tool. In such cases, the surgeons' ability can be augmented by techniques such as virtual fixtures (VF). Virtual fixtures, which have been discussed previously in the literature for both telerobotic and cooperative robots [15, 16, 17, 18, 19, 20], are algorithms which provide anisotropic behavior to surgeons motion command besides filtering out tremor to provide safety and precision.

An important case of virtual fixtures are forbidden regions, where the surgical tool is restricted to a certain region in the workspace. Davies *et al.* [15] set active constraints to constrain the robot to cut the femur and tibia within a permitted region for prosthetic knee surgery. Park *et al.* [16] developed sensor-mediated virtual fixtures that constrain the robot's motion or create haptic feedback directing the surgeon to move the surgical instruments in a desired direction. They applied a virtual wall based on the location of the internal mammary artery obtained from a preoperative CT scan to guide a surgeon's instrument during teleoperated coronary bypass. The recent work at JHU by Okamura, Hager *et al.* on virtual fixtures [17, 18, 19] used admittance control laws to implement vision-based guidance virtual fixtures for retinal vein cannulation. These works are based either on a specific robot type or on a specific task.

This chapter presents a new method to implement virtual fixtures for surgical assistant robots. We extend the work of Funda *et al.* [21] by applying the method to generate complicated virtual fixtures based on the human's input for surgical assistant robots. Funda presented an optimal motion control method to control both redundant and deficient robotic systems based on motion constraints. Our approach [22] uses a weighted, linearized multi-objective optimization framework to formalize a library of virtual fixtures for task primitives. Our paradigm covers the implementation of guidance virtual fixtures, forbidden region virtual fixtures and combinations of both for generating spatial motion constraints to control the robotic assistant. It is independent of manipulator characteristic and can be used for admittance or impedance type.

22.2 Constrained Motion Control for Virtual Fixtures

Virtual fixtures are task-dependent computer-generated constraints that limit the robot's movement into restricted regions and/or influence its movement along desired paths. The goal of the virtual fixture algorithm is to generate a sequence of incremental motion commands for the robot such that certain task-specific constraints are satisfied. To keep the system intuitive, the incremental motion should be proportional to the user input. For a surgical assistant robot, it is very important to be able to place absolute bounds on the spatial motion of the different parts of the instrument.

Often, surgical robots are designed to be kinematically redundant for providing dexterous assistance. At the same time, certain tasks such as passing a tool through a cavity place certain requirements and constraints on the robot motion and restrict dexterity. Indeed, some special purpose designs for minimally invasive surgery, such as the IBM LARS [23] and the JHU Steady Hand robot [24] provide such constraints through mechanism design. Other robots such as the Intuitive daVinci [3] and Endorobotics [25] combine a kinematically constrained remote center of motion (RCM) mechanism with a kinematically redundant wrist. Thus, it is important for the robot control algorithm to be able to accommodate unique, special purpose mechanical designs (such as kinematically redundant or deficient mechanisms). In this section we present an overview of constrained optimization approach, followed by certain approximations that allow us to execute the algorithm in real-time.

22.2.1 Constrained Optimization Approach

We begin by defining different task frames that specify a point and/or direction of interest associated with different parts of the instrument. For each of the task frames, we define actual state variables $\mathbf{x}$ and desired state variables $\mathbf{x}^d$. The state, $\mathbf{x} = \mathbf{x}(\mathbf{q} + \Delta\mathbf{q})$ is a function of joint variables $\mathbf{q}$ and joint incremental motion $\Delta\mathbf{q}$. The desired state, $\mathbf{x}^d = \mathbf{x}^d(\boldsymbol{\tau}, \mathbf{q})$ is a function of human's input $\boldsymbol{\tau}$, joint variables $\mathbf{q}$.

We can formulate a constrained optimization problem to generate the constrained motion for a certain task frame. The most general formulation for this problem is:

$$\begin{aligned} \Delta\mathbf{q}_{cmd} &= \arg\min_{\Delta\mathbf{q}} \; C(\mathbf{x}(\mathbf{q}+\Delta\mathbf{q}), \mathbf{x}^d) \\ s.t. \;\; & A(\mathbf{x}(\mathbf{q}+\Delta\mathbf{q})) \leq \mathbf{b} \end{aligned} \tag{22.1}$$

where $C(\mathbf{x}(\mathbf{q}+\Delta\mathbf{q}), \mathbf{x}^d)$ is the objective function associated with the difference between the actual state variables $\mathbf{x}$ and the desired state variables $\mathbf{x}^d$. The inequality, $A(\mathbf{x}(\mathbf{q}+\Delta\mathbf{q})) \leq \mathbf{b}$ represents the constraint conditions. These constraints are used to force the solution vector $\Delta\mathbf{q}_{cmd}$ to satisfy certain critical requirements, such as restricting the motion of a part of the instrument within a strict motion envelope.

We can combine the constrained motions on different task frames for generating complicated constrained motions. For an example, assume the virtual fixture for task frame $\{i\}$ is

$$\begin{aligned} \Delta\mathbf{q}_{cmd} &= \arg\min_{\Delta\mathbf{q}} \; C_i(\mathbf{x}_i(\mathbf{q}+\Delta\mathbf{q}), \mathbf{x}_i^d) \\ s.t. \;\; & A_i(\mathbf{x}_i(\mathbf{q}+\Delta\mathbf{q})) \leq \mathbf{b}_i \end{aligned} \tag{22.2}$$

Then the complicated virtual fixtures generated by constraining on task frames $\{i, \, (i = 1, ..., N)\}$ can be formulated as

$$\begin{aligned} \Delta\mathbf{q}_{cmd} &= \arg\min_{\Delta\mathbf{q}} \sum_{i=1}^{N} w_i C_i(\mathbf{x}_i(\mathbf{q}+\Delta\mathbf{q}), \mathbf{x}_i^d) \\ s.t. \;\; & A_i(\mathbf{x}_i(\mathbf{q}+\Delta\mathbf{q})) \leq \mathbf{b}_i \\ & i = 1, \cdots, N. \end{aligned} \tag{22.3}$$

where w_i specifies the relative importance of minimizing the objective function error for different task frames. The combination of a weighted objective function and an additional set of task constraints allow us to exploit the geometry of a particular task space motion and effectively trade off the various performance criteria. Our formulation could easily integrate any behavior, such as asserting joint limits, resolving redundancy or incorporating haptic information to the control strategy.

In this work we discuss virtual fixtures for five task primitives that form the basis of a virtual fixture library. The nomenclature used is presented in the table below.

$\mathbf{x}_i$	Cartesian state of task frame $\{i\}$. $\mathbf{x}_i \in \Re^6$. Subscript i can be omitted for compactness.
$\mathbf{x}_i^d$	Desired Cartesian state of task frame $\{i\}$.
$\mathbf{x}_{p,i}$	Translational component state of task frame $\{i\}$. $\mathbf{x}_{p,i} \in \Re^3$.
$\mathbf{x}_{r,i}$	Rotational component state of task frame $\{i\}$. $\mathbf{x}_{r,i} \in \Re^3$.
$\Delta\mathbf{x}$	Incremental Cartesian motion. $\Delta\mathbf{x} \in \Re^6$.
$\Delta\mathbf{q}$	Incremental joint motion. $\Delta\mathbf{q} \in \Re^n$, n is number of robot joints.
$J(q), J$	Jacobian relating the instantaneous kinematics to joint motion.
$\boldsymbol{\delta}$	Signed distance error between desired and current task frame. $\boldsymbol{\delta} \in \Re^6$
$\boldsymbol{\delta}_p$	Translational component of error.
$\boldsymbol{\delta}_r$	Rotational component of error.
$\hat{\mathbf{l}}$	Orientation of task frame. $\hat{\mathbf{l}} \in \Re^3$. E.g. $\hat{\mathbf{z}}$ component of tool tip rotation matrix.
$\hat{\mathbf{l}}^d$	Desired or reference orientation of task frame.
$\boldsymbol{\tau}$	User input, from force sensor or joystick or master.
$\hat{\mathbf{d}}$	Predefined direction specified by user. $\hat{\mathbf{d}} \in \Re^3$. E.g normal to plane, direction of path.
ϵ_i	Small positive numbers. $i = 1, 2, \ldots$

Using the formulation shown in (22.3), one or more of these primitives applied to one or more task frames can be combined to create complex fixtures. Complicated surgical tasks can then be composed from a sequence of these customized virtual fixtures. The names and descriptions of these task primitives are listed below.

- **Stay on a point:** Keep the tool position represented by $\mathbf{x}_p$ fixed on the reference position $\mathbf{x}_p^d$.
- **Maintain a direction:** Keep the tool orientation represented by $\hat{\mathbf{l}}$ aligned with the reference direction $\hat{\mathbf{l}}^d$.

- **Move along a line:** Keep the tool position, $\mathbf{x}_p$ on line L which has the direction $\hat{\mathbf{d}}$ and passes through point $\mathbf{L}_0$. At the same time, the tool should move along L proportional to the human's input $\boldsymbol{\tau}$.
- **Rotate around a line:** Keep the tool orientation, $\hat{\mathbf{l}}$ perpendicular to line L which has the direction $\hat{\mathbf{d}}$ and passes through point $\mathbf{L}_0$. At the same time, the tool should rotate around L proportional to the human's input $\boldsymbol{\tau}$.
- **Stay above a plane:** Keep the tool position, $\mathbf{x}_p$ stay above a plane Π which has the normal direction $\hat{\mathbf{d}}$ pointing to the free half space and passes through point $\mathbf{P}_0$. At the same time, the tool should move proportional to the human's input $\boldsymbol{\tau}$.

We define a desired nominal behavior together with constraints specifying how far actual behavior can differ from the nominal for each of these primitives. The terms in the objective function of (22.3) are used to relate the desired motion to user input, while the constraints place an absolute bound on the motion. In Sec. 22.3 we elaborate on constraints for each of the five primitive virtual fixtures.

22.2.2 Linearly Constrained Control

The general form of the optimization problem desribed by (22.3) has many variants, both for the objective function and the constraints. As in Funda's work [21], we specialize (22.3) to produce a quadratic optimization problem with linear constraints. We use linear constraints because of the efficiency and the robustness of the computation. The objective function is a two-norm of motion error in different task frames. Because the velocity of a surgical robot is relative low, we can use robotic instantaneous kinematics to map the different task frames to joint variables. The incremental motion of a certain task space is approximated as $\Delta\mathbf{x} = J(\mathbf{q})\Delta\mathbf{q}$. Then we set an optimization problem over incremental joint motion $\Delta\mathbf{q}$. The cost function for task frame $\{i\}$ has the form $\left\|J_i(\mathbf{q})\Delta\mathbf{q} - \Delta\mathbf{x}_i^d\right\|_2^2$, and the constraint has the form $A_i \cdot J_i(\mathbf{q})\Delta\mathbf{q} \leq \mathbf{b}_i$. This inequality is linear in terms of our variables $\Delta\mathbf{q}$. The matrix A_i and vector $\mathbf{b}_i$ are based on the virtual fixture primitive for task frame $\{i\}$.

Then the complicated virtual fixtures generated by combining task frames $\{i, (i = 1, ..., N)\}$ can be expressed as

$$\begin{bmatrix} A_1 & & 0 \\ & \ddots & \\ 0 & & A_N \end{bmatrix} \begin{bmatrix} J_1(\mathbf{q}) \\ \vdots \\ J_N(\mathbf{q}) \end{bmatrix} \Delta\mathbf{q} \leq \begin{bmatrix} \mathbf{b}_1 \\ \vdots \\ \mathbf{b}_N \end{bmatrix} \tag{22.4}$$

where $J_i(\mathbf{q})$ $(1 \leq i \leq N)$ is the Jacobian matrix that maps Cartesian velocities of frame $\{i\}$ to the joint space. If the constraints are placed on the rotational (with subscript r) and translational (with subscript p) components of the same frame, the virtual fixtures can be expressed as

$$\begin{bmatrix} A_p \\ A_r \end{bmatrix} J(\mathbf{q})\Delta\mathbf{q} \leq \begin{bmatrix} \mathbf{b}_p \\ \mathbf{b}_r \end{bmatrix} \tag{22.5}$$

Our control algorithm need not to be modified for singularity, but a safety check is important for actual robot control. Our optimization algorithm forms a least squares problem with linear inequality constraints (algorithm LSI in [26]). At each control loop, we check the solution of LSI before we command robot to move. If the singularity is reached, then there will be no solution for LSI. If the robot is very close to singularity, the solution for the LSI will be extremely large. In both cases, we should stop the robot. Moreover an objective term $||W_q \Delta q||_2^2$ can be added to the overall objective function, where W_q is a diagonal matrix specifying the relative weights between different joints. Such an objective function ensures reasonable and smooth joint velocities when the robot is near singular points. The weights are chosen empirically and we typically use weights in the range 1×10^{-3} to 1×10^{-9}.

Additonally, it is straightforward to incorporate a per joint rate limit by using additional inequality constraints of the form

$$\begin{bmatrix} I \\ -I \end{bmatrix} \Delta \mathbf{q} \leq \begin{bmatrix} \mathbf{q}_{max} - \mathbf{q} \\ \mathbf{q} - \mathbf{q}_{min} \end{bmatrix} \tag{22.6}$$

Besides this, the norm of the cost function gives an indication if an appropriate solution that would satisfy all the constraints is possible. A large norm implies no appropriate solution exists for the given set of constraints. Under such conditions the safest behavior for a surgical robot would be to stop motion. Alternatively some constraints can be relaxed if possible, but we suggest that this should be done only through some user intervention.

22.3 Basic Geometric Constraints for the Virtual Fixture Library

Next we present linearized approximations of constraints for five task primitives. That is, we provide a method to set A and $\mathbf{b}$ in the inequality linear constraints (22.4). We model the robot i^{th} task frame as a purely kinematic Cartesian device with the task frame position $\mathbf{x}_p \in \Re^3$ and the task frame orientation given by unit vector $\hat{\mathbf{l}} \in \Re^3$. Given a reference target, we define the signed distance error $\boldsymbol{\delta} = \left[\boldsymbol{\delta}_p{}^T, \boldsymbol{\delta}_r{}^T\right]^T \in \Re^6$ from the reference target frame to the task frame. The incremental motion for each computational loop is $\Delta\mathbf{x} = \left[\Delta\mathbf{x}_p{}^T, \Delta\mathbf{x}_r{}^T\right]^T \in \Re^6$. We denote translational components by subscript p, and rotational components expressed in Rodriguez angles by subscript r. We assume that both the distance error $\boldsymbol{\delta}$ and the incremental motion $\Delta\mathbf{x}$ are very small and that for small angles, $\boldsymbol{\delta}_r$ and $\Delta\mathbf{x}_r$ approximate Euler Angles. $\epsilon_{1,2,3,4,5}$ are small positive values.

22.3.1 Stay on a Point (VF1)

The first basic geometric constraint we describe is to keep the task frame position on a given target point $\mathbf{x}_p^d \in \Re^3$. The signed errors are then set as $\boldsymbol{\delta} \equiv [\boldsymbol{\delta}_p^T, \boldsymbol{\delta}_r^T]^T = [(\mathbf{x}_p - \mathbf{x}_p^d)^T, \mathbf{0}^T]^T$. We require that after the incremental motion, the task frame position $\mathbf{x}_p + \Delta\mathbf{x}_p$ to be as close to the target point $\mathbf{x}_p^d$ as possible. As shown in Fig. 22.1(a), the constraint can be expressed as

$$\|\boldsymbol{\delta} + \Delta\mathbf{x}\|_2 = \|\boldsymbol{\delta}_p + \Delta\mathbf{x}_p\|_2 \leq \epsilon_1 \tag{22.7}$$

which implies that the various projections of vector $\boldsymbol{\delta}_p + \Delta\mathbf{x}_p$ on the pencil through $\mathbf{x}_p^d$ are less than ϵ_1. We approximate the sphere of radius ϵ_1 by considering a polyhedron with $n \times m$ vertices, and rewrite (22.7) by linear inequalities.

$$\begin{gathered}\begin{bmatrix}\cos\alpha_{1i}\cos\beta_{1j} & \cos\alpha_{1i}\sin\beta_{1j} & \sin\alpha_{1i} & 0 & 0 & 0\end{bmatrix} \cdot (\boldsymbol{\delta} + \Delta\mathbf{x}) \leq \epsilon_1, \\ i = 0, 1, \cdots, n-1; \quad j = 0, 1, \cdots, m-1.\end{gathered} \tag{22.8}$$

where $\alpha_{1i} = \frac{i2\pi}{n}$ and $\beta_{1j} = \frac{j2\pi}{m}$. Then we set A and $\mathbf{b}$ as

$$A = \begin{bmatrix} \cos\alpha_{11}\cos\beta_{11} & \cos\alpha_{11}\sin\beta_{11} & \sin\alpha_{11} & 0 & 0 & 0 \\ & & \cdots & & & \\ \cos\alpha_{1n}\cos\beta_{1m} & \cos\alpha_{1n}\sin\beta_{1m} & \sin\alpha_{1n} & 0 & 0 & 0 \end{bmatrix}, \quad \mathbf{b} = \begin{bmatrix} \epsilon_1 \\ \cdots \\ \epsilon_1 \end{bmatrix} - A\boldsymbol{\delta}. \tag{22.9}$$

Note that (22.7) and (22.8) are equivalent only if $m = n = \infty$. For finite values of m and n, (22.8) results in a polyhedron. As the value of $n \times m$ increases, the volume of polyhedron reduces and the polyhedron approaches the inscribed sphere with radius ϵ_1. Therefore, the linearized conditions of (22.8) are a better approximation to (22.7) for larger values of $n \times m$. However, more constraints require more time to solve the optimization problem. From (22.8), the minimum value for $n \times m$ to obtain a symmetrical polyhedron is 4×4, though 3×3 gives a bounded polyhedron with least value of $n \times m$. On a Pentium IV, 2.0GHz, 512MB computer the average time for each computational loop is 4.1 ms to solve the problem with 8 constraints and 7 decision variables, 7.2 ms if the number of constraints is 16 and 14.3 ms if the number of constraints is 32.

22.3.2 Maintain a Direction (VF2)

This basic geometric constraint is to maintain the task frame orientation $\hat{\mathbf{l}}$ along a given direction $\hat{\mathbf{l}}^d$. The signed errors are then set as $\boldsymbol{\delta} \equiv [\boldsymbol{\delta}_p^T, \boldsymbol{\delta}_r^T]^T = [\mathbf{0}^T, (\hat{\mathbf{l}}^d \times \hat{\mathbf{l}})^T]^T$. We require that after the incremental motion, the angle between the new task frame orientation $\hat{\mathbf{l}}'$ and $\hat{\mathbf{l}}^d$ is close to zero. As shown in Fig. 22.1(b), we can approximate this constraint for a small angle assumption as

$$\|\boldsymbol{\delta}_r + \Delta\mathbf{x}_r\|_2 \leq \epsilon_2 \tag{22.10}$$

where ϵ_2 defines the size of the range that can be considered as the desired direction. Applying the approach described in Sec. 22.3.1 to the angular components, we set A and $\mathbf{b}$ as

$$A = \begin{bmatrix} 0 & 0 & 0 & \cos\alpha_{21}\cos\beta_{21} & \cos\alpha_{21}\sin\beta_{21} & \sin\alpha_{21} \\ & & & \cdots & & \\ 0 & 0 & 0 & \cos\alpha_{2n}\cos\beta_{2m} & \cos\alpha_{2n}\sin\beta_{2m} & \sin\alpha_{2n} \end{bmatrix}, \quad \mathbf{b} = \begin{bmatrix} \epsilon_2 \\ \cdots \\ \epsilon_2 \end{bmatrix} - A\boldsymbol{\delta}. \tag{22.11}$$

where $\alpha_{2i} = \frac{i2\pi}{n}$ and $\beta_{2j} = \frac{j2\pi}{m}$.

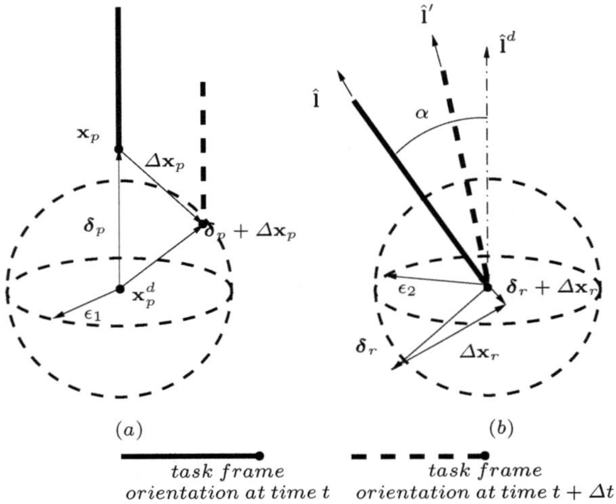

Fig. 22.1. Geometric relation for (a) "stay on a point" and (b) "maintain a direction"

22.3.3 Move Along a Line (VF3)

The next basic geometric constraint is to guide the task frame position to move along a reference line in 3D space, given by $L : L(s) = \mathbf{L}_0 + \hat{\mathbf{d}} \cdot s,\ s \in (-\infty, \infty)$. We require that after each incremental motion, the translational component of the task frame $\mathbf{x}_p + \Delta\mathbf{x}_p$ to be along (or close to) the reference line (see Fig. 22.2(a)). If the actual position is off the path because of some external disturbance, the control algorithm should drive the task frame back to the line. The geometric constraint should envelop the reference line and absorb the disturbance.

From the given line L, we can compute the closest point, $\mathbf{x}_p^{cl}$ to $\mathbf{x}_p$ on L. The signed errors are then set as $\boldsymbol{\delta} \equiv [\boldsymbol{\delta}_p^T, \boldsymbol{\delta}_r^T]^T = [(\mathbf{x}_p - \mathbf{x}_p^{cl})^T, \mathbf{0}^T]^T$. We define a vector $\mathbf{u}_p$ as the projection of vector $\boldsymbol{\delta}_p + \Delta\mathbf{x}_p$ on the plane Π which is perpendicular to line L. As shown in Fig. 22.2(a), our requirement is equivalent to $\|\mathbf{u}_p\|_2$ be close to zero, which can be written as

$$\|\mathbf{u}_p\|_2 \leq \epsilon_3 \tag{22.12}$$

To determine $\mathbf{u}_p$ from $\boldsymbol{\delta}_p + \Delta\mathbf{x}_p$ we need to compute a rotation matrix R_3, which would transform plane Π to the XY plane of the world (or robot) coordinate frame. Though R_3 is not unique, to compute R_3, we first define an arbitrary vector, which is not aligned with $\hat{\mathbf{l}}$, then we generate two unit vectors that span the plane Π. Any unit vector with arbitary angle α_3 in the plane Π with o as origin can be written in the world coordinate frame as

$$R_3 \begin{bmatrix} \cos\alpha_3 \ \sin\alpha_3 \ 0 \end{bmatrix}^T \tag{22.13}$$

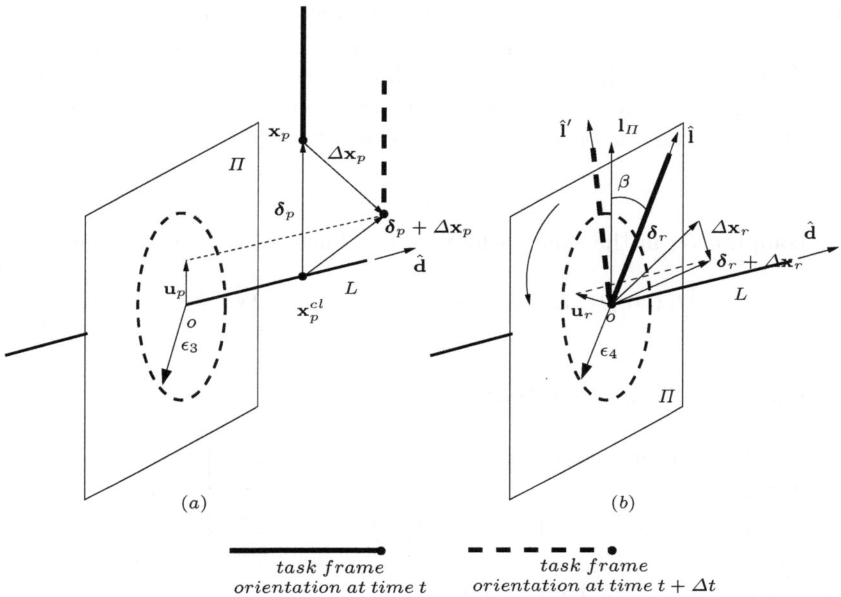

Fig. 22.2. Geometric relation for (a) "move along a line" and (b) "rotate around a line"

Form (22.12) implies that the projections of $\mathbf{u}_p$ on the pencil at o in the plane Π be less than ϵ_3. We approximate the circle of radius ϵ_3 by considering a polygon with n vertices centered at the origin, and rewrite (22.12) as

$$\begin{gathered}\left[R_3 \left[\cos\alpha_{3i} \; \sin\alpha_{3i} \; 0\right]^T \; 0 \; 0 \; 0\right] \cdot (\boldsymbol{\delta} + \Delta\mathbf{x}) \le \epsilon_3, \\ i = 0, 1, \cdots, n-1.\end{gathered} \tag{22.14}$$

where $\alpha_{3i} = \frac{i2\pi}{n}$. We can set A and $\mathbf{b}$ as,

$$A = \begin{bmatrix} R_3 \left[\cos\alpha_{31} \; \sin\alpha_{31} \; 0\right]^T \; 0 \; 0 \; 0 \\ \cdots \\ R_3 \left[\cos\alpha_{3n} \; \sin\alpha_{3n} \; 0\right]^T \; 0 \; 0 \; 0 \end{bmatrix}, \quad \mathbf{b} = \begin{bmatrix} \epsilon_3 \\ \cdots \\ \epsilon_3 \end{bmatrix} - A\boldsymbol{\delta}. \tag{22.15}$$

22.3.4 Rotate Around a Line (VF4)

Given a line L, with direction $\hat{\mathbf{d}}$, the geometric constraint is to rotate the task frame orientation around the lines while keeping the orientation of the task frame on the plane Π that is perpendicular to L. Even if an external disturbance changes the orientation away from the plane Π, our virtual fixture is required to drive it back.

As shown in Fig. 22.2(b), $\hat{\mathbf{l}}_{\Pi}$ is the unit vector of the projection of the orientation of the task frame $\hat{\mathbf{l}}$ on plane Π. The signed errors are set as $\boldsymbol{\delta} \equiv [\boldsymbol{\delta}_p^T, \boldsymbol{\delta}_r^T]^T = [\mathbf{0}^T, (\hat{\mathbf{l}}_{\Pi} \times \hat{\mathbf{l}})^T]^T$. We define a vector $\mathbf{u}_r$ as the projection of vector $\boldsymbol{\delta}_r + \Delta\mathbf{x}_r$ on the plane Π. Our constraint that after the incremental rotation the task frame is on plane Π is equivalent to $\|\mathbf{u}_r\|_2$ being close to zero:

$$\|\mathbf{u}_r\|_2 \leq \epsilon_4 \tag{22.16}$$

We compute R_4 in the same fashion as R_3. Then we write (22.16) as

$$\begin{gathered} \left[0\ 0\ 0\ R_4 \left[\cos\alpha_{4i}\ \sin\alpha_{4i}\ 0\right]^T\right] \cdot (\boldsymbol{\delta} + \Delta\mathbf{x}) \leq \epsilon_4, \\ i = 0, 1, \cdots, n-1. \end{gathered} \tag{22.17}$$

where $\alpha_{4i} = \frac{i2\pi}{n}$. Then A and $\mathbf{b}$ are set as:

$$A = \begin{bmatrix} 0\ 0\ 0\ \ R_4 \left[\cos\alpha_{41}\ \sin\alpha_{41}\ 0\right]^T \\ \cdots \\ 0\ 0\ 0\ \ R_4 \left[\cos\alpha_{4n}\ \sin\alpha_{4n}\ 0\right]^T \end{bmatrix}, \quad \mathbf{b} = \begin{bmatrix} \epsilon_4 \\ \cdots \\ \epsilon_4 \end{bmatrix} - A\boldsymbol{\delta}. \tag{22.18}$$

22.3.5 Stay Above a Plane (VF5)

This basic constraint is to prevent the task frame position from penetrating the given plane Π. From the given plane $\Pi(s)$, we can easily compute $\mathbf{x}_p^{cl}$ on $\Pi(s)$ which is the closest point to $\mathbf{x}_p$. The signed errors are set as $\boldsymbol{\delta} \equiv [\boldsymbol{\delta}_p^T, \boldsymbol{\delta}_r^T]^T = [(\mathbf{x}_p - \mathbf{x}_p^{cl})^T, \mathbf{0}^T]^T$.

As shown in Fig. 22.3, this constraint can be expressed by

$$\hat{\mathbf{d}}^T \cdot (\boldsymbol{\delta}_p + \Delta\mathbf{x}_p) \geq 0 \tag{22.19}$$

where $\hat{\mathbf{d}}$ is the unit normal direction of $\Pi(s)$ and points to the free half space. Then A and $\mathbf{b}$ are set as

$$\begin{gathered} A = \left[-\hat{\mathbf{d}}^T\ 0\ 0\ 0\right], \\ \mathbf{b} = -A\boldsymbol{\delta} \end{gathered} \tag{22.20}$$

If we further want to confine the task frame position on the plane, we can add constraints

$$\hat{\mathbf{d}}^T \cdot (\boldsymbol{\delta}_p + \Delta\mathbf{x}_p) \leq \epsilon_5 \tag{22.21}$$

where ϵ_5 is a small positive number, which defines the range of error tolerance. Then A and $\mathbf{b}$ are set as

$$A = \begin{bmatrix} -\hat{\mathbf{d}}^T\ 0\ 0\ 0 \\ \hat{\mathbf{d}}^T\ \ 0\ 0\ 0 \end{bmatrix}, \quad \mathbf{b} = \begin{bmatrix} 0 \\ \epsilon_5 \end{bmatrix} - A\boldsymbol{\delta}. \tag{22.22}$$

All the small values $\epsilon_{1,2,3,4,5}$ specify how much the robot can drift away from the reference direction. If they are small (close to zero), then the user can only move the manipulator along the desired direction. If they are relatively large, then the user has more freedom to deviate from the programmed virtual fixture.

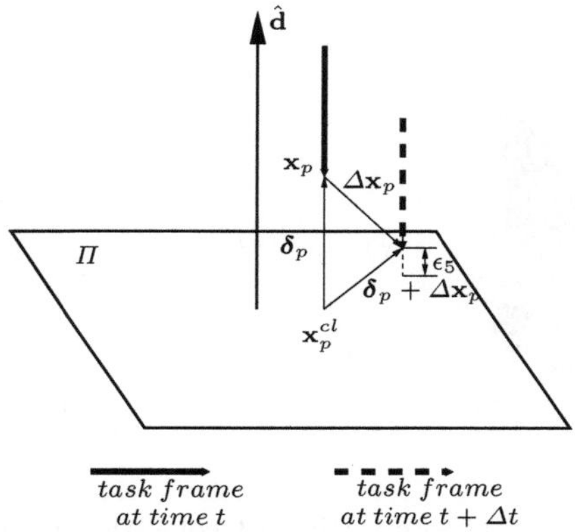

Fig. 22.3. Geometric relation for "stay above a plane"

22.4 Applications and Experiments

In this section we demonstrate customized virtual fixtures generation using the virtual fixture library on task primitives [22, 27]. As a preliminary test-bed we have used a "remote center of motion" JHU Steady Hand robot [24], which is equipped with a tool holder and a 6-DoF force-torque sensor (ATI Nano43 F/T transducer) mounted on the tool handle. The Optotrak (Northern Digital Inc, Waterloo, CA) infrared optical position tracking system was used for robot calibration. Our control algorithm is independent of manipulator characteristic. Moreover, the desired user input can be obtained either from a master robot, a joystick or a force sensor attached to the robot.

22.4.1 Application Task 1: Path-Following

Numerous surgical situations require surgeons to follow a predetermined path, while maintaining certain other constraints. One such example is the cutting procedure for a knee prosthetic implant. Knee prosthetic implants are used to replace the bearing surfaces of the knee. Normally when done by manual technique, the surgeon positions cutting blocks (jigs) on the bone and then cuts the femur and tibia to the required shape using a hand held saw to mount the prosthetic components in position. In the robotic procedure adopted by systems such as Robodoc [28], Acrobat [29] and CASPAR [30], a mill is used for the cutting procedure, and the blade is required to cut along the planned path on the bone. Meanwhile, the cutting edge of the tool should be kept perpendicular to the cutting plane in order to provide more efficient force.

Modeling of Task

We model the femur cutting task as a task to guide the tip of a long straight tool following a 2D b-spline curve C_1 in plane Π while keeping the tool shaft

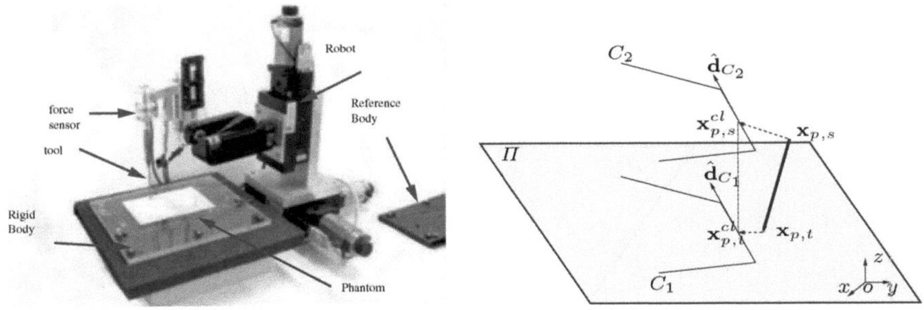

Fig. 22.4. (left) Experimental setup, (right) Geometric relation for the task "path-following"

perpendicular to the plane. The geometric relation is shown in Figure 22.4. We assume that the path C_1 and the cutting plane are known in the robot coordinate frame by using an appropriate registration method. During the procedure, the tip of the tool (task frame $\{t\}$) is allowed to move along the planned path C_1. At the same time, a point, $\mathbf{x}_{p,s}$ on the tool shaft (task frame $\{s\}$) is only allowed to move along the second path C_2, which is a translation of C_1 above the target plane. $\mathbf{x}^{cl}_{p,t}$ is the closest point to the tip of the tool on C_1 and $\mathbf{x}^{cl}_{p,s}$ is the projection of $\mathbf{x}^{cl}_{p,t}$ on C_2. We use each of these points and the tangent at these points to create two sets of constraints according to formulation VF3 in Sec. 22.3.3.

Experimental Results

We mounted a straight tool on the robot end-effector. We drew a set of line segments on a flat plastic plate, assumed to be a plane, and attached Optotrak LEDs to the plate. We used a digitizer to gather sample points on the line segments and then we generated a 5th degree b-spline curve in the target coordinate frame by interpolating these sample points. We used an Optotrak to record the tool tip position and the tool orientation. The tip position error is defined as the distance from the tool tip position to the reference b-spline curve.

The average tip position error of five trials is $0.32 \pm 0.19\,mm$. The trajectory of the tool tip with respect to the b-spline curve and the error profile of a trial are shown in Figure 22.5. The large errors occur at the sharp turnings. The time for each loop is around $150\,ms$, in which more than $140\,ms$ is for communication between the robot and the Optotrak reading via a local network. The communication delay contributes to the large errors on the sharp turnings where the tangent direction changes dramatically.

To evaluate the effect of the communication delay, we compared the tip error with different time intervals for the control loop. We removed the communication between the robot and Optotrak from the loop, only robot encoders and kinematics were used to record the tool tip motion. As shown in Fig. 22.6, the error

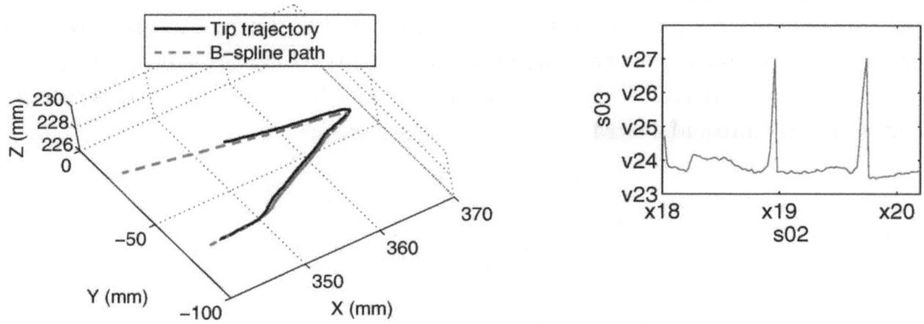

Fig. 22.5. (left) The trajectory of the tool tip with respect to the reference b-spline curve (right) The magnitude of the tool tip position error measured by Optotrak in "path-following" task

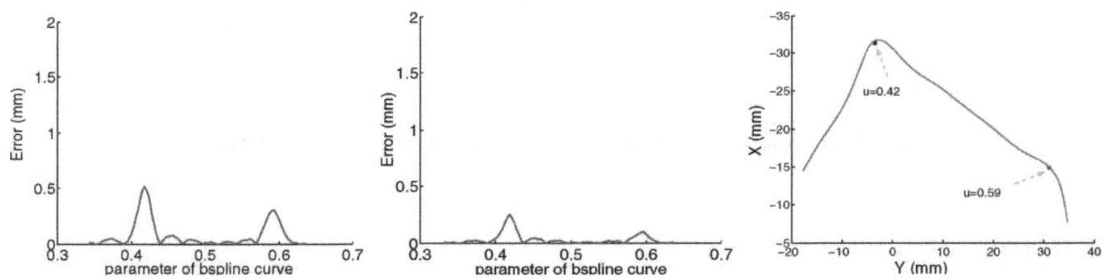

Fig. 22.6. The magnitude of the tool tip position error measured by the robot encoders and kinematics with different time interval for each control loop in "path-following with fixed orientation" task (left) around 150ms each loop, (middle) around 40ms each loop, (right) bspline curve and the position on which the large errors occur.

at the sharp turning decreased in the case that the time interval of each loop is shorter. In this case, the velocity is updated more frequently. Especially at the sharp turning, before the tool tip goes too far, the new velocity is computed and applied. The prompt action of the robot reduces the error. With the same time interval (150 ms), the error recorded by the Optotrak is larger than that by the robot encoders and kinematics. This is due to the system registration error and the accuracy of the Optotrak.

22.4.2 Application Task 2: Virtual RCM

In percutaneous needle insertions and also in robotic-assisted minimally invasive surgery, the surgical tools are inserted into the human body through a port. It is highly desireable to limit the motion of the tools at the entry port, and provide sufficient degrees of freedom for manipulation of tools inside the body. Some surgical robots [3, 23, 25] constrain the tool motion by providing a mechanical isocenter mechanism also known as remote center of motion (RCM). This task demonstrates a virtual remote center of motion (RCM) configuration, which provides an isocentric motion that is fundamental to percutaneous procedures. Our virtual fixture paradigm can implement virtual RCM on any given position other than mechanical RCM. Moreover, it can also be used in robots which do not provide a mechanical RCM.

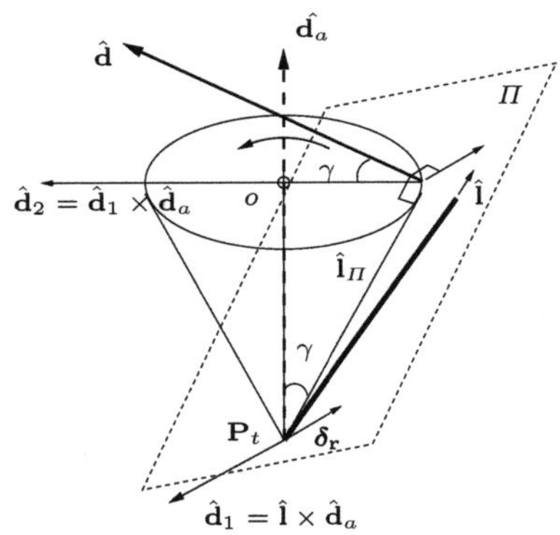

Fig. 22.7. Geometric relation for the task "Virtual RCM"

Modeling of Task

A surgical procedure using the virtual RCM can be modeled as to keep a selected position on the tool staying on a given insert port on a patient skin, while moving the orientation of the tool to follow the preplanned trajectory. The position and tool trajectory can be defined by the surgeon. In our work, we simplify the task as to keep the tip of a straight tool pivoting on a given point $\mathbf{P}_t$ (which is other than any mechanical RCM of the given robots) while rotating the tool shaft around a given direction $\hat{\mathbf{d}}_a$ with a fixed angle γ. The cone shape in Fig. 22.7 shows the desired trajectory of the tool shaft. We have combined the motion primitives VF1 and VF4 as presented in sections 22.3.1 and 22.3.4 respectively to create the virtual fixture for this task.

Experimental Results

For the experiment we set the given direction as $\mathbf{d}_a = [0\ -0.2\ 1]^T$, the fixed angle as $\gamma = 15\,\text{deg}$ and the pivot point as $\mathbf{P}_t = [0\ 0\ -10]^T(mm)$ with respect to the robot mechanical RCM. The values (ϵ_i) for positional and angular tolerance was selected as $0.1mm$ and $0.2deg$ respectively.

The trajectory of the tool and the error profile, which are measured by the robot encoders, are shown in Figure 22.8. The average pivot point position error for five trials is $0.01 \pm 0.01\ mm$ measured by the robot encoders. The average tool orientation angle error is $0.03 \pm 0.02\,\text{deg}$.

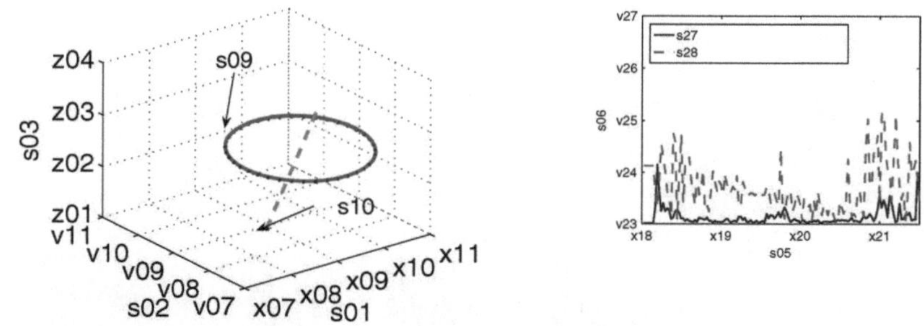

Fig. 22.8. Experimental results of the "virtual RCM" task measured by the robot itself. (left)The trajectories of a point on the tool shaft and the virtual RCM points. The circle shows the actual trajectory of a point on the tool shaft, the dots close to the circle represents the desired trajectory of the point. (right) The error profile for both the pivot point position error and the tool orientation angle error.

22.4.3 Application Task 3: Suturing

Suturing is considered to be one of the most difficult and time consuming minimally invasive surgical procedures. The surgeon faces the challenge of the limited and constrained motion as well as the loss of direct visualization. The suturing task was observed and analyzed as performed in training videos. This task involves the following steps 1) (Select) Determine a suitable entry and exit point for the suture needle leaving sufficient room from the edge to be sutured together. 2) (Align) Grasp the needle, move and orient it such that the tip is aligned with the entry point. 3) (Bite) Entry and exit "bites" are made such that the needle passes from one tissue to the other. 4) (Loop) Create a suture loop to tie a knot. 5) (Knot) Secure the knot under proper tension.

In this application, we address the align and bite steps of the suturing process, where the primary challenges are manipulation of a curved needle under non-ideal haptic conditions using a robot with complex kinematics. In the align step, the goal is to move the robot to align the position and the orientation of the suture needle such that it pierces the tissue correctly, while minimizing extraneous motion of the needle and robot. The goal of the bite step is to move the needle tip from the entry point to the exit point with minimum damage to the tissue through which the needle passes.

Task Modeling

We assume that the entry and the exit points are known in the robot coordinate frame. These could be specified by the surgeon using a tracked instrument or by using a computer vision system registered to the robot coordinate frame, to determine suitable points on the surface based on distance from the edge to be sutured together. Fig. 22.9 (top left) shows the various task frames $\{i\}$ associated with the suturing task.

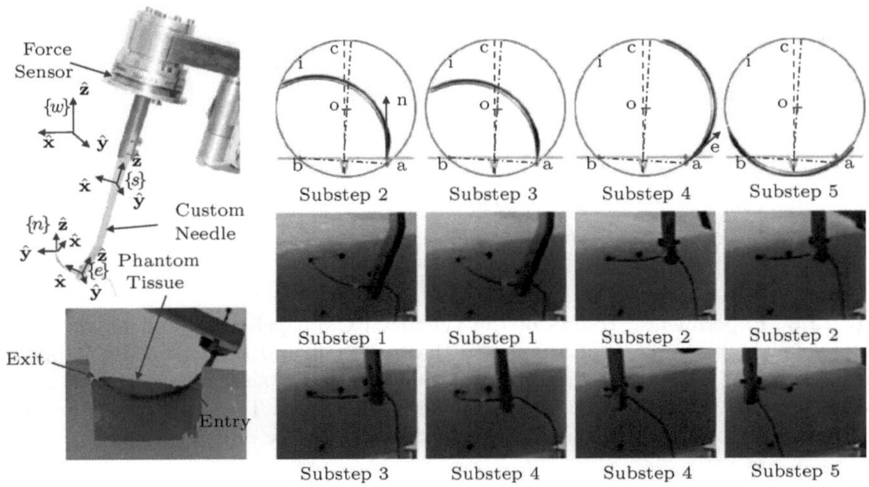

Fig. 22.9. (top left) Custom needle holder, needle and assigned frames. (bottom left) Actual path of needle. Top of phantom was removed after the expermient for verification. (top right) Sketch of sequence of substeps. a - Entry Point, b - Exit Point, c - Center line of wound, e - Entry direction, i - Ideal path of needle, n - Surface normal, o - Center of rotation. (bottom right) Sequence of substeps in phatom.

In our approach the required VF constraints for each substep are analyzed and broken into a combination of one or more of the basic constraints. We make use of the common structure between the different substeps and construct generalized constraints that take the desired target into consideration. Furthermore we utilize the sequential nature of the task to switch between different substeps. The switch could be triggered when the error between the current value and target decreases below a threshold. Fig. 22.9 shows these substeps, along with the entry, and the exit points on a phantom tissue.

Align Step. *(Substep 1)* First the needle tip is allowed to move in a straight line such that the needle tip coincides with the desired entry point; at the same time its orientation is allowed to change only about an axis such that this motion will result in the tangent at the needle tip being coincident with the normal to the surface at the entry point by using primitives VF3 and VF4. *(Substep 2)* In the next substep, the orientation of the normal to the needle plane is allowed to change, such that the needle plane coincides with the line joining the entry and exit points. Assistance is provided by not allowing any motion of the needle tip or the tangent at the needle tip by using the using primitives VF1 and VF2 respectively. *(Substep 3)* Once the desired orientations are reached we allow the surgeon to penetrate the tissue by a small distance, *(Substep 4)* followed by motion constraints that would let the surgeon bring the tangent at the needle tip to coincide with the desired entry direction without changing the plane normal and tip position by using primitives VF2 and VF1 repectively. The align step is completed once the desired orientation is reached, which is computed using the entry and exit points specified by the surgeon and the needle radius. In all these substeps only those motions that bring the needle closer to the desired position and orientation are allowed.

Bite Step. *(Substep 5)* Once the entry and exit points are determined, and the radius of needle is known, clearly the trajectory of the needle tip that would cause minimum damage to the tissue lies on a circle with the entry and exit points as points on a chord and with radius equal to the needle radius. To ensure sufficient depth of penetration in the tissue we ensure that the needle plane is parallel to the line joining the entry and exit points and the surface normal at the entry point. In this step our constrained motion algorithm permits only those motions that satisfy these constraints by using primitives VF1 and VF2 for needle center and a normal to the needle plane respectively.

Experimental Results

For these experiments we selected a 3/8 circle $30mm$ cutting needle from Ethicon (needle diameter $1mm$). We recorded the encoder readings of the robot joints and used direct kinematics of the robot to verify our algorithm by measuring the errors between the ideal target path and that followed by the robot. Fig. 22.9 (bottom right) shows the progression of different substeps for one of the trials. Fig. 22.9 (bottom left) shows the phantom with a portion cut out so that the actual path taken by the needle is visible, the entry and exit points are $13.5mm$ apart. As seen in Fig. 22.9, we have selected an angle that places limits on performing the suture manually, to emphasize the ability of our algorithm to assist in non-favorable orientations.

Table 22.1. The error (mm) in ideal and actual points as measured by the Optotrak

	Entry	Exit
Robot	0.6375	0.7742
Manual	-	2.1

Fig. 22.10 shows the errors between actual and ideal robot motion as measured by the robot encoders and kinematics for different substeps. The values (ϵ_i) for positional and angular tolerance were selected as $0.5mm$ and $0.25deg$. We also demonstrate our algorithm using a phantom tissue. Since the phantom is opaque, the measurements available are the entry and exit points of the needle. Table 22.1 presents the differences between the user specified targets and the actual ones as measured by the Optotrak.

As expected, the errors measured by the Optotrak are higher than measured by encoders alone, because this represents the overall accuracy of the system, which also includes errors arising from calibration of the needle and accuracy of the Optotrak ($0.1mm$). The residual calibration errors appear as errors in

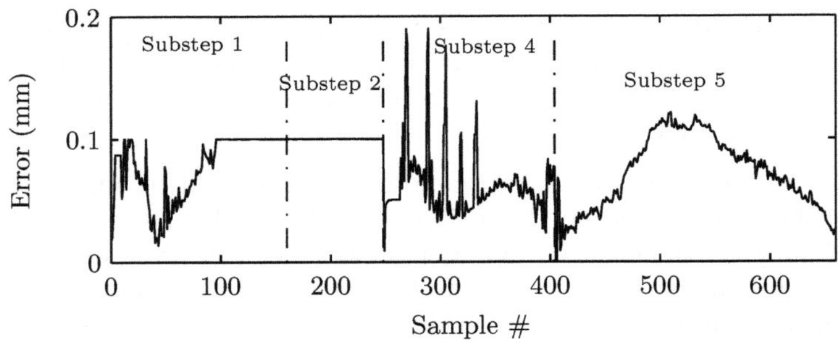

Fig. 22.10. The magnitude of error in the needle tip with respect to ideal path as measured by robot encoders

the entry point errors in Table 22.1. Average errors for free hand suturing as performed by four users (5 trials each), using the same needle holder and without robot assistance, are presented in Table 22.1. We believe that robot assistance can improve accuracy especially in a constrained environment such as that of endoscopic surgery. Moreover, robot assisted motions did not require multiple trials and large undesirable movements of tissue, which is often the case in free hand suturing.

22.5 Conclusion

This chapter described a new method to generate spatial motion constraints for surgical robots that provide sophisticated ways to assist surgeons. Our approach is based on the optimized constrained control. We set the objective function based on the user input that can be obtained through a force sensor, joystick or a master robot. We set the linearized subjective function based on five basic geometric constraints. The combinations of one or more basic geometric constraints for the same or different task frames could generate customized virtual fixtures for complicated surgical tasks. Theoretically, different virtual fixtures can be implemented by using this method if we know the instantaneous kinematics of the manipulator and the geometric constraints [31].

Our approach provides the link between surgeon-understandable task behaviors and low level control for surgical assistance robots. The strength of this approach is that it is extensible to include additional constraints that are important in robotic assisted surgery, such as collision avoidance, anatomy-based constraints and joint limits by using the instantaneous kinematic relationship between the task variables and the robot joints. In [32, 33], we extended our algorithm to create anatomy-based motion constraints for a path-following task in a constrained workspace. We integrated a 3-D geometric model of the workspace to generate virtual fixtures to guide the tool tip along the paths while preventing the tool shaft from entering forbidden regions for sinus surgery.

References

1. P. Berkelman, P. Cinquin, J. Troccaz, J. Ayoubi, C. Letoublon, and F. Bouchard. A compact, compliant laparoscopic endoscope manipulator. In *Proc. IEEE Int. Conf. Robotics and Automation*, pages 1870–1875, 2002.
2. J. Rosen, J. D. Brown, L. Chang, M. Barreca, M. Sinanan, and B. Hannaford. The bluedragon - a system for measuring the kinematics, the dyanmics of minimally invasive surgical tools in-vivo. In *Proc. IEEE Int. Conf. Robotics and Automation*, pages 1876–1881, 2002.
3. G. S. Guthart and J. K. Salisbury. The intuitive telesurgery system: Overview, application. In *Proc. IEEE Int. Conf. Robotics and Automation*, pages 618–621, 2000.
4. M. Chodoussi, S. E. Butner, and Y. Wang. Robotic surgery-the transatlantic case. In *Proc. IEEE Int. Conf. Robotics and Automation*, pages 1882–1888, 2002.
5. M. C. Cavusoglu, W. Williams, F. Tendick, and S. S. Sastry. Robotics for telesurgery: second generation berkeley/ucsf laparoscopic telesurgical workstation, looking toward the future applications. *Industrial Robot, Special Issue on Medical Robotics*, 30(1):22–29, 2003.
6. P. S. Schenker, H. Das, and R. T. Ohm. Development of a new high-dexterity manipulator for robot-assisted microsurgery. In *Proc. SPIE - The International Society for Optical Engineering: Telemanipulator and Telepresence Technologies*, 2351:191–198, 1995.
7. P. S. Jensen, K. W. Grace, R. Attariwala, J. E. Colgate, and M. R. Glucksberg. Toward robot-assisted vascular microsurgery in the retina. *Graefes Archive for Clinical and Experimental Ophthalmology*, 235(11):696–701, 1997.
8. M. Mitsuishi, Y. Iizuka, H. Watanabe, H. Hashizume, and K. Fujiwara. Remote operation of a micro-surgical system. In *Proc. IEEE Int. Conf. Robotics and Automation*, pages 1013–1019, 1998.
9. R. H. Taylor, H. A. Paul, P. Kazanzides, B. D. Mittelstadt, W. Hanson, J. F. Zuhars, B. L. Musits B. Williamson, E. Glassman, and W. L. Bargar. An image-directed robotic system for precise orthopaedic surgery. *IEEE Transactions on Robotics and Automation*, 10(3):261–275, 1994.
10. J. Wurm, H. Steinhart, K. Bumm, M. Vogele, C. Nimsky, and H. Iro. A novel robot system for fully automated paranasal sinus surgery. In *International Congress Series*, 1256:633–638, 2003.
11. I. W. Hunter, T. D. Doukoglou, S. R. Lafontaine, P. G. Charette, L. A. Jones, M. A. Sagar, G. D. Mallinson, and P. J. Hunter. A teleoperated microsurgical robot, associated virtual environment for eye surgery. *Presence*, 2(4):265–280, 1993.
12. M. Mitsuishi, T. Watanabe, H. Nakanishi, T. Hori, H. Watanabe, and B. Kramer. A tele-micro-surgery system with co-located view, operation points, a rotational-force-feedback-free master manipulator. In *2nd Int. Symp. Medical Robotcis and Computer-Assisted Surgery (MRCAS)*, pages 111–118, 1995.
13. M. Mitsuishi, H. Watanabe, H. Nakanishi, H. Kubota, and Y. Iizuka. Dexterity enhancement for a tele-micro-surgery system with multiple macro-micro co-located operation point manipulators, understanding of the operator's intention. In *3rd Int. Symp. Medical Robotcis and Computer-Assisted Surgery (MRCAS)*, pages 821–830, 1997.

14. S. E. Salcudean and G. Bell S. Ku. Performance measurement in scaled teleoperation for microsurgery. In *1st Int. Symp. Medical Robotics and Computer-Assisted Surgery (MRCAS)*, pages 789–798, 1997.
15. B. L. Davies, S. J. Harris, W. J. Lin, R. D. Hibberd, R. Middleton, and J. C. Cobb. Active compliance in robotic surgery - the use of force control as a dynamic constraint. In *Proc. Inst. Mech. Eng. H*, 211(4):285–292, 1997.
16. S. Park, R. D. Howe, and D. F. Torchiana. Virtual fixtures for robotic cardiac surgery. In *Proc. Medical Image Computing and Computer Assisted Intervention (MICCAI)*, pages 1419–1420, 2001.
17. A. Bettini, P. Marayong, S. Lang, A. M. Okamura, and G. D. Hager. Vision assisted control for manipulation using virtual fixtures. In *IEEE Transactions on Robotics*, 20(6):953–966, 2004.
18. R. Kumar, G. D. Hager, A. Barnes, P. Jensen, and R. H. Taylor. An augmentation system for fine manipulation. In *Proc. Medical Image Computing and Computer Assisted Intervention (MICCAI)*, pages 956–965, 2000.
19. P. Marayong, M. Li, A. M. Allison, and G. D. Hager. Spatial motion constraints: theory, demonstrations for robot guidance using virtual fixtures. In *Proc. IEEE Int. Conf. Robotics and Automation*, pages 1954–1959, 2003.
20. D. Aarno, S. Ekvall, and D. Kragic. iadaptive virtual fixtures for machine-assisted teleoperation tasks. In *Proc. IEEE Int. Conf. Robotics and Automation*, pages 1151–1156, 2005.
21. J. Funda, R. H. Taylor, B. Eldridge, S. Gomory, and K. G. Gruben. Constrained cartesian motion control for teleoperated surgical robots. *IEEE Transactions on Robotics and Automation*, 12(3):453–465, 1996.
22. M. Li, A. Kapoor, and R. H. Taylor. A constrained optimization approach to virtual fixtures. In *Proc. IEEE/RSJ Int. Conf. Intelligent Robots and Systems*, pages 2924–2929, 2005.
23. R. H. Taylor, J. Funda, B. Eldridge, K. Gruben, D. LaRose, S. Gomory, M. Talamini, L. R. Kavoussi, and J. Anderson. A telerobotic assistant for laparoscopic surgery. *IEEE Eng. Med. Biol. Mag.*, 14:279–287, 1995.
24. R. H. Taylor, P. Jensen, L. L. Whitcomb, A. Barnes, R. Kumar, D. Stoianovici, P. Gupta, Z. Wang, E. deJuan, and L. Kavoussi. A steady-hand robotic system for microsurgical augmentation. *International Journal of Robotics Research*, 18(12):1201–1210, 1999.
25. M. Cavasoglu, F. Tendick, M. Cohn, and S. Sastry. A laparoscopic telesurgical workstation. *IEEE Transactions on Robotics and Automation*, 15(4):728–739, 1999.
26. C. Lawson and R. Hanson. *Solving Least Squares Problems*. Prentice-Hall, 1974.
27. A. Kapoor, M. Li, and R. H. Taylor. Spatial motion constraints for robot assisted suturing using virtual fixtures. In *Proc. Medical Image Computing and Computer Assisted Intervention (MICCAI)*, pages 89–96, 2005.
28. U. Wiesel and M. Boerner. First experiences using a surgical robot for total knee replacement. In *Proc. Computer Assisted Orthopaedic Surgery (CAOS Intl)*, pages 143–146, 2001.
29. S. J. Harris, K. L. Fan, R. D. Hibberd, and B. L. Davies. Experiences with robotic systems for knee surgery. In *Proc. 3rd Int. Conf. Medical Robotics and Computer Assisted Surgery*, pages 757–766, 1997.
30. S. Mai and W. Siebert. Planning and technique using the robot system 'caspar' for tkr. In *Proc. Computer Assisted Orthopaedic Surgery (CAOS Intl)*, pages 278–288, 2001.

31. A. Kapoor, N. Simaan, and R. H. Taylor. Suturing in confined spaces: constrained motion control of a hybrid 8-dof robot. In *Proc. IEEE Int. Conf. Advanced Robotics*, pages 452–459, 2005.
32. M. Li and R. H. Taylor. Optimum robot control for 3d virtual fixture in constrained ent surgery. In *Proc. Medical Image Computing and Computer Assisted Intervention (MICCAI)*, pages 165–172, 2003.
33. M. Li and R. H. Taylor. Spatial motion constraints in medical robot using virtual fixtures generated by anatomy. In *Proc. IEEE Int. Conf. Robotics and Automation*, pages 1270–1275, 2004.

23

Technologies for a Telesurgery Laboratory Implementation

Jesús Manuel Gómez-de-Gabriel, Víctor Fernando Muñoz-Martinez, Francisco Jesús Domínguez-Fernández, and Javier Serón-Barba

University of Málaga
Dept. Ingeniería de Sistemas y Automática,
Plaza de El Ejido s/n. 29013 Málaga, Spain
{degabriel,victor,javierseron}@ctima.uma.es,
frandominguez@isa.uma.es

Summary. This chapter describes the telerobotic surgery developments and technologies involved in the implementation of a telesurgery research laboratory. These developments cover from communications software, with real-time audio and video transmission, to the use of a dual bilateral master-slave teleoperation system, with force-feedback, considering also the use of existing tools. As it will be described, one of the special characteristics of this system, that makes it different from other commercial approaches, is that it may be able to replace the local surgeon, with a standard surgical equipment and non stereoscopic commercial cameras, handled by a custom robot, and standard surgical tools handled by industrial manipulators. This characteristics make it accessible from small laboratories or research groups.

23.1 Introduction

The goal of robotic telesurgery is to develop robotic tools to augment or replace hand instruments used in surgery [1]. In this case robotic tools are not automated robots but teleoperated systems under direct control of the surgeon.

The purpose of the medical robotics research groups is the development of telesurgery applications and techniques. For this it is necessary to have an advanced experimental laboratory or operating room equipped with a large variety of different technologies. This chapter describes the technologies involved in the implementation of an experimental system, its equipment, and applications that were specifically developed in order to perform experimental remote surgery on live animals.

Minimally invasive surgery (MIS) is a surgical technique performed with instruments inserted through small incisions rather than making an open large incision to provide access to the operation point inside the patient [2]. These instruments have only four degrees of freedom (DOF) through the entry port, losing the ability to arbitrarily orient the instrument tip. In order to perform MIS, at least three surgical instruments are required, although the usual number is four. One of these instruments is an endoscope that provides the video feedback. The other two are grippers/scissors with electric scalpel functions for use on the patient's tissues. These last may provide secondary feedback in the form of tactile or force sensations [3].

M. Ferre et al. (Eds.): Advances in Telerobotics, STAR 31, pp. 403–414, 2007.
springerlink.com © Springer-Verlag Berlin Heidelberg 2007

In this chapter we are using a custom laparoscopic camera positioning system (the ERM Robot) [4, 5] developed by the Medical Robotics Research Group from the University of Málaga, that was designed to assist the surgeon, and now is controlled remotely and give the reference position to locate the point of view of the endoscopic camera and to transform the movement coordinates of the manipulator robots, that handles the surgical instruments. Current research, using commercial solutions approved for human surgery is not accessible for most of the research groups and it is necessary to have a basic system that allows to get relevant and transferables results to the medicine and industry. In this results are included studies and comparatives of different communication systems for video transmission. The object of the experimental telesurgery system is to develop and test new communication and control technologies to remotely control these surgical instruments without any loss of operating effectiveness, thus enabling the surgeon to operate from distant places while using standard surgical instruments.

Sec. 23.2 describes the overall system and the telerobotic architecture. Sec. 23.3 details the robotic system describing the force-feedback problem in telesurgery systems and the approach used for this implementation. Section 23.4 describes the solutions for real time video transmission and Sec. 23.5 is on development tools available to researchers. Finally in Sec. 23.6 and 23.7 the experiments realized, and obtained conclusions can be found.

23.2 Description of the Telerobotic System

The principal telesurgery system is based on remote communication of the surgeon's intentions via a computer system with a client-server architecture using a teleoperation work station and a teleoperation server (see Fig. 23.1).

Thus, the teleoperation server provides control of different robots in this case three operating robots are used. Two of them are responsible for carrying out the surgeon's commands, and the third robot controls the laparoscopic camera. To control these robots, the teleoperation workstation has two haptic devices that sense the surgeon's movements and transmit them through the infrastructure to the operating robots. As an important part of the system, a series of video imaging sources that allow the surgeon to obtain remote, real-time feedback on the status of the operation, as well as to interact with the healthcare personnel in the operating room.

23.2.1 The Teleoperation Workstation

This is the human-machine interface (HMI) that will allow the surgeon to interact remotely. It is formed by a high performance computer system dedicated mainly to processing and visualizing the video signals received via a TCP/IP network. In order to ensure that the surgeon can have complete and permanent information on the status of the operation, the system needs to be able to display large amounts of information, reason for which two display screens have been incorporated (see Fig. 23.2).

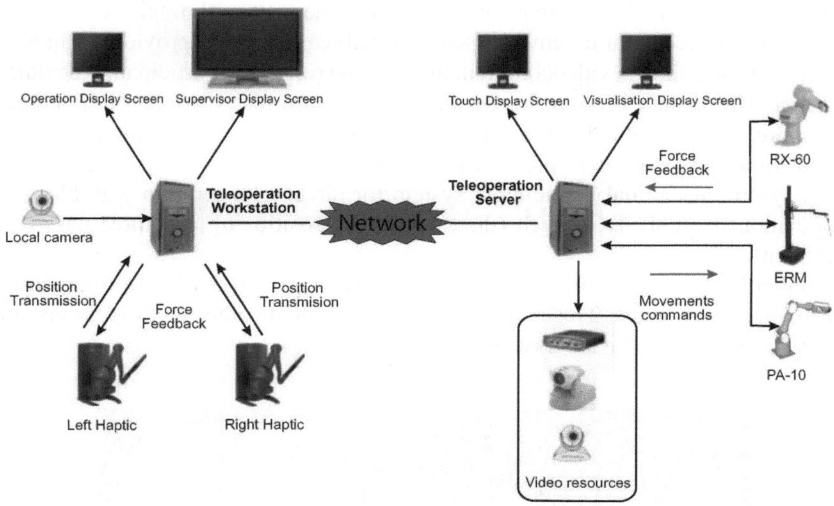

Fig. 23.1. General diagram of the telesurgery system architecture

The teleoperation workstation's missions are to receive the audio and video information from the remote operating room; supply, through a pair of haptic devices, the feedback force detected at the tips of the surgical instruments to both hands of the surgeon; return the position of these, as positional references, to the robots in the operating room, and transmit an image of the remote surgeon to facilitate interaction with the local surgeon. In order to do all this, the workstation is equipped with a computer and display screens, a video-conference camera, and two force-reflecting master manipulators.

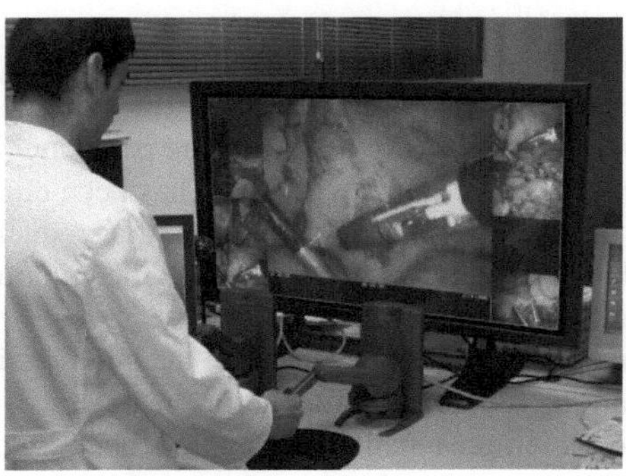

Fig. 23.2. Telesurgery workstation with force feedback master manipulators

The main display screen, or supervisor screen, a 42" plasma screen, will be dedicated to monitoring the video signals coming from the operating room and for video feedback in direct teleoperation when using the haptics. The basic idea is that the surgeon does not miss any detail of what is happening, both inside the patient (through the endoscopic camera) and in the operating environment, through cameras that provide a panoramic view of the room, and via videoconferencing with the rest of the participating healthcare personnel. The system is based on customized software that allows a high degree of video source management flexibility, as it can work with up to eight different sources at the same time and these can be presented in several ways as can be seen in Fig. 23.3. The current system has a dual haptic input system for direct teleoperation with bi-manual manipulation, with six DOF each (three active for positioning and three passive for orientation).

Together with this information, a status bar is also displayed in which the surgeon can view the operation status. This area can be used to provide other kinds of information to the surgeon like remote ECG reading.

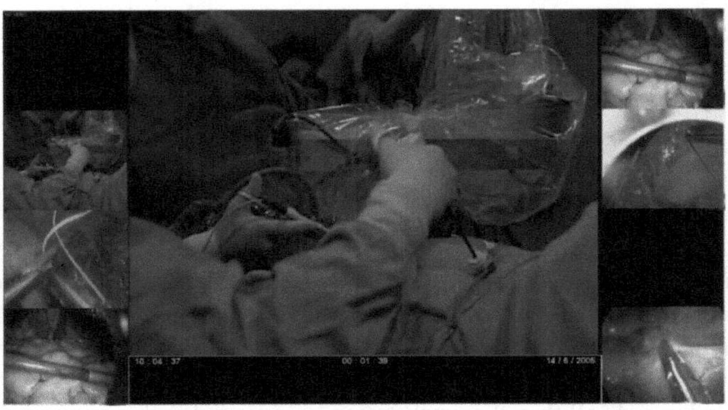

Fig. 23.3. Large display (42") showing multiple (external and internal) views to the workstation supervisor

The second screen, a standard 17" TFT monitor, will display an user interface in which it will be possible to control, intuitively, all the system's functionalities, with particular emphasis on interaction with the healthcare personnel in the operating room. For the user control interface, an uniform interface has been chosen so that all the devices to be controlled present a similar user interface, thus reducing the time needed to learn how to control the system. This system provides the recommended image quality for medical applications [6].

This HMI is divided basically into two areas. On the left there is a toolbar that allows the surgeon to assess all the functions available for the device, whilst on the right side, occupying most of the screen area available, the associated video source is displayed (see Fig. 23.4). This display usually shows real-time video sources obtained from the endoscopic camera. A toolbar allows the user to move the ERM robot remotely. Furthermore, the functionality of this tool has been completed with telementoring and telestration capabilities.

By means of Telementoring, an expert surgeon at a remote location can mentor a second surgeon in an operating room anywhere. The expert surgeon can control the field of view (move the camera arm), from set up and port placement, including telestration, through the entire procedure.

Telestration is an illustrative technique which allows a remote surgeon to make marks on the local surgeon's video monitor [7]. Both the remote surgeon and the surgeon with the patient have the exact same view of the surgical site. The remote surgeon can draw to show where to make an incision or can highlight a tumour mass, for example.

It also has an interface that enables remote control of pan-tilt-zoom cameras and allows the surgeon to have a full view of the operating room. To simplify managing the computer environment, two drop-down lists have been positioned at the top of the control windows from which the surgeon can select the device to be controlled or the video display option for the supervisor monitor.

The most important advantages of this HMI are its usability and simplicity for non engineering staff, so that the surgeon can pay attention to the surgical operation itself instead to the application interface.

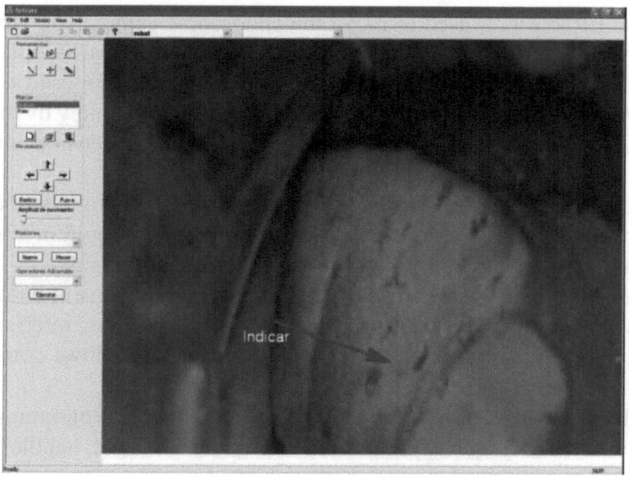

Fig. 23.4. Operation display user interface showing telementoring/telestration capabilities

23.2.2 The Telesurgery Server

The telesurgery server, located at the surgical place, is the operational core of the system and it is where all the routines are implemented for controlling the elements that make up the operating room's robotic system and the services that enable remote surgery to be carried out. It is comprised of the controls for the operating robots, the laparoscopic camera positioning robot, and the communications and video server (see Fig. 23.5).

The server has an interface that allows the healthcare personnel in the operating room to interact with the system, as, despite the fact that the objective is to carry out remote surgical operations, it is the operating room personnel that carry out the preliminary preparation of the patient for the operation. The display screen is used exclusively to display images captured by the camera attached to the endoscope and will provide the local surgeon the local video feedback for the operation. To increase interaction between the two medical teams (local and remote) bidirectional telestration has been implemented by means of a 14" TFT touch-screen. This HMI also includes local control and management of the surgical robots and videoconferencing with the remote surgeon.

This system is the global operating room server, and the responsible of the communication with the different robots and sensors in order to achieve the requested movements, and returning status and sensorial information to its client. The server also provides other services, such as remote surgeon authentication, the service and resource discovering system that informs on the resources available in the operating room to carry out the operation.

23.3 Surgical Robots

Apart from the ERM robot, used to move the endoscopic camera [6], which has been tested in human surgery clinical trials as assistant in local laparoscopic surgery, the operating manipulator robots have an important role within the proposed system, as they are responsible for performing locally the actions requested remotely by the surgeon. In general terms, we could say that the robots represent a remote extension of the surgeon's hands.

In this case, two industrial robots have been used, which, though hard to implement in a real operating room scenario because its power supply requirements, noise level and volume, are an extremely useful tool in the experimental stages.

Its great flexibility and accuracy allows testing several movement control strategies and trajectory generation. This industrial manipulators have user interfaces for quick development and provides an easy way to monitor it, which allows comparing easily the different movement control strategies used.

A Stäubli RX60, with a proprietary control system, VAL+ programming language, six DOF, and an external trajectory control method called ALTER, handles the left-hand instrument.

On the other side, a seven DOF Mitsubishi PA-10 robot, manages the right-hand instrument, with a proprietary low-level controller with a PC-based supervisor. The main advantage of using a robot with a seven DOFs is its redundancy; its infinite articulation configurations to reach the same point gives it more movement flexibility. This is particularly interesting taking into account that there will be three robots working at the same time in a confined workspace as can be seen in Fig. 23.5.

The instrument compliance to the fulcrum point, can be implemented in two ways. The ERM with a passive instrument (camera) has a two passive DOF, and a special kinematic structure for keeping the image vertically. The other two active instruments need to be full controlled and the compliance is made by means of a wrist force-torque sensor, one on each industrial robot, and a mixed position-compliance control system described in [8].

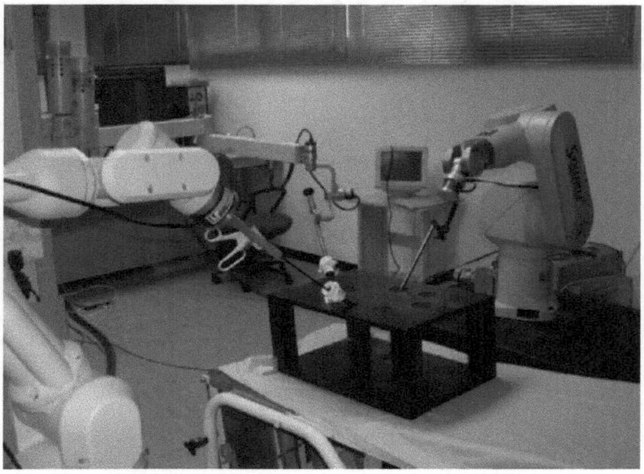

Fig. 23.5. Experimental operating room showing the surgical robots working on a patient simulator. Form left to right: RX-60, ERM, and PA-10.

23.3.1 Force-Feedback and Coordinate Transformation in MIS

One of the requirements of telesurgery systems is that surgeons can obtain, remotely, the same sensations they would have if they were in the operating room. As already mentioned, the teleoperation workstation has two devices that allow transmission of force and tactile sensations to the user.

These forces are measured by industrial 6-DOF force sensors that return forces and pairs in local coordinates. These sensors have been positioned in the wrists of the operating robots and processing is required to scale and transform the measurements in accordance with the surgeon's reference system (point of view) before being transmitted to the teleoperation workstation. All of these will be associated through a global fixed reference system, situated in the operating table. The movements carried out by the master manipulator (haptic) have to be translated into movement of the tool tip attached to the slave manipulator and to obtain the position augmentations of the tool tip, it is necessary to associate the master manipulator reference system with that of the camera and this latter with that of the slave manipulator. The existing transformation matrices between the different reference systems are used.

Minimally invasive surgery applications are basically vision guided, by means of an endoscopic camera that often changes its position and orientation. This coordinate transformations, for movements and forces, are strictly necessary to maintain the eye-hand coordination.

Specifically, the ${}_{H}T^{C}$ transformation is applied on the vector originated from the position variation of the haptic end-effector with respect to the reference system of its base, obtaining the position variation of the reference system associated to the camera; multiplying the result by the ${}_{C}T^{ERM}$ transformation gives the position variation referred to the reference system of the ERM base; the transformation ${}_{ERM}T^{G}$ refers to the position variation to the global system; finally the ${}_{G}T^{PA10}$ or ${}_{G}T^{RX60}$ matrix transformation is applied (see Fig. 23.6), depending on the slave manipulator used, obtaining

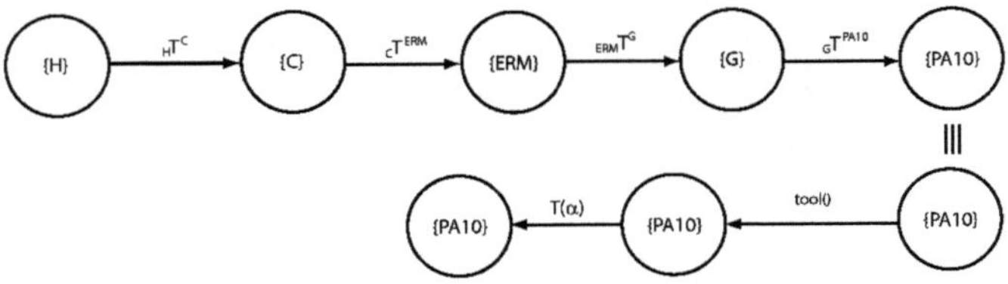

Fig. 23.6. Diagram of the coordinated systems association

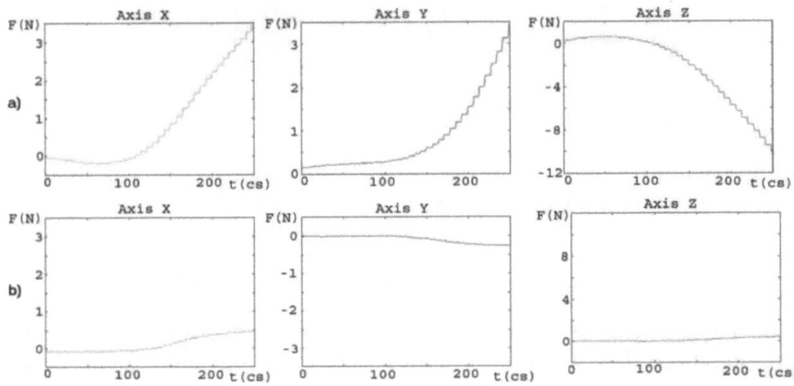

Fig. 23.7. Force measurements. a) Wrong fulcrum point. b) Right fulcrum point.

the position variation of the haptic end-effector in relation to the reference system of the slave manipulator base. As this is the position variation that the tool-tip should reproduce, this is the point that gives the end-tool position in relation to the reference system of the slave manipulator base.

The position and orientation of the reference system associated to the end-effector of this manipulator need to be calculated from the aforementioned vector. Due to the fact that the tools are inserted through an incision point, the slave manipulator should pivot around this point. For this, each tool-tip position augmentation is translated into rotational and translational movement in the reference system associated to the aforementioned end-effector of the manipulator.

This system makes the operation easier because the surgeon directly moves the tool end effector, avoiding the necessity of training and adaptation of the surgeon to the movement inversion caused by the fulcrum point of each hand tool.

The rotational axis is the vector perpendicular to the plane defined by the pivot point, the current position of the tool-tip and the objective position; the application point of this vector is the pivot point. Ideally, by pivoting around the point of entry, no force is exercised on the patient at this point. Fig. 23.7 shows the force measurements in a pivoting operation with a wrong and with a right fulcrum point.

23.4 Real-Time Video Feedback

In laparoscopic surgery the video stream captured by the laparoscopic camera is an essential element. Therefore, in order to achieve a system that enables remote surgery, as is the objective of this system, the remote workstation needs to be able to access these images with minimum delay and maximum quality.

To ensure that direct teleoperation can be carried out correctly, the video transmission has to fulfil some specifications. It must provide good quality images in terms of resolution, color depth and frame rate, but also the transmission delay must be reduced. A study was carried out on existing video transmission and reproduction libraries [9], in which the most important multimedia libraries were analyzed, both for Windows platforms and for other operating systems, such as JMF. None of the multimedia libraries analyzed offered a complete solution that allowed for fulfilment of the aforementioned specifications.

To solve this transmission problem, we developed a custom system, based on DirectShow, which, using commercial video codecs based on MPEG-4, made it possible to capture images and transmit them in real-time with an average delay time of 80 ms, using a low bandwidth. A drawback of this transmission system is that it was implemented upon the UDP protocol, which means that in the event of network overload, packets are discarded by the routers, resulting in a temporal image distortion until a new full key-frame is received.

A commercial video server, based on MJPEG compression, has been also used, which affords the possibility of capturing up to four video sources in real-time and video image transmission. The only negative aspect we have found with this server is that the server's development library, which enables video reception in our application, is quite weak and only allows reception.

23.5 Development Tools and Resources

As the purpose of this telesurgery laboratory is the research, the system must provide utilities for easy data analysis and implementation of different control methods.

The early development stages have been carried out in local mode, with one of the haptics as master manipulator, the PA-10 robot as slave manipulator and with a JR3 force sensor attached to it.

Two specific libraries have been developed to help researchers in the development of the control system. A library for Matlab in order to test control systems prototypes, and another for LabVIEW in order to test user interaction. The Matlab library is composed of the full PA-10 programming interface plus the force sensor reading functions, and running on the same robot controller. The LabVIEW library has been developed to build interactive Real-Time applications (see Fig. 23.8) in order to check user interaction; it includes functions for Haptic reading and writing (Phantom Desktop) and to send cartesian references to the slave robots.

In the global teleoperation system integration stage, all the functions developed on Matlab or LabVIEW have been implemented in C++ using Visual Studio. For the

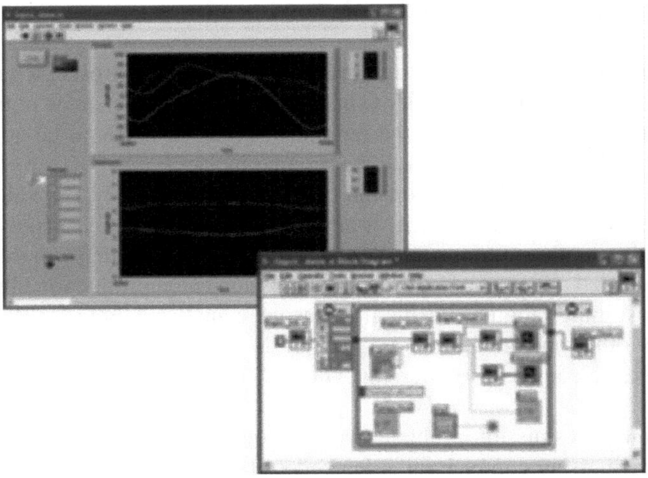

Fig. 23.8. Example of quick Real-Time application development in LabVIEW

routines developed in C++ to have control over the operating robot's movements, the movement types chosen were real-time movement (RMRC) for the PA-10 and ALTER mode for the RX60. These movement modes allow a position update rate for the PA-10 and RX60's controllers every 15 ms and 16 ms, respectively.

23.6 Experiments

In order to check the system performance, in-vitro and in vivo experiments have been made. The in-vitro experiments were devoted to verify the movement accuracy the coordinate transformation and force feedback quality under different conditions. Network Bandwidth limitation (10 Mbps) made it impossible to achieve a desired 1 ms update rate for the bilateral force-feedback loop, so a 3 ms rate has been selected. The experiments have been done with a local ethernet network and the teleoperation station was in the same building as the operating room so the delays were not very significant with an average round trip time of 5 ms. Nevertheless, the effective position loop delay may rise up to 100 ms including the slowest robot update period (16 ms) and the video streaming delay (80 ms). Regarding the first point, it has been realized path tracking experiments with both operating robots. In order to test the force-feedback feel it has been studied the sensation generated by haptic when the instrument take on touch with soft and hard surfaces. For each kind of surface, it has been tested with two kinds of tools mounted on the robot wrist: i) an experimental rigid tool, ii) a standard laparoscopic surgery tool.

In-vivo experiments are currently being carried out with local control systems, and the obtained developments are being progressively moved from local functions to the telesurgery workstation.

23.7 Conclusions

The special Teleoperation Workstation of this telerobotic laboratory contributes a series of advantages compared to other solutions:

- There is no need for stereoscopic displays as the system uses standard surgical equipment for MIS. Non-stereoscopic camera provides suitable video feedback, as does for manual surgery.
- It can work simultaneously with different video streaming systems such as Webcams, IP-Cameras, Camera servers, including a proprietary video streaming software developed by the author's research group. It provides simultaneous display and quick source selection (acting as a control center video-wall) allowing collective decision-making.
- It has a simultaneous dual HMI: Big screen for direct teleoperation and collective decision making and small screen for surgeon-to-surgeon telementoring/telestration.
- The software integrates the tools needed for datalogging and procedures to characterize the teleoperation process delays.
- It has been developed with an Object Oriented Programming methodology that gives modularity for easy reconfiguration and quick integration with input devices as Joysticks, Haptics, Force-Balls or mouse.
- Its client-server architecture makes the Telesurgery Workstation independent from the kind of surgical robots used (kinematics or operating systems).

The use of a video server constitutes a compact and easy to use solution that avoids having to have a computer equipment dedicated exclusively to video transmission, with the consequent problem of system integrity. The video server is based on the Linux operating system, and up to the present, it has provided high level of robustness and stability. As regards the transmission system developed on DirectX, its only negative aspect is that it only allows image transmission using MJPEG, with the resulting increase in bandwidth.

Using high-level language programming and powerful development environments allowed shorter experiment cycles and a better understanding of the results thanks to the easy graphic representation of the different signals even in run-time.

One of the challenges of teleoperation systems in general, as well as for telesurgery systems, is to achieve force-feedback avoiding the effects of the communication delays. This is a classic problem [10] in which there is a large scope for research and for finding technological solutions for the problems to which these give rise, such as the instability of the basic manual control. This implementation focuses on efficiency but there are some factors as proprietary robots with slow update rate an the delays for digital video encoding/decoding that introduce undesirable delay. Furthermore, direct teleoperation may not be possible with the variable delays of TCP/IP networks. For this reason the implementation of teleautonomous tasks will be necessary to ensure the system stability in the presence of significative/variables delays.

Acknowledgements. This project is the result of the efforts of all the members of the Medical Robotics Research Group of the Department of Automatic Control and Systems Engineering of the University of Málaga and the 1st Surgery Chair, directed by

Dr. Carlos Vara Thorbeck and financed by the FIS00/0050-01, FIS00/0050-02 and CICYT DPI2003-08263 Projects.

References

1. M. C. Cavusoglu, W. Williams, F. Tendick, and S. S. Sastry. Robotics for telesurgery: Second generation berkeley/ucsf laparoscopic telesurgical workstation and looking towards the future applications. *Industrial Robot, Special Issue on Medical Robotics*, 30(1):22–29, 2003.
2. L. W. Way, S. Bhoyrul, and T. Mori. *Fundamentals of Laparoscopic Surgery*. Churchill Livingstone, 1995.
3. V.F. Muñoz, C. Vara-Thorbeck, J.M. Gómez de Gabriel, J. Fernández-Lozano, E. Sanchez-Badajoz, A. García-Cerezo, R. Toscano, and A. Jiménez-Garrido. A medical robotic assistant for minimally invasive surgery. In *IEEE Int. Conf. on Robotics and Automation*, pages 2901–2902, 2000.
4. V.F. Muñoz, J.M. Gómez de Gabriel, J. Fernández-Lozano, I. García-Morales, R. Molina-Mesa, C. Pérez del Pulgar, J. Serón-Barba, and M. Azouaghe. Design and control of a robotic assistant for laparoscopic surgery. In *Proc. of the* 9th *Int. Symp. on Inteligent Robotic Systems*, 2001.
5. V.F. Muñoz, J.M. Gómez de Gabriel, J.J. Fernández-Lozano, I. García-Morales, C. Vara, C. Pérez del Pulgar, and M. Azouaghe. Development of a telerobotic camera manipulator for laparoscopic surgery. In *Proc. of the* 15th *IFAC World Congress on Automatic Control*, 2002.
6. S.I. Brown, C. White, K. Wipat, G. B. Hanna, T. G. Frank, and A. Cuschieri. Characterizing the "gold standard" image for laparoscopic surgery. *Surgical Endoscopy*, 18:1192–1195, 2004.
7. L.H. Eadie, A.M. Seifalian, and B R. Davidson. Telemedicine in surgery. *British Journal of Surgery*, 90(6):647–658, 2003.
8. V.F. Muñoz, J.M. Gómez de Gabriel, I. García-Morales, J.J. Fernández-Lozano, and J. Morales. Pivoting motion control for a laparoscopic assistant robot and human clinical trials. *Advanced Robotics*, 19(6):695–713, 2005.
9. J.M. Gómez de Gabriel, V. F Muñoz Martínez, Francisco J. Domínguez, and J. Serón Barba. Sistemas de transmisión de video para telemedicina. *SISAB 2005*, 2005.
10. J.M. Gómez de Gabriel. *Contribuciones a la teleoperación con retardos de comunicación*. PhD thesis, Universidad de Mlaga, Málaga, Spain, 1999.

Proprio and Teleoperation of a Robotic System for Disabled Persons' Assistance in Domestic Environments

Carlos Balaguer[1], Antonio Giménez[1], Alberto Jardón[1], Raúl Correal[1], Santiago Martínez[1], Angelo M. Sabatini[2], and Vincenzo Genovese[2]

[1] University Carlos III de Madrid
Depto. de Ingeniería de Sistemas y Automática
Avenida de la Universidad, 30. 28911 (Leganés) Madrid, Spain
{balaguer,agimenez,ajardon,rcorreal,smartinez}@ing.uc3m.es
[2] Scuola Superiore SantAnna
ARTS Lab
Piazza Martiri della Libertá, Pisa 56127, Italy
a.sabatini@arts.sssup.it, genovese@mail-arts.sssup.it

Summary. The chapter describes a teleoperation system for assisting disabled and elderly people in their lives and work environments. The developed system ASIBOT (assistive robot) is part of the EU 5th FP project MATS (IST 2001-32080). The goal of the project is to come up with a new concept in teleoperated robotics systems that would help people in their daily domestic activities such as eating, drinking, shaving,grooming or simply retrieving objects from shelves or from the floor, etc. The best feature of the ASIBOT system is represented by a snake-like robot arm that is capable of moving around serially from a wall-mounted or table- mounted docking stations. The robot is also able to "jump" to or from a wheelchair. One of the important properties of proprio and teleoperation of assistive robots is the fact that the operator could be at the same time the patient,the user or the target of the task.Due to disability, the operator has limited capability on controlling the master and also considerable delays in closing the teleoperation loop. Depending on the level of the operators disability, different types of HMI are needed to be used. Some HMI are commanded by voice,simple switches or a joystick.

24.1 Introduction

Teleoperated robotics system is commonly formed by two different scenarios: the operator site where the master and the human operator are located, and the remote site where the robot, performs the remote task. It clearly shows that the human is "isolated" from the working environment and is to be safe at every moment. The present chapter introduces a new concept of robotics teleoperation, called proprio & teleoperation, where sometimes both areas, the operator and remote environment are the same, but not at all times. The human operator teleoperates the robot whose working environment includes himself or herself. Humans in general could not be safe in the master environment as such area could be at the same time the remote environment.

Human factors are important not only for the teleoperation itself but also for safety reasons. The present work also describes the teleoperation architecture of the ASIBOT

M. Ferre et al. (Eds.): Advances in Telerobotics, STAR 31, pp. 415–427, 2007.
springerlink.com © Springer-Verlag Berlin Heidelberg 2007

system where two different control loops (internal and external feedbacks) automatically adjust their role in an overall control strategy. This adjustment depends on the level of motion impairment from the operator/patient.

Finally, the chapter presents the experimental results of the ASIBOT robot applications in serving people in their domestic environment. The main conclusions and operator/patient preferences are also presented.

24.2 Proprio and Teleoperation Control Architecture

Lots of data are sent between the user and the workspace of a slave robot in most teleoperated systems. [1, 2]. Devices such as joysticks, keypads, haptic devices, etc. are used for sending different commands from the user to a computer. Consequently, communication between these two computers comes with delay due to distance and/or complexity in calculation. These problems are of special relevance during teleoperation, which has generated a lot of advancement in the study of this problem [3, 4, 5], and presented here in previous chapters.

In a teleoperated system, slave environments have different sensors, such as stereoscopic cameras, microphones, etc. and the slaves also carry force/torque sensors to determine force reflection. All these elements allow users to better know their workspace and obtain good telepresence. The user receives information, forces, images, and audio from the slave environment through a communication link. All of these data are received on displays, speakers, etc.

Up to now, previous teleoperated systems have been modelled with high degree of accuracy. There are different proposed solutions depending on the slave work-environment, sensorial system, and distance between the master and the slave. Most of these applications have been teleoperated by skilled people who know the system and are usually experienced in such tasks.

In an assistive system, users do not have same skills as described above. So, not all classic teleoperation devices are appropriate for disabled people. It is necessary to develop new tools, or different HMI appropriate to the users level of disability. Depending on their disability, teleoperators will have a slower time response in most cases. This means that a delay can arise due to users behaviour, and not due to a communication delay between the master and slave. This fact requires changes in the value of transference function of the user, proposed by McRuer [20, 6]. Furthermore, the operators ability to sense reacted forces reflected from the slave in the telecontrol operation mechanism is lessened.

The proposed system as shown in Fig. 24.1 is a special teleoperated system since the user is its slave workspace and its target for carrying out the required task. It is the environment where the robot must work. Sometimes, it is not necessary to use cameras or other sensors to provide telepresence for the user. As the user is located at the centre of the task, the robot can be clearly seen and thereby estimate its distance.

Due to such factors,the functionality of the teleoperated system has to be changed in this type of applications. The robot must be autonomous for certain tasks (Fig. 24.2, as the system cannot wait for any response from the user. Several tasks such as eating or drinking are better performed when they are pre-programmed. It will allow a

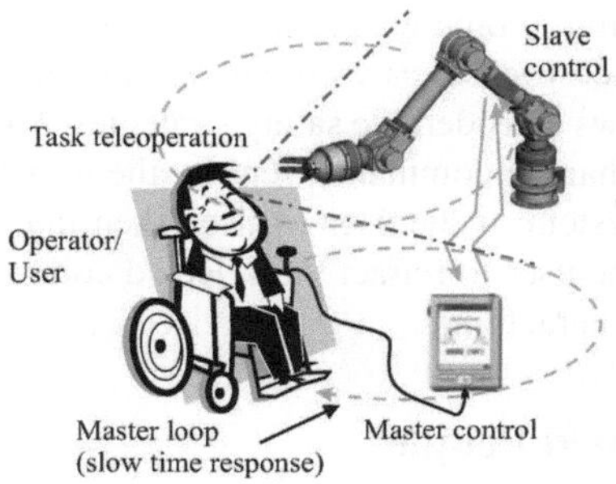

Fig. 24.1. Proprio & teleoperation scheme

Fig. 24.2. Teleoperation

non-skilled user to simply push a button, move a stick, or send voice command considering the adapted users HMI, i.e. the user pushes the meal button, and then the robot fills the spoon and gives the food to the user. In such manner, it is not necessary for the user to guide the robot directly.

In other applications (gaming) it is more suitable for the user to command the robot directly. In such case, the teleoperation is mostly discontinuous in order to reach a high safety level and avoid time delays.

24.2.1 Teleoperation Control Strategy

Teleoperated systems have a strong control loop between the remote and operator sites. The user can command all slave movements and trajectories with main priority in the control loops of the remote and operator sites. Both control loops try to get good telepresence from the operator.

The teleoperation architecture of the ASIBOT system has two different control loops (master and slave feedbacks) which automatically adjust their role in the overall control strategy. This adjustment depends on the level of operator/patients motion impairment, such as shown in figure 24.1.

In such kind of applications, it is not necessary to send data from the remote site to the operator site, because the non-skilled user is located at the task area. In this case the patient is involved in the both control loops, master and slave. In the master loop, there is a slow time response, because the user does not have the same dexterity as of an operator in a classical teleoperated system. In the slave loop, as the robot has to work next to the user/patient, this control loop is designed so as not to hurt him/her.

Patients with minor degree of disability can use a control strategy very similar to classical systems. When the degree of users disability is higher, the ASIBOT system increases the priority of the slave loop. If the robot finds a flaw after a command has been sent, the user may have no time to abort the order, however, the robot is still able to control its safety level. Safety and reliability are particularly important in applications where user's welfare is involved and especially in motion-impaired or cognitive disabled users. Since the system's full safety and precision is unfeasible, [7, 8], it is thereby required to establish the maximum cost over risk acceptance ratio.

The design of this kind of assistive systems must come with redundancy procedures that assure no user control is missed by the user, even in cases of sub-system failure. Appropriate design in the control interface allows considerable safety strategies. Moreover, "error recovery", which is the ability to handle commands sent by the user that could affect the welfare and security of the system or the user is also taken into account. The system must "forgive" and allow the user to retract the selected command with minimal penalty in time loss and system interaction.

24.3 Control Architecture for Disabled People

Fig. 24.3 shows the overall control architecture of the ASIBOT system. Three different levels of computational tasks are considered, and are implemented in the following subsystems:

(i) The Human-Machine-Interface (HMI)
(ii) The Room Controller (RC)
(iii) The Arm Controller (AC)

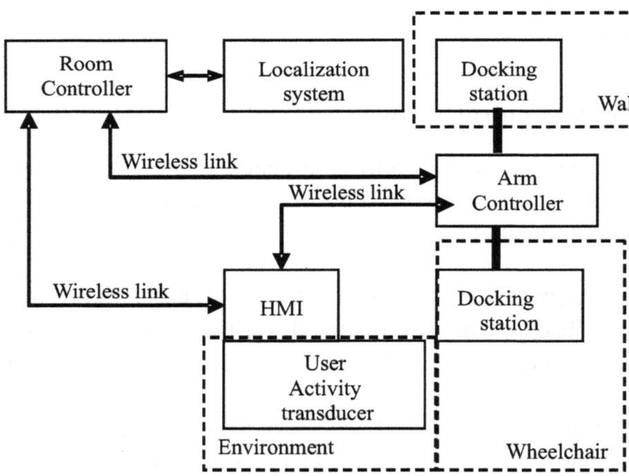

Fig. 24.3. The control architecture of the ASIBOT robot

The HMI is the device available to the user: a) to command the arm's working; b) to be informed about the state of the device or the task that the arm is involved in; c) to benefit from navigational feedback offered during transfer maneuvers from wall-mounted docking station to the wheelchair docking station and vice-versa; d) to get access to standard application software, including Internet browser and e-mailer.

The RC is a computer whose main functions are: a) to perform path-planning activities, so that the arm can be moved to its optimum within the network of available docking stations from a given starting point to the specified target configuration; b) to select and send the list of motion commands needed by the AC to move the robot arm; c) to perform wheelchair localisation via a monocular vision-based system built around web-cam image sensors (sensor-based assistance to docking for management of arm transfer procedures) [21].

The AC is embedded within the robot arm structure. The main functions of the AC are: a) Communication protocol (to interact with the client, HMI) Commands interpreter, b) Kinematics transformations (Direct kinematics and Inverse kinematics), c) Path-planning (for straight line movements), d) Connection to the amplifiers, and e) Commands to digital inputs and outputs, (i.e.: open and close the grippers).

Peculiar to the ASIBOT-approach is the consideration of different interaction situations between the disabled person and the robot:

(i) **Proprio & telerobotics**, the robot and the user are in close spatial relationships, e.g., when the user eats or drinks.
(ii) **Telerobotics**, the robot is controlled by the user, however the robot and the user are in different spatial locations, e.g., the system is selecting the specific tool needed for the function to be accomplished: electric razor, tooth-brush, etc.
(iii) **Autonomous robotics**, the robot might or might not be in close spatial proximity, however no interaction between the user and the robot is required - the arm moves in an autonomous way through different points in the network of docking stations.

These different interaction situations combine spatial relationships, situational context, and modality of interaction which, in using the robot system, can change over time. The bridge between the different interaction situations can be built by providing users with a mobile control and command platform.

24.3.1 Human Factors of Disabled Operators

Human factors are the main requirements of the design of the operator-site, especially in this application. The key part in the control architecture of any assistive robot is the usability of the HMI, because overall performance is HMI dependent. Interaction devices address several mutually exclusive design trade-offs and complications. Users, by nature of their potential benefit from an assistive robotic device, are also very limited in the manner in which they are able to interact with the device. Simultaneously, device specification is variable. However, direct control is good in avoiding uncertainty but task execution is tedious. Executing a pre-programmed task is much faster, yet such systems cannot meet some of the user's requirements and the effort required to program a task has been criticized. A need has been perceived for a non-technically oriented person to be provided with easy tools for performing or programming tasks [9]. The conflicting constraints are to maximize flexibility while minimizing the length of time that requires to perform a task and minimizing the cognitive load placed on the user [10].

In order to design an interface for an assistive robot which allows the user to be 'in the loop' as the main part of the interaction architecture, the ASIBOT robot takes into account the following considerations:

- The HMI device must be portable, and preferably wearable. This defines some physical characteristics of input device including size, range of motion and strength required for activation, and whether the device is a joystick, single switch, or other mode of input. The interface must be updatable and expandable, in order to easily add new devices which allow adaptation of the overall system to the progressive degeneration of the user's residual capabilities. A PDA is a suitable device to obtain these requirements.

- Flexibility and connectivity are needed in order to perform communication with the robot and the environment. It's necessary to pay special attention to issues such as how different users are when using the interface, and how the interface fits into the each user's environment.

- High degree of usability is required. Non-skilled users and the cognitive handicapped must be able to use all the functions of the systems without much effort or heavy mental load.

- The interface system must reduce the mental load on the user, showing only real relevant information and performing adequate sensor data fusion to free the user from doing such. This allows the user to concentrate on the problems related to task execution and not on handling the interface itself.

- The interface and control architecture must allow modality of interaction in execution time (related to the degree of autonomy) to be changed The HMI has to allow scalability in the implication of the user inside the control loop. In such manner, the user decides how to use the robot, by direct control or acts as an observer while the system performs an automated task like robot connector transference.

A thorough analysis of several HMI techniques can be found in the literature. [11]. Nevertheless, table 24.1 shows a list of interface devices vs. kinds of disability, from different motion impaired levels and residuals. Each column shows a group of target users and the rows show the usability of several kinds of interface devices. Residual capacities are ordered from left to right in ascendant order of disability, from those users that are able to move lower limbs, to those with high degree of motion impairment. The second column refers to the output format of the device actuated by the user. The 'C' represents a command type output, generated by the software running in the PC or PDA, and in general for any mechatronic device able to communicate with the robot. The letter 'O' refers to any simple device like switches, licorns, push button, physically connected to a control unit like a PDA or other complex system, and associated to a screen or voice menu that allows selecting the desired action by the user. This is a popular system to interface severely disabled people, [14, 12]. The letter 'P' mentions all analogue transductor-based devices, like joysticks, activated by one hand, chin, back of the neck, foot, etc. in which a proportional control requires dextrous control of the related movement.

User response analysis and characterization provide the basis for defining the architecture and behaviour of any assistive system. One of the simplest and most straightforward user modelling method is the Model Human Processor (MHP). It is based on the segmentation of the user response time in three aspects that are independent among them: first, the time in perceiving an event; second, the time for processing information and deciding upon a course of responsive action; and, finally, the time to perform the appropriate response. Consequently, total response time to stimuli can be described by (24.1):

$$\textit{Total time} = A \cdot \tau_p + B \cdot \tau_c + C \cdot \tau_m \tag{24.1}$$

where A, B and C are integers and τ_p, τ_c y τ_m correspond to the times for single occurrences of the perceptual, cognitive and motor functions [13].

Table 24.1. Interface vs. disability classification; A: Output Type; B: No legs mobility; C: No upper and lower limbs; D: No head, neck or feet mobility; E: Totally motion impaired only vision, hearing, and voice; F: No voice or very difficult vocalization only vision, and hearing

%	A	B	C	D	E	F
PC, keyboard, and mouse	C	Yes	No	No	No	No
PDA + pointer	C	Yes	No	No	No	No
PDA + tactile screen	C	Yes	No	No	No	No
Joysticks hand activated, *Space Mouse 3D*	P/C	Yes	No	No	No	No
Tactile input/ haptic output devices	P/C	Yes	No	No	No	No
Single switch handled screen interfaces	O	Yes	Yes	No	No	No
Gesture recognition, head, shoulder or hand movements	C	Yes	Yes	Yes	No	No
Head/shoulders activated Joysticks	P/C	No	Yes	No	No	No
EMG, *Eyes or gaze tracking*	P/C	Yes	Yes	Yes	Yes	Yes
EEG-BCI	C	Yes	Yes	Yes	Yes	Yes
Face recognition, facial command generation	C	Yes	Yes	Yes	Yes	Yes

Although this is a very simple model, it was selected because it is very easy to understand and to observe deviations from predicted behaviour. Other articles related to the analysis of motion-impaired users and comparatives with able-bodied users are e.g. [17].

24.3.2 ASIBOT HMI

Every group of users have different characteristics, abilities and possibilities. However, most of them have been considered with mobility problems and restricted to a wheelchair.

The device that has been chosen to serve as a user interface is a PDA (Pocket PC) due to several reasons. One reason is its small size and weight. It becomes more portable, which can be carried easily by any user or be attached to a wheelchair and visible to the user. It also consumes very little power. Another characteristic is its versatility and ease of use. There is a screen on the front of the device, which offers tactile use as will be later on explained.

Different ways using the PDA for controlling the robot have been developed. These possibilities are: tactile screen using a pointer or a finger, a scanning system, a button for option selection, a joystick connection and a voice recognition system.

Tactile: Users can choose most suitable interface depending on the users ability to control the robot.For users that could move their hands or a hand, a graphic interface using a PDA in its typical manner, which is by a pointer to select different options from a screen similar to a conventional PC program. Control application has been designed to be used with ease. It lets an unskilled user to adequately carry out desired task and operate the robot.

The graphic interface is based on windows. The goal is to keep it easy because of its similarity to standard programs that use typical window selection, buttons or text boxes. If the user cannot move his/her hands but can move a finger, the application can be controlled in the same manner, due to the screen's tactile feature. The buttons can therefore be pressed by using the finger to point it to the desired option on the screen.

Scan system: The goal of this system is to ease the way a user selects among different options offered by an HMI through its graphic interface. It is done by rotationally highlighting different menus and possible choices on the screen. It means that option 1 is highlighted for a number of seconds, then option 2, later option 3, etc. This allows users to select a highlighted option by simply pressing a button that has been installed on the wheelchair or connected to the PDA. This method has been planned for most severe disabled users.

Joystick: A joystick that was specially designed for this project (Fig. 24.4), it has been developed to control the robot faster and is connected for multiple purpose uses. One of its uses is to move from one application to another or from one screen to the other by simply using the joystick as a pointer in selecting desired options from a screen or a menu.

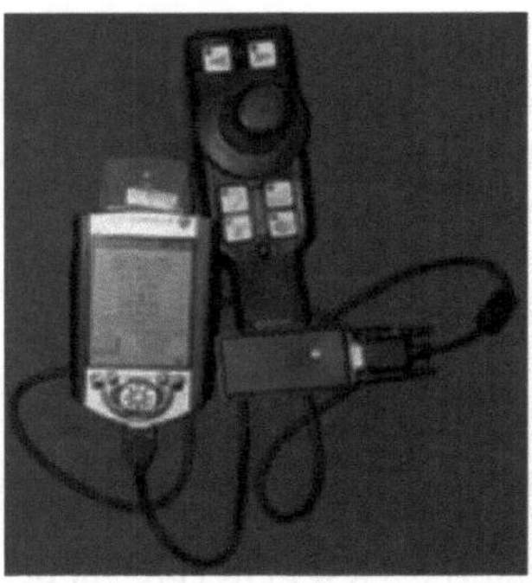

Fig. 24.4. Joystick-activated HMI prototype

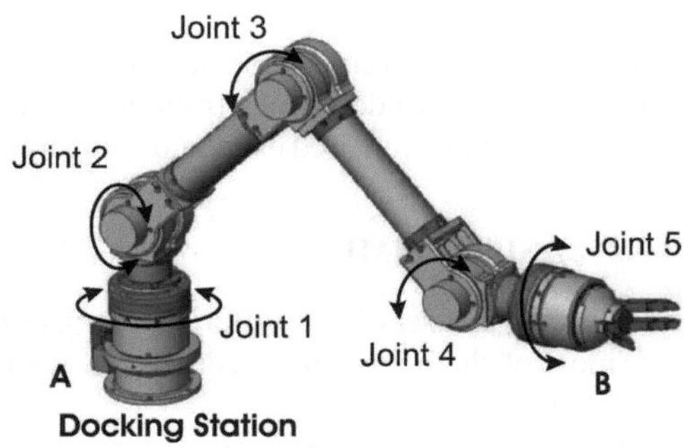

Fig. 24.5. ASIBOT robot design

Another purpose is to move the robot using this device. It allows users to move the joystick in one direction so as to direct the robot to the desired direction. When the joystick comes back to rest position, the robot stops. It also has the option of changing the speed of the robot depending on the pressure users exert on the handle. This lets the user control the system using his/her hands, achieving a more realistic sense of moving the robot.

This joystick could be the one the wheelchairs joystick, so making it unnecessary to have two different joysticks. By pressing a button, the function of this joystick could be switched from wheelchair control to robot control.

Voice recognition: Unfortunately, there are users who are unable to move their arms, hands, fingers or neck. Those users are not able to control the robot using any of the methods described previously. Considering such, another way of control has been developed by using a voice recognition system . This system is connected to the PDA so as to listen for the user's orders.

A wireless headphone with a microphone can be used and connected to the PDA via Bluetooth; this allows the user to speak anywhere through a wireless connection in the room without having the PDA in front of him/her.

24.3.3 Assistive Robot ASIBOT

The ASIBOT robot has five degrees of freedom, and is divided into two parts: the tips, which have a docking mechanism (DS) to connect the robot to the wall, or a wheelchair, and a gripper. The body has two links that contain the electronic equipment and the control unit of the arm. In this manner, the robot is self-constrained, being portable with overall weight of 11 Kg. It is important to note that the robot is symmetric, and due to this, it is possible to attach the arm at any of its ends. It is made of aluminium and carbon fiber. The actuators are torque DC motors, and the gears are flat Harmonic-Drive. Power supply is taken from the connector that is placed in the centre of the docking station. The range and position of the different joints can be seen in Fig. 24.5.

ASIBOT is designed to be modular and capable of fitting into any environment. This means that the robot can move accurately and reliably in between rooms and up or downstairs. It can be transfered from/to a wheelchair [9]. For this purpose the environment is equipped with serial docking stations which make the transition of the robot from one to another possible. This degree of flexibility has significant implications for the care of disabled and elderly people with special needs. Modularity makes the system able to grow as the users degree of disability changes.

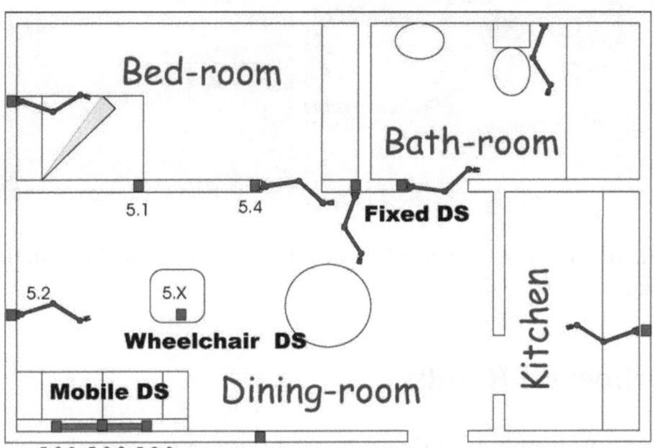

Fig. 24.6. ASIBOT robot in domestic environment and types of DS

There are three different kinds of DS (Fig. 24.6):

- Fixed DS. These kinds of mechanisms are fixed to a walls wherever needed to perform special tasks. It could be fixed on a table for putting plates into the dishwasher.
- Mobile DS. When the robot is required to move a long distance between two DS, moving at high speed is best desired. It can be done by moving on a rail attached to the wall or the table.

- DS inside the wheelchair. It is a special DS, located inside the wheelchair. There is a special DS in the room which allows the transition between the room DS and wheelchair.

24.4 ASIBOT Robot Applications

Main applications of the robot are focused on domestic tasks. A high degree of precision during these motions is not necessary, unless upon moving between two DS. During the design process, it was decided that eating and shaving tasks the only action that the robot must do, is to move the spoon, the shaver, or the toothbrush to the user. Fig. 24.7 shows several working environments where ASIBOT robot operates in performing several domestic tasks, such as shaving and drinking. Likewise, it also shows the control strategy being explained. During these tasks, control, speed and accelaration of the different trajectories for the arm are very important. It is such due to the proximity of the user to the robot being operated. If the robot is moving a spoon with food, it will be crucial to control the orientation of its outermost part in order to avoid dropping the food.

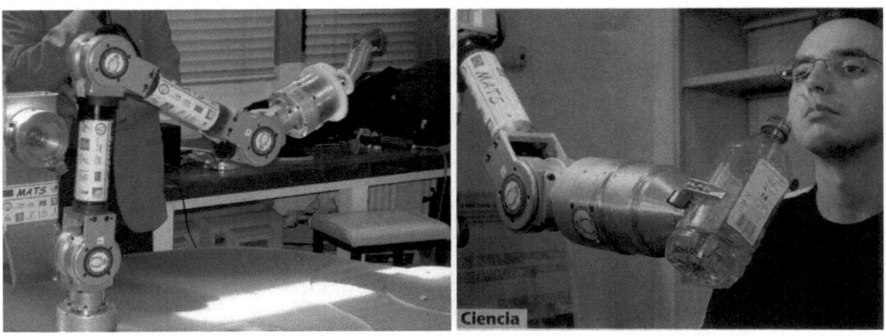

Fig. 24.7. The ASIBOT robot connected to two different fixed docking stations during shaving and drinking tasks

24.5 Experimental Results

Directly seeking out disabled user's opinion about domestic and workplace application motivated the users trials. Our intention was to focus on the detection of acceptance level, identity of prejudice and fear, as well as uncovered needs and expectations.

The protocol followed was applied in two different scenarios. First, doing live demonstrations in the laboratory with users from rehabilitation centres, and second, performing demonstration via teleconference with patientslocated remotely from users who control the robots. In both cases,the demonstration has been divided into two stages:

Six scenarios or tasks assisted by the ASIBOT have been selected for users evaluation: eating, drinking, shaving, applying make-up (Fig. 24.8, picking up and placing objects, and arm-operating from a wheelchair. A brief explanatory report of the

Fig. 24.8. Making-up task assistance by ASIBOT

system was given to the user. Information was collected by an examiner via an open and close-ended questionnaire. Results obtained had to be correlated with the nature of the user's pathologies, culture, residual motor abilities, etc. An exhaustive explanation for the groups of users tested is beyond the scope of this chapter.

24.5.1 Performance of the Proprio and Teleoperation

After gathering data and its analysis, some results have been obtained. Among them are user contributions on how to improve its functions. However some proposals are contradictory and others seem to be closer to fiction than reality. For example, size reduction and at the same time increasing the distance between dockings stations. The following are the main reasons for its little acceptance: too big, lack of use, risk of isolation, reduction of communication, bad appearance, frightening, too slow, etc.

Overall the subject group responded positively to the demonstration. They felt that the robot could constitute a welcome change to their lives. Of the additional comments received 89% were positive. It was a concern that not being able to actually use the robot would mean that the subject would have difficulty relating the robot to their real, everyday situation. This does not appear to be the case because while some of them were not able to relate its use to their situation, majority felt that they could. When asked to express free ideas, the most popular tasks identified were food preparation, household tasks, and grasping high and low objects.

Slightly more than half the subjects felt that the robot would have an effect on the level of the needed care/help. Only 9.5% felt negatively about this effect.

The size of the robot was thought to be the most significant factor. Further work is needed in order to understand exactly how changes in physical size would influence this. Time constraints for the final user evaluation have resulted in the condition profile of the subject sample being biased towards spinal injury (75% spinal injury). This population is more likely to be driven towards greater levels of independence. This could account for the relatively large number who felt that a reduction in care levels was positive. This in itself is an important result, but more work is needed before generalisation across a wider spectrum of conditions is possible.

The most positive tasks (ranked interesting or above) were: wheelchair transfer / gripping and releasing objects - over 75%, drinking - 65%, the largest area thought to be definitely not of interest was eating (approx 30%). Physical size and speed of movement of the robot are likely to have had an effect on this result.

This illustrates the complex nature of evaluating this type of equipment and points towards the importance of a more experiential evaluation than has been possible at this time [18].

There was significant support for some measure of direct control of the robot (as well as with pre-programmed), the use of a joystick / chin control was the most popular. The remote subjects were not able to perceive any possible difficulties of directing the end effector in 3 dimensional space from a two dimensional system such as a joystick.

24.6 Conclusions

The above features of this robot contribute to robotics research by adding a new concept of robotics teleoperation, proprio & teleoperation, which describes a new scenario where both areas, master and remote, are the same. The human operator teleoperates the robot whose working environment includes him- or herself. Humans in general do not feel safe in the master area as it has been the same to the remote environment.

In this sense, the chapter has described the use of this new concept, a proprio & teleoperation system for assistance to disabled and elderly people in their lives and work environments. This robot helps people in their daily domestic activities such as eating, drinking, shaving, applying make-up, toothbrushing, retrieving an object from a shelf or from the floor, etc.

Depending on the degree of users disability, different types of HMI have been presented. Human factors are not only important for the teleoperation itself but also are crucial for safety. The actual tests have demonstrated the feasibility of the system. During the initial trials, it was highly accepted by the users.

References

1. M. Buss and G. Schmidt. Multi-modal telepresence. In *Advances in control, Hightlights of the 5th European Control Conference ECC'*, 1999.
2. T.B. Sherindan. Telerobotics and Human Supervisory Control. *The MIT Press*, 1992.
3. P. Arcara and C. Melchori. Control Schemes for Teleoperation with Time Delay: A Comparative Study. *Robotics and Autonomous Systems*, Vol. 38, 2002.

4. C. E. Garcia, R. Carelli, J. F. Postigo, and B. Morales. Time Delay Compensation Control Structure for a Robotic Teleoperation system. In *Proceeding 4th IFAC International Symposium on Intelligent Components and Instruments for Control Applications.*, 2000.
5. P. Arcara and C. Melchori. Control Schemes for Teleoperation with Time Delay: A Comparative Study. *Robotics and Autonomous Systems*, Vol. 38, 2002.
6. D. McRuer. Human Dynamics in Man-Machine Systems *Automatica*, Vol. 16, 1980.
7. W. Harwin and T. Rahman. Safe software in rehabilitation mechatronic and robotics design In *RESNA 15th Annual Conference*, pages 100–102, 1992.
8. H.F.M. Van Der Loos, D.S. Lees, and L.J. Leifer. Safety considerations for rehabilitative and human service robot systems In *RESNA 15th Annual Conference*, pages 322–324, 1992.
9. A. Gimnez, A. Jardn, R. Correal, R. Cabas, and C. Balaguer. A portable light-weight climbing robot for personal assistance applications In *8th International Conference on Climbing and Walking Robots (Clawar'05)* 2005.
10. C. Balaguer, A. Giménez, and A. Jardón. Climbing Robots Mobility for Inspection and Maintenance of 3D complex Environments. *Autonomous Robots*, Vol. 18. No. 3. pages 157–169.
11. R. Rammoun, J.M. Dtrich, and F. Lauture. The new MASTER man-machine interface In *International Conference on Rehabilitation Robotics*, 1994.
12. Z. Han, H. Jiang, P. Scucces, S. Robidoux, and Y. Sun. PowerScan: a Single-Switch Environmental Control System for Persons with Disabilities In *Proceedings of the IEEE Bioengineering Conference 8-9th*, pages 171–172.
13. M.J. Topping, H. Helmut, and G. Bolsmjo. An Overview Of the BIOMED 2 RAIL Robotic Aid to Independent Living Project. In *International Conference on Rehabilitation Robotics ICORR'97*, pages 23–26. Vol. 37 No. 5, pages 591–598, 1997.
14. A. Craig, Y. Tran, P. McIsaac, and P. Boord. The efficacy and benefits of environmental control systems for the severely disabled. *Med Sci Monit*11(1): RA32-39,PMID: 15614204, 2004.
15. J. Angelo. Factors affecting the use of a single switch with assistive technology devices, Journal or Rehabilitation. *Research and development*, Vol. 37 No. 5, pages 591–598, 2000.
16. S.K. Card, T.P. Moran, and A. Newell. The Psychology of Human-Computer Interaction. *Lawrence Erlbaum Associates*, 1983.
17. S. Keates, P.J. Clarkson, and P. Robinson. Investigating the applicability of user models for motion-impaired users In *Proceedings of ACM ASSETS 2000, Arlington*, VA. pages 129–136, 2000.
18. C. Balaguer, A. Giménez, A. Jardón, R. Cabas, and R. Correal. Live experimentation of the service robot applications elderly people care in home environments In *IEEE/RSJ. International Conference on Intelligent Robots and Systems (IROS'2005)*
19. A.M. Sabatini, V. Genovese, and E.S. Maini. Be-Viewer: vision-based navigation system to assist motor-impaired people in docking their mobility aids In *Proc. IEEE International Conference onRobotics and Automation (ICRA 2003)*, pages 1318–1323.
20. D. McRuer and E.S. Krendel. Mathematical Models of Human Pilot Behavior. *ARGARD-AG-188*, 1974.

25

Internet-Based Tele-Laboratory: Remote Experiments Using the SNRP Distributed Network Architecture

Raúl Marín[1], Raúl Wirz[1], Pedro J. Sanz[1], and Josep Fernández[2]

[1] Universidad Jaume I.
Dept. d'Enginyeria i Ciéncia dels Computadors
Campus de Riu Sec. 12006 Castellon, Spain
{rmarin,wirz,sanzp}@icc.uji.es
[2] Technical University of Catalonia (UPC)
Automatic Control and Computer Engineering Department
Campus Nord. 08034 Barcelona, Spain
josep.fernandez@upc.edu

Summary. This chapter describes the way students at the UJI University (Jaume I University of Castellon) are able to design remote visual servoing experiments from any computer connected to the Internet. These experiments are able to control a set of real robotic devices such as manipulators, cameras, and sensors. It means the experiments are finally executed on real robots instead of 3D simulations.

In fact, the experiments presented in this chapter focus on the on-hand visual servoing problem, which has been possible thanks to the images obtained from a camera mounted on the gripper of an educational robot.

Moreover, the Distributed Network Architecture is presented, which permits any external algorithm to have access to the state and services of several network robots (e.g. robots control, network cameras, object recognition, etc) [17] [16]. We call this architecture SNRP (Simple Network Robot Protocol), due to the fact that commanding a robot via Internet follows a very simple Web Service protocol.

Simplicity is maybe the most important challenge of a network robotics architecture, due to the fact that it must be possible for a very broad range of devices to be part of it. In fact, thanks to this simplicity we were able to implement a prototype of SNRP Network Camera using an FPGA.

The last part of the article explains how the system can execute remote experiments that use several SNRP robots in a concurrent and synchronized manner. As an example two educational manipulators and a mobile robot are used to simulate an industrial task.

Keywords: Distributed Systems, Internet, Network Robot, Multirobot Programming, Remote Visual Servoing.

25.1 Introduction

The concept of Network robot recently came up at the Workshop "Network Robots" within the IEEE ICRA 2005 World Conference[1]. During this event it was realised that

[1] IEEE International Conference on Robotics and Automation, Barcelona, Spain, 2005.

M. Ferre et al. (Eds.): Advances in Telerobotics, STAR 31, pp. 429–444, 2007.
springerlink.com © Springer-Verlag Berlin Heidelberg 2007

many interesting issues related to networking and distributed systems are still open and unresolved, that is when the device that communicates is able to sense, move, cooperate, learn, and react (i.e. network robot) [3] [4].

A "networked robot" is a robotic device connected to a communications network such as the Internet or LAN. The network could be wired or wireless, and based on any of a variety of protocols such as TCP, UDP, or 802.11. Many new applications are now being developed ranging from automation to exploration. There are two subclasses of Networked Robots:

(i) Tele-operated, where human supervisors send commands and receive feedback via the network. Such systems support research, education, and public awareness by making valuable resources accessible to broad audiences.
(ii) Autonomous, where robots and sensors exchange data via the network. In such systems, the sensor network extends the effective sensing range of the robots, allowing them to communicate with each other over long distances to coordinate their activity. The robots in turn can deploy, repair, and maintain the sensor network to increase its longevity, and utility. A broad challenge is to develop a scientific background that couples communication to control enabling such new capabilities.

A very important part of a system that permits the control of a remote network robot (e.g. telerobotic system) is the user interface. Predictive displays [26] can be used in order to avoid the time delays effects [22]. Moreover, augmented reality user interfaces enhance very much the information provided to the user from the cameras located at the robot place [25].

In October 2000 the UJI Online robot was connected to the web for research purposes. It consisted of an educational manipulator robot with three cameras, which enabled a user to remotely control pick and place operations of objects located on a board. Experiments about distributed systems, object recognition, virtual reality, augmented reality, speech recognition, and telemanipulation were accomplished in order to enhance the way people interacted with the system. Again, the students were provided with the possibility to program more sophisticated pick and place operations, using this time both, the offline and the on-line robots.

Nowadays, the system presents a more sophisticated experimental setup, including a multirobot configuration. Two educational robot manipulators (see Fig. 25.1) and seven cameras are presented: two taking images from the top of each scene, one pan-tilt camera from the side, two cameras situated on the grips and two more cameras from the front. Each one of these devices can be accessed and programmed concurrently by a remote algorithm.

The top cameras are calibrated, and used as input to the automatic object recognition module and 3D-model construction. The other five cameras (two cameras for each manipulator plus a pan-tilt camera) give different points of view to the user when a teleoperation mode is necessary in order to accomplish a difficult task (e.g. manipulating overlapped objects). For the results presented in this paper an Intranet at 10Mbps connection was used.

It was considered necessary to enhance the system in order, not only to let students control the robot from a user interface, but also to program it by using any standard programming language. In April 2003 the UJI TeleLab project came up, which involved the

Fig. 25.1. Current Robotic Tele-Lab

design of a Java library called "Experiments" that let students program their own control algorithms from any computer connected to the Internet and perform them with the real robot. (See Fig. 25.2) Some pilot experiments have been performed with researchers and students since then. The interest in the design of Internet-based Tele-Laboratories is increasing enormously, and this technique is still very new. A very good example of already existing experiments in this area can be found in [24] [23].

25.2 The Experiments Library

The Tele-Laboratory can be programmed in two ways. The first one is sending SNRP commands as a web service (e.g. "http://mentor1.robots.uji.es /move /left /10"). This SNRP commands follow the REST convention (Representational State Transfer) and are very convenient for controlling the devices remotely from any programming language like for example Matlab, Visual Basic, Java, etc.

On the other hand, the Tele-Laboratory provides the Java specific Experiments Library, wich enables any Java program to get access the to remote Java objects of the server side. This second method is more convenient because it provides a higher set of possibilities and as well it lets the student upload the experiment to the server side instead of executing it on the client computer. This library already includes templates that are examples of simple experiments that manage the remote robots and the cameras (see Fig. 25.3).

To facilitate implementation of the experiments even more, an "Experiment_" "template" is provided that already inherits from the "Experiment" class and presents the structure of a typical Remote Programming experiment, which is: Extending the

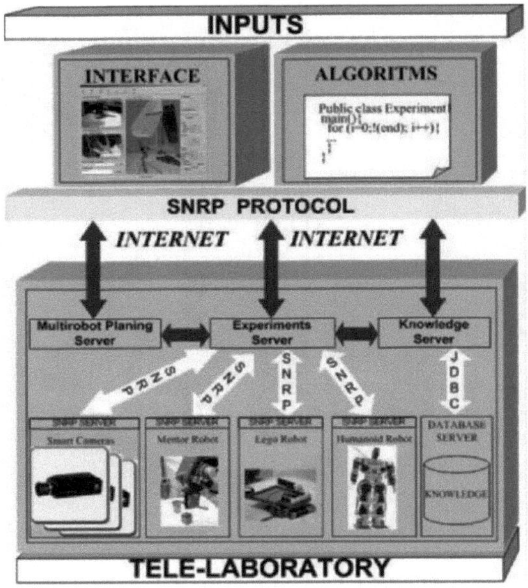

Fig. 25.2. Simplified architecture of the UJI TeleLab

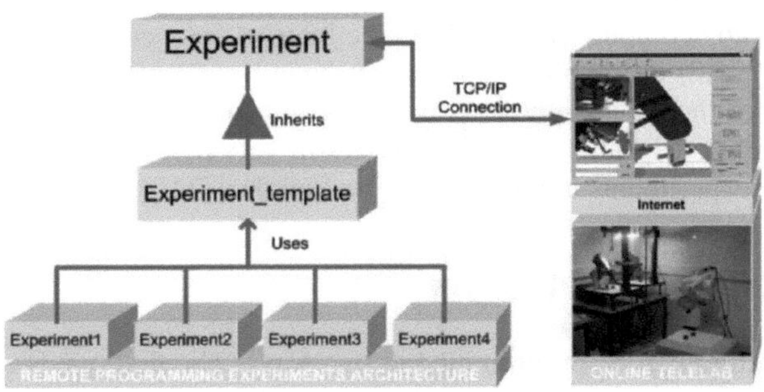

Fig. 25.3. The Experiments Library

"Experiment" class, creating an instance of the experiment, calling the "getSceneManagerSer" to obtain the serialized objects from the TeleLab, executing the corresponding actions on the Telelaboratory, and closing the connection (see Fig. 25.4).

25.3 The Experiments in Action: Programming Visual Servoing Algorithms

Visual serving [10] involves the use of one or more cameras and a computer vision system to control the position of the robot's end-effector relative to the work piece as

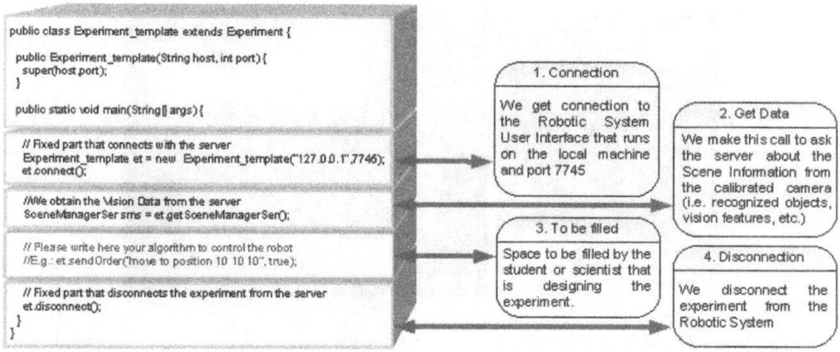

Fig. 25.4. Remote Programming Template

required by the task. It is a multi-disciplinary research area spanning computer vision, robotics, kinematics, dynamics, control and real-time systems

Remote Visual Servoing techniques are normally used for teleoperation using a real-time communication bus [20]. When using the Internet as communication media the challenge is bigger, due to the fact that unpredictable time-delays and bandwidth limitations are introduced. Moreover, if the system is designed in a distributed way and allows the concurrent control of multiple network robots, the challenge is even bigger.

The idea is to introduce the Internet communication channel inside the visual servoing loop, and then to enable the remote programming of a real Internet tele-laboratory to test those algorithms in a simple way.

The remote experiments that have been tested are the following:

- **Experiment 1:** Remote Visual servoing using one fixed top camera. The client side (user's program) does the feature extraction and the pose determination. It means obviously that the Tele-Laboratory gives the actual camera image to the user and is the experiment that performs the computer vision and the control law.
- **Experiment 2:** Remote Visual servoing using one fixed top camera. In this case it is the server side (Tele-Lab) that provides the user with the feature extraction and the pose determination. This second experiment is oriented to the experimentation of simple visual servoing techniques within an introductory program of Robotics at the University. The student or researchers only has to resolve the control law function.
- **Experiment 3:** Remote Visual servoing using one on-hand camera (See Fig. 25.5). For this experiment the student gets as input an image extracted from a camera mounted on the robot arm. This image is transferred from the remote camera to the user via Internet. The gripper of the robot can be seen from this mounted camera. The feature extraction and the pose determination are performed by the user experiment in the client side. It corresponds to the Experiment 1 using in this situation a camera mounted on the gripper instead of a fixed one.

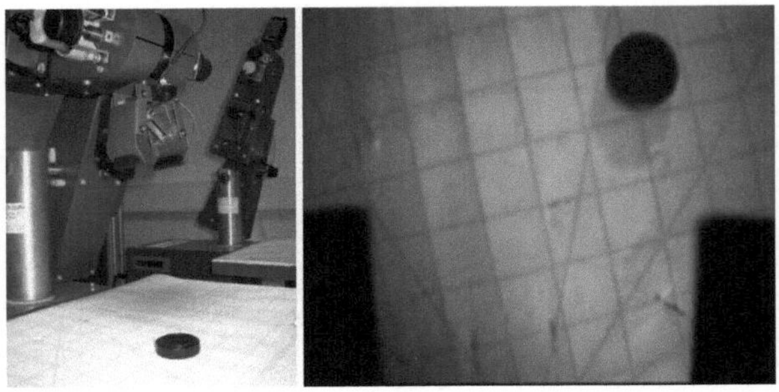

Fig. 25.5. Eye-In-Hand visual servoing experiment within the UJI Educational Telelaboratory

Experiment 1 and experiment 2 do the same, but the difference between both experiments is which side does the feature extraction and the pose determination, the client side (experiment at the user place) or the server side (Tele-Lab). More information and results of experiment 1 and experiment 2 can be found in [12].

The following is a description in more detail of the actions performed by both the client and the server side for the proposed remote visual servoing experiment 3. For that, the Interface provided by the Experiments Library, which implements the following commands, must be used:

- Transform Point (pixel1 pixel2): It transforms a point in camera coordinates (pixel1 pixel2) to its related robot coordinates (X Y Z). This order is used for both experiments.
- Move to position (X Y Z): It moves the robot to the position (X Y Z) in world coordinates. The server uses the robot's inverse kinematics to transform this point in joint movements. This order is used for both experiments.
- GetSceneManager (camera): It returns a Java serialized object that has access to every detail of the Computer Vision procedure (Binarization, Segmentation, objects' details, etc.).
- GetRobotPosition: It returns the position of the robot in the camera view (pose determination) by recognizing the pink label attached to the robot arm.
- Get Image (camera): It sends the actual camera image to the client. This order is only used in Experiment 1.

The common procedure for every experiment is: First, the position of the object to be grasped is obtained in camera coordinates, as well as the position of the robot in the camera, which is calculated by the Tele-Laboratory. This camera coordinates are transformed into robot coordinates by means of the "Experiments" command "transform Point". Then, the resulting error in camera coordinates is calculated and the control law in order to get a better approximation to the object on the next iteration is applied. The control law provides the movement that must be applied to the robot in order to stabilize the error (see Fig. 25.6).

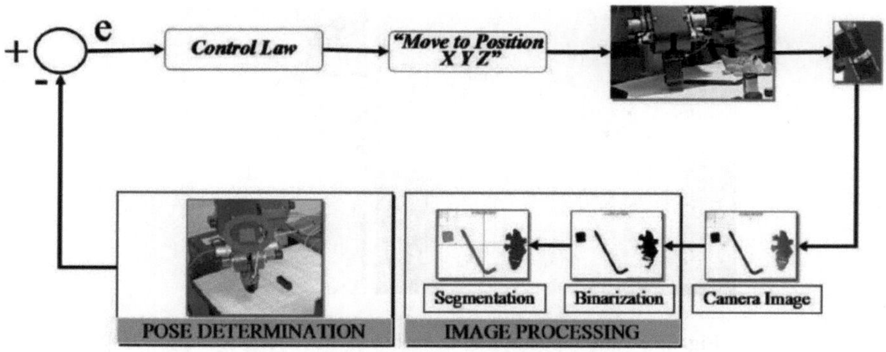

Fig. 25.6. Remote Visual Servoing Programming Details

25.4 Experimental Results

In the following the time delays due to the communication are analysed, through a general-purpose network. The temporal cost of each one of the stages in the experiment 3 is then evaluated.

Every image is in color format (RGB 352x288 pixels), and its size is 11679 bytes. The image obtained through the HTTP protocol requires 26 IP packets of 516 bytes average packet size.

The experiment 3 is a little bit more sophisticated because the camera is moving while the robot goes from one position to another. It means that, from the camera point of view, the objects in the scene are in movement too. In this situation, the whole scene must be binarized and segmented every time the robot moves, in order to acquire the required mathematical features for every object.

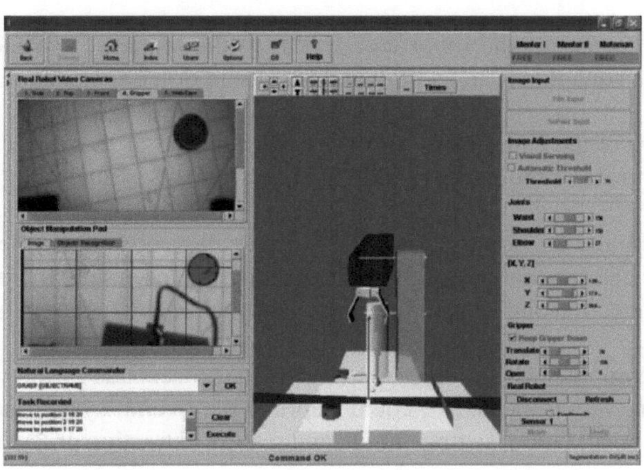

Fig. 25.7. Initial position of the robot as shown by the educational Tele-Lab user Interface

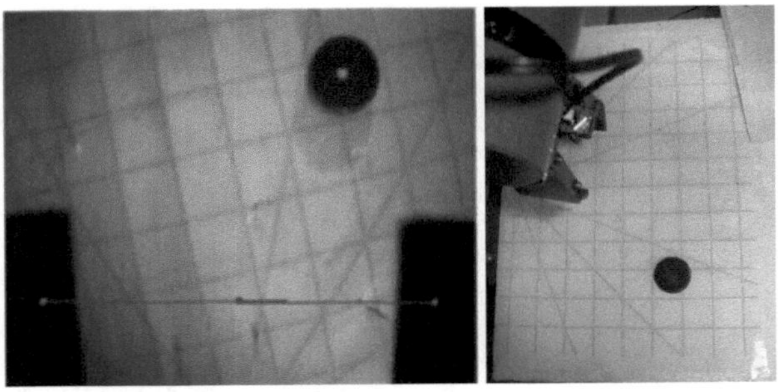

Fig. 25.8. Initial position of the robot at the robotic scenario

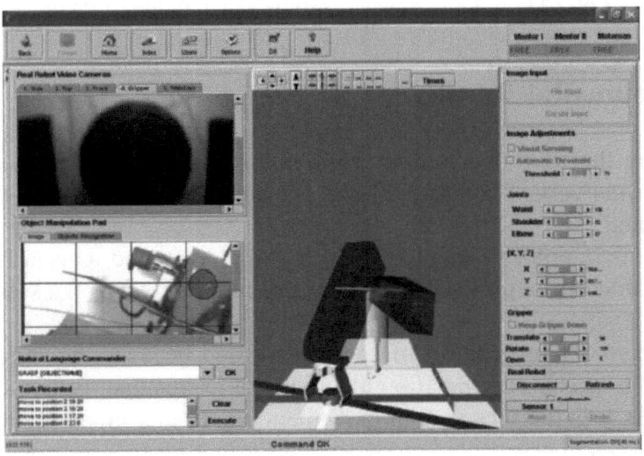

Fig. 25.9. Final position of the robot as shown by the Tele-Lab user Interface

The gripper fingers of the robot will appear in the scene as two independent objects, so that the student has an additional difficulty in order to calculate the next robot movement.

The objective of the experiment is to bring the centroid of the object (Fig. 25.7 and 25.8) to the centroid of the gripper by iterative movements of the robot(Figs. 25.9 and 25.10).

In Figs. 25.11 and 25.12 we can see the times employed for both, the client and the server side. We can see that most of the time is invested in sending the images through the Internet connection and as well by waiting for the robot to accomplish the required movement.

The whole computer vision process takes about 2.5 seconds of the whole manipulation operation, which means we could improve the overall system performance by optimizing the binarization, segmentation, and feature extraction procedures.

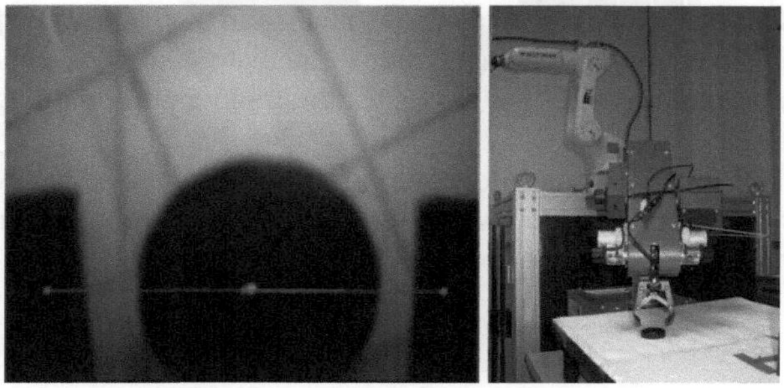

Fig. 25.10. Final position of the robot at the robotic scenario

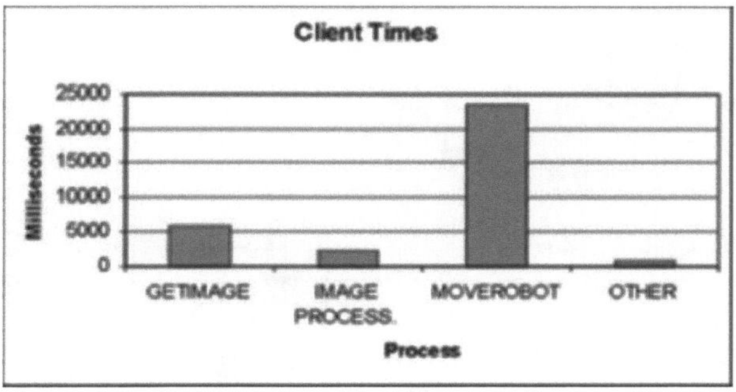

Fig. 25.11. Time employed by the client

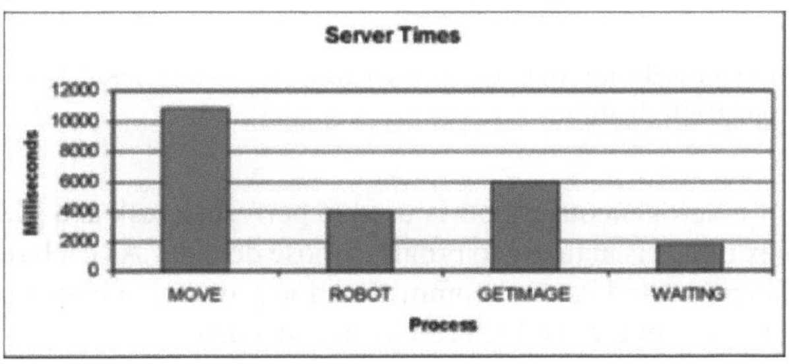

Fig. 25.12. Time employed by the server

25.5 Enhancing the System: The Multirobot Architecture

After having shown the way the system allows a student/researcher to program remotely a single robot provided with cameras and sensors, now, our challenge goes further because it is required that the user is able to program multirobot collaborative tasks in a concurrent manner.

Due to the fact that every robot has its own hardware and software specification, as well as its own programming language, our immediate goal is to implement a hardware and software architecture that facilitates the remote programming of multirobot synchronized tasks.

As seen in Fig. 25.13, the multirobot architecture is organized in several networks, where a router offers access to the whole set of devices (i.e. robots, cameras, etc.) that stay in the same location (i.e. same tele-laboratory).

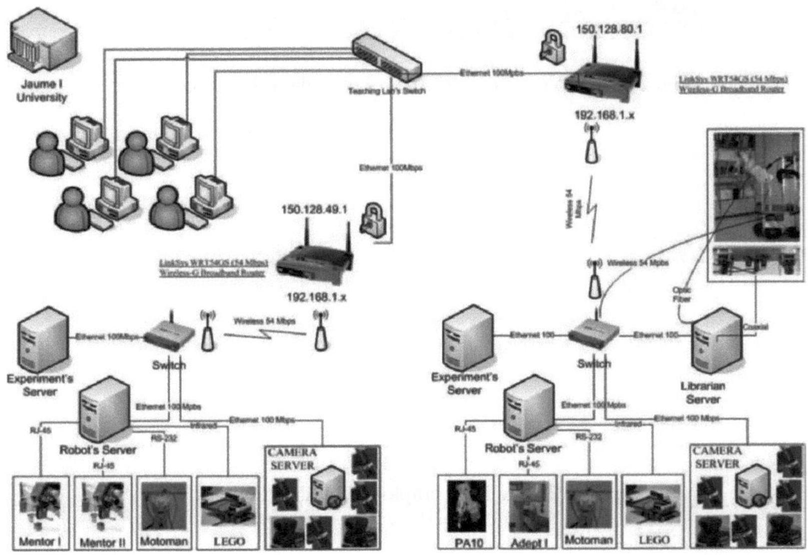

Fig. 25.13. Multirobot Hardware Architecture including the teaching tele-laboratory (left-bottom) and the research one (right-bottom)

When a set of heterogeneous robots is used to perform a task at a given time, one of the most complex things is actually to program those devices. As each robot has its own programming language and its own communication protocol, a platform that assures a reliable and efficient cooperation between them is needed.

To accomplish this, the experiments interface has been improved by including the possibility of having a set of behaviours (concurrent tasks) associated to every experiment. In fact, behaviour has the responsibility for programming a given robot, camera or sensor. These behaviours work in a synchronized and concurrent manner (See Fig. 25.14). To let students and researchers program the Tele-Laboratory in a simple and reliable manner, we provide a Java library called "Experiments" that already manages the implementation details of these behaviours using our SNRP interface (see next subsection). This library already includes templates that are examples of simple experiments that manage the remote robots and the cameras.

As explained in [23], Networked Robotics is an emerging research area for creating intelligent robotic architectures that integrate embedded systems, sensor and

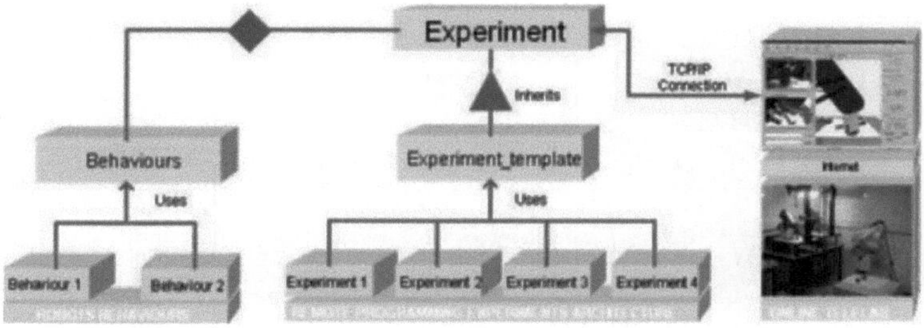

Fig. 25.14. Experiment class with behaviours

actuator networks. The challenge is defining a software and network architecture within the network robotics area that provides the following features: (1) simple, (2) open, (3) flexible, (4) dynamic, (5) robust, (6) scalable, (7) efficient, (8) secure, (9) platform independent. Simplicity is maybe the most important challenge of a network robotics architecture, due to the fact that it must be possible for a very broad range of devices to be part of it. In fact, as we will describe later, thanks to this simplicity we were able to implement a prototype of SNRP Network Camera using an FPGA. In the scientific literature several works can be found that propose different ways and architectures to organize task-oriented applications of multiple network robots. Some of these architectures are focused on internet software frameworks (e.g. web services) and have been extended from previous works in single-robot telerobotics. Other works focus on the internet network protocols themselves and study internet transport protocols that enable real-time control and teleoperation of network robots over IP. In fact, as explained in [19], solutions can be found to cope with the problems associated to the Internet in order to control networked robots: Time-varying transmission delay, and not-guaranteed bandwidth.

In the Fig. 25.15 we can see the software architecture of the SNRP framework which provides the following modules:

- SNRPRobot: Every robot/device in the SNRP framework would provide an SNRP_" "Robot network interface, which allows any client (e.g. user experiment) to use a service provided by it (e.g. "motoman. service. moveToPosition (x, y, z)". Examples of this interfaces are "SNRPConveyorBelt", "SNRPMotoman", and "SNRPFPGAVision".
- SNRPRobotsGroup: An SNRP robot can be the union of several SNRP robots (e.g. a Mobile manipulator is the union of a mobile robot, an arm). Moreover, the SNRP module for the arm can be the union of two modules, the one for the gripper and the one for the arm itself. Thus, SNRPRobotsGroup permits defining new services for the several networks robots that work together as if they were a unique robot.
- SNRPNamingService: An SNRP network robot can register to a naming service in order to select a name (e.g. UJI/telelabs/industrial/motoman) and inform other peers of which IP and port it is listening to.
- SNRPServiceHolder: The services provided by an SNRP robot can be programmed in a static manner within the SNRP Module itself, or on the other hand, they can be added dynamically in runtime. For that, an SNRP service that follows a given

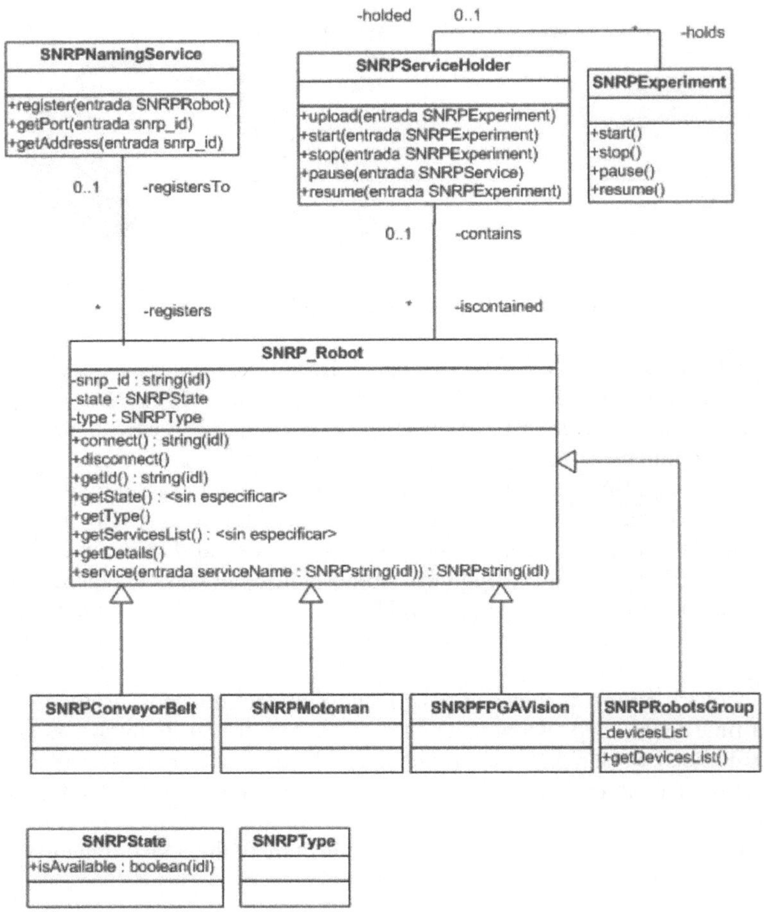

Fig. 25.15. Tele-Laboratory's software architecture: UML Diagram

interface most be uploaded into the SNRPServiceHolder. At the moment of writing the industrial telelaboratory has only one SNRPServiceHolder for the whole system. Anyway, the architecture would permit having a holder in every SNRP robot.

- SNRPExperiment: An SNRP Experiment is a robot service that can be allocated into a service holder. In fact, the experiments that we are performing in this moment provide a unique service holder for the telelaboratory that is located in the Experiment's server computer. Further experiments could be defined as the union of several SNRP services (agents) that are running concurrently on different service holders and that all together provides a certain robotic task.

Once we have seen the software architecture for the SNRP framework, now we are going to focus on the SNRP protocol itself that permits the communication between SNRP experiments, holders, naming services and robots.

First of all, as we want to enable the devices to be accessed through the internet, they should be able to manage the IP protocol (See Fig. 25.16). On top of it, the SNRP framework enables the device to accept TCP and UDP connections. As explained before, UDP and TCP are not the best solutions to perform remote control through the Internet, so the SNRP framework provides the possibility to transport the internet datagrams through other protocols like "trinomial" [18] or RTP (real-time transport protocol) among many others.

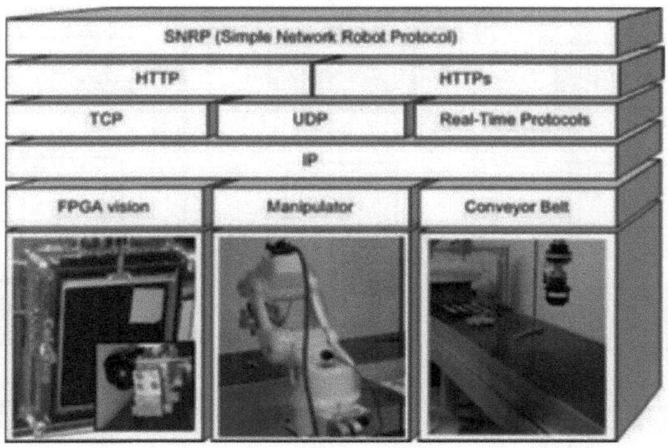

Fig. 25.16. SNRP network architecture

In order to make the SNRP simple to use and implement, it uses the HTTP protocol as basis, which give it even more interoperability and flexibility. However, for this kind of situation the HTTP does not provide the following features: event notification and support for structured information. These two characteristics are very important to design the SNRP framework in the industrial robotics area. To accomplish this, we have incorporated into the SNRP protocol the REST model [21], which permits the implementation of state-oriented applications and a simple scenario to design event notification and structured information features.

25.5.1 Example of Multirobot Experiment

To validate the proposed architecture, a simple multirobot experiment has been implemented. In this situation three robots (two manipulators called Mentor, and one mobile LEGO robot) are used. The task's challenge to let the first manipulator pick the objects in his environment and put them one by one on the mobile LEGO robot. Once the LEGO detects that an object has been put on its platform, it transports it to the second manipulator. And finally, the last manipulator picks the object up from the LEGO platform, classifies it (i.e. applies object recognition), and drops it in the correct classification grid.

This experiment was implemented using our SNRP protocol (Fig. 25.17) using three concurrent agents, one for every robot in the scenario. First of all, the Mentor 1 robot waits until an object is dropped on the work-area (WaitObjectinField) and, once it detects it, the Mentor 1 robot picks it up (TakeObject). After that, the Mentor 1 robot waits

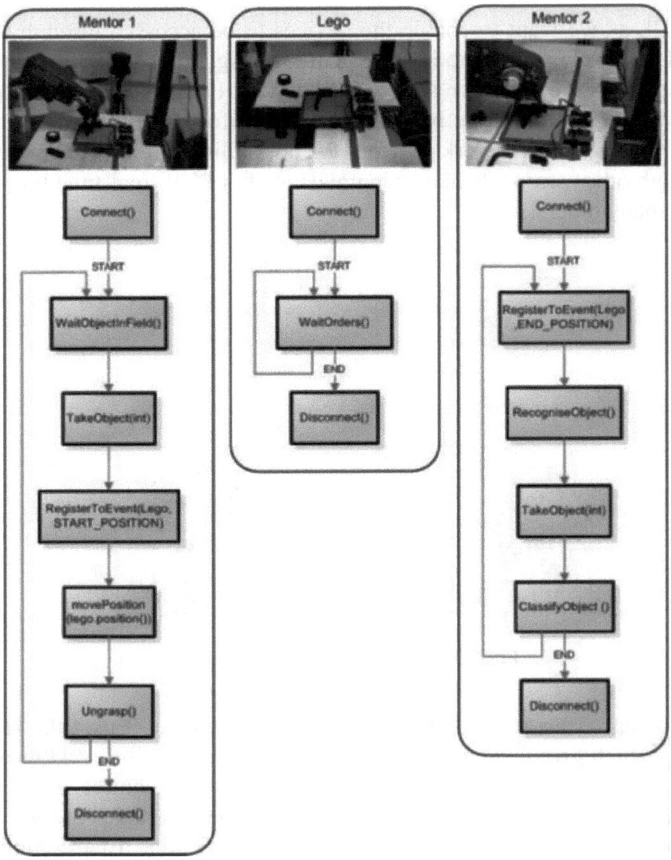

Fig. 25.17. SNRP Diagram

until the Lego robot is situated in the start position (RegisterToEvent start_position). Once it happens, the Mentor1 robots is moved on top of the Lego Robot (MovePosition(lego.position()), and drops the object (Ungrasp). After that, it tells the Lego robot to move to the end position at the Mentor2 workarea (GoFront). When the Lego robot arrives at the end position, the event "end_position" is activated. As the Mentor 2 robot is waiting for this event, it is able to proceed accordingly. When the Mentor 2 robot detects the event, it first recognizes the object (RecogniseObject), picks it up (TakeObject), classifies it (ClassifyObject) and orders the Lego robot to return to its start position (GoBack). Finally, Mentor 2 robot goes to its home position and waits until a new end_position event gets activated (RegistertoEvent end_position). When the Lego Robot arrives at the start position, the start position event (start_position) is activated, so the Mentor 1 Robot starts the same loop again.

25.6 Conclusions

This chapter has presented a real education and training experience that allows students to program remotely a set of networked robots, cameras and sensors in a simple way. The SNRP protocol has been described that follows a Web Service Architecture and enables the devices to be controlled over any platform (e.g. Linux) and using any programming language.

The Multirobot architecture has been validated using two experiments, one related to controlling a robotic manipulation using visual servoing loops over the network, and the second programming a whole multirobot experience that simulates an industrial classification problem.

Next efforts will be oriented to improve the SNRP protocol by including specific network congestion capabilities for teleoperation. In fact, TCP/IP is not very convenient for these situations due to the fact that it implies many fluctuations that are not good for a smooth manual control over the internet.

Acknowledgements

This work has been partially funded by the Spanish Ministry (MEC) under Grants TIC2003-08496, DPI2004-01920, TSI2004-05165-C02-01, by the Fundació Caixa Castelló under Grants P1-1B2002-07, P1-1B2003-15, and P1-1A2003-10, and by the Generalitat Valenciana under grant GV04A-698.

References

1. R. Marín, P.J. Sanz, and A.P. del Pobil. The UJI Online Robot: An Education and Training Experience. *Autonomous Robots*, vol 15, Number 3, 2003.
2. R. Marín, and P. Sanz. Grasping Determination Experiments within the UJI Robotics Telelab. *Journal of Robotic Systems, Internet and Online Robots for Telemanipulation Special Issue (Part 2)*, vol 22, Number 4, 2005.
3. B. K. Kim et al. Web Services Based Robot Control Platform for Ubiquitous Functions. In *Proc. of the IEEE Int. Conf. On Robotics and Automation (ICRA)*. Barcelona, Spain, 2005.
4. D. Lee and M. W. Spong. Bilateral Teleoperation of Multiple Cooperative Robots over Delayed Communication Networks: Theory. *In Proc. of the IEEE Int. Conf. On Robotics and Automation (ICRA)*. Barcelona, Spain, 2005.
5. Jesus Manuel Gomez de Gabriel, Anibal Ollero Baturone, and Alfonso Jose Garcia Cerezo. Teleoperacion y telerrobotica. *Pearson Education*, cop. Madrid, 2006.
6. J.M. Azorin, O. Reinoso, R. Aracil, and M. Ferre. Generalized control method by state convergence for teleoperation systems with time delay. *AUTOMATICA 40* (9): 1575–1582, SEP 2004.
7. N. Turro, O. Khatib, E. Coste-Maniere. Haptically augmented teleoperation. In *Proc. IEEE International Conference on Robotics and Automation (ICRA)*, 2001.
8. G. T. McKee, D. I. Baker, and P. S. Schenker. Robot Scpaces, Module Networks and Distributed Robot Architectures. In *Proceedings of the IROS 2004 Workshop on Networked Robotics:issues, architectures and applications*, Sendai, Japan, 2004.

9. G. T. McKee, D. I. Baker, and P. S. Schenker. Network robotics: Dynamic reconfigurable architectures. In *Proceedings SPIE Intelligent Robots and Computer Vision XXII: Algorithms, Techniques and Active Vision*, 2004.
10. S. Hutchinson, G.D. Hager, and P.I. Corke. A tutorial on visual servo control. *IEEE Trans. on Robotics and Automation*, 12(5):651-670, 1996.
11. C.S. Kim, W.H. Seo, S.H. Han, and O. Khatib. Fuzzy logic control of a robot manipulator based on visual servoing. In *Proc. Industrial Electronics (ISIE)*, 2001.
12. R. Wirz, R. Marin, and P. J. Sanz. Remote programming over multiple heterogeneous robots: a case study on distributed multirobot architecture. *Industrial Robot: An International Journal*. Vol33 Issue 6, 2006.
13. R. Wirz, R. Marín, and E. S. Quintana-Orti. Distributed System for Remote Programming of Multiple Network Robots: System Performance & Parallization Issues. In *CEDI 2005 Ist Spanish Conference on Computer Science CEDI 2005, Workshop on Parallelization (JP 2005)*, Granada, 2005.
14. P. X. Liu, M. Q. H. Meng, and S. X. Yang. Data Communications for Internet Robots. *Autonomous Robots*, vol 15, 2003.
15. P. X. Liu, M. Q. H. Meng, P. R. Liu, S. X. Yang. An End-to-End Transmission Architecture for the Remote Control of Robots Over IP Networks. *IEEE Transactions on Mechatronics*, vol 10, Number 5, 2005.
16. R. Marin and P. J. Sanz. Grasping determination experiments within the UJI robotics telelab. *J. Robot. Syst.*, vol 22, Number 4,203–216, 2005.
17. R. Marin, P. J. Sanz, and A. P. Del Pobil. The UJI Online Robot: An Education and Training Experience. *Auton. Robots*, vol 15, Number 3,283–297, 2003.
18. P. X. Liu, M. Q. H. Meng, P. R. Liu, and S. X. Yang. An end-to-end transmission architecture for the remote control of robots over IP networks. *IEEE/ASME Transactions on Mechatronics*, vol 10, Number 5, 560–570, 2005.
19. P. X. Liu, M. Q. H. Meng, and S. X. Yang. Data Communications for Internet Robots. *Auton. Robots*, vol 15, Number 3, pages 213–223, 2003.
20. C. S. Kim, W. H. Seo, S. H. Han, and O. Khatib. Fuzzy logic control of a robot manipulator based on visual servoing. In *Proceedings. ISIE 2001. IEEE International Symposium on Industrial Electronics*, pages 1597–1602, 2001.
21. R. T. Fielding and R.N. Taylor. Principled design of the modern Web architecture. In *Proceedings of the 2000 International Conference on Software Engineering*, pages 407–416, 2000.
22. J. M. Azorin, O. Reinoso, R. Aracil and M. Ferre. Generalized control method by state convergence for teleoperation systems with time delay. *Automatica*, vol. 40(9) 1575–1582, 2004.
23. D. I. Baker, G. T. McKee, and P. S. Schenker. Network robotics: dynamic reconfigurable architectures. In *Proc. of SPIE. Intelligent Robots and Computer Vision XXII: Algorithms, Techniques, and Active Vision*, pages 360–369, 2004.
24. G. McKee and D. Baker. Robot Spaces, Module Networks and Distributed Robot Architectures. In *Networked Robotics: Issues, Architectures and Applications (IROS-2004)*, pages 360–369, 2004.
25. N. Turro and O. Khatib. Haptically Augmented Teleoperation. *ISER '00: Experimental Robotics VII*. Edit. Springer-Verlag, 2001.
26. J. M. Gomez de Gabriel, A. Ollero Baturone, and A. J. Garcia Cerezo. Teleoperacion y Telerrobotica. (in Spanish), *Edit. Prentice-Hall, series Automatica Robotica*, 2006.

26

Force Reflecting Teleoperation Via IPv6 Protocol with Geometric Constraints Haptic Guidance

Emmanuel Nuño, Adolfo Rodríguez, and Luis Basañez

Technical University of Catalonia (UPC)
Institute of Industrial and Control Engineering
Av. Diagonal 647, 11th floor, 08028 Barcelona, Spain
{emmanuel.nuno,adolfo.rodriguez,luis.basanez}@upc.edu

Summary. When dealing with teleoperated systems, several important aspects have to be considered: unstructured environment, communication delay, human operator uncertainty, and safety at the remote site, amongst others. The main contribution of this work, that tackles some of the aforementioned issues, is a system that combines a force feedback teleoperation scheme with geometric constraints and haptic guidance. The allowed motion space of a robot can be reduced by specifying a set of geometric relations between the robot tool and the workcell's fixed objects. These relations are processed by a geometric reasoning module that generates a compatible motion subspace. Restriction forces are then fed to the operator via a haptic interface in order to guide its movements inside this subspace. The communication channel between the local center and the remote cell is implemented using high speed networks with the novel IPv6 protocol. The slave robot control is based on position or velocity. Experimental results validate the proposed approach.

26.1 Introduction

The trajectories to be described by a robot end-effector –either in free space or in contact with other objects– depend strongly on the task to be performed and on the topology of the environment with which it is interacting. As common examples, peg-in-hole insertions require the alignment between the peg and the hole (in [1] a virtual execution of this task is presented), spray painting tasks require maintaining the nozzle at fixed distance and orientation with respect to the surface to be painted, and assembly tasks often involve the alignment or coincidence of faces, sides and vertices of the parts to be assembled.

One of the main reasons for the increasing use of robots instead of humans for performing certain tasks, besides economics, is their high accuracy, speed and repeatability. Shon and McMains [2] describe experiments for evaluating speed and accuracy when drawing 3D objects with a haptic device, and they conclude that, if the operator is provided with a guidance method, the drawings are clearly better. With the geometric constraints haptic guidance approach the operator defines a set of geometric constraints which restrict the motion of the haptic device [3] from the 3D space with 6 DOF to a subspace with less DOF which satisfies the aforementioned geometric constraints.

In order to assist humans while performing different tasks some approaches have been developed. They can be divided in two groups, depending on how the motion

M. Ferre et al. (Eds.): Advances in Telerobotics, STAR 31, pp. 445–458, 2007.
springerlink.com © Springer-Verlag Berlin Heidelberg 2007

restrictions are created, either by software or by hardware. To the first group belongs the work by Turro et al. [4] that have implemented three types of constraints for the operator movements: constrained movement along a line, virtual obstacle avoidance using a potential field force and geometric cube constraint in order to limit the robot workspace. A drawback of this approach is the need of reprogramming when a new restriction must be introduced. In [5] constrained teleoperation has been develop using predictive control techniques. The constraints act in the nominal path of the robot end effector, but on the master side motion guidance is not implemented. It means that if the operator moves the robot away from the nominal path, the robot will keep moving along the desired path.

An often used method is to provide the obstacles with a repulsive force potential field. Thus, the operator will not make the robot collide with the obstacles. This method has been used in [6] with a mobile robot, where the force generated by the obstacles is fed back to the operator. The work described in [7] is one of the first to add geometric restrictions in the robot workspace. In this case a stiff virtual wall is modeled as a spring-damper system. Several authors propose the use of hardware to guide motion, for example guide-rails [8] and sliders with circled rails [9]. Mechanical guides such that only translation is needed, would make it easy to move into a restricted space [10].

The main contribution of this paper is the teleoperation framework in which motion restrictions can be easily defined and modified by the operator according to the task needs. Deviations from the restrictions are translated into forces via a haptic device, providing the operator with an intuitive interface to ensure movements inside the restricted subspace, thus improving task performance and immersiveness.

The paper contents are as follows: the system architecture of the proposed teleoperation system is laid out in Sec. 26.2; in Sec. 26.3 the functional description of each subsystem is presented; Sec. 26.4 discusses how the geometric constraints are calculated; Sec. 26.5 refers to the haptic guidance forces; Sec. 26.6 describes the experimental validation performed, and in Sec. 26.7 the conclusions and future work are outlined.

26.2 System Architecture

A scheme depicting the proposed system architecture is shown in Fig. 26.1. Three main subsystems are pointed out in the diagram: the local command center, where information concerning the restrictions and guidance are computed; the communication channel, which manages the information flow; and the remote robotic cell, where the actual task is performed.

- The *local command center* hosts the two modules that permit interaction with the operator: the Geometric Reasoning Module, responsible for the geometric constraint processing, and the Force Guidance Module, which handles forces that have to be fed to the operator as well as the integration of position/velocity with the restricted subspace data.

  In addition to the main control loop a video stream provides video feedback from cameras located at the remote robotic cell, whose zoom and orientation can be remotely actuated.

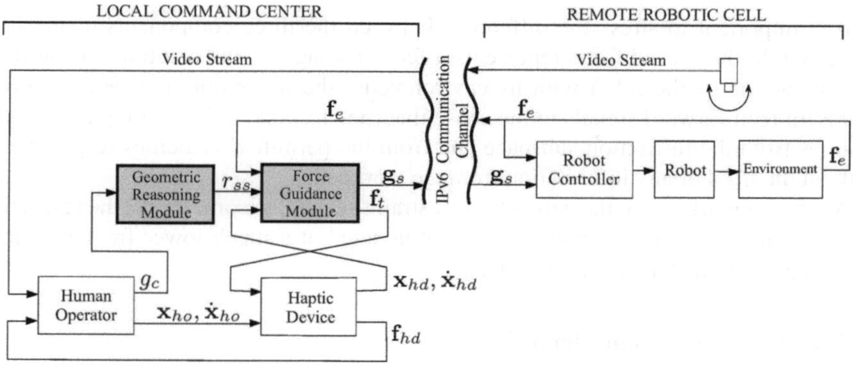

Fig. 26.1. System architecture. The novelty of the scheme resides on the use of the Geometric Reasoning Module, the Force Guidance Module and the IPv6 communication channel.

- The *communication channel* is responsible for the management of the data flow between the local command center and the Remote Robotic Cell. It is served by a high-speed Local Area Network (LAN) with a client-server application structure. These structures have been implemented using a socket based configuration with TCP/UDP and IPv6 protocols.
- The *remote robotic cell* is composed of a robot, its controller, a force-torque sensor, and a video server with two 3 DOF cameras. The subsystem sends information about the interaction of the robot with the environment to the local command center.

26.3 Subsystems Description

26.3.1 Local Command Center

The sequence of events that take place in the geometric constraints haptic guidance is the following:

(i) The operator defines a set of geometric constraints g_c.
(ii) The Geometric Reasoning Module computes a set of restricted subspaces that satisfy the input constraints.
(iii) The Force Guidance Module computes the restriction force $\mathbf{f}_r$ that must be exerted to maintain the position of the end-effector inside the currently selected restricted subspace r_{ss}, as well as the viscous force $\mathbf{f}_v$ that prevents the velocity of the end-effector from becoming too large for the robot to follow the master.
(iv) The restriction force $\mathbf{f}_r$ and viscous force $\mathbf{f}_v$ are combined with the raw force measurement $\mathbf{f}_e$ coming from the Remote Robotic Cell to generate the total force $\mathbf{f}_t$, which is fed to the operator via the haptic device.
(v) The guidance signal $\mathbf{g}_s$ is sent to the remote robotic cell via the communication channel.
(vi) Steps 3 to 5 are repeated while the operator sends new position/velocity $(\mathbf{x}_{ho}, \dot{\mathbf{x}}_{ho})$ data and the geometric constraints remain unchanged.

It is important to stress the difference between the three components of the total force. While the sensed force represents a feedback signal –the reaction arising from the interaction of the robot with its environment– the restriction and viscous forces represent feed forward signals in the sense that they respond to known inputs –the deviations from the restriction subspace and from the permitted velocities respectively– without the need of any information from the workcell.

Another remark is that the geometric constraint set and, in some cases, the restriction subspace are symbolic information and are updated at a much lower frequency than other signals in the teleoperation scheme.

26.3.2 Communication Channel

The overall structure of the client-server application uses the IPv6 protocol due to its Quality of Service (QoS) benefits [11]. Additional information about IPv6 can be found in reference [12]. Amongst the new implementations of IPv6 applications over next generation networks all over the world (as an example see [13]), telerobotics have a great potential to develop.

Comparative studies between using TCP or UDP as the transport layer protocol [14, 15, 16] state that TCP provides a point to point channel for applications that require reliable communication while UDP provides communication that is not guaranteed. This is because TCP is a confirmation based protocol and UDP is not. However, TCP has the drawback that it has an unpredictable data arrival time because it retransmites lost packets after a timeout of any acknowledge message of the transmitted packet. Since UDP does not require an acknowledgment message, the network delay can be substantially lower. In this work sockets are compatible with both transport layer protocols. In [17] a protocol for internet robots is presented which is based on UDP and in [14] a new protocol featuring the benefits of both TCP and UDP is developed. Other Internet based robots have used higher level protocols as http [18]. In [19] and [20] examples of internet robots are described.

When dealing with a teleoperated system one must take into account that delay plays a critical role in the system stability. High-speed networks with an increased QoS can reduce the delay by using communications based on priorities rather than the usual best effort networks. In the teleoperation scheme of Fig. 26.1 the guidance and the force signals have the highest priority and the video signal the lowest.

26.3.3 Remote Robotic Cell

The controller inputs are either position or velocity commands, sent from the master site. According to the nature of the task the controller can receive the restrictions from the Geometric Reasoning Module and the variables corresponding to the allowed DOF.

When position or velocity commands are selected from the local command center depending on the task, the robot can move strictly in the restriction subspace ($\mathbf{x}_r$) or with a deviation from it ($\mathbf{x}_{hd}$), allowed by the stiffness and damping implemented in the Force Guidance Module. In Fig. 26.2, vector $\mathbf{d}$ represents the deviation of the position or velocity command produced by the operator.

The position control scheme of the remote robotic cell is stable. Then, if the input references (position/velocity) of the controller are bounded the overall system is also stable. The drawback of this straightforward approach is that in some cases transparency of the overall system is sacrificed for the sake of system stability.

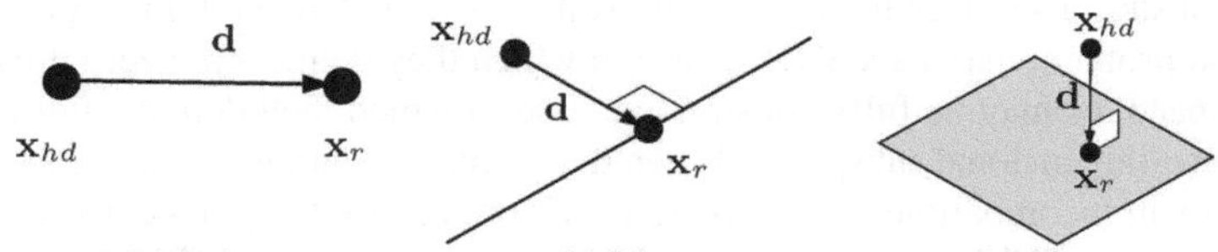

(a) Point (b) Line (c) Plane

Fig. 26.2. Motion with restrictions

26.4 Geometric Reasoning Module

Most geometric constraint solvers come from the CAD world and deal with the general problem of positioning multiple objects in such a way that they satisfy a set of predefined geometric constraints. Due to the high complexity of such problems, existing solvers often limit their scope to planar cases and rely heavily on numeric and graph-searching techniques to find the solutions [21].

The Geometric Reasoning Module is a geometric constraint solver that addresses the problem of positioning a mobile object in the 3D space, such as the robot tool or the grasped object, with respect to its fixed environment given a set of geometric constraints. The nature of this particular problem, where the only unknown is the position of the mobile object, renders its analytical possible solution for most cases that arise in practice.

The solver implemented in the Geometric Reasoning Module is mainly based on the methodology described in [22]. Its computational simplicity and efficient implementation makes it fast enough to comply with the real time demands of a teleoperated system.

The solution process can be summarized as follows:

(i) The operator defines as input a set of pairwise constraints between geometric elements of the mobile object and the fixed environment. This implies that the positions of the objects comprising the environment of the robot must be known (i.e. a virtual model of the workcell).

Currently points, lines and planes are considered as valid input elements. Each constraint can be –depending on the selected elements– coincidence, parallelism, angle, perpendicularity or distance. It is important to note that the constraints set does not need to be well defined, since the solver detects redundant and incompatible constraints and deals with them appropriately.

(ii) The input constraints are decomposed into a simpler set which will be referred to as the fundamental constraint set. If the rotational and translational parts of the constraint can be decoupled, pure rotational and translational constraints are derived. As an example, a line-line coincidence constraint can be decomposed into one parallelism constraint between the line direction vectors and one point-line coincidence constraint.

(iii) The fundamental constraints are combined in pairs to obtain implicitly defined ones in order to restrict the solution space as much as possible.

(iv) The solutions –if any– are analytically computed based on the resulting fundamental constraint set. Depending on how well constrained the system is, the solutions to the problem may have DOFs, case in which they define a nonzero dimensional subspace; or may be fully constrained, case in which they define a finite number of zero-dimensional subspaces. When the solver does not find a solution, it can be due to an incompatible input constraint set, or because the input constraint set defines an unhandled problem. When either of these occurs, the situation is detected and notified to the operator.

(v) A validity test is performed to verify whether the obtained solutions satisfy the input constraint set.

Fig. 26.3 shows the steps followed to solve the case of two point-point coincidence constraints between identical cubes. This problem illustrates how a rotation constraint is obtained from pure translational input constraints. The initial configuration of the cubes is shown in Fig. 26.3a. The input constraints are

$$\begin{aligned} P_{m1} &= P_{f1} \\ P_{m2} &= P_{f2} \end{aligned}$$

where P_{m1}, P_{m2} are points of the mobile cube and P_{f1}, P_{f2} are points of the fixed cube. Since the distance $d(P_{m1}, P_{m2}) = d(P_{f1}, P_{f2})$ it follows that the initial constraints are compatible and can be substituted by one point-point coincidence constraint and one parallelism constraint (Fig. 26.3b)

$$\begin{aligned} P_{m1} &= P_{f1} \\ \mathbf{u_m} &\parallel \mathbf{u}_f \end{aligned}$$

where $\mathbf{u}_m = \overrightarrow{P_{m2}, P_{m1}}$ and $\mathbf{u}_f = \overrightarrow{P_{f2}, P_{f1}}$. The resulted restriction subspace (Fig. 26.3c) has dimension one and corresponds to rotations along the direction of $\mathbf{u}_f$.

26.5 Force Guidance Module

The main objectives of this module are to send the guidance signal $\mathbf{g}_s$ to the remote robotic cell through the communication channel and to generate the total force $\mathbf{f}_t$ the haptic device will fedback to the operator. The information to be sent to the remote robotic cell from the local command center follows two different schemes, explained in Sec. 26.5.1, depending on the task and on the operator knowledge of the remote environment.

The total force (at an instant k) that is fed back to the operator is:

$$\mathbf{f}_t = \mathbf{f}_s + \mathbf{f}_r + \mathbf{f}_v$$

where $\mathbf{f}_s$ is the contribution of the force sensor, $\mathbf{f}_r$ is the restriction force (the effect of the restriction subspace) and $\mathbf{f}_v$ is the viscous force. Sec. 26.5.2 shows how these forces are calculated.

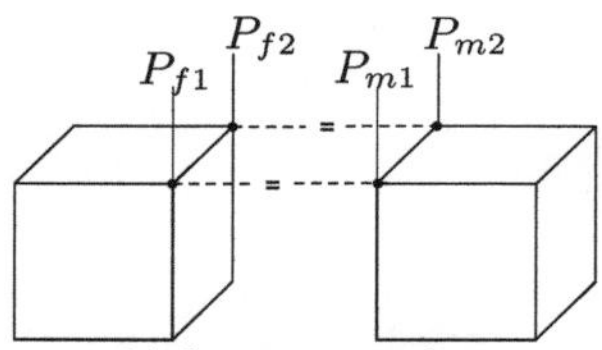

(a) Input constraints

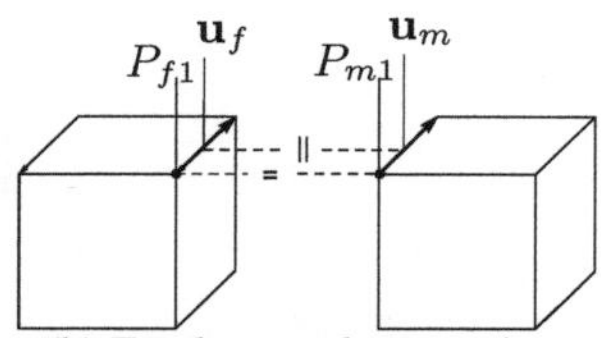

(b) Fundamental constraints

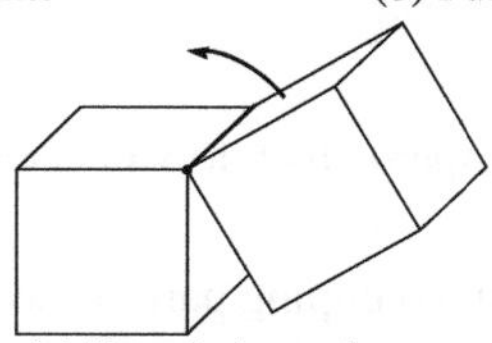

(c) Restriction subspace

Fig. 26.3. Geometric Reasoning Module example

26.5.1 Master Guidance Signal Schemes

- *Scheme 1.* There are two possible combinations of position/velocity of the master in this scheme. Depending on the task, the operator must decide whether the real position/velocity of the master or the projection into the restricted subspace r_{ss} of these values is send to the slave. In the first case the slave motion is strictly within the restricted subspace ($\mathbf{g}_s = \mathbf{x}_r$ or $\mathbf{g}_s = \dot{\mathbf{x}}_r$), and, in the second, it moves along the actual values of the master ($\mathbf{g}_s = \mathbf{x}_{hd}$ or $\mathbf{g}_s = \dot{\mathbf{x}}_{hd}$) meaning that the master's position/velocity could deviated by a certain value $|\mathbf{d}|$ from the restricted subspace (Fig. 26.2).
- *Scheme 2.* In this scheme, instead of holding the master's position/velocity data, the guidance signal carries the restriction subspace information along with the current position/velocity in the subspace, $\mathbf{g}_s = (r_{ss}, \mathbf{p}_m)$ or $\mathbf{g}_s = (r_{ss}, \dot{\mathbf{p}}_m)$. Since the movements only take place in r_{ss}, it is not necessary to send all six position/velocity components but only the ones concerning the available DOF.

26.5.2 Force Feedback Generation

- *Sensor force.* The raw force measurement $\mathbf{f}_e$ that comes from the sensor's data is filtered in the local command center at a cutoff frequency of 500 Hz. The resulting sensor force $\mathbf{f}_s$ is calculated as follows:

$$\mathbf{f}_s = \mathbf{T}_s \mathbf{f}_e$$

  where $\mathbf{T}_s$ is a transformation between the force sensor frame and the haptic frame.
- *Restriction force.* This is an attractive force that tends to fix the haptic position to the restriction subspace. The direction of this force is calculated by vector $\mathbf{d}$ (Fig. 26.2) and its magnitude is given by:

$$\mathbf{f}_{r_k} = K_P \mathbf{e}_k + \mathbf{D}_k$$

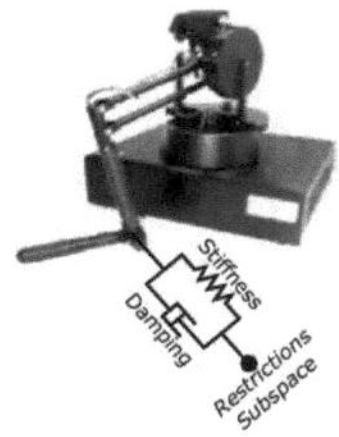

Fig. 26.4. Approach of the virtual restrictions

where $\mathbf{D}_k$ is the corresponding damping part of the controller and it is given by:

$$\mathbf{D}_k = K_D\left(\mathbf{e}_k - \mathbf{e}_{k-1}\right)$$

where $\mathbf{e}_k$ is the position error: $\mathbf{e}_k = \mathbf{x}_r - \mathbf{x}_{hd}$ with $\mathbf{x}_r$ the reference point that lies on the restriction subspace and $\mathbf{x}_{hd}$ is the haptic device position at an instant k. $\mathbf{f}_r$ is the force contribution of the applied restriction. The value of $\mathbf{f}_r$ will be zero if no restriction is set. Currently, only the force feedback corresponding to the translational DOF of the restriction subspace have been implemented.

Fig. 26.4 shows an intuitive interpretation of the restriction force in order to visualize the concept. K_P and K_D are chosen to set the stiffness and damping of the restriction.

- *Viscous force.* If velocity mode is selected and the velocity of the master is high enough, the slave may not be able to follow the velocity commands. To avoid this situation an additional restriction has been implemented: above a certain velocity value, which depends on the maximum velocity achievable by the slave, the motion restricting force is a function of the master velocity $\dot{\mathbf{x}}_{hd} = \hat{\mathbf{v}}$, and this force is zero below that value. The resulting force of this effect $\mathbf{f}_v$ at instant k is given by:

$$\mathbf{f}_{v_k} = K_v\hat{\mathbf{v}}_k$$

where K_v is a gain that fits the needs to restrict velocity. With

$$\begin{aligned}\hat{\mathbf{v}}_k &= b_0\mathbf{v}'_k + b_1\mathbf{v}'_{k-1} + a_0\mathbf{v}_{k-1}\\ \mathbf{v}'_k &= \mathbf{v}_k - \mathbf{v}_{k-1}\\ \mathbf{v}_k &= \tfrac{1}{T}\left(\mathbf{x}_k - \mathbf{x}_{k-1}\right)\end{aligned}$$

the expression of $\hat{\mathbf{v}}_\mathbf{k}$ corresponds to a 1st order Butterworth filter with coefficients b_0, b_1, a_0 calculated at a frequency ratio (sample freq / cutoff freq) of 10, and T is the sample period (see [23] for more details). In the computation of the viscous force, the noisy position signal must be filtered with a low-pass frequency component, from which a velocity estimate can be reliably derived.

26.6 Experimental Testbed

The experimental testbed consists of a TX-90 Stäubli robot, with a CS8-C Stäubli controller, and a JR3 force-torque sensor, a PHANToM 1.5™ 6 DOF haptic device from

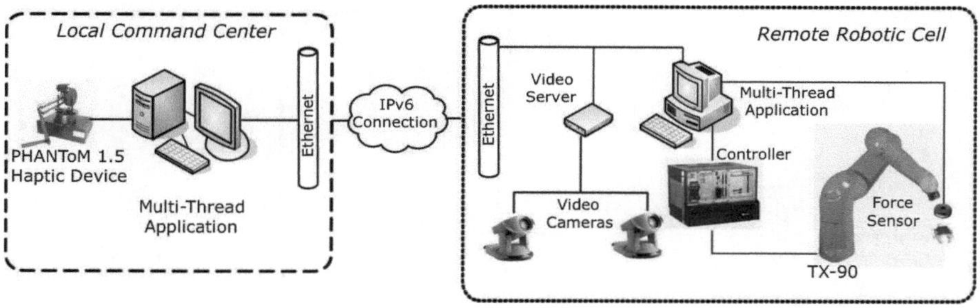

Fig. 26.5. Physical system architecture

Sensable Technologies, and two CANON VC-C4 video cameras with an AXIS 2400 video server which provides a 10-20 fps motion JPEG video stream. A sketch of the system is shown in Fig. 26.5.

On the software side, the interaction with the haptic device is done with Sensable Technologies' GHoST™ libraries. Since the haptic's control loop runs at 1kHz, forces must be calculated within the millisecond time window. All software is written in C++ using sockets and POSIX threads. Graphics User Intefaces were developed with Trolltech's QT library and 3D virtual scenes are managed by Systems in Motion's COIN 3D library, an open source implementation of Open Inventor (SGI).

Fig. 26.6 shows a window of the Geometric Reasoning Module GUI. The interface features two main components, on the right a 3D virtual model of the robot workcell is displayed, and on the left there is a control panel that permits the definition of geometric constraints (top-left), diplay solutions to the current problem (middle-left), and if the selected solution has at least one DOF, as many knobs as DOFs will become active and permit the motion of the constrained object along its unrestricted directions (bottom-left).

Fig. 26.7 depicts the Force Guidance Module GUI window. The user can define the restriction subspace manually or by using the Geometric Reasoning Module (top-right), choose one of the two master guidance signal schemes –described in section 26.5.1– (middle-right), and configure protocols, addresses and ports of the communication channel (bottom-right). Additionally, there is a visual representation of the measured forces and torques (top-left), and a text-box that outputs relevant information to the user (bottom-left).

In order to validate the teleoperation scheme, a peg-in-hole task was remotely performed using the proposed teleoperation architecture. Although this is a simple task, it was chosen amongst others because it reflects in a straightforward manner the benefts of Geometric Constraints Haptic Guidance, namely the performance improvement due to the added precision and repeatability. Such benefits can be greater when dealing with more complex tasks that involve restricted motion along higher-order curves and surfaces, as well as sequences of multiple restricted movements, like assembly tasks.

The peg-in-hole insertion has the following characteristics:

- Since both peg and hole have a circular cross-section, a line-line coincidence constraint is defined between their revolution axes (Fig. 26.8). By setting this constraint

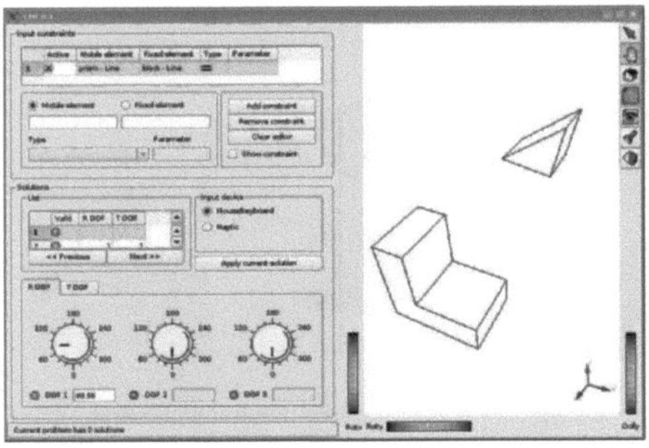

Fig. 26.6. Geometric Reasoning Module GUI

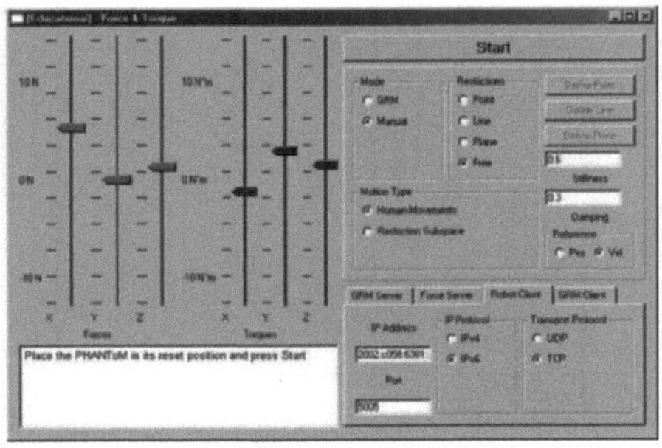

Fig. 26.7. Force Guidance Module GUI

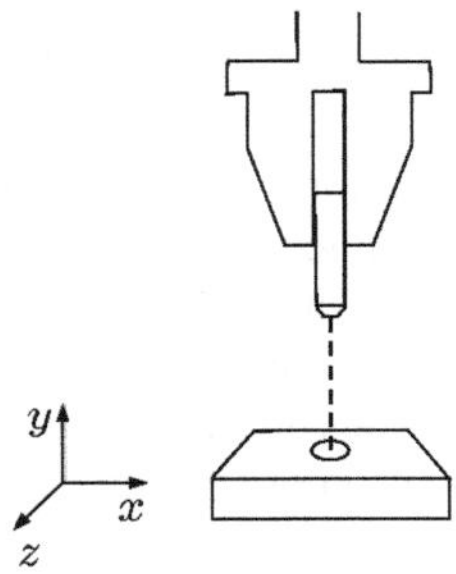

Fig. 26.8. Peg-in-hole scheme

the motion of the peg is restricted to the hole's centerline, providing a natural guide towards the task goal.

- The forces coming from the robot sensor provide information about the peg contact status, and since the peg is chamfered, these forces can help its guidance if contacts with the hole's edge occur.
- The information carried by the guidance signal corresponds to the velocity of the haptic device ($\mathbf{g}_s = \dot{\mathbf{x}}_{hd}$), according to *Scheme 1*, in Sec. 26.5.1.
- Packets have been transmitted using TCP/IPv6 sockets with the scheme of a classical client-server application. The round trip time (RTT) per packet is bounded in the 0.5ms - 1.5ms time interval.
- Since the current implementation does not reflect rotational DOF force data, the robot orientation has been fixed parallel to the hole's axis direction.

Figs. 26.9 – 26.11 plot the time evolution of positions and forces along the x, y, and z directions. The force plots show the three components of the total force $\mathbf{f}_t$: the restriction force $\mathbf{f}_r$, the viscous force $\mathbf{f}_v$, and the sensed force $\mathbf{f}_s$. The insertion direction is along the y axis.

From 0 s to 1.5 s no geometric constraint has been set ($\mathbf{f}_r = 0$ in this time interval), so the robot moves freely in space.

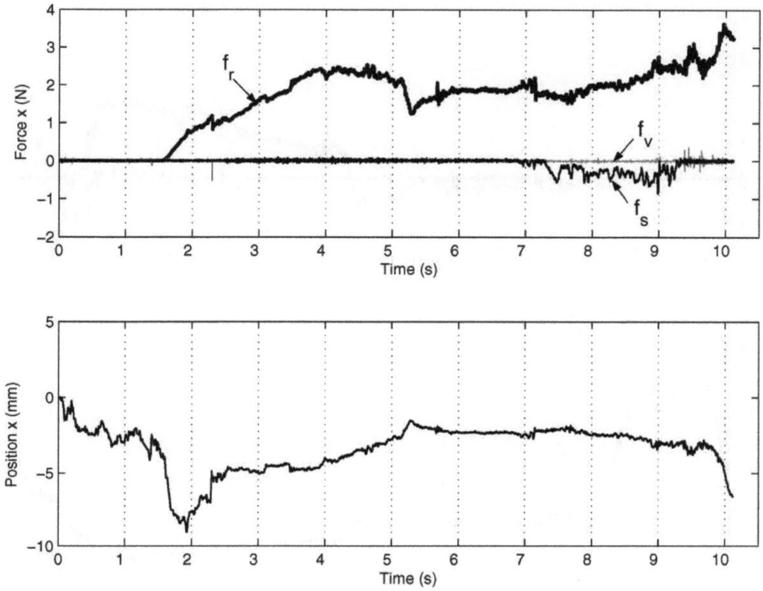

Fig. 26.9. Position and force in the x axis, where f_r is the restriction force, f_v the viscous force and f_s the force feedback, along this axis

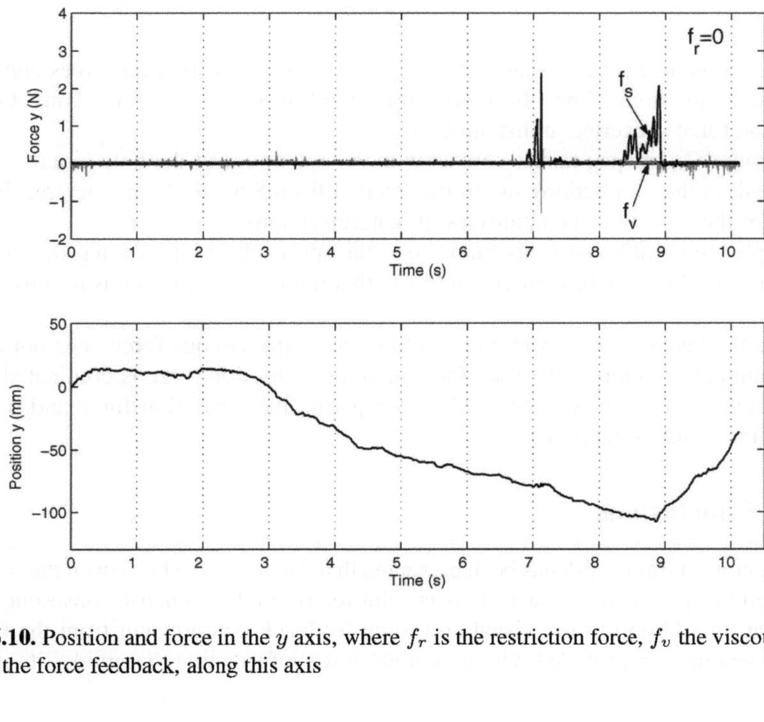

Fig. 26.10. Position and force in the y axis, where f_r is the restriction force, f_v the viscous force and f_s the force feedback, along this axis

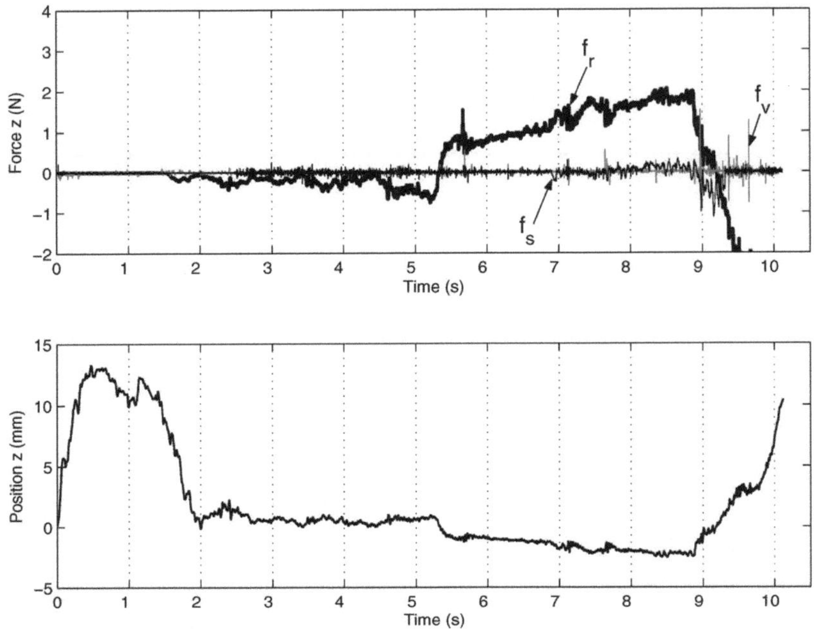

Fig. 26.11. Position and force in the z axis, where f_r is the restriction force, f_v the viscous force and f_s the force feedback, along this axis

Once the geometric constraint is set at time 1.5 s, the restriction forces appear on the x and z directions. Since the y axis corresponds to the unconstrained direction, no restriction force is exerted in this direction.

At around 7 s the peg makes contact with the outer edge of the hole (see the sensed force peak in the y direction) and in the interval from 8.5 s to 9 s the peg reaches the bottom of the hole, which contains a soft material (foam).

The position plot in the y direction shows the approach and insertion processes, that take place until the 9 s time mark, and from that moment on, the peg is removed from the hole.

Since the task has been performed at low speed, the viscous force does not have a significant contribution to the total force, and the restriction force is dominated by its spring component. This can be verified comparing the restriction force and position plots in the x and z directions.

26.7 Conclusions

A teleoperation framework has been presented that can lower the burden on the operator while remotely executing a task. This is achieved through geometric constraint haptic guidance. In addition to the visual and force feedback that are sent from the remote site, the operator is provided with additional force information that guides its motion according to some predefined geometric constraints between the robot tool and its environment. The IPv6 protocol was used to handle communications in an efficient manner, enabling important data such as control signals to be transmitted with higher priority than less relevant and bandwidth-hungry signals like video feeds. The benefits of this approach were demonstrated through a peg-in-hole task.

References

1. B.J. Unger, A. Nicolaidis, P.J. Berkelman, A. Thompson, S. Lederman, R.L. Klatzky, and R.L. Hollis. Virtual peg-in-hole performance using a 6-dof magnetic levitation haptic device: comparison with real forces and with visual guidance alone. In *Proceedings of the 10th IEEE Symposium on Haptic Interfaces for Virtual Environment and Teleoperator Systems*, pages 263–270, March 2002.
2. S. Shon and S. McMains. Evaluation of drawing on 3d surfaces with haptics. In *IEEE Computer Graphics and Applications*, 24(6):40–50, Nov. - Dec. 2004.
3. T.H. Massie and J.K. Salisbury. The phantom haptic interface: A device for probing virtual objects. In *Proceedings of the ASME Winter Annual Meeting, Symposium on Haptic Interfaces for Virtual Environment and Teleoperator Systems*, 55(1):295–300, Nov. 1994.
4. N. Turro, O. Khatib, and E. Coste-Maniere. Haptically augmented teleoperation. In *Proceedings of the IEEE International Conference on Robotics and Automation*, 1:386–392, May 2001.
5. A. Casavola and M. Sorbara. Towards constrained teleoperation for safe long–distance robotic surgical operations. In *Proceedings of the IEEE International Conference on Robotics and Automation*, pages 697–702, 2005.
6. N. Diolaiti, G. Niemeyer, F. Barbagli, J.K. Salisbury, and C. Melchiorri. The effect of quantization and coulomb friction on the stability of haptic rendering. In *First Joint Eurohaptics Conference and Symposium on Haptic Interfaces for Virtual Environment and Teleoperator Systems*, pages 237–246, March 2005.
7. J.E. Colgate, P.E. Grafing, M.C. Stanley, and G. Schenkel. Implementation of stiff virtual walls in force-reflecting interfaces. In *Proceedings of the IEEE Virtual Reality Annual International Symposium*, pages 202–208, 1993.
8. P. Pan, K.M. Lynch, M.A. Peshkin, and J.E. Colgate. Static single-arm force generation with kinematic constraints. In *Proceedings of the IEEE International Conference on Robotics and Automation*, 3:2794–2800, 2004.
9. T. Tickel, D. Hannon, K.M. Lynch, M.A. Peshkin, and J.E. Colgate. Kinematic constraints for assisted single-arm manipulation. In *Proceedings of the IEEE International Conference on Robotics and Automation*, 2:2034–2041, 2002.
10. E.S. Boy, E. Burdet, C.L. Teo, and Colgate. Motion guidance experiments with scooter cobot. In *Proceedings of the 11th IEEE Symposium on Haptic Interfaces for Virtual Environment and Teleoperator Systems*, 20(4):63–69, March 2003.
11. E.B. Fgee, J.D. Kenney, W.J. Phillips, W. Robertson, and S. Sivakumar. Comparison of qos performance between ipv6 qos management model and intserv and diffserv qos models. In *Proceedings of the 3rd IEEE Annual Communication Networks and Services Research Conference*, pages 287–292, May 2005.
12. IPv6. Ipv6 information page. `http://www.ipv6.org/`, 2005.
13. V. Srivastava, C. Wargo, and S. Lai. Aviation application over ipv6: performance issues. In *Proceedings of the 2004 IEEE Aerospace Conference*, 3:1661–1670, March 2004.
14. P.X. Liu, M. Meng, and S. Yang. Data communications for internet robots. *Kluwer Academic Autonomous Robots*, 15(3):213–223, 2003.

15. R. Oboe. Web-interfaced, force-reflecting teleoperation systems. *IEEE Transactions on Industrial Electronics*, 48(3):1257–1265, Dec. 2001.
16. S. Munir and W.J. Book. Internet-based teleoperation using wave variables with prediction. *IEEE/ASME Transactions on Mechatronics*, 7(2):124–133, June 2002.
17. P.X. Liu, M. Meng, Y. Xiufen, and J. Gu. An UDP-based protocol for internet robots. In *Proceedings of the IEEE 4th World Congress on Intelligent Control and Automation*, 1:59–65, June 2002.
18. R. Safaric, S. Sinjur, B. Zalik, and R.M. Parkin. Control of robot arm with virtual environment via the internet. In *Proceedings of the IEEE*, 91(3):422–429, March 2003.
19. K. Taylor and B. Dalton. Internet robots: a new robotics niche. *IEEE Robotics & Automation Magazine*, 7(1):27–34, March 2000.
20. R.C. Luo, K.L. Su, S.H. Shen, and K.H Tsai. Networked intelligent robots through the internet: issues and opportunities. In *Proceedings of the IEEE*, 91(3):371–382, March 2003.
21. C.M. Hoffman and R. Joan-Arinyo. A brief on constraint solving. *Unabridged version at* `http://www.cs.purdue.edu/homes/cmh/distribution/papers/Constraints/ThailandFull.pdf`*; abridged version to appear in CAD&A*, pages 873–881, 2005.
22. E. Celaya. LMF: A program for positioning objects using geometrical relationships. In *VII International Conference on Applications of Artificial Intelligence in Engineering. Elsevier Applied Science*, pages 873–881, 1992.
23. F. Janabi-Sharifi, V. Hayward, and C-S.J Chen. Discrete-time adaptive windowing for velocity estimation. *Control Systems Technology, IEEE Transactions on*, 8(6):1003–1009, Nov. 2000.

27

Telerobotics for Aerial Live Power Line Maintenance

Rafael Aracil and Manuel Ferre

Universidad Politécnica de Madrid
Dpto. de Automática, Ingeniería Electrónica e Informática Idustrial
C/. José Gutierrez Abascal, 2. 28006 - Madrid, Spain
aracil@etsii.upm.es, m.ferre@upm.es

Summary. This chapter describes a telerobotic system for maintenance of aerial electrical live-power lines. Maintenance-related tasks are of great relevance for utilities companies, as their goal of achieving a good service quality depends to a great extent on the performance of maintenance and inspection issues. A telerobotic system called *ROBTET*, has been developed by the Universidad Politécnica de Madrid, in collaboration with Iberdrola and Cobra, a Spanish utility company and contractor, respectively. The *ROBTET* system has been in operation in the Spanish electricity network since 2002, which attests to the success of the prototype and also allows for the incorporation of improvements.

27.1 Introduction

The electrical sector plays an important role in the technological and industrial progress of a country. There are great possibilities for the application of robotic systems in this field [1]. In recent decades, the fast growth of industrialised countries has been made possible by the contribution of utilities companies. These companies have reached an agreement with the industry to supply the demanded energy, along with the proper service quality.

Maintenance tasks on live lines always involve a certain risk, despite worker skill. Risk usually appears with the occurrence of unexpected events. The most frequent tasks to be carried out in electrical networks are: changing insulator sets, opening and closing switches, establishing new connections, changing line equipment, inspecting line equipment. All these types of tasks are executed manually by highly qualified workers. They use techniques that have been carefully studied and designed to avoid workers' risks and also to increase efficiency and minimise task execution time. The two widely spread techniques used in manual live-line maintenance are: distance works (working indirectly) and potential works (working directly). In the first technique, the worker manipulates the line using different kinds of insulated hot-sticks. The operator works close to the line, tying up to the perfectly insulated pole. In the case of the second technique, workers remain in touch with the line. Certain areas are covered in advance with insulated accessories in order to prevent electrical shocks, and the contact with different line elements is executed with the appropriate rubber gloves. This technique can also be done with the worker in an insulated bucket placed on the top of a boom and close

M. Ferre et al. (Eds.): Advances in Telerobotics, STAR 31, pp. 459–469, 2007.
springerlink.com © Springer-Verlag Berlin Heidelberg 2007

Fig. 27.1. Classical techniques for aerial live line maintenance: distance works (left) and potential works (right)

to the line. Some other techniques include the use of helicopters to hold the bucket with the worker inside and close to the line. This technique is quite expensive. Fig. 27.1 shows workers using hot-sticks (distance works) and working from a bucket (potential works).

The possibility of outage-free maintenance for aerial distribution and transport lines (without interrupting the electric energy on the line) depends on the legislation of each country and the features of the electrical network. In some countries, it is possible to perform different maintenance tasks on aerial lines without having previously disconnected the energy. In some other countries, a total disconnection must be done before the worker starts maintenance works. The risks associated with these tasks on hot lines are: electrical shocks, exposure to electromagnetic fields and falling from the high working place. A potential task to be carried out by telerobotics is live-power line maintenance but at present only a few prototypes have been developed and their applications have not extended beyond the developing companies. Therefore, no commercial system is currently found on the market. Live power line maintenance is characterised by the following main features:

- Works are carried out over elements with different degrees of power. It is therefore necessary to avoid short circuits by preventing contact between different power elements, i.e. two lines or a line and the tower cannot be touched simultaneously by the same manipulator.
- Although it is possible to know details about the remote work environment in advance, this is not known for the positions of the elements and the relation among them.

- The location of the tower and lines is a determining factor in maintenance works. Access to them often presents some difficulties, and thus a proper vehicle and the corresponding power supplies for these tasks are required.
- Local features of the electrical network in each country, such as isolation of aerial lines or potentials use for distributions lines.

These features should be taken into account in the development of any telerobotic system. All proposed systems are based on teleoperated robots where the human operator monitors, in different supervisory degrees, the task execution. For these cases, telerobotic systems are considered a powerful tool from which workers in the sector can benefit a great deal. In accordance with the previous features, teleoperated robot applications on live-power line maintenance shows the following advantages: *Security and comfort increase* for workers, as tasks are carried out by robots, thus reducing worker exposure to the risks of manipulation, *cost reduction* in maintenance works since teams of 4-5 people can be reduced to 2-3, and *higher degree of standardisation* and systematisation of maintenance tasks due to the programmed procedures for robots. Although these advantages fully justify the implementation of this system, the realities of economics, finances and work environments are having a limiting effect on the introduction of new teleoperated prototypes in the sector.

27.1.1 Systems Developed for Live-Power Lines

In the 1980s, the electrical sector used teleoperation technology to develop new procedures for maintenance tasks on aerial transport and distribution lines. Some interesting semiautomatic robotic systems have been designed around the world. Countries like the US, Japan, France, Canada and Spain have developed telerobotic prototypes for these tasks. Most of them are comprised of robots which are teleoperated from a cabin on the ground or at the top of a boom close to the electrical line.

At the end of 1980s, Japan developed some telerobotics systems for live-power lines. These systems were designed to operate on their distribution lines up to 6,6 kV. In the early 1990s Aichi Inc. [2] designed a system with two hydraulic teleoperated robots. Robots were guided by the operator situated at the top of the boom. Shikoku Electric Power Co. developed a hydraulic arm controlled from a bucket on the top of a boom to execute maintenance tasks; the system also incorporated a complete set of tools to support the line manipulation. Kyushu Electric Power Co. (KEPCO) is the company with most robotic units in use for this purpose [3, 4, 5]. Among their systems, worth special mention is Phase I which includes two electric robots teleoperated by a joystick, from an aerial bucket placed on the top of a boom. At the end of 1993, Phase II system was developed incorporating two robots teleoperated from a cabin on a vehicle deck. The system has several cameras located on the platform to show the operator all information related to the task's environment. There is also a man-machine interface with a touching panel display and a voice recognition system to control the camera movements. Phase II system incorporates a very useful device: the automatic tool changer (ATC). The latest research at KEPCO focuses on the system called Phase III, the objective of which is a fully automatic robot, recently developed by Takaoka [6]. Most of Japanese developments are focused on distribution lines for urban areas. These lines are significantly different to European or American ones, since Japanese aerial lines are usually isolated by plastic covers and potential in the distribution lines is low ($\leq 10kV$). Therefore, developments from Japan are dificult to directly be applied in networks from other countries.

Since the mid-1980s, the Electric Power Research Institute (EPRI) in the United States has been working on a system called TOMCAT [7]. The first prototype consisted of a single hydraulic robot teleoperated from a cabin located on the ground. The system was designed to work with lines ranging from 50 to 345 kV. The system is capable of working in very high aerial lines. The next stage in this system is the TOMCAT 2000, which incorporates the possibility of developing live-line maintenance tasks under conditions of freezing rain and darkness. This system is focused on aerial distribution and transport lines. It is a very complex system designed for high potentials; therefore, this technology is difficult to be applied in urban areas.

In Canada, the Hydro-Quebec Research Institute [8,9] developed a system made up of two hydraulics robots teleoperated from an aerial cabin placed on the top of a boom and an auxiliary arm capable of a very long access. The system may operate in distribution lines of 25 kV. Two projects for live-line maintenance have been started in France, but they have not yet produced results. The first was supported by EDF in the 1990s, but it was cancelled later (Soler93). At the end of 1990s a consortium called TST 2000 was started, supported by the French Thomson-CSF and the Japanese Nissho-Iwai and Yaskawa, but no prototype has been developed at this time. Both systems involve a medium degree of complexity which can be found between Japanese and American developments.

27.2 ROBTET: Telerobot for Aerial Live Line Maintenance

The ROBTET system has been designed to carry out maintenance tasks on the Spanish power supply network up to 69 kV. All components are installed on a 4*4 truck [10,11]. This truck is autonomous, it holds a hydraulic pump unit and an electric generator of 10 kW, both power units are used for the power supply of all components. Fig. 27.2 shows a general view of the system.

The ROBTET system works in a teleoperation control mode with the operator sending commands from the cabin on the truck, and receiving information from the remote working environment. Telemanipulators and tools are placed on a remote platform located on an isolated telescopic boom. This configuration provides the operator with completely safe conditions. Efficient performance should also achieve proper interaction between the operator and the remote working environment, in order to obtain the most teleproprioception as possible. In response to these requirements, two workplaces have been developed: the operator's working site placed on the cabin and the remote working site placed on the remote platform. Communication between these sites takes place through optical fibre cable to guarantee electrical isolation.

The main goal of the telerobotic system is to give the operator as much telepresence feeling as possible in order to perform the teleoperated task in an easy, safe and comfortable way. Under these conditions, the teleoperated system should be as powerful as the traditional manual system. Both systems can be compared according to parameters like productivity, time execution and safety.

Fig. 27.2. General view of Robtet system (left), robot working site (top-right) and operator working site (bottom-right)

The objectives in the ROBTET design are twofold: firstly, to optimize operator interaction in order to achieve as much telepresence as possible; and secondly to manage a virtual model of the remote enviroment in order to provide operator advanced teleoperation tools.

Operator interaction is based on his senses stimulation and his command processing. Both information flows allow the operator to close the task control loop. Three operator's senses are stimulated to report him the teleoperated task: vision through stereoscopic and monoscopic video cameras and graphics, touch through force feedback, and hearing to receive alarm information. The operator generates commands in two different ways: (i) hand movements to guide manipulators and (ii) voice to control the overall camera and the interface configuration.

Operator's teleproprioception depends on the relation between operator commands and the displayed result of his action, i.e. a movement to the right in the master should have the corresponding displayed right movement. Operator's cabin and platform have a similar configuration, the overall camera is placed at the manipulators back to simulate the operator head position, and its image is shown on the main display in front of him.

A virtual model of the remote environment is used during the execution of the teleoperated task. When the task is starting up this model have to be calibrated. The virtual model allows the implementation of advanced teleoperation helps, such as high level of voice command, collision detection, virtual force reflection, and graphic simulation. The main problem to properly use the virtual model of the remote environment is its calibration with the real world. Information about the object manipulated by the telerobots are defined before starting the teleoperated task. It is assumed that all the geometric elements of the intervening model are known but the spatial location of the electrical line is not defined. Two different processes for the calibration have been developed. The first process was based on a computer vision system. This computer vision system is described in [10]. A manual calibration by the operator is the alternative to the vision system. Hereby, the operator must indicate the object location in the remote environment. Before starting, the operator moves the manipulators at key positions of the remote site to calibrate the computer model. The key positions are usually related to the isolator chain ends, several points at the live lines and predefined tower points.

Operator can easily check the computer information. Video images from the robot working site and simulation images using the computer model are blended. This image blending is described in [12]. The blending image system is used by the operator to decide if the information of the computer model is valid. Object position errors close to 5 cm are tolerated by the programs that use this information. It is advisable to know that computer model information is used for control a pan/tilt camera and collision detection. However, this information can not be used by automates for object manipulation; therefore, all manipulation are carried out by the operator.

27.2.1 Operator's Working Site

The operator's working site is placed in a cabin on a truck. It contains the operator interface devices and some control computers. The function of operator interface devices is two-fold :(1) to show the operator the status of the executing task and, (2) to send the operator's commands to the remote environment. The operator interface contains the following devices: two haptics master arms, a graphical display, a stereoscopic display, and a voice recognition system.

The **two haptics master-arm** devices with force feedback are developed by Kraft Telerobotics. The masters have six degrees of freedom (dof) and are used to guide the manipulators and to reflect forces to the operator. Two kind of forces are reflected to

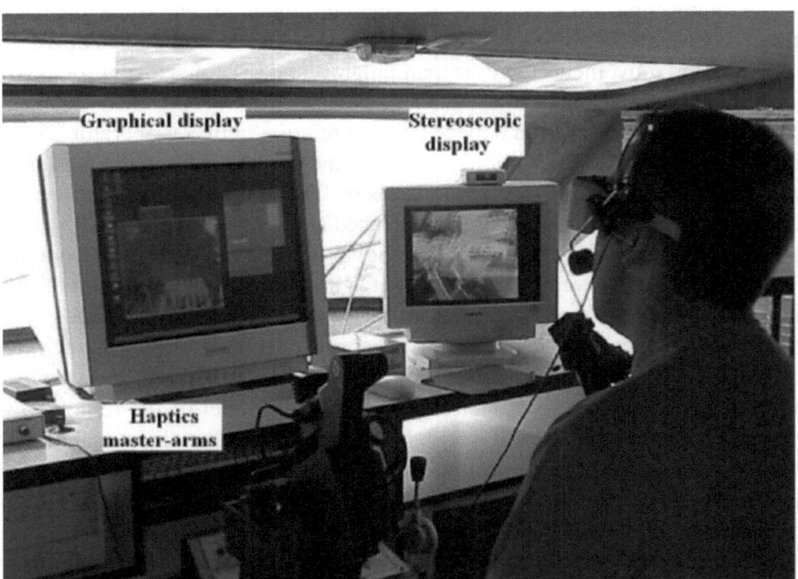

Fig. 27.3. Operator cabin

the operator: (1) Reflected forces from the manipulation interaction. They are obtained by applying the measured torque of the slave joint actuators-reduced by a scale factor-through the corresponding joint motors of the master. (2) Virtual forces from the check collision module. When a manipulator is near to a different potential site this module generates a repulsion force in order to avoid a short circuit.

The **graphical display** shows a monoscopic video image, a graphic simulation and some information about the task. The video image is taken by the panoramic camera. The camera has three dof - pan, tilt and zoom. This camera shows general and detailed images providing the operator with panoramic scenes and specific views. These images provide the operator with a high degree of teleproprioception since he can perceive the relation between the different elements at the remote working site. The operator can also see a graphical simulation of the task and information about the executed step of the task procedure.

The **stereoscopic display** shows images from the stereoscopic cameras placed at each slave manipulator. These cameras show detailed images and provide an excellent depth perception to the operator. The operator must use special glasses to see the stereoscopic images. The operator can only see one stereoscopic image from the two stereoscopic video cameras located at the slaves, and therefore must select the appropriate image. The selection between the stereoscopic cameras depends on the point of view that the operator wishes to have. He can change the camera articulating the proper voice command. The stereoscopic images are very useful for the operator as he guides the manipulator in precise movements such as the insertion of screws.The perception of depth is a very important factor when it comes to implementing a precise manipulation. If the operator does not have the proper spatial information, he may execute several movements before achieving the desired position, usually a zigzag trajectory is described. As a result, task execution time increases and the performance of teleoperated system decreases.

The **voice recognition** is used during task performance to control the overall camera, to indicate the step of the procedure executed, to select the stereoscopic image shown on the stereoscopic display, and to enable the external communication for disable. The voice interface is a powerful tool for the operator, because it allows him to use only his hands for guiding the manipulators, while the rest of the operator commands are sent by voice.

27.2.2 Robot's Working Site

The remote platform is located on the top of an isolated telescopic boom, as shown in Fig. 27.4. The platform has pan and tilt movements, controlled by a pendant placed in the cabin. It holds two slave manipulators, an auxiliary jib, the panoramic camera, tools and isolated power supplies with 12 h-autonomy electronic components, such as cameras and slaves communications. The communication between the platform and the cabin is carried out by several optic fibre cables inside the boom. The hydraulic power for manipulators and jib is taken from the truck using isolated oil and hoses. The components are: two slave manipulators, two stereoscopic video cameras, a panoramic video camera, and an auxiliary jib and tools. These devices can have different configurations so far some factors are taken into account like power isolation and operator

Fig. 27.4. Platform placed at the boom tip

teleproprioception. Power isolation provides different potentials at the remote platform and therefore each slave can manipulate a different live line.

Two **slave manipulators**, developed by Kraft Telerobotics, are hydraulically powered and have six dof plus the grip action. The maximum payload of each arm is 45 kg and the net weight 60 kg. The master arm movements guide slaves and the joint torques is reflected on the master device. Manipulators are mounted on special insulators in order to provide different potentials to each manipulator. Each manipulator has an independent power supply that guarantees its electrical insulation. The degree of power supply autonomy can be a critical aspect for long lasting tasks.

Two stereoscopic video cameras are mounted at the end of each manipulator. The stereoscopic cameras provide two synchronised video signals for the stereoscopic image computer. The images are calibrated in order to obtain the best depth perception at 50 cm. The distance to the parts manipulated by the slaves is the usual one.

The **panoramic video camera**, model EVI by Sony, has three dof: pan, tilt and zoom; all movements are monitored through operator's voice commands. The camera is located behind both slaves to have a good general view. From this point of view, the operator can perceive manipulators as if they were his/her arms. It provides an excellent teleproprioception when the task begins. Within hot line manipulations, more detailed images are needed. For this purpose it is possible to use the camera zoom or the stereoscopic cameras.

The **auxiliary jib** is placed next to the right slave manipulator. It is hydraulic powered and has three dof: pan, tilt and telescopic extension. The telescopic link is isolated up to 69 kV. Therefore, the jib can hold up the live line while the manipulators are working with a different potency. The jib is specially designed for the ROBTET and constitutes a powerful tool for the telemanipulators. The use of a jib, as shown in

Fig. 27.4, has simplified the procedures and consequently reduced the execution time. Several tools are placed on a special box before starting up task execution.

27.3 Performance of ROBTET System

The main contribution of this telerobotic system is the achievement of the same productivity as the classical methods for live-line maintenance. It implies the definition of proper procedures for each task achieving a similar execution time. Next two sections describe the features of ROBTET procedures; operator's training is later explained.

27.3.1 Procedures Execution

The procedures for developing hot line maintenance tasks follow strict rules in order to prevent hazardous risks. Depending on the kind of task to be implemented on the line, a certain procedure must be applied. These procedures aim at minimising the risks to the operators and also at reducing the task operation time and increasing efficiency. Another important factor to be considered is to reduce the effort exerted by the operator who remotely works on the line. The ROBTET operation is defined by the corresponding well-established procedures. They are based on the traditional manual procedures, so they are modified according the new facilities of the telerobotic system; the most important one is the worker safety working on the ground, in a very comfortable way with total security.

The most frequent maintenance tasks carried out on a distribution line are insulator set change, the attachment of a jumper cable between two points of the line, opening switching units on the line and the replacement of fuse elements. These works are performed by using the previous well-developed manual procedures for the direct work method. An example of procedure executed by the telerobot is described in [10].

Two main conclussions have been drawn from the development of new execution procedures for telerobotics live-line task:

- Operator does not guide both telemanipulator at time, usually he moves one and halts the other telemanipulator. Consequently, procedures have to take into account the following: when a manipulator is executing a step the other one plays a cooperative role. For example, a manipulator is removing a plate and the other is getting the rest of the plate chain. Telemanipulators have a cooperative behaviour; one is active while the other supports the action. This cooperation reduces significantly execution time. In other cases, more complex procedures and/or tools should be implemented.
- Specific tools have to be designed in order to adapt traditional tools to robot features. In many cases, traditional tools only required small changes to be used by the telemanipulators, but such adaptations reduce significantly the procedure execution performance.

This telerobot system requieres two operators to execute a complete task. One of them will operate inside the cabin, control the platform, the boom and will perform the task; the other one has to carry out the corresponding steps outside the cabin and works also as a supervisor reporting the operator during and about the task execution. Traditional methods usually require 3 or 4 operators to execute the same task. It represents a significant increasing of productivity.

27.3.2 Operator's Training

The introduction of a new technology represents a very important fact for companies. Some points must be taken into account: operator's interface must be friendly, computer starting up process have to run automatically, devices used by operators should be comfortable, and so on.

A lot of interesting information has been obtained from the training phase. Fig. 27.5 shows a typical training evolution. Initially, the operators spends a lot of time because they need to get used to the system. The dexterity in the use of the master devices increases gradually reducing the execution time. Finally, the interface is skilfully managed and the time to execute the procedure remains around a final execution time. Usually, it is necessary to repeat the procedure over 20 times to reach the final execution time.

The final spent time is similar for many operators. To reduce this time, it is necessary to use better tools, or introduce more automation in the procedure. Sometimes, new tools are easy to be developed, but more automation involves new computers and sensors, including more complexity to the teleoperation system.

Operator behaviour during task execution evolves as follows. First of all, he/she becomes a general working scene view which allows him/her to localize telemanipulators as to the objects within the same enviroment. Images come from a panoramic camara on which operator can monitor pan, tilt and zoom. Once the operator is aware of his/her actions regarding telemanipulators in the remote environment, he/she makes then use of the stereoscopic camaras in order to properly manipulate remote objects. At the time, force reflection provides him/her with approximate information about forces involved in the task execution.

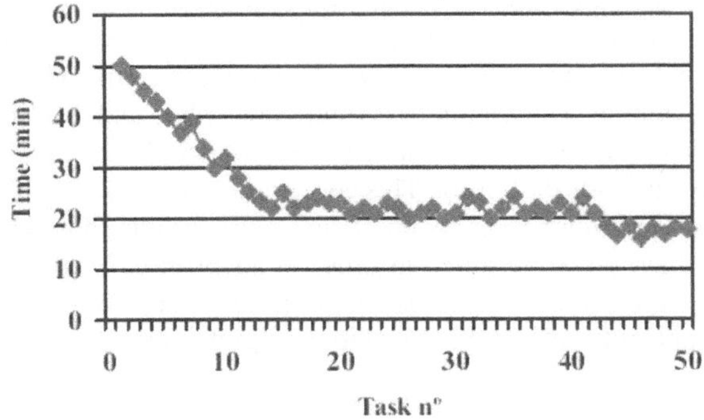

Fig. 27.5. Operator's training evolution

27.4 Conclusions

Telerobotic systems have been developed for outagefree maintenance tasks. Technical and working factors must be taken into account for its future implementation in order to assure its viability. New and better teleoperated systems for live-power lives should be promoted to achieve a step forward in the telerobotics field.

The new telerobotic systems will be conceived as powerful tools for electrical network workers. They represent a breakthrough for the security and standardisation of live-power line maintenance tasks. It has become clear that there are two aspects which are key to the successful functioning of ROBTET: the interaction with the operator and the variety of assistance forms for teleoperation (voice recognition, collision detection, virtual forces, calibration or the remote environment). Thanks to the stereoscopic system and the force reflection, the operator can manipulate elements with dexterity and safety. These features have proved a significant increase in the ROBTET productivity, as they drastically reduce the time used for teleoperated tasks. These results have been ratified by the works which have been taking place on the Spanish electrical lines over the last years.

References

1. L.E. Parker and J.V. Draper. *Handbook of industrial robotics*. Editorial Shimon Nof, 1999.
2. Aichi Inc. *AICHI manipulator system design feature*. Report of AICHI Development Planning Department, 1993.
3. Y. Maruyama. Robot applications for hot-line maintenance. *Industrial Robot: An International Journal*, 27 (5):357-365, 2000.
4. Y. Maruyama et al. A hot-line manipulator remotely operated by the operator on the ground. In IEEE, editor, *Proc. of Fifth IEEE international conference on transmission and distribution construction, operation and live-line maintenance ESMO-93*, 1993.
5. Y. Maruyama et al. Mv overhead hot-line work robot. In ICOLiM, editor, *Proc. of Third international conference on line maintenance, ICOLiM-96*, volume 1, 1996.
6. L. Takaoka et al. Development of the fully-automatic live-line maintenance robot-phase iii. In IEEE, editor, *Proc. of the 4th IEEE International Symposium on Assembly and Task Planning*, 2001.
7. Electric Power Research Institute (EPRI). Live-line repair with tomcat. *EPRI Journal*, 1:1–29, 1987.
8. M. Boyer. Telerobotics for maintenance of distribution lines. In ICOLiM, editor, *Proc. of the international conference on line maintenance*, 1994.
9. J. Coté and M. Peletier. Telemanipulator design and optimization software. In SPIE, editor, *Proc. of the SPIE. Telemanipulator and telepresence technologies*, 1995.
10. R. Aracil, M. Ferre, M. Hernando, E. Pinto, and J.M. Sebastian. Telerobotic system for live-power line maintenance:robtet. *Control Engineering Practice*, 10:1271–1281, 2002.
11. R. Aracil, L.F. Penin, M. Ferre, and A. Barrientos. Robtet: Robot for live-line maintenance. In ICOLiM, editor, *Proc. of the Int. Conference on live maintenance (ICOLIM96)*, 1996.
12. M. Ferre, R. Aracil, M. Navas, and J.A. Escalera. Real time video image processing for teleoperation: Image blending and stereoscopy. In IEEE, editor, *Proc. of the 9th IEEE International Conference on Emerging Technologies and Factory Automation. ETFA2003*, pages 539–544, 2003.

Advanced Telerobotics: Dual-Handed and Mobile Remote Manipulation

Martin Buss, Kwang-Kyu Lee, Norbert Nitzsche, Angelika Peer, Bartlomiej Stanczyk, and Ulrich Unterhinninghofen

Technische Universität München
Institute of Automatic Control Engineering (LSR)
D-80290 Munich, Germany
{mb,kk.lee,angelika.peer,stanczyk,
ulrich.unterhinninghofen}@tum.de, norbert.nitzsche@mytum.de

Summary. This work presents an advanced dual-handed, mobile telerobotic system developed at the High-Fidelity Telepresence and Teleaction Research Centre, Munich, Germany. To the authors' best knowledge, it is the first attempt to integrate mobile and multi-robot strategies in one physical and logical framework. In order to exploit human manipulation capabilities, a high fidelity telemanipulation system was developed. It consists of two redundant human-scaled anthropomorphic telemanipulator arms controlled by two redundant haptic interfaces providing a large, convex workspace and force feedback in a wide range of human perception. To provide a multi modal immersion, the haptic modality is augmented by 3D visual and audio channels. The main research issues are the control of devices with dissimilar kinematics, redundancy resolution methods, and six DOF compliance control. To extend the accessible workspace in remote environments, mobile robots are used as transporting platform extending the functionality of both the input devices and the telerobot. Mechatronic design topics and experimental results of six degree of freedom telemanipulation tasks and mobile telemanipulation are presented. The motion compression concept is exploited to cover large remote environments on a relatively small local area. Finally, architectures for collaborative telemanipulation are classified and corresponding interaction schemes are discussed.

28.1 Introduction

A telepresence system can be used to perform tasks in a distant, dangerous, or inaccessible environment. In recent years, several specialized telemanipulation systems have been developed for space and underwater exploration, plant maintenance, construction, and surgery. In order to operate in highly variable, unstructured, unknown, or dynamic environments, an advanced telerobotic system with increased adaptability and human like manipulation capabilities is required. This implies the possibility for dual-handed and mobile telemanipulation augmented with appropriate visual, auditory, and haptic information.

Existing telepresence systems are often limited to visual and auditory feedback, thus rendering real immersion into the remote environment impossible. Providing a combination of visual, auditory, and haptic information allows to establish a multi-modal telepresence system which guarantees a fully immersive representation. Moreover,

M. Ferre et al. (Eds.): Advances in Telerobotics, STAR 31, pp. 471–497, 2007.
springerlink.com © Springer-Verlag Berlin Heidelberg 2007

existing solutions with force feedback are often restricted to a few degrees-of-freedom (DOF), low interaction forces, or a strongly limited workspace. In this chapter, a multimodal telepresence system which allows telemanipulation with two arms in full 6 DOF is presented. Intuitive operation is achieved by tracking the motions of a human user at the operator site, replicating these motions by a telemanipulator at the remote site, and transferring back sensor data such as visual, auditory, and haptic information (see Fig. 28.1).

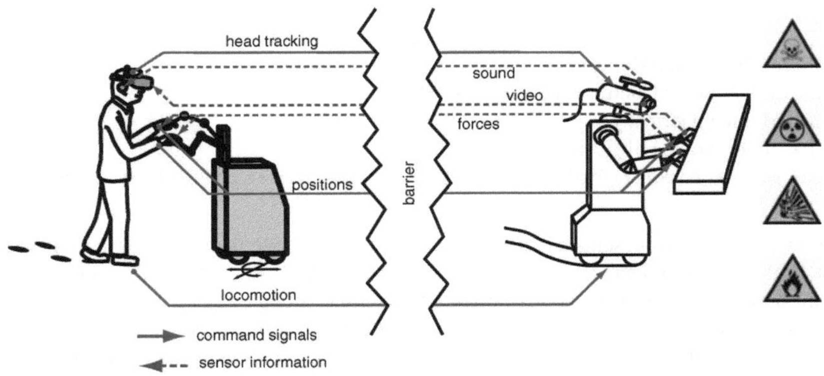

Fig. 28.1. Teleoperation robotic system

Visual Feedback: Visual data is collected by a stereo camera head and presented to the operator via a head-mounted display (HMD), see [1]. The gaze direction of the robotic head is adjusted according to the orientation of the operator's head measured by a tracking system. Using this setup, the user can intuitively look around in the remote environment just by turning his/her own head.

Auditory Feedback: The robotic head is equipped with two microphones placed on each side of the head. The audio data is transmitted to the operator site and played back through stereo headphones.

Haptic Feedback: In manipulation tasks, the sensation of interaction forces is crucial to allow complex operations. This is achieved by use of haptic displays which enable bidirectional human system interactions via the sense of touch. The haptic display acquires the position of the operator's hand and relays it to the telemanipulator. In turn, the resulting contact forces in the remote environment are transmitted back and displayed to the operator.

In order to allow natural telepresence and teleaction in a wide remote environment, a mobile telemanipulation system is employed. It comprises a telemanipulator mounted on a mobile platform such that it can move around and operate in an arbitrarily large remote environment. The use of a mobile human-system-interface guarantees a realistic feeling of locomotion which is necessary for accurate path integration and also contributes to the immersiveness of the telepresence system.

In Sec. 28.2 an advanced teleoperation system capable to perform dual-handed manipulations in full six degrees of freedom is presented. After showing the mechatronic design topics and control algorithms for the haptic display and the telemanipulator, the overall control structure of the telemanipulation system is discussed. Some experimental results concerning the redundancy resolution of the telemanipulator and the performance of the proposed telemanipulation system conclude this section. In Sec. 28.3 this system is extended by a mobile platform which enables to operate in an arbitrarily large remote environment. The control algorithm for the mobile platform as well as the idea of motion compression is presented. Finally, Sec. 28.4 deals with possible architectures for collaborative telemanipulation systems and classifies types of collaborative interactions. Challenges and requirements on the control of collaborative telemanipulation systems are discussed.

28.2 Bimanual Telepresence System

When accomplishing complex tasks, human beings are used to perform bimanual manipulations. This allows increasing dexterity as well as loading and handling capability. Manipulations become safer and more robust. In order to make use of these advantages, a telepresence system should provide a dual-handed human-system-interface.

Most haptic displays for bimanual interaction are limited to the use in virtual environments. Examples are the commercially available Haptic Workstation [2], the linkage based haptic display SPIDAR [3], as well as some medical simulation and training systems [4]. Some specialized dual-handed teleoperation systems with real remote environments can be found in the field of tele-surgery [5] and aerospace [6, 7]. More universal systems are the bimanual teleoperation system applied to the disposal of explosive ordnances, see [8, 9], as well as the humanoid teleoperator controlled by an exoskeleton presented in [10].

But most of these systems suffer from limitations to a few degrees of freedom, a small workspace of the haptic display, or the necessity to index in order to adjust the robot motion to a human scaled workspace. In the following sections an advanced teleoperation system capable to perform dual-handed manipulations in full six degrees of freedom is presented. Manipulator redundancy plays an important role because it enables to operate in a singularity and collision free human scaled workspace without indexing.

28.2.1 Haptic Interface

In order to enable an intuitive telemanipulation, the hyper redundant haptic display ViSHaRD10 (**Vi**rtual **S**cenario **Ha**ptic **R**endering **D**evice with **10** actuated DOF) is used as a human system interface. Its main characteristics are a very large workspace free of singularities, a high payload capability to accommodate various application specific end-effectors, foreseen redundancy to avoid kinematic singularities and user interferences, as well as the possibility for dual-arm haptic interaction with full 6 DOF. Further information about the design and control of this device can be found in chapter 2.

28.2.2 Telemanipulator

The superior manipulation-dexterity of humans is a result of the kinematic redundancy of human arms and the ability to adapt their compliance to the current task. As many technical design solutions being inspired by nature, the bi-manual telemanipulator is also anthropomorphically designed. Although the kinematical structure of the human operator and the slave robot do not need to match exactly, a 7 DOF structure is chosen as a trade-off between system complexity and performance. Investigations of possible 7 DOF structures result in the design shown in Fig. 28.2, which is proven in [11] to be optimal in the sense of "elimination of singularities, mechanical realizability, kinematic simplicity, and workspace shape".

Design of Telemanipulator — The telemanipulator consists of two identical human-scaled arms. A schematic view and physical construction are shown in Fig. 28.2. Each arm consists of two spherical joints with 3 DOF at shoulder and wrist, each, and one revolute joint at the elbow, which results in 7 DOF. Comparing to a conventional 6 DOF arm with spherical wrist, it is simply extended by adding a revolute joint for an elbow.

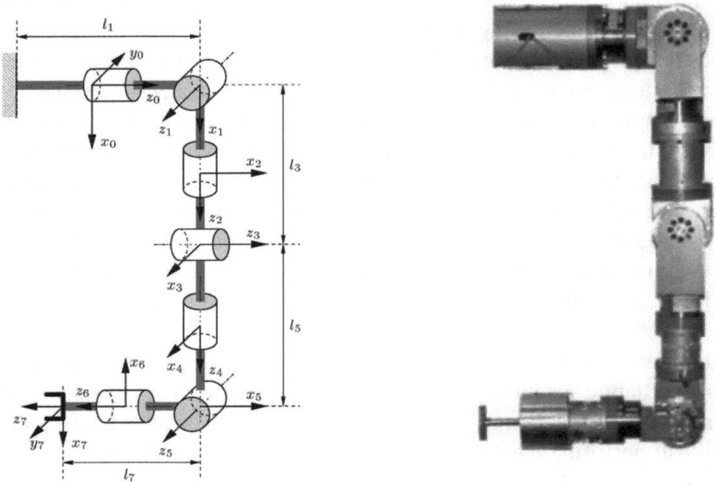

Fig. 28.2. Schematic view and link coordinate systems of 7 DOF arm (left) and physical view of 7 DOF arm (right)

Control of Telemanipulator — In teleoperation applications the desired trajectory of a telemanipulator is directly generated by a human operator in the Cartesian space, which is measured by the haptic display, and sent to the telemanipulator, placed at the remote site. During telemanipulation the telemanipulator may have to handle interactions with unstructured rigid environments. For such reasons, a task space control algorithm is desired, which guarantees compliant behavior during contact situation. On the other hand, to resolve the redundancy of the telemanipulator, a suitable constraint or a secondary task should be devised.

Extended Kinematics — Common methods to control redundant manipulators are typically based on the pseudo-inverse of the Jacobian although it has well known drawbacks: a) the null space motion can only be controlled in open-loop-fashion; b) nonrepeatability of the joint trajectory, i.e., a closed task space trajectory may not result in

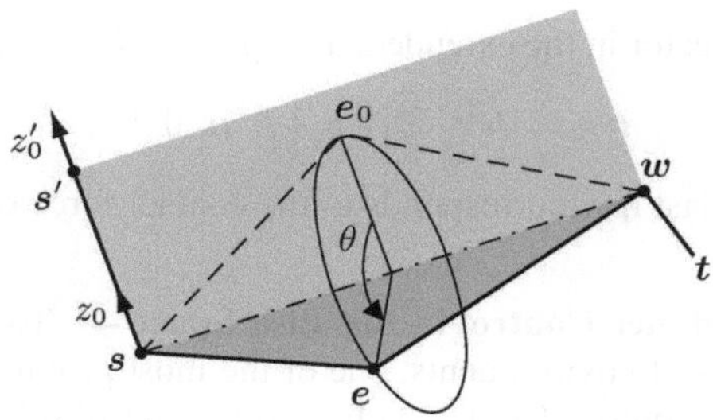

Fig. 28.3. Arm angle θ definition

a closed loop joint space trajectory. To solve this, the configuration control presented in [12] makes the Jacobian square and uses the augmented Jacobian in the control law as in control of a nonredundant manipulator. The extended task space vector in our case is defined as follows:

$$\boldsymbol{x}_E = \left[\boldsymbol{p}^T(\boldsymbol{q})\, \boldsymbol{Q}^T(\boldsymbol{q})\, \theta(\boldsymbol{q})\right]^T, \tag{28.1}$$

where $\boldsymbol{p} = [p_x\ p_y\ p_z]^T$ is the position vector of the end effector and $\boldsymbol{Q} = [\boldsymbol{\epsilon}^T\ \eta]^T$ is the *unit quaternion* representing the orientation of the end effector in which η and $\boldsymbol{\epsilon}$ denote scalar and vector part. Using unit quaternions to represent orientation errors allows global parametrization of orientation not suffering from representation singularities [13].

The arm angle θ proposed in [11, 12] represents the orientation of the arm plane determined by the centers of the shoulder $\boldsymbol{s}$, elbow $\boldsymbol{e}$, and wrist $\boldsymbol{w}$, see Fig. 28.3. It is a kinematic function of the joint angle vector $\boldsymbol{q}$, which gives a measure of the following physical mobility: if we hold the shoulder $\boldsymbol{s}$, the wrist $\boldsymbol{w}$, and the end-effector $\boldsymbol{t}$ in fixed positions, the elbow $\boldsymbol{e}$ is still free to swivel about the axis from the shoulder $\boldsymbol{s}$ to wrist $\boldsymbol{w}$. The arm angle on the circle can be defined by an interior angle between the planes $\triangle \boldsymbol{s}\boldsymbol{e}_0\boldsymbol{w}$ and $\triangle \boldsymbol{sew}$. The reference position $\boldsymbol{e}_0$ of elbow is chosen such that $\boldsymbol{e}_0$ is on the plane which is spanned by $\boldsymbol{s}$, $\boldsymbol{w}$ and the other shoulder position $\boldsymbol{s}'$ of the dual arm manipulator so that θ is equal to zero when $\boldsymbol{e} = \boldsymbol{e}_0$. In this way, the null space motion of the manipulator is uniquely parameterized.

The velocity vector $\dot{\boldsymbol{x}}_E$ in the extended task space is defined as:

$$\dot{\boldsymbol{x}}_E = \left[\dot{\boldsymbol{p}}^T(\boldsymbol{q})\, \boldsymbol{\omega}^T(\boldsymbol{q})\, \dot{\theta}(\boldsymbol{q})\right]^T, \tag{28.2}$$

where $\dot{\boldsymbol{p}}$ and $\boldsymbol{\omega}$ denote the translational and rotational velocity of the end effector, and $\dot{\theta}$ denotes the arm angle (change) velocity. The differential kinematics in the extended task space is given as

$$\dot{\boldsymbol{x}}_E = \boldsymbol{J}_E(\boldsymbol{q})\dot{\boldsymbol{q}} = \begin{bmatrix} \boldsymbol{J}(\boldsymbol{q}) \\ \boldsymbol{J}_\theta(\boldsymbol{q}) \end{bmatrix} \dot{\boldsymbol{q}}, \tag{28.3}$$

where $\boldsymbol{J}$ is the (6×7) manipulator geometric Jacobian and $\boldsymbol{J}_E$ is the (7×7) augmented Jacobian extended by the (1×7) elbow Jacobian $\boldsymbol{J}_\theta = \frac{\partial \Theta(\boldsymbol{q})}{\partial \boldsymbol{q}}$.

The contact force vector in the extended task space is defined as

$$\boldsymbol{h}_E = [\boldsymbol{h}^T \quad 0]^T = [\boldsymbol{f} \ \boldsymbol{\mu} \ 0]^T \ , \tag{28.4}$$

where the zero in the last row corresponds to the contact force of elbow motion set to zero.

Position Based Impedance Control in the Task Space — To ensure stable interaction with various types of environments, one of the most promising compliant control schemes is impedance control. Unlike explicit force control schemes that handle contact force directly the impedance control regulates the dynamics between the measured contact force and the position trajectory. The desired trajectory, which is generated by the human operator during teleoperation, is modified according to the current measured force $\boldsymbol{f}$ and moment $\boldsymbol{\mu}$ such that the end-effector shows the target impedance. Further, the modified trajectory can be considered as reference value for the inner position controller, which is responsible for the actual end-effector position. The target impedance can be written as

$$\boldsymbol{M}_T \Delta\ddot{\boldsymbol{p}}_{dc} + \boldsymbol{D}_T \Delta\dot{\boldsymbol{p}}_{dc} + \boldsymbol{K}_T \Delta\boldsymbol{p}_{dc} = \boldsymbol{f}, \tag{28.5}$$

$$\boldsymbol{M}_R \Delta\dot{\boldsymbol{\omega}}_{dc} + \boldsymbol{D}_R \Delta\boldsymbol{\omega}_{dc} + \boldsymbol{K}_R \Delta\boldsymbol{\epsilon}_{dc} = \boldsymbol{\mu}, \tag{28.6}$$

where $\boldsymbol{M}$, $\boldsymbol{D}$ and $\boldsymbol{K}$ are virtual inertia, damping, and stiffness matrices, which characterize the desired target impedance of the telemanipulator, and subscripts 'T' and 'R' denote translational and rotational direction, respectively. Denoting the desired task frame Σ_d, in which the desired trajectory is described, and the compliant frame Σ_c, in which the modified trajectory by the impedance equation is defined, then $\Delta\boldsymbol{x}_{dc}$ is the amount of modification of the desired trajectory according to the measured contact force $\boldsymbol{h}$. It comprises the translational modification $\Delta\boldsymbol{p}_{dc}$ and the rotational modification $\Delta\boldsymbol{\epsilon}_{dc}$ which is the vector part of the unit quaternion representing the orientation error. Furthermore, the following relations are exploited in (28.6):

$$\boldsymbol{K}_R = 2\boldsymbol{E}^T(\eta, \boldsymbol{\epsilon})\, \boldsymbol{K}'_R, \tag{28.7}$$

$$\boldsymbol{E} = \eta \boldsymbol{I} - \boldsymbol{S}(\boldsymbol{\epsilon}), \tag{28.8}$$

where $\boldsymbol{K}'_R$ is the virtual stiffness matrix in Cartesian space and $\boldsymbol{S}(\,\cdot\,)$ denotes the skew-symmetric matrix based on an argument. The rotational velocity can be calculated from the following quaternion propagation rule [13]:

$$\dot{\eta} = -\frac{1}{2}\boldsymbol{\epsilon}^T \boldsymbol{\omega}, \tag{28.9}$$

$$\dot{\boldsymbol{\epsilon}} = \frac{1}{2}\boldsymbol{E}(\eta, \boldsymbol{\epsilon})\, \boldsymbol{\omega}. \tag{28.10}$$

For the inner control loop, common motion control algorithms such as PD control with gravity compensation can be implemented either in the joint space or in the task space. However, the latter is preferable because the arm self-motion compliance can be controlled independently from the end-effector compliance by means of the inner motion control loop. The end-effector compliance is governed by the outer "impedance control" loop, see Fig. 28.4. Therefore, the inner control loop should be designed in such a way that it does not deteriorate the performance of the outer impedance control loop. Hence, considering that the transition from free space motion to interaction with a rigid environment is a high dynamic process, a static model based control scheme such as PD control with gravity compensation is less preferable to a dynamic model based control like inverse dynamic control, which is often called resolved acceleration control.

The telemanipulator impedance control, which is implemented based on the resolved acceleration control as an inner position control, is shown in Fig. 28.4. Here, $\boldsymbol{x}_{Ed}$ denotes the desired motion trajectory of the end effector in the extended task space; $\boldsymbol{D}_K$ and $\boldsymbol{K}_K$ are proportional and derivative gain of the inner position controller; FK denotes the forward kinematics; $\boldsymbol{J}_E^{-1}$, $\boldsymbol{M}$ and $\boldsymbol{n}$ denote the inverse matrix of the square augmented Jacobian, the manipulator inertia matrix, and the sum of the nonlinear term such as Coriolis-centrifugal torque, gravity torque, and friction torque, respectively. As the last row of the extended modification vector $\Delta\boldsymbol{x}_{Edc} = [\Delta\boldsymbol{p}_{dc}\ \Delta\epsilon_{dc}\ 0]^T$ and its derivatives are zero, the arm angle and its derivative are not modified by means of the target impedance relation, see (28.5), (28.6).

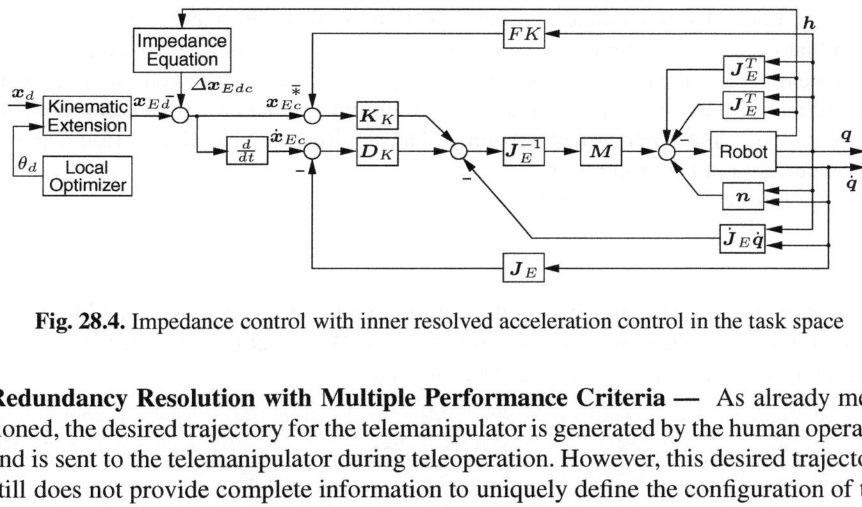

Fig. 28.4. Impedance control with inner resolved acceleration control in the task space

Redundancy Resolution with Multiple Performance Criteria — As already mentioned, the desired trajectory for the telemanipulator is generated by the human operator and is sent to the telemanipulator during teleoperation. However, this desired trajectory still does not provide complete information to uniquely define the configuration of the telemanipulator with respect to the extended task space vector $\boldsymbol{x}_E$ defined in (28.1). To make it complete, the arm angle presented in the previous section should be appropriately defined. To remedy this, there are two viable methods. One is to measure the arm angle of the human operator directly and send it with the other desired trajectories. An alternative one, which has been more typically used to resolve kinematic redundancies, is obtained using the so-called *Moore-Penrose pseudoinverse* of the Jacobian defined as

$$\boldsymbol{J}^{\#} = \boldsymbol{J}^T(\boldsymbol{J}\boldsymbol{J}^T)^{-1}. \tag{28.11}$$

Using this pseudoinverse of the Jacobian inverse kinematic solution can be written at the velocity level as

$$\dot{\boldsymbol{q}} = \boldsymbol{J}^{\#}\dot{\boldsymbol{x}} + (\boldsymbol{I} - \boldsymbol{J}^{\#}\boldsymbol{J})\dot{\boldsymbol{q}}_0, \tag{28.12}$$

where $\dot{\boldsymbol{q}}$ and $\dot{\boldsymbol{x}}$ denote the joint velocity vector and the derivative of the task space vector $\boldsymbol{x}$ without augmentation; $\dot{\boldsymbol{q}}_0$ is an arbitrary vector which can be chosen as a gradient of a potential (objective) function that is optimized as a secondary task (often called *gradient projection method*). The first term of the right hand side in (28.12) describes the primary task of the end effector which minimizes $\|\dot{\boldsymbol{q}} - \boldsymbol{J}^{\#}\dot{\boldsymbol{x}}\|$. The second term defines self motion of the telemanipulator in which the matrix $(\boldsymbol{I} - \boldsymbol{J}^{\#}\boldsymbol{J})$ projects an arbitrary vector $\boldsymbol{q}_0$ onto the null space of the manipulator Jacobian $\mathcal{N}(\boldsymbol{J})$ such that the secondary task does not affect the primary task of the end-effector. Using this method, the desired angle θ_d is calculated by defining an appropriate objective function to control the null-space motion of the manipulator arm.

Typical objective functions are avoiding singularities, obstacles, and structural limitations (e.g., joint angle limits), carrying out reasonable actions (e.g., low-energy-consuming motion), and so on. However, most research in redundant robotics has mainly concentrated on a single performance criterion [14, 15, 16, 17] while relatively little attention has been paid to redundancy resolution with multiple performance criteria [18, 19]. In the following, redundancy resolution schemes with multiple performance criteria, implemented on the telemanipulator, are discussed. Before going into details, it should be noted that redundancy resolution with multiple performance criteria does not mean that these criteria are optimized at the same time, but that they are considered at the same time, i.e., the most dominant (crucial) performance criterion out of concerning performance criteria may be optimized, or the self motion can be settled compromising between two or more performance criteria. The significance of each criterion depends on the current joint configuration.

The solution (28.12) can be rewritten as

$$\dot{\boldsymbol{q}} = \dot{\boldsymbol{q}}_p + \alpha\dot{\boldsymbol{q}}_h, \tag{28.13}$$

where α is a real scalar and $\dot{\boldsymbol{q}}_p$, $\dot{\boldsymbol{q}}_h$ denote the minimum-norm solution and homogeneous solution for the joint velocities, respectively. When taking s number of multiple criteria into account, then $\dot{\boldsymbol{q}}_h$ can be determined as

$$\dot{\boldsymbol{q}}_h = \sum_{i=1}^{s} c_i \dot{\overline{\boldsymbol{q}}}_{hi}, \tag{28.14}$$

where $\dot{\overline{\boldsymbol{q}}}_{hi}$ is a normalized homogeneous solution of $\dot{\boldsymbol{q}}_{hi}$ and c_i is a weighting factor, which takes usually the form of a sigmoidal function of the ith normalized performance criterion $\overline{w}_i$

$$c_i = \begin{cases} 0 & \text{for } \overline{w}_M \leq \overline{w}_i, \\ 2\overline{w}'^3 - 3\overline{w}'^2 + 1 & \text{for } \overline{w}_m \leq \overline{w}_i < \overline{w}_M, \\ 1 & \text{for } \overline{w}_i < \overline{w}_m. \end{cases} \tag{28.15}$$

Thereby, $\overline{w}_m$ and $\overline{w}_M$ describe predefined minimum and maximum activation threshold, respectively, and $\overline{w}'$ is defined as

$$\overline{w}' = \frac{\overline{w}_i - \overline{w}_m}{\overline{w}_M - \overline{w}_m}. \tag{28.16}$$

Because performance criteria are of different functional forms and magnitudes, they must be normalized to allow reliable comparison and combination. With normalization, criteria with different numerical ranges are altered to comparable magnitudes so that, when combined, no single criterion can excessively dominate the solution. After normalization the performance criterion $\dot{\boldsymbol{q}}_{hi}$ can be written as

$$\dot{\boldsymbol{q}}_{hi} = \left(\boldsymbol{I} - \boldsymbol{J}^{\#}\boldsymbol{J}\right)\frac{\partial \overline{w}_i}{\partial \boldsymbol{q}}. \tag{28.17}$$

It should be noted that effective normalization of performance criteria is not trivial because it involves knowing the normalizing factors (e.g. criteria's maximum values) of all the criteria. However, the normalization of the ith homogeneous solution $\dot{\boldsymbol{q}}_{hi}$ before weighting compensate the effect of various normalization of performance criteria.

Once $\dot{\boldsymbol{q}}_h$ is calculated as above, the velocity of self-motion (the arm angle rate) can be obtained as

$$\dot{\theta}_d = \boldsymbol{J}_\theta \dot{\boldsymbol{q}}_h. \tag{28.18}$$

Finally, the arm angle θ_d and the arm angle rate $\dot{\theta}_d$ are added to the desired trajectory generated by a human operator such that the resulting extended task space vector takes the form of (28.1).

It should be noted that this redundancy resolution brings many attractive characteristics at the expense that more complicated control algorithms are required to control the additional mobility. As already mentioned, this algorithm requires no measurement of the θ angle and opens the possibility to control the null space motion in closed-loop fashion.

28.2.3 Telemanipulation with Dissimilar Kinematics

Fig. 28.5 shows a diagram of the implemented teleoperation control, which combines the control of haptic display and telemanipulator to a two-channel architecture. While the control algorithm of the haptic display is implemented as mentioned in chapter 2, the teleoperator is controlled according to the control scheme presented in Sec. 28.2.2. Desired positions are sent to the teleoperator and the measured interaction forces are fed back to the operator site. In order to reduce complexity, Fig. 28.5 can be simplified considering perfect position tracking ($\boldsymbol{x}_m = \boldsymbol{x}_r$, $\boldsymbol{x}_s = \boldsymbol{x}_{Ec}$) of both devices, which results in the structure depicted in Fig. 28.6.

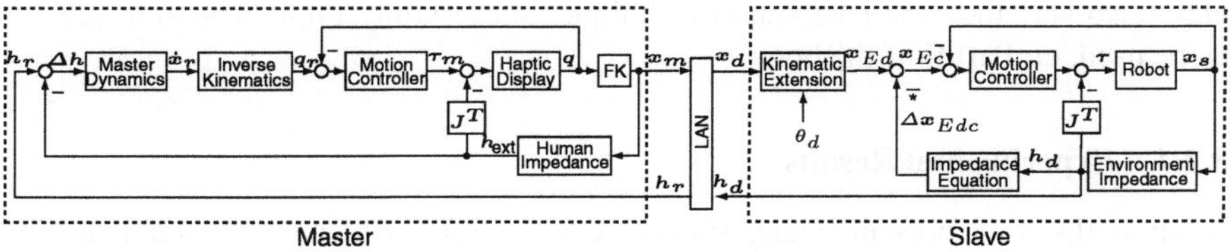

Fig. 28.5. Teleoperation control architecture

As reported in [20,21], using impedance control instead of a pure position control enhances stability of the overall telemanipulation system if the impedance parameters are chosen in an adequate manner. This allows to operate in stiffer environments than possible using a pure position controller. On the other hand, one can see that in contact, as a result of the impedance control, the reference position of the teleoperator is modified according to (28.5) and (28.6). Consequently, transparency is influenced by the parameter settings of the slave impedance control. The stiffer the impedance control of the slave the smaller the position deviation and the better the real environmental impedance is reproduced. As already mentioned, this stays in contrast to the stability behavior of the system. As a result, the impedance parameters must be chosen as tradeoff between stability and performance.

Transparency is also affected by the master control, because perfect transparency requires the master mass to be set to zero. However, this is not possible because the minimum target inertia of the haptic display is bounded by stability. Thus, in free space motion, a minimal mass and inertia, necessary for stability of the master control, can be felt by the human operator.

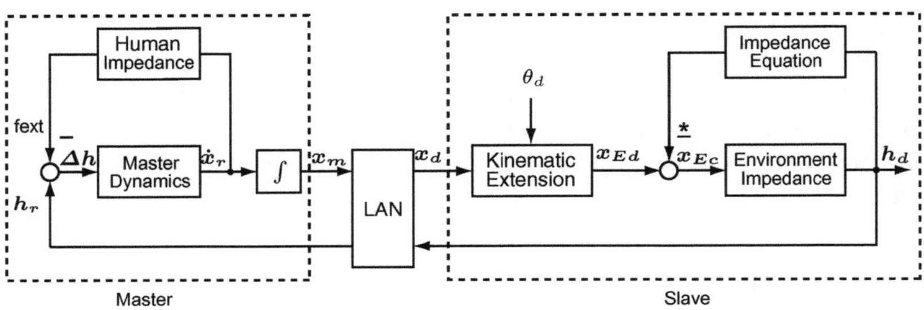

Fig. 28.6. Simplified teleoperation control architecture

The stability of the overall system is guaranteed in conformity to the passivity theorem, see [22]. The mechanical impedance/admittance of the manipulators as well as the environment used in the experiments are passive. It is also assumed that the human operator behaves in a cooperative i.e. passive way. The only potentially non passive element is the communication network. However, in the presented experiment, the delay introduced by the internal lab network may be neglected. For a case of considerable time delay (also variable), the passivation techniques successfully implemented in [22, 23] can be used, cf. Part II of the book.

28.2.4 Experimental Results

The strengths of the presented telepresence system are its sophisticated redundancy resolution schemes and its ability to provide a comprehensive teleoperation environment comprising dual-handed manipulation with visual, auditory, and haptic feedback. In the following section, firstly, experimental results for the implementation of a redundancy resolution scheme with multiple objective functions are presented. Secondly, the performance of the telepresence system for exploration and manipulation tasks is assessed.

Redundancy Resolution with Multiple Performance Criteria — For dual-handed telemanipulation it is crucial to ensure collision avoidance between both arms because they share the same workspace. Some types of collisions can be avoided through utilization of self-motion. Moreover, it is desirable to keep the current joint configuration as far as possible from its joint limits without disturbing the primary task of the end effector. Therefore, joint limit avoidance and collision avoidance between both arms were chosen as multiple performance criteria.

A corresponding objective function for the joint limitation avoidance can be easily found in the literature. According to [17] follows:

$$w_{JL}(\boldsymbol{q}) = \frac{1}{4}\sum_{i=1}^{n} \frac{(q_{iM} - q_{im})^2}{(q_{iM} - q_i)(q_{im} - q_i)}, \tag{28.19}$$

where w_{JL} denotes the objective function for the joint limit avoidance, n denotes the number of joints, and q_{iM} and q_{im} denote the maximum and the minimum joint limits for the *i*-th joint angle q_i, respectively. Considering that $w_{JL} = -n$ if all the joint angles are placed in the middle of their joint limits and goes to minus infinity if one of the joint angles approaches either of the limits, the normalized objective function can be defined as

$$\overline{w}_{JL}(\boldsymbol{q}) = e^{(n+w_{JL}(\boldsymbol{q}))}. \tag{28.20}$$

The normalized objective function $\overline{w}_{JL} = e^0 = 1$ is equal to one when all joint variables stand in the middle of their feasible ranges and goes to zero ($\overline{w}_{JL} = e^{-\infty} = 0$) at either limit.

A typical objective function for collision avoidance can be defined as the shortest distance between manipulator and obstacle. For the purpose of collision avoidance between both arms, it can be written in the form

$$d(\boldsymbol{q}) = \|\boldsymbol{p}_L(\boldsymbol{q}_L) - \boldsymbol{p}_{ELR}(\boldsymbol{q}_R)\|, \tag{28.21}$$

in which redundancy resolution for the right hand is assumed. Thereby, $\boldsymbol{p}_L$ and $\boldsymbol{p}_{ELR}$ denote the position vector of the point on the left hand nearest to the right arm's elbow and the position vector of the right arm's elbow with respect to the global reference frame, respectively. Practically, however, it is not necessary to take this objective function into account if there is no danger of collision. Let assume that if the distance is greater than d_M there is no danger of collision. Otherwise ($0 < \overline{w}_{CA}(\boldsymbol{q}) < 1$), there is a danger of collision and the optimizing process needs to be activated. In that case the normalized objective function for collision avoidance $\overline{w}_{CA}$ may be written as

$$\overline{w}_{CA}(\boldsymbol{q}) = \begin{cases} 1 & \text{for } d > d_M \\ \frac{d-d_m}{d_M-d_m} & \text{for } d_m \le d \le d_M \\ 0 & \text{for } d < d_m \end{cases}, \tag{28.22}$$

where d_M and d_m are design parameters of the normalized objective function for collision avoidance.

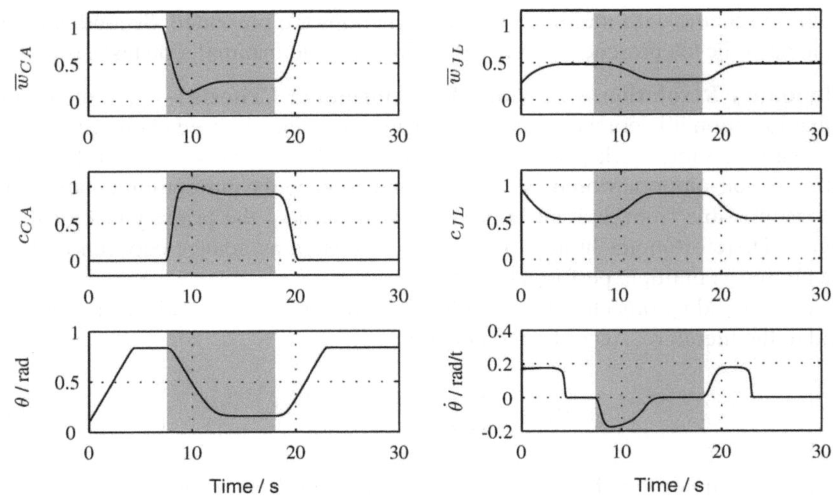

Fig. 28.7. Optimization of a multiple objective function: Collision avoidance with joint limitation avoidance

These two objective functions are combined to define the arm angle optimally as described in Sec. 28.2.2. The experimental results are illustrated in Fig. 28.7, whereby the shaded area denotes dominance of the collision avoidance criterion. Initially, the left arm is placed far enough from the elbow of the right arm. After 7 s the left arm starts to move toward the elbow of the right arm, and at the time of 9 s it starts to move back. Therefore, the whole process can be divided into three phases. In the first phase (left side of the shaded area in the figures), there is no danger of collision so that the criterion for joint limit avoidance is dominant. Therefore, the criterion for joint limitation avoidance $\overline{w}_{JL}$ is optimized through the null space motion. However, as collision possibilities are induced in the second phase due to the approach of the left arm, the criterion for collision avoidance ($\overline{w}_{CA}$) becomes dominant, and the self motion is resolved to avoid the collision with the left arm while $\overline{w}_{JL}$ is reduced. In the third phase, after the left arm moved back, the $\overline{w}_{JL}$ becomes dominant such that the settled arm angle in the first phase is recovered. It should be noted that these performance criteria can be integrated into the overall objective function either with same or different priorities.

High-Fidelity Telemanipulation — The experimental setup consists of the haptic input device VisHaRD10 [24, 25], the 7DOF slave telemanipulator, and a stereo vision system. According to the bilateral control structure (see Sec. 28.2.3), the motion of the operator is read by the master device and sent as desired positions to the slave impedance controller. It should be noted that during these experiments no redundancy resolution based on the optimization of multiple objective functions is applied, and therefore the angle θ_d is kept at a constant value. The measured contact forces are sent back as the input to the master admittance controller. The devices communicate using UDP (in the lab LAN) with a sampling rate of 1kHz, which is the same as for the local loop control.

The vision system consists of two CCD cameras placed on a 3DOF camera head. The recorded video streams are displayed on the head mounted display (HMD) carried by the operator. The HMD has a built in tracker, which is used for controlling the motion of the camera head. Such a setup provides the operator with a realistic visual information about the location of the objects, the environment, and the telemanipulator. Here, the anthropomorphic construction of the telemanipulator plays an important role: the operator can drive it as if it were his/ her own arm. The visual information is useful not only for motion generation but also for handling the contact and minimizing effects of the impact.

The experiment consists of three tasks:

- tracking of free space motion
- haptic exploration of different materials (soft and stiff), see Fig. 28.8a
- driving a screw with an aluminium tool, see Fig. 28.8b. This last experiment consists of three phases: contact with extreme stiff materials, a classic peg-in-hole operation and manipulation in a constrained environment.

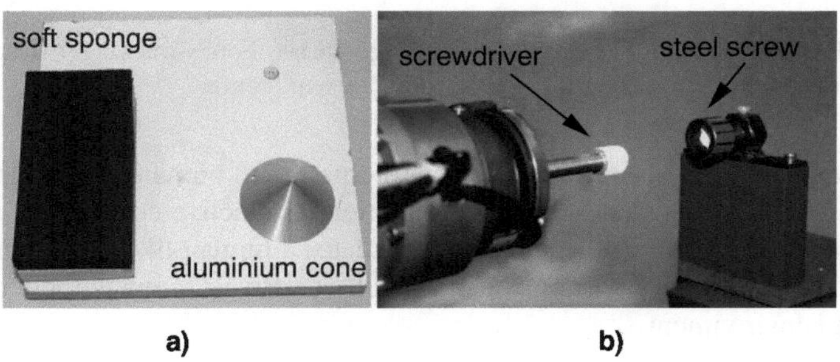

Fig. 28.8. Slave side: a) objects for haptic exploration b) screw and screwdriver

Fig. 28.9 and Fig. 28.10 show the position and force tracking performance during haptic exploration of different materials (see Fig. 28.8a). The shaded areas indicate the several contact phases. One can see that during free space motion, the position tracking of the slave arm works very well while in the contact situation, as a consequence of the implemented impedance controller, the slave position differs from the master position. Please note that, as the force tracking is very good, this position displacement influences the displayed and felt environmental impedance in such that hard objects are perceived softer then they are. As the master controller is of admittance type, which reacts on the human force input, non zero forces (forces depend on the minimal master dynamics) are necessary during free space motion to change the actual end-effector position.

Similar results can be achieved in the screw tightening experiments, see Fig. 28.11 and Fig. 28.12. Although screw tightening differs from the simple exploration scenario because more than one translational and rotational constraint is active at the same time, a quite small position displacement and a very good force tracking in all translational and rotational directions can be achieved. As the y-direction represents the actual screw

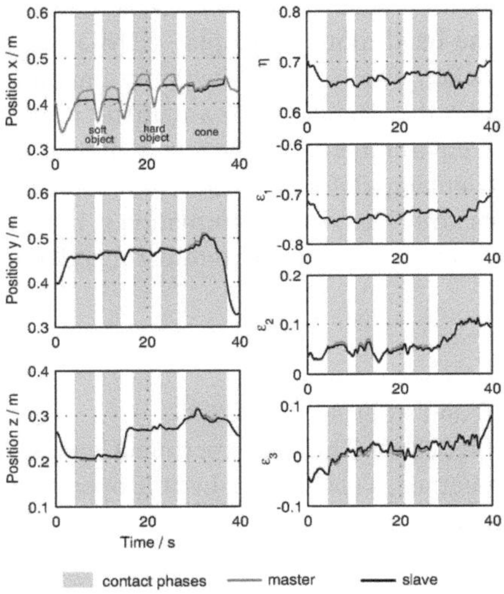

Fig. 28.9. Position tracking (Slave) during haptic exploration

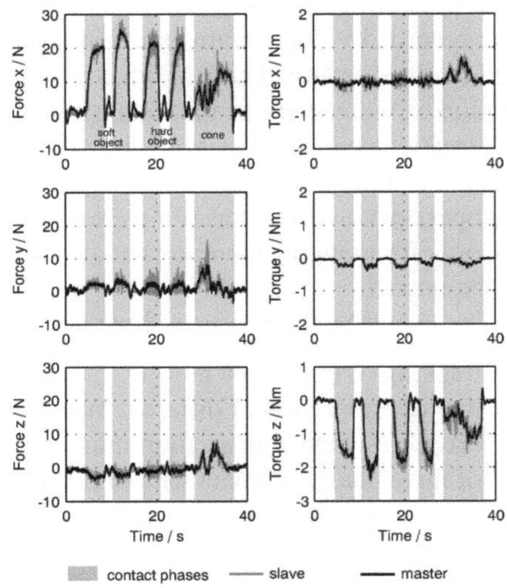

Fig. 28.10. Force tracking (Master) during haptic exploration

tightening axis, torques around it should be interpreted as human torque inputs necessary to change the robot's end-effector orientation. The active compliance introduced by the impedance control of the slave arm emulates a human like compliant behavior when interacting with the environment and enables to drive a screw without destruction of screw and environment.

28.3 Extensive Telepresence

Telepresence systems are commonly operated in unstructured, hazardous remote environments. These environments are often far larger than the workspace of a human arm or a stationary telemanipulator. To extend the accessible workspace, a mobile teleoperator (MTO) which is able to freely move around in the remote environment can be employed. In order to maintain the feeling of presence in the remote environment, it is highly important for extensive telepresence systems to convey a natural sensation of locomotion in addition to visual, auditory, and haptic feedback. This is achieved by tracking motions and gaze direction of the human operator and replicating them by the MTO. Thereby, a feeling of presence in a wide remote environment is realized [26,27].

Current telepresence systems provide either realistic force-feedback or intuitive feeling of locomotion but not both at the same time. In the former case, a stationary haptic device is used to track the position of operator hands and reflect forces back to them. In the latter case, the position of the user is tracked by some kind of tracking device, but no forces can be applied to the user's hands. In this section we present a mobile haptic interface (MHI) which comprises a haptic interface mounted on a mobile base. The motion of this base is controlled in such a way that it maximizes the manipulability

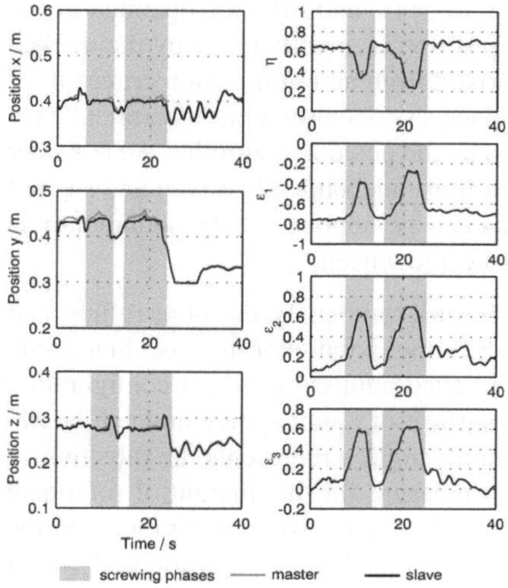

Fig. 28.11. Position tracking (Slave) during screw tightening

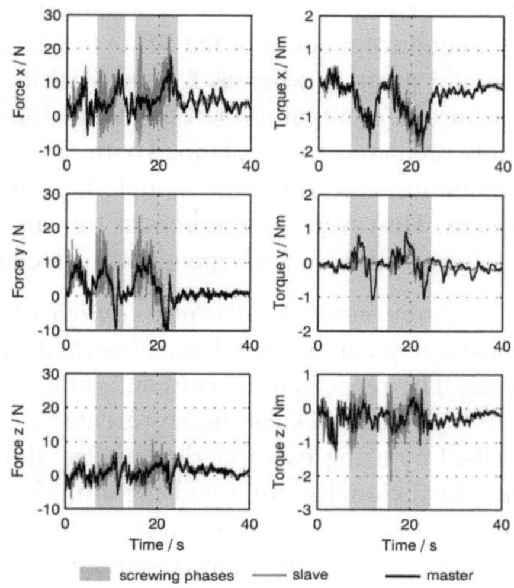

Fig. 28.12. Force tracking (Master) during screw tightening

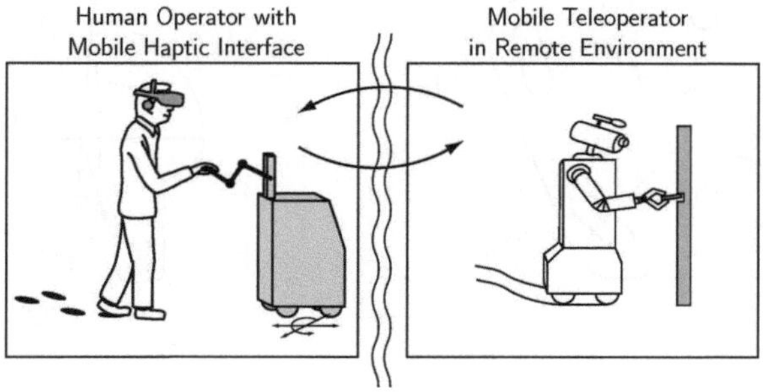

Fig. 28.13. Extensive telepresence with a human operator mastering a mobile teleoperator (MTO) through a mobile haptic interface (MHI)

of the haptic interface. The MHI allows the operator to freely move around while being in contact with the haptic interface. Thereby, realistic force-feedback and sensation of locomotion is achieved at the same time (Fig. 28.13).

28.3.1 Mobile Platform

The human legs allow the operator to move in three DOFs (two translational, one rotational). As the MTO has to mimic all motions of the operator, its mobile platform must have at least the same DOFs. The same holds for the MHI, which must follow the path of the operator in order to maintain manipulability.

Position Control — Both the mobile platforms of MTO and MHI are operated in position control mode. For the MTO the desired position is directly given by the tracked operator position whereas for the MHI it is derived from an optimization criterion regarding the configuration of the haptic input device. As operator motions cannot be reliably predicted, no preplanned path can be used by the position controller. We consider a nonholonomic, omnidirectional platform with four independently driven and steered wheels. However, the theoretical fundamentals can be applied to any nonholonomic platform with $n > 2$ independently driven and steered wheels.

1) Representation of wheel configuration: In order to satisfy the pure rolling and nonslipping condition, all wheel normals must be either parallel or intersect in a single point. The respective wheel configurations are called admissible wheel configuration (AWC). All AWCs can be represented on the surface of a unit sphere and can be described by using two spherical angles: the azimuth angle η represents the direction of the translational motion and the altitude ζ is a measure for the amount of rotational motion. In Fig. 28.14b the unit sphere model is illustrated. All configurations on the equator ($\zeta = 0$) correspond to pure translational motion while configurations at one of the poles ($\zeta = \pm\pi/2$) represent pure rotational motion.

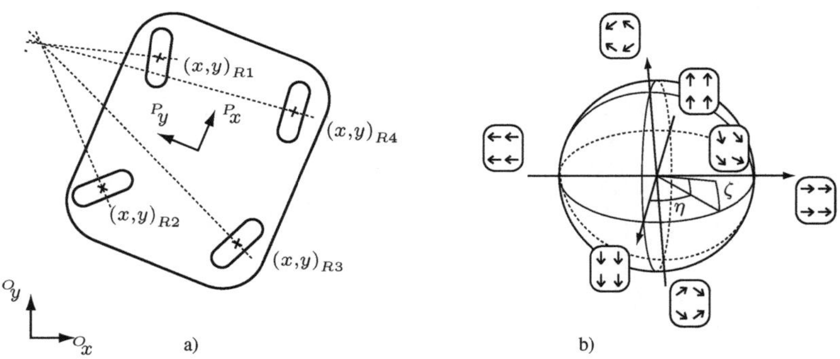

Fig. 28.14. (a) Kinematic model of the mobile platform (b) Unit sphere representation of wheel configurations

As an AWC does not specify the absolute speed of the platform motions, a third variable, the generalized velocity ω is introduced. From the AWC (η, ζ) the unit vector $\boldsymbol{e}(\eta, \zeta)$ can be calculated. Including the generalized velocity ω yields the platform velocity in the Cartesian platform coordinate system:

$$\begin{pmatrix} {}^P\dot{x} \\ {}^P\dot{y} \\ {}^P\dot{\psi}/\kappa_G \end{pmatrix} = \omega \begin{pmatrix} \cos\zeta\, \cos\eta \\ \cos\zeta\, \sin\eta \\ \sin\zeta \end{pmatrix} = \omega \boldsymbol{e}(\eta, \zeta), \tag{28.23}$$

where κ_G is a scaling factor of dimension m^{-1}.

For any given Cartesian velocity ${}^P\dot{\boldsymbol{x}}$, two solutions $\boldsymbol{e_1}$, $\boldsymbol{e_2}$ exist. These are located at two diametrically opposed points on the surface of the unit sphere with $\omega_1 = -\omega_2$.

Both solutions yield the same motion because the corresponding steering angles of each wheel differ by $\pm\pi$ while their revolution speeds have opposite signs.

The actual steering angles φ_i and wheel velocities v_i of the four wheels $i \in \{1 \dots 4\}$ can be obtained from the triple (η, ζ, ω) by solving

$$v_i \begin{pmatrix} \cos\varphi_i \\ \sin\varphi_i \end{pmatrix} = \omega \underbrace{\begin{pmatrix} \cos\zeta\,\cos\eta - \kappa_G y_{Ri} \sin\zeta \\ \cos\zeta\,\sin\eta - \kappa_G x_{Ri} \sin\zeta \end{pmatrix}}_{\boldsymbol{x}_{Ri}}, \tag{28.24}$$

where (x_{Ri}, y_{Ri}) are the coordinates of wheel i with respect to the platform coordinate frame.

For every configuration with $\boldsymbol{x}_{Ri}^2 \neq 0$, (28.24) can be uniquely solved by introducing the following constraint:

$$\text{sgn}\,(v_i)s_i = \text{sgn}\,(\omega). \tag{28.25}$$

In this equation $s_i \in \{-1, 1\}$ determines the sign of the wheel velocity. The use of s_i is discussed below.

If $\boldsymbol{x}_{Ri}^2 = 0$, no unique solution for (28.24) exists. The corresponding wheel configuration is called a singular wheel configuration. In this case, the steering angle φ_i is undetermined and the wheel velocity v_i is zero. This occurs when the instantaneous center of rotation of the mobile platform is coincident with wheel i.

2) Controller design: The position controller consists of three major parts: a linear compensator, which calculates Cartesian velocities from position errors, the mapping from desired platform velocities to an appropriate wheel configuration, and the wheel controllers. The complete structure is presented in Fig. 28.15.

The desired Cartesian velocities in the absolute coordinate frame are calculated from the Cartesian control error $(x_{err}, y_{err}, \psi_{err})$ by applying a pure linear P-control law: After transformation into the platform coordinate system, an AWC (η, ζ) and a generalized velocity ω is calculated from this velocity using (28.23). Due to the nonholonomic nature of the platform, solving (28.23) and (28.24) leads to infinitely high steering velocities. Therefore, the velocity controller has to limit steering velocities by balancing between tracking accuracy and steering effort.

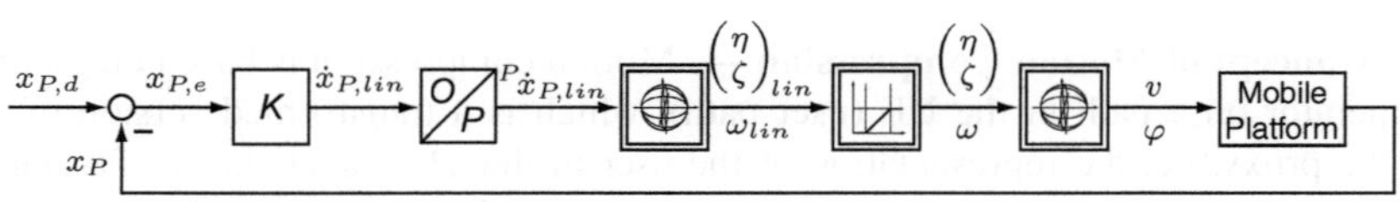

Fig. 28.15. Position control of the mobile platform

3) Overall control structure: The complete structure of the haptic telepresence system for telemanipulation in extensive remote environment is shown in Fig. 28.16. At the remote site, contact forces are measured and send to the haptic display. Forces from the operator site are used as input to the manipulator control. The position of the mobile platform is controlled according to the tracker signals received from the operator site. At the operator site, an analogous architecture is used. The desired position of the mobile platform, however, is updated from the current manipulator configuration.

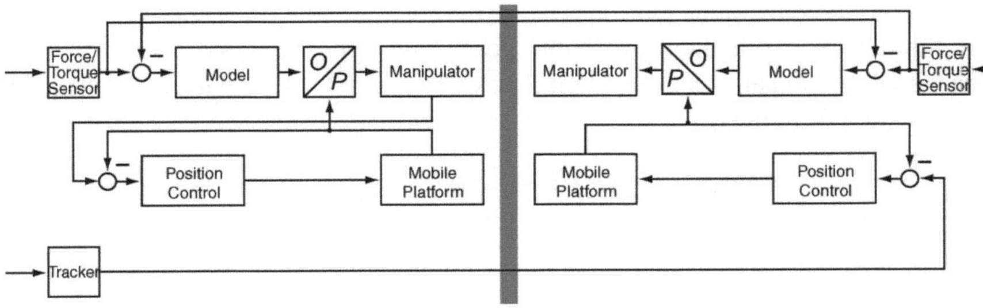

Fig. 28.16. Overall control structure of an extensive telepresence system

28.3.2 Motion Compression

A natural feeling of locomotion is of high importance for fully immersive exploration of real remote environments as well as for visiting virtual environments. The sensation of moving around in the target environment (TE) can be achieved by tracking the position of the operator and presenting according visual, auditory, and haptic feedback to him/her. The feedback can be either provided by sensors from a teleoperator or can be generated from a virtual world model. In both cases this approach is bounded by the size of the user environment (UE). In most practical applications, the coverage of the tracking system imposes additional constraints.

To allow intuitive locomotion and navigation in TEs which are far larger than the available user space, various techniques have been investigated. But all methods like indexing, scaling etc. considerably reduce the feeling of presence in the TE and deteriorate the navigation capabilities of the operator.

In the following section, we present a technique termed *Motion Compression* which is based on curvature deviations between paths in the TE and the UE. An optimization strategy is used to determine the user path with minimum curvature deviation which fits inside the boundaries of the UE. As humans are unable to detect small curvature deviations, motion compression does not affect the experience of locomotion in the target environment.

Basic Concept of Motion Compression — Motion compression is based on guiding the operator on a path in the UE (user path), which is a transformed version of the path the proxy, i.e. the representation of the user in the TE, travels in the TE (target path). The user is presented visual feedback which reflects motions of the proxy in the target environment correctly. Proprioceptive sensation (vestibular and kinesthetic feedback), however, matches motions of the operator in the UE. It has been shown that humans can tolerate a certain degree of inconsistency between visual and proprioceptive sensations [28, 29].

User guidance is achieved by exploiting human path tracking behavior, which is based on both, visual and proprioceptive information. While proprioception cannot be altered by external devices, the visual information can be arbitrarily generated and presented to the operator via a head-mounted display. The resulting deviations between visual and proprioceptive stimuli should be kept at a minimum in order to avoid confusion of the operator.

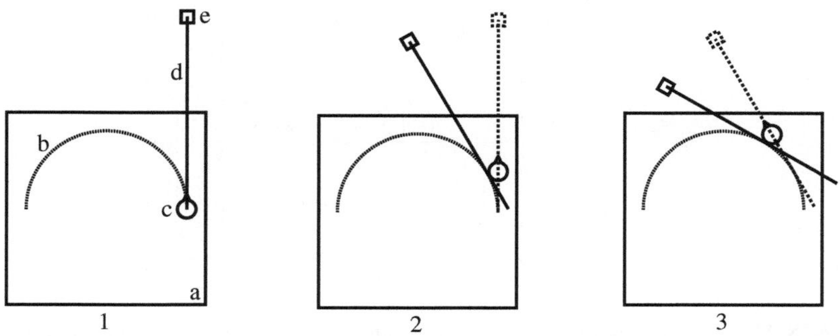

Fig. 28.17. User guidance

The principle of user guidance is illustrated in Fig. 28.17. In this example, the path in the TE is a straight line while its transformed version in the UE is a semi-circle:

(i) The user (c) is standing in the user environment (a) at the beginning of the user path (b). The object (e) is displayed to the user via a HMD. The user decides to walk towards the object on the target path (d).
(ii) The user has moved a short distance straight ahead in the user environment. As this motion reflects a circular trajectory in clockwise sense on the target path a deviation between actual and intended path occurs. The object is shifting a little to the left on the HMD.
(iii) The user has turned to the left by a small amount to correct the deviation. Thereby, he/she is guided on the user path.

The three major parts of motion compression are: path prediction, i.e. estimation of the intended path in the target environment; path transformation, i.e. finding a transformed path which fits to the user environment; and user guidance, i.e. applying the transformation to transform positions and orientations from user environment to target environment and vice versa.

In the following, these components of motion compression are described in more detail for a simple static case and the general dynamic case.

Motion Compression for Target Environments with a Finite Number of Discrete Predefined Paths — As discussed in the previous section, a single straight line in the TE can be mapped to a circle in the UE. Fig. 28.18 shows a more complex environment, which consists of an equally spaced grid of lines. A real world example would be the aisles between bookshelves in a library.

Path Prediction: For this scenario also a static mapping of all target path into the user environment exists, which can be calculated offline. Therefore, the path prediction step can be omitted.

Path Transformation: In Fig. 28.18 all horizontal lines in the TE are mapped to the lower left circle in the UE whereas all vertical lines are mapped to the upper right circle. As the distance between the lines equals the perimeter of the circles, the mapping is distance preserving. The intersections of horizontal and vertical lines in the TE are all

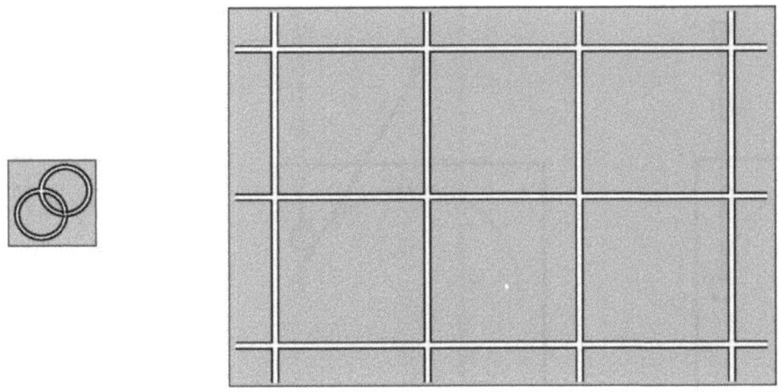

Fig. 28.18. Motion Compression with static path transformation. Left: Paths in the user environment (UE), right: Paths in the target environment (TE).

mapped to the upper left intersection of the circles in the UE. The mapping is also (locally) angle preserving because the tangents of the two circles are perpendicular to each other at this intersection.

User Guidance: User guidance works in the same way as described in the previous section.

Motion Compression for General Target Environments — It is a key feature of Motion Compression that the transformation between coordinates in the user and target environments is distance and angle preserving. As this preserving condition cannot be globally fulfilled in the case of a general target environment, a dynamic mapping has to be calculated in real-time. This mapping is valid in a subspace of the TE which is located around the current user position and the intended target path.

Path Prediction: A correct assumption on the intended target path is essential for calculating a transformed user path which neither exceeds the boundaries of the UE nor exhibits great curvature deviations. Two different methods of path prediction can be distinguished depending on whether information about the content of the TE are taken into account or not. Path prediction without knowledge of the TE relies on extrapolating past user motions and tracking the gaze direction of the user. In a very simple implementation, the future target path is assumed to be a straight line coincident with the gaze direction of the user. This method is well-suited for unstructured environments like unknown remote environments. However, this path prediction method has a strongly limited time horizon and produces bad results in case of abrupt turns. In contrast to that, an approach including knowledge of the TE can be chosen when interesting sites in the TE have been identified beforehand. A model of the environment can be used to estimate the intended goal and path of the user. The required knowledge of the TE can be easily acquired for static virtual environments whereas for telepresence systems it requires complicated feature extraction algorithms.

Path Transformation: The predicted path has to be transformed from TE to UE in such a way that it fits into the available user space. To account for unpredicted user motion, the user path should have a certain clearance from the boundaries of the UE. This is particularly important for intermediate points on the target path, where an estimation of future motion direction is very unreliable. These points should be mapped to locations close to the center of the UE.

Path transformation can be understood as dynamic optimization problem with the continuous path variable s as the independent variable. The predicted path is given by its curvature function

$$\kappa_T = \kappa_T(s) \quad 0 \leq s \leq s_E, \tag{28.26}$$

where κ_T is the target path curvature, s the distance along the path, and s_E the total length of the path. Starting position and orientation of the target path are not required for the transformation. The optimal user path to be found is given by the function

$$\kappa = \kappa(s) \quad 0 \leq s \leq s_E \tag{28.27}$$

together with its starting position and orientation

$$x_0, y_0, \text{and } \varphi_0. \tag{28.28}$$

An intuitive objective functional for minimizing overall curvature deviation is

$$J_1 = \int_0^{s_E} \frac{1}{2}(\kappa - \kappa_T)^2 \, ds. \tag{28.29}$$

The user path is subject to the following equality constraints describing the relationship of curvature, orientation, and position:

$$\frac{dx}{ds} = x' = \cos\varphi, \tag{28.30}$$

$$\frac{dy}{ds} = y' = \sin\varphi, \tag{28.31}$$

$$\frac{d\varphi}{ds}\varphi' = \kappa, \tag{28.32}$$

and a set of inequality constraints characterizing the feasible space in the user environment,

$$\underline{\mathcal{G}}(x, y) \geq 0. \tag{28.33}$$

If the terminal position of the target path is known, it can be used as an additional constraint to increase the clearance of the corresponding terminal point in the UE from the boundaries. This is achieved by introducing a penalty term Φ in the objective function:

$$J_2 = J_1 + \Phi(x(s_E), y(s_E), \varphi(s_E)). \tag{28.34}$$

The general optimization problem given in (28.26) through (28.34) can be solved numerically by using the DIRCOL framework [30].

User Guidance: User position and orientation are transformed from the UE to the TE and vice versa according to the calculated path transformation. The transformed user position and orientation, i.e. position and orientation of the proxy, are used to generate visual feedback to the user. Because of human path tracking control, this visual feedback leads to guiding the user on the user path.

Some important characteristics of the transformation between UE and TE are:

- Positions on the user path are mapped to positions on the target path where the distance the user has moved along the user path equals the distance the proxy has moved on the target path.
- Positions off the user path are mapped to positions off the target path with the perpendicular distance from the user to the user path equal to the distance of the proxy to the target path.
- User orientation tangential to the user path is mapped to proxy orientation tangential to the target path; the angle between user orientation and the tangent of the user path equals the angle between proxy orientation and target path.

These properties ensure that the transformation from UE to TE is locally distance and angle preserving.

Experimental Validation — Experimental tests with untrained subjects have shown that motion compression does not reduce the localization capabilities of a human as long as curvature deviations are kept below the perception threshold. For motions along a straight line in the TE, motion compression becomes typically noticeable if the corresponding circle in the UE is smaller than 10 m radius. For detailed experimental data, the reader is referred to [29, 31].

28.4 Collaborative Telemanipulation

In collaborative environments humans multiply their output by sharing facilities and capabilities which are necessary to perform a common task or achieve a common goal. Combining this ability with classical telemanipulation, a collaborative telemanipulation system results.

28.4.1 Collaborative Telemanipulation Architectures

As usual in the literature, collaborative telemanipulation systems can be divided into:

- multiple operator, multiple robot systems[1](MOMR),
- single operator, multiple robot systems[1](SOMR), and
- multiple operator, single robot systems[1](MOSR).

Referring to this classification, five different collaborative telemanipulation architectures can be constructed, see Tab. 28.1. The first architecture represents a MOMR system. In order to control a robot located in a remote environment, each operator interacts with a human system interface. Depending on the given task visual, auditory, and haptic feedback are provided. To accomplish a common task, the robots can share the same workspace. In such a case, a collision free coordinated motion can be considered as the simplest form of collaboration. More complicated situations result when robots interact directly or via an object. Applications for such an architecture are situations when

[1] It should be noted that in the context of this report the term "robot" describes collaborating entities at the remote site which can be a robot but also a human.

more than one person are required to perform a given task, e.g. while transportation or assembling of heavy, bulky, or fragile objects. Direct contact between the robots can be necessary to guide the motion of the collaborating operator or to call his attention. Also, social conventions as hand-shaking could be possible.

The second architecture describes a similar situation with similar interaction and collaboration possibilities, but in this case only one operator is responsible for multiple robots. E.g. bimanual telemanipulation can be treated as such an architecture whereby each human system interface are controlled by one human hand. Even possible is an architecture whereby the human operator controls multiple robots by using only one human system interface. In these cases the telerobots possess local intelligence, which can be used to avoid collisions between other participants and the environment, to commonly grasp and/or hold an object, or to achieve a superior mission, which cannot be obtained by a single telerobot because of limited time or missing tools. Therefore, such an architecture could be suitable for inspection or transshipment tasks whereby also the transportation of heavy, bulky, or fragile objects could be considered. In the latter case, multiple robots can grasp an object and form a closed chain with it. While local force controller guarantee the contact with the object, the human operator is responsible for the positioning of the object.

The third architecture shows a MOSR situation. In this case multiple operators are simultaneously responsible for one robot that is controlled according to an average or sum of all operator inputs. Applications for such an architecture could be found in the entertainment sector as well as in student/teacher scenarios, whereby the teacher can interact with the student, correct him, or give instructions. Modifying this architecture in such a way that only one operator controls the robot which is located remotely, but the information provided by the human system interface are available for more than one operator, this architecture can be used to teach motion sequences and procedures. Such an architecture, described by the fourth scheme, can also be used to increase the situation awareness of an operator before the control of a robot is switched from one operator to the other.

Finally, the fifth architecture shows a SOMR system, whereby a human collaborates with a robot controlled by a remote human operator. As in case of the first architecture, the simple case of coordinated motion as well as the more complicated case of direct contact and interaction over an object can be considered. Such an architecture could be of interest when the task requires an expert which is not located at the remote site. In this case the expert can operate remotely assisted by a local human operator.

28.4.2 Interactions in Collaborative Telemanipulation Systems

As usual in the literature, constrained and unconstrained cases of collaboration can be distinguished. Constrained collaboration refers to situations where multiple individuals are in contact with the same object or with each other such that the motion of each participant is tightly associated with the other's motion. In the unconstrained case, each individual is controlled independently, but in order to perform a common task they share the same workspace and need to avoid collisions.

While in unconstrained telemanipulation systems, the telerobots are in contact with different objects and, therefore, tasks can be described by decoupled parallel linkages,

Table 28.1. Architectures for collaborative telemanipulation

N°	Type	Architecture
1	MOMR	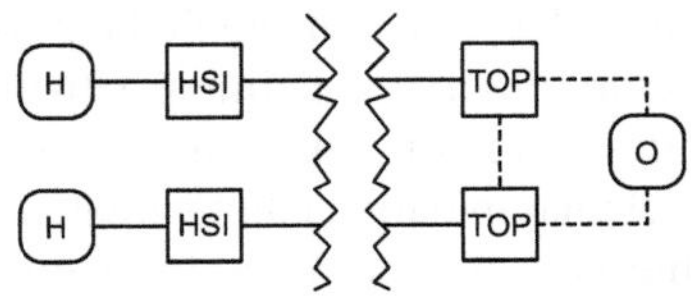
2	SOMR	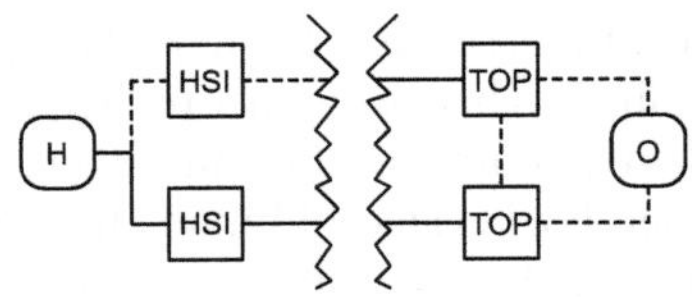
3	MOSR	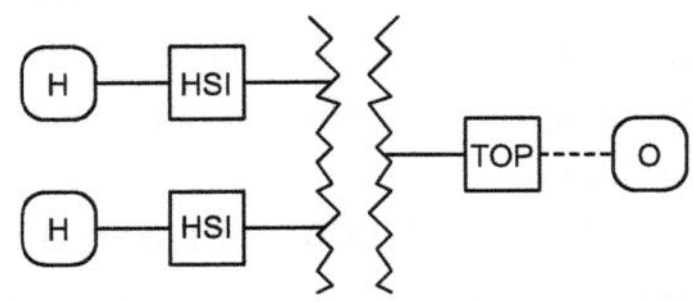
4	MOSR	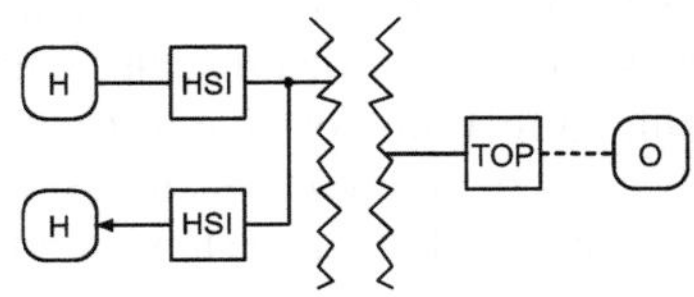
5	SOMR	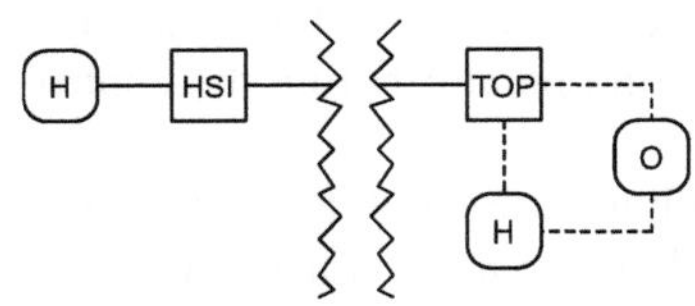

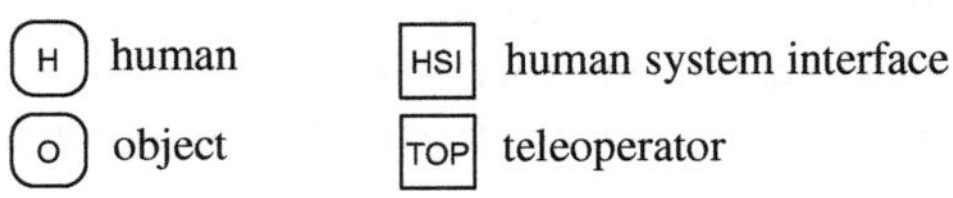

in constrained telemanipulation systems, also the following two configurations must be considered:

- direct coupled linkage between the robots: telerobots are in direct contact with their end-effectors; the task can be described by a direct coupled linkage, and
- indirect coupled linkage via an object: the end-effectors of the telerobots are coupled via a common object; the task can be described as indirect coupled linkage.

These additional configurations enable an increased dexterity, loading and handling capability, as well as an enhanced efficiency and feasibility. Beside these advantages, also increased safety (due to the distributed energy) and improved robustness (which results from the manipulation redundancy) can be achieved.

But in order to reap the benefits of such a system, stable control algorithms for all these possible configurations must be implemented. This means that depending on the actual configuration, control algorithms have to deal with uncertainties in the environment, the specific task trajectory and the object manipulation, as well as constraints on the motion of the robots and the manipulated object. As the manipulation is accomplished remotely and all signals have to be sent over the internet, algorithms should also be robust against non-ideal network conditions. Finally, also properties as transparency and quality of tele-coordination should be addressed. In order to implement an appropriate control algorithm which guarantees a synchronous operation, efficient coordination, and safe manipulation, all these mentioned effects and requirements have to be considered.

28.5 Summary and Conclusions

A multi-modal telepresence system which allows dexterous teleoperation in large remote environments was presented. Thereby, the combination of two redundant devices for teleoperator and haptic display enables an intuitive manipulation in 6 DOF. The requirements for such a manipulation may be summarized as follows: accurate position and force control, dextrous free space motion, singularity robust kinematic transformations, small contact forces, and stable environment contact. To satisfy these requirements, adequate control algorithms were implemented on the master and slave side. A suitable task space definition was chosen and implemented. To assure the global parametrization of the rotational motion and the environment interaction, the unit quaternion representation was applied. The redundancy of the master and slave manipulators was efficiently utilized to fulfill additional kinematical or dynamical task, e.g. to avoid singularities, joint limits, and collisions. Experiments have shown the suitability of the developed control algorithms performing different manipulations which require all 6 DOF.

As an approach to extend the applicability of telepresence to wide remote environments, an extensive telepresence system was described. The main design objective was to maintain a high degree of immersion including a natural sensation of locomotion. This objective was successfully addressed by mounting telemanipulator and haptic interface on mobile bases. The mobile teleoperator replicates all operator motions including changes of gaze direction whereas the mobile haptic interface is controlled in such a way that the manipulability of the haptic interface is guaranteed. Motion compression was presented as a method to map arbitrarily large remote environments to user spaces of limited size with almost no degradation of the feeling of presence.

Architectures for collaborative telemanipulation systems and a classification of possible types of interactions were presented. The assignment of these interaction types to the proposed architectures for collaborative telemanipulation allows the formulation of challenges and requirements on the control of such systems.

Future research aims at developing algorithms for constrained bimanual interactions and enlarging the existing system for collaborative telemanipulation tasks. This requires further the development of control algorithms for a mobile human-system-interface for bimanual manipulations. Furthermore, the telemanipulator arms will be equipped with robotic hands, which will make more complex assignments e.g. grasping tasks possible.

References

1. H. Baier, F. Freyberger, and G. Schmidt. A high fidelity interactive stereo vision. In *Proceedings of the 2001 Workshop on Advances in Interactive Multimodal Telepresence Systems*, pages 33–42, Munich, Germany, 2001.
2. Immersion Corp. Two-handed cyberforce system. http://www.immersion.com/3d/products/haptic_workstation.php, 2002.
3. S. Walairacht, M. Ishii, Y. Koike, and M. Sato. Two-handed multi-finger string-based haptic interface device. In *IEICIE Trans. on Information and Systems*, (84):365–373, 2001.
4. K.J. Waldron and K. Tollon. Mechanical characterization of the immersion corp. haptic, bimanual, surgical simulator interface. In B. Siciliano and P. Dario, editors, *Proc. of the 8th Int. Symposium on Experimental Robotics (ISER'02)*, pages 106–112. Springer, 2003.
5. M.C. Cavusoglu, F. Tendick, W. Winthrop, and S.S. Sastry. A laparoscopic telesurgical workstation. *IEEE Trans. on Robotics and Automation*, 15(4):728–739, 1999.
6. S. Hayati, T. Lee, K. Tso, and P. Backes. A testbed for a unified teleoperated-autonomous dual-arm robotic system. In *Proc. of the IEEE Trans. on Robotics and Automation*, pages 1090–1095, 1990.
7. W.K. Yoon, Y. Tsumaki, and M. Uchiyama. An experimental system for dual-arm robot teleoperation in space with concepts of virtual grip and ball. In *Proc. of the Int. Conf. on Advanced Robotics*, pages 225–230, 1999.
8. A. Kron and G. Schmidt. Bimanual haptic telepresence system employed to demining operations. In *Proc. of the EuroHaptics*, pages 490–493, 2004.
9. A. Kron, G. Schmidt, B. Petzold, M. Zäh, E. Steinbach, and P. Hinterseer. Disposal of explosive ordnances by use of a bimanual haptic telepresence system. In *Proc. of the IEEE Int. Conf. on Robotics and Automation, ICRA*, New Orleans, LA, 2004.
10. S. Chang, J. Kim, J.H. Borm, and C. Lee. Kist teleoperation system for humanoid robot. In *Proc. of the IEEE/RSJ Int. Conf. on Intelligent Robots and Systems*, pages. 1198–1203, 1999.
11. J.M. Hollerbach. Optimum kinematic design for a seven degree of freedom manipulator. *Int. J. Robotics Research*, pages 341–349, 1984.
12. H. Seraji. Configuration control of redundant manipulators: Theory and implementation. *IEEE Transactions on Robotics and Automation*, 5(4):472–490, 1989.
13. F. Caccavale, B. Siciliano. Quaternion-based kinematic control of redundant spacecraft/manipulator systems. In *IEEE Int. Conf. on Robotics and Automation*, pages. 435–440, 2001.
14. Y. Nakamura and H. Hanafusa. Optimal redundancy control of robot manipulators. *Inter. J. of Robotics Research*, 6(1):34–42, 1987.
15. T. Yoshikawa. Analysis and control of robot manipulators with redundancy. In *First Inter. Symp. on Robotics Research*, 6(2):3–15, 1987.
16. T.F. Chan, R.V. Dubey. A weighted least norm solution based scheme for avoiding joint limits for redundant manipulators. In *IEEE Int. Conf. on Robotics and Automation*, 3:395–402, 1993.
17. H. Zghal, R.V. Dubey, J.A. Euler. Efficient gradient projection optimization for manipulators with multiple degrees of redundancy. In *IEEE Inter. Conf. on Robotics and Automation*, pages 1006–1011, 1990.

18. K. Cleary and D. Tesar. Incorporating multiple criteria in the operation of redundant manipulators. In *IEEE Inter. Conf. on Robotics and Automation*, 1:618–624, 1990.
19. J.A. Pemanes and S. Zeghloul. Optimal placement of robotic manipulators using multiple kinematics criteria. In *IEEE Inter. Conf. on Robotics and Automation*, pages 933–938, 1991.
20. A. Kron. *Beiträge zur bimanuellen und mehrfingrigen haptischen Informationsvermittlung in Telepräsenzsystemen*. PhD thesis, Technische Universität München, 2004.
21. A. Kron and G. Schmidt. Stability and performance analysis of kinesthetic control architectures for bimanual telepresence systems. *Journal of Intelligent Robotic Systems*, 46(1):1–26, 2006.
22. S. Hirche, B. Stanczyk, and M. Buss. Transparent Exploration of Remote Environments by Internet Telepresence. In *Proceedings of Int. Workshop on High-Fidelity Telepresence and Teleaction jointly with the conference HUMANOIDS*, Munich, Germany, 2003.
23. S. Hirche. *Haptic Telepresence in Packet Switched Communication Networks*. PhD thesis, Technische Universität München, 2005.
24. M. Ueberle, N. Mock, A. Peer, C. Michas, and M. Buss. Design and Control Concepts of a Hyper Redundant Haptic Interface for Interaction with Virtual Environments. In *Proceedings of the IEEE/RSJ International Conference on Intelligent Robots and Systems IROS, Workshop on Touch and Haptics*, Sendai, Japan, 2004.
25. M. Ueberle. *Design, Control, and Evaluation of a Family of Kinesthetic Haptic Interfaces*. PhD thesis, Technische Universität München, 2006.
26. N.H. Bakker, P.J. Werkhoven, and P.O. Passenier. The effect of proprioceptive and visual feedback on geographical orientation in virtual environments. *Presence: Teleoperators and Virtual Environments*, 8:36–53, 1999.
27. N. Nitzsche and G. Schmidt. A mobile haptic interface mastering a mobile teleoperator. In *Proceedings of the IEEE/RSJ International Conference on Intelligent Robots and Systems (IROS)*, pages 3912–3917, Sendai, Japan, 2004.
28. W. Becker, G. Nasios, S. Raab, and R. Jürgens. Fusion of vestibular and podokinesthetic information during self-turning towards instructed targets. *Exp. Brain Res.*, 144(4):458–474, 2002.
29. N. Nitzsche, U.D. Hanebeck, and G. Schmidt. Motion compression for telepresent walking in large target environments. *Presence: Teleoperators and Virtual Environments*, 13:44–60, 2004.
30. O. von Stryk. Numerische Lösung optimaler Steuerungsprobleme: Diskretisierung, Parameteroptimierung und Berechnung der adjungierten Variablen. 1995.
31. N. Nitzsche. *Weiträumige Telepräsenz: Unbeschränkte Fortbewegung und haptische Interaktion*. PhD thesis, Technische Universität München, 2005.

Index

Architecture 12
ASIBOT 418
Auditory interfaces 19
Augmented reality 140

Bimanual teleoperation 473
Body modeling 125

Carving 126
Collaborative telemanipulation 492
Contact models 213

Disabled people 419

Force based control 123
Force-position 182, 297
Force-velocity 185, 297
Fuzzy techniques 113

Geometric constraints 386
Geometric reasoning 449

Hammerstein model 217
Haptic interfaces 20, 25, 27
Human gestures 121, 130
Human intention 121
Human interface 404, 421, 464
Human-robot interaction 123, 129
Hunt-Crossley model 219

Image correspondence 96
Image distorsion 95
Impedance estimation 212
Impedance reflection 212
Internet protocols 435, 448

Just noticeable difference 194

Laparoscopic 411
Live-line maintenance 461

Micro-telemanipulation 148
Mobile platform 485
Model estimation 218
Multi-modal 11, 15
Multirobot systems 437

Natural language 108
NEPTUNE 351

Online estimation 218
Operator's training 468

Packet losses 246, 267
Passivity theory 258
Port-Hamiltonian 235, 239
Position-position 178
Power scaling 242, 244
Predictive simulator 307

Re-configurable controller 291
ROBTET 462
ROTEX 326

Sense of hearing 18
Sense of touch 19
Sense of vision 15
SFS 125, 126
Shape from silhouette 125, 126
Space telerobotics 323
State convergence 278
State space 273

Stereoscopic interface 81, 100
Stereovision 125
Stochastic techniques 116

TECSAS 339
Tele-laboratory 431
Teleoperator synchronization 262
Telepresence 12, 21, 154, 484
Teleprogramming 305
Telesurgery 381, 404
Time delay 185, 198, 244, 267, 280
Transparency 193, 195

Underwater telerobotics 347

Virtual environments 144
Virtual fixtures 392
ViSHaRD 31, 36
ViSHaRD10 473
Visual interfaces 17
Visual servoing 432

Springer Tracts in Advanced Robotics

Edited by B. Siciliano, O. Khatib and F. Groen

Vol. 31: Ferre, M.; Buss, M.; Aracil, R.; Melchiorri, C.; Balaguer C. (Eds.)
Advances in Telerobotics
500 p. 2007 [978-3-540-71363-0]

Vol. 30: Brugali, D. (Ed.)
Software Engineering for Experimental Robotics
490 p. 2007 [978-3-540-68949-2]

Vol. 29: Secchi, C.; Stramigioli, S.; Fantuzzi, C.
Control of Interactive Robotic Interfaces – A Port-Hamiltonian Approach
225 p. 2007 [978-3-540-49712-7]

Vol. 28: Thrun, S.; Brooks, R.; Durrant-Whyte, H. (Eds.)
Robotics Research – Results of the 12th International Symposium ISRR
602 p. 2007 [978-3-540-48110-2]

Vol. 27: Montemerlo, M.; Thrun, S.
FastSLAM – A Scalable Method for the Simultaneous Localization and Mapping Problem in Robotics
120 p. 2007 [978-3-540-46399-3]

Vol. 26: Taylor, G.; Kleeman, L.
Visual Perception and Robotic Manipulation – 3D Object Recognition, Tracking and Hand-Eye Coordination
218 p. 2007 [978-3-540-33454-5]

Vol. 25: Corke, P.; Sukkarieh, S. (Eds.)
Field and Service Robotics – Results of the 5th International Conference
580 p. 2006 [978-3-540-33452-1]

Vol. 24: Yuta, S.; Asama, H.; Thrun, S.; Prassler, E.; Tsubouchi, T. (Eds.)
Field and Service Robotics – Recent Advances in Research and Applications
550 p. 2006 [978-3-540-32801-8]

Vol. 23: Andrade-Cetto, J,; Sanfeliu, A.
Environment Learning for Indoor Mobile Robots – A Stochastic State Estimation Approach to Simultaneous Localization and Map Building
130 p. 2006 [978-3-540-32795-0]

Vol. 22: Christensen, H.I. (Ed.)
European Robotics Symposium 2006
209 p. 2006 [978-3-540-32688-5]

Vol. 21: Ang Jr., H.; Khatib, O. (Eds.)
Experimental Robotics IX – The 9th International Symposium on Experimental Robotics
618 p. 2006 [978-3-540-28816-9]

Vol. 20: Xu, Y.; Ou, Y.
Control of Single Wheel Robots
188 p. 2005 [978-3-540-28184-9]

Vol. 19: Lefebvre, T.; Bruyninckx, H.; De Schutter, J. Nonlinear Kalman Filtering for Force-Controlled Robot Tasks
280 p. 2005 [978-3-540-28023-1]

Vol. 18: Barbagli, F.; Prattichizzo, D.; Salisbury, K. (Eds.)
Multi-point Interaction with Real and Virtual Objects
281 p. 2005 [978-3-540-26036-3]

Vol. 17: Erdmann, M.; Hsu, D.; Overmars, M.; van der Stappen, F.A (Eds.)
Algorithmic Foundations of Robotics VI
472 p. 2005 [978-3-540-25728-8]

Vol. 16: Cuesta, F.; Ollero, A.
Intelligent Mobile Robot Navigation
224 p. 2005 [978-3-540-23956-7]

Vol. 15: Dario, P.; Chatila R. (Eds.)
Robotics Research – The Eleventh International Symposium
595 p. 2005 [978-3-540-23214-8]

Vol. 14: Prassler, E.; Lawitzky, G.; Stopp, A.; Grunwald, G.; Hägele, M.; Dillmann, R.; Iossifidis. I. (Eds.)
Advances in Human-Robot Interaction
414 p. 2005 [978-3-540-23211-7]

Vol. 13: Chung, W.
Nonholonomic Manipulators
115 p. 2004 [978-3-540-22108-1]

Vol. 12: Iagnemma K.; Dubowsky, S.
Mobile Robots in Rough Terrain – Estimation, Motion Planning, and Control with Application to Planetary Rovers
123 p. 2004 [978-3-540-21968-2]

Vol. 11: Kim, J.-H.; Kim, D.-H.; Kim, Y.-J.; Seow, K.-T.
Soccer Robotics
353 p. 2004 [978-3-540-21859-3]

Vol. 10: Siciliano, B.; De Luca, A.; Melchiorri, C.; Casalino, G. (Eds.)
Advances in Control of Articulated and Mobile Robots
259 p. 2004 [978-3-540-20783-2]

Vol. 9: Yamane, K.
Simulating and Generating Motions of Human Figures
176 p. 2004 [978-3-540-20317-9]

Vol. 8: Baeten, J.; De Schutter, J.
Integrated Visual Servoing and Force Control – The Task Frame Approach
198 p. 2004 [978-3-540-40475-0]

Vol. 7: Boissonnat, J.-D.; Burdick, J.; Goldberg, K.; Hutchinson, S. (Eds.)
Algorithmic Foundations of Robotics V
577 p. 2004 [978-3-540-40476-7]

Vol. 6: Jarvis, R.A.; Zelinsky, A. (Eds.)
Robotics Research – The Tenth International Symposium
580 p. 2003 [978-3-540-00550-6]

Vol. 5: Siciliano, B.; Dario, P. (Eds.)
Experimental Robotics VIII – Proceedings of the 8th International Symposium ISER02
685 p. 2003 [978-3-540-00305-2]

Vol. 4: Bicchi, A.; Christensen, H.I.; Prattichizzo, D. (Eds.)
Control Problems in Robotics
296 p. 2003 [978-3-540-00251-2]

Vol. 3: Natale, C.
Interaction Control of Robot Manipulators – Six-degrees-of-freedom Tasks
120 p. 2003 [978-3-540-00159-1]

Vol. 2: Antonelli, G.
Underwater Robots – Motion and Force Control of Vehicle-Manipulator Systems
268 p. 2006 [978-3-540-31752-4]

Vol. 1: Caccavale, F.; Villani, L. (Eds.)
Fault Diagnosis and Fault Tolerance for Mechatronic Systems – Recent Advances
191 p. 2003 [978-3-540-44159-5]

Printing: Mercedes-Druck, Berlin
Binding: Stein + Lehmann, Berlin

MIX
Papier aus verantwortungsvollen Quellen
Paper from responsible sources
FSC® C105338

If you have any concerns about our products,
you can contact us on
ProductSafety@springernature.com

In case Publisher is established outside the EU,
the EU authorized representative is:
Springer Nature Customer Service Center GmbH
Europaplatz 3, 69115 Heidelberg, Germany

Printed by Libri Plureos GmbH
in Hamburg, Germany